Mathematical Sciences Research Institute Publications

15

Mathematical Sciences Research Institute Publications

Volume 1	Freed and Uhlenbeck: Instantons and Four-Manifolds
Volume 2	Chern (ed.): Seminar on Nonlinear Partial Differential Equations
Volume 3	Lepowsky, Mandelstam, and Singer (eds.): Vertex Operators in Mathematics and Physics
Volume 4	Kac (ed.): Infinite Dimensional Groups with Applications
Volume 5	Blackadar: K-Theory for Operator Algebras
Volume 6	Moore (ed.): Group Representations, Ergodic Theory, Operator Algebras, and Mathematical Physics
Volume 7	Chorin and Majda (eds.): Wave Motion: Theory, Modelling, and Computation
Volume 8	Gersten (ed.): Essays in Group Theory
Volume 9	Moore and Schochet: Global Analysis on Foliated Spaces
Volume 10	Drasin, Earle, Gehring, Kra, and Marden (eds.): Holomorphic Functions and Moduli I
Volume 11	Drasin, Earle, Gehring, Kra, and Marden (eds.): Holomorphic Functions and Moduli II
Volume 12	Ni, Peletier, and Serrin (eds.): Nonlinear Diffusion Equations and their Equilibrium States I
Volume 13	Ni, Peletier, and Serrin (eds.): Nonlinear Diffusion Equations and their Equilibrium States II
Volume 14	Goodman, de la Harpe, and Jones: Coxeter Graphs and Towers of Algebras
Volume 15	Hochster, Huneke, and Sally (eds.): Commutative Algebra
Volume 16	Ihara, Ribet, and Serre (eds.): Galois Groups over Q

M. Hochster C. Huneke J.D. Sally
Editors

Commutative Algebra

Proceedings of a Microprogram Held June 15-July 2, 1987

Springer-Verlag
New York Berlin Heidelberg London Paris Tokyo

cat/sep

M. Hochster
Department of Mathematics
University of Michigan
Ann Arbor, MI 48109-1003, USA

C. Huneke
Department of Mathematics
Purdue University
West Lafayette, IN 47907, USA

J.D. Sally
Department of Mathematics
Northwestern University
Evanston, IL 60201, USA

Mathematical Sciences Research Institute
1000 Centennial Drive
Berkeley, CA 94720, USA

MATH

The Mathematical Sciences Research Institute wishes to acknowledge support by the National Science Foundation.

Mathematics Subject Classification: 13A15, 13B20, 13C10, 13C13, 13C15, 13D05, 13E05, 13F20, 13H05, 13H10, 13H15, 14B12, 14M10

Library of Congress Cataloging-in-Publication Data
Commutative algebra: proceedings from a microprogram held in June 15–July 2, 1987/Melvin Hochster, Craig Huneke, Judith D. Sally.
p. cm.—(Mathematical Science Research Institute publications; 15)
ISBN 0-387-96990-X (alk. paper)
1. Commutative algebra—Congresses. I. Hochster, Melvin, 1943– . II. Huneke, Craig. III. Sally, Judith D. IV. Series
QA251.3.C655 1989
512'.24—dc20

Printed on acid-free paper

Camera-ready text prepared by the Mathematical Sciences Research Institute.
Printed and bound by Edwards Brothers, Inc., Ann Arbor, Michigan.
Printed in the United States of America.

9 8 7 6 5 4 3 2 1

ISBN 0-387-96990-X Springer-Verlag New York Berlin Heidelberg
ISBN 3-540-96990-X Springer-Verlag Berlin Heidelberg New York

Preface

During late June and early July of 1987 a three week program (dubbed "microprogram") in Commutative Algebra was held at the Mathematical Sciences Research Institute at Berkeley. The intent of the microprogram was to survey recent major results and current trends in the theory of commutative rings, especially commutative Noetherian rings. There was enthusiastic international participation.

The papers in this volume, some of which are expository, some strictly research, and some a combination, reflect the intent of the program. They give a cross-section of what is happening now in this area. Nearly all of the manuscripts were solicited from the speakers at the conference, and in most instances the manuscript is based on the conference lecture. The editors hope that they will be of interest and of use both to experts and neophytes in the field.

The editors would like to express their appreciation to the director of MSRI, Professor Irving Kaplansky, who first suggested the possibility of such a conference and made the task of organization painless. We would also like to thank the MSRI staff who were unfailingly efficient, pleasant, and helpful during the meeting, and the manager of MSRI, Arlene Baxter, for her help with this volume. Finally we would like to express our appreciation to David Mostardi who did much of the typing and put the electronic pieces together.

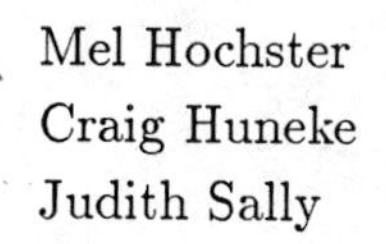

Mel Hochster
Craig Huneke
Judith Sally

Commutative Algebra

TABLE OF CONTENTS

Representations, Resolutions and Intertwining Numbers

KAAN AKIN AND DAVID A. BUCHSBAUM

Introduction.

It may, at first glance, seem inappropriate to be talking about representations of the general linear group at a conference on commutative algebra. We would like, therefore, to offer some observations: one, historical; one, personal; and one, perhaps presumptuous, that might reconcile this apparent anomaly.

Probably most mathematicians would agree that the paper, [**13**], in which Hilbert proved (among other important results) the Hilbert Basis Theorem and the Hilbert Syzygy Theorem, was fundamental for the development of commutative algebra. What some of us occasionally lose sight of is that this paper really was dealing with basic problems in invariant theory. Once we recall that fact, it shouldn't be too surprising that there can be a substantial interplay between representation theory and commutative algebra.

Let us look, now, at a slightly updated version of a part of the problem Hilbert was considering in the above paper [**13**]. Let F, G, and V be finite-dimensional vector spaces over a field R, of dimensions m, n and $p-1$ respectively. Then the general linear group of V, $\mathrm{GL}(V)$, acts on the symmetric algebra $S(F\otimes V\oplus V^*\otimes G)$ over R, and we have the ring of invariants $S' = S(F\otimes V\oplus V^*\otimes G)^{\mathrm{GL}(V)}$. Question: Is S' a finitely generated algebra over R? Answer: Yes (First Fundamental Theorem of Invariant Theory). In fact, the following is shown. The natural isomorphism $\mathrm{Hom}_R(V,V) \approx V\otimes_R V^*$ gives us the element $c_V \in V\otimes V^*$ corresponding to the identity map in $\mathrm{Hom}_R(V,V)$, and we have the map $\phi : S(F\otimes G) \to S(F\otimes V\oplus V^*\otimes G)$ associated to the map $F\otimes G \to F\otimes V\otimes V^*\otimes G$ which is defined by $x\otimes y \mapsto x\otimes c_v\otimes y$. The image of ϕ is the ring S', so S' is finitely generated. Now let I_p be the kernel of ϕ. Question: Is I_p a finitely generated ideal,

The greater part of this article consists of a lecture given by the second-named author at the Mathematical Sciences Research Institute in Berkeley for a microprogram on Commutative Algebra, and is therefore mainly expository. The results on intertwining numbers in the last section, however, have not appeared elsewhere.

and what are its generators? Answer: Yes, and its generators are the minors of order p of the generic $m \times n$ matrix (Second Fundamental Theorem of Invariant Theory). (To make sense out of this answer, we choose bases $\{f_1, \ldots, f_m\}$, $\{g_1, \ldots, g_n\}$ of F and G. Then the matrix $(f_i \otimes g_j)$ is defined over $S(F \otimes G)$, and I_p is the ideal in $S(F \otimes G)$ generated by the $p \times p$ minors of this matrix.) The next question that could naturally be asked is: What is a minimal free resolution of $S(F \otimes G)/I_p$ over $S(F \otimes G)$? This question was not asked in [**13**], but the Syzygy Theorem was proved to ensure that a finite minimal resolution existed in situations of this kind (one wanted a method of computing the number of linearly independent invariant forms of degree k for each k; i.e. the Hilbert function). The answer to the above question, extended to the case when R is a commutative ring, has been sought by one of the authors on and off since around 1957, and by the other since around 1978. For $p = \min(m, n)$, a complete answer has been known for many years [**8,11**]. In 1977, A. Lascoux [**14**] found the key to the solution for all p, provided R contains the rational numbers. Namely, assuming that the terms of such a resolution should be $\mathrm{GL}(F) \times \mathrm{GL}(G)$-modules, he used classical representation theory (i.e. characteristic zero theory) to determine what these terms should be. Subsequent work by Nielsen [**15**] and Pragacz-Weyman [**16**] nailed all the arguments down, and provided definitions of the boundary maps of the resolutions. For arbitrary R, the case $p = \min(m, n) - 1$ was settled in [**6**], and involved the extension and modification of representation-theoretic techniques to arbitrary commutative rings. Much more will be said about this in the later sections, but for now this brief description of the attempt to find resolutions of determinantal ideals should serve to explain the personal involvement in the relationship between commutative algebra and representation theory.

Finally, we offer this exposition of a small part of characteristic-free representation theory to suggest that the introduction of homological techniques in this area (and our contribution represents only a small part of these) may prove as fruitful as the introduction of such methods to commutative algebra. Remember that the regularity of localization of regular local rings, and the factoriality of regular local rings were known only in the geometric case for many years before the 1950's. The application of homological methods made it possible to prove and extend these results free of characteristic considerations. Work in the past twenty-five years, while at times very rooted to positive characteristic, has nevertheless been focused on the homological conjectures. In a similar vein, one can hope that spectral sequence and filtration arguments may be strong enough to replace some of the classical counting arguments of characteristic zero theory, and that universal coefficient theorems (or change of rings arguments generally) may serve to enrich some of the modular theory.

In the foregoing apologia, we mentioned that we had to extend and modify the classical representation theory of the general linear group to arbitrary commutative ground rings. The major part of this paper will present some of the basic definitions and describe some of the central problems and techniques that have been evolving over the past several years, as well as some partial solutions of the problems. The exposition will be discursive rather than formal, as all but the very last material on intertwining numbers has appeared elsewhere both in original and expository form. The original papers are [**2,3,5,7**] and the expository one to which the reader may want to refer is [**4**].

1. Basic definitions and questions.

If F is a finite-dimensional vector space over a field R of characteristic zero, the irreducible representations of $\mathrm{GL}(F)$ are modules $L_\lambda(F)$ where $\lambda = (\lambda_1, \ldots, \lambda_k)$ is any partition with $\lambda_1 \leq \dim_R F$. Recall that λ is a partition if λ_i are integers with $\lambda_1 \geq \lambda_2 \geq \cdots \geq \lambda_k \geq 0$. The shape, or diagram, of λ is:

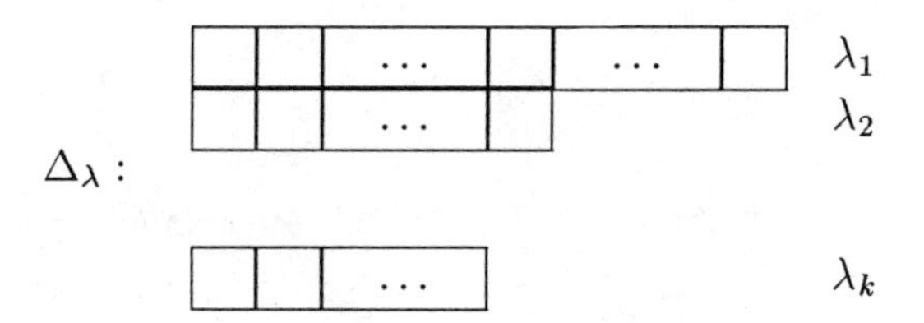

where the i-th row has λ_i boxes. If $\{x_1, \ldots, x_n\}$ is a basis for F, the basis for $L_\lambda(F)$ consists of all standard tableaux of shape Δ_λ in the basis $x_1, \ldots, x_n$. That is, we fill in the boxes of the diagram of λ with basis elements x_i (getting a tableau), and we say the tableau is standard if the entries in each row are strictly increasing, while the entries in each column are weakly increasing. For instance, if $\lambda = (3,3,2,1)$ is a partition, and

$\{x_1, \ldots, x_n\}$ is a basis for F, the tableau

$$\begin{array}{|c|c|c|}\hline x_2 & x_3 & x_5 \\ \hline x_2 & x_4 & x_5 \\ \hline x_4 & x_5 \\ \cline{1-2} x_4 \\ \cline{1-1}\end{array}$$

is standard, but

$$\begin{array}{|c|c|c|}\hline x_2 & x_3 & x_5 \\ \hline x_1 & x_4 & x_5 \\ \hline x_4 & x_4 \\ \cline{1-2} x_3 \\ \cline{1-1}\end{array}$$

is not.

Thus we see that there is a requirement of skew-symmetry in the rows, and symmetry in the columns. Taking this as our point of departure, we make the following definitions.

As $s \times t$ matrix $A = (a_{ij})$ is a *shape matrix* if $a_{ij} = 0$ or 1 for $1 \leq i \leq s$, $1 \leq j \leq t$. We set $\alpha_i = \sum_{j=1}^{t} a_{ij}$, $\beta_j = \sum_{i=1}^{s} a_{ij}$.

To the shape matrix A, and any free R-module of finite type F over the commutative ring R, we associate the *Schur map* $d_A(F)$:

$$d_A(F) : \wedge^{\alpha_1} F \otimes \cdots \otimes \wedge^{\alpha_s} F \to S_{\beta_1} F \otimes \cdots \otimes S_{\beta_t} F,$$

where $\wedge$ denotes exterior and S denotes symmetric power, as the composition of the following maps:

$$\begin{aligned} u : \quad & \wedge^{a_1} F \otimes \cdots \otimes \wedge^{a_s} F \\ & \qquad \to (\wedge^{a_{11}} F \otimes \cdots \otimes \wedge^{a_{1t}} F) \otimes \cdots \otimes (\wedge^{a_{s1}} F \otimes \cdots \otimes \wedge^{a_{st}} F) \end{aligned}$$

$$\begin{aligned} v : \quad & (\wedge^{a_{11}} F \otimes \cdots \otimes \wedge^{a_{1t}} F) \otimes \cdots \otimes (\wedge^{a_{s1}} F \otimes \cdots \otimes \wedge^{a_{st}} F) \\ & \qquad \to (S_{a_{11}} F \otimes \cdots \otimes S_{a_{1t}} F) \otimes \cdots \otimes (S_{a_{s1}} F \otimes \cdots \otimes S_{a_{st}} F) \end{aligned}$$

$$\begin{aligned} \theta : \quad & (S_{a_{11}} F \otimes \cdots \otimes S_{a_{1t}} F) \otimes \cdots \otimes (S_{a_{s1}} F \otimes \cdots \otimes S_{a_{st}} F) \\ & \qquad \to (S_{a_{11}} F \otimes S_{a_{21}} \otimes \cdots \otimes S_{a_{s1}} F) \otimes \cdots \otimes (S_{a_{1t}} F \otimes \cdots \otimes S_{a_{st}} F) \end{aligned}$$

$$\begin{aligned} w : \quad & (S_{a_{11}} F \otimes \cdots \otimes S_{a_{s1}} F) \otimes \cdots \otimes (S_{a_{1t}} F \otimes \cdots \otimes S_{a_{st}} F) \\ & \qquad \to S_{\beta_1} F \otimes \cdots \otimes S_{\beta_t} F \end{aligned}$$

where u is the indicated diagonalization on $\wedge F$, v is the isomorphism identifying $\wedge^{a_{ij}}$ with $S_{a_{ij}}F$, θ is the indicated rearrangement of factors, and w is the appropriate multiplication in SF.

The image of $d_A(F)$ is denoted by $L_A(F)$, and is called the *Schur module of shape A of F.*

If $\lambda = (\lambda_1, \ldots, \lambda_k)$ is a partition, we set A_λ to be the $k \times \lambda_1$ matrix defined by

$$a_{ij} = \begin{cases} 1 & \text{if } 1 \le j \le \lambda_i \\ 0 & \text{if } \lambda_i < j \le \lambda_1. \end{cases}$$

More generally, if $\lambda = (\lambda_1, \ldots, \lambda_k)$ and $\mu = (\mu_1, \ldots, \mu_k)$ are two partitions with $\mu_i \le \lambda_i$ for $i = 1, \ldots, k$, we set $A_{\lambda/\mu}$ to be the $k \times \lambda_1$ matrix defined by

$$a_{ij} = \begin{cases} 1 & \text{if } \mu_i \le j \le \lambda_i \\ 0 & \text{otherwise.} \end{cases}$$

We then set $d_{\lambda/\mu}(F) = d_{A_{\lambda/\mu}}(F)$,and $L_{\lambda/\mu}(F) = L_{A_{\lambda/\mu}}(F)$. The shape $A_{\lambda/\mu}$ is called a *skew-shape*, and $L_{\lambda/\mu}(F)$ is called a *skew-Schur module.* (Note that $L_\lambda(F) = L_{\lambda/(0)}(F)$.)

If λ is a partition, A_λ the associated shape matrix and $\alpha_i = \sum_j a_{ij}$, $\beta_j = \sum_i a_{ij}$ the corresponding row- and column-sums, it is easy to see that $(\alpha_i, \ldots, \alpha_s) = \lambda$ and that $(\beta_1, \ldots, \beta_t)$ is also a partition. The partition $(\beta_1, \ldots, \beta_t)$ is denoted by $\tilde{\lambda}$ and is called the transpose partition of λ. If $A_{\lambda/\mu}$ is a skew-shape, then $A_{\tilde{\lambda}/\tilde{\mu}} = (A_{\lambda/\mu})^*$, where X^* denotes the transpose of the matrix X.

If R is a field of characteristic zero, the modules $L_\lambda(F)$ are precisely the irreducible representations mentioned at the beginning of this section. For an arbitrary commutative ring R, it is clear that for any shape matrix A, $L_A(F)$ is a GL(F)-module, since $d_A(F)$ is a GL(F)-equivariant map.

For fields of characteristic zero, the symmetric algebra $S(F)$ and the divided power algebra $D(F)$ are isomorphic. However, over Z, or $Z/(p)$, or an arbitrary commutative ring, they are not, and this leads us to make the following additional definitions.

For A a shape matrix as above, with α_i and β_j defined as before, we define the *Weyl map*

$$d'_A(F) : D_{\alpha_1}F \otimes \cdots \otimes D_{\alpha_s}F \to \wedge^{\beta_1}F \otimes \cdots \otimes \wedge^{\beta_t}F$$

by diagonalizing, identifying $D_{a_{ij}}F$ with $\wedge^{a_{ij}}F$, rearranging, and multiplying as we did with the Schur map. The image of $d'_A(F)$ is denoted by

$K_A(F)$, and is called the *Weyl module of shape A of F*. When $A = A_{\lambda/\mu}$, we denote $K_{A_{\lambda/\mu}}(F)$ by $K_{\lambda/\mu}(F)$.

A few easy examples that are worth noting are:

a) $\lambda = (p)$. Then $L_\lambda(F) = \wedge^p F$, $K_\lambda(F) = D_p F$.

b) $\lambda = (\overbrace{1,\ldots,1}^{p})$. Then $L_\lambda(F) = S_p F$, $K_\lambda(F) = \wedge^p F$.

c) $\lambda = (p, \overbrace{1,\ldots,1}^{q-1})$. Then $L_\lambda(F) = \mathrm{Im}(\wedge^p F \otimes S_{q-1}F \to \wedge^{p-1}F \otimes S_q F) = \mathrm{Ker}(\wedge^{p-1}F \otimes S_q F \to \wedge^{p-2}F \otimes S_{q+1}F)$ and $K_\lambda(F) = \mathrm{Im}(D_p F \otimes \wedge^{q-1}F \to D_{p-1}F \otimes \wedge^q F) = \mathrm{Ker}(D_{p-1}F \otimes \wedge^q F \to D_{p-2}F \otimes \wedge^{q+1}F)$.

The map $\wedge^p F \otimes S_{q-1}F \to \wedge^{p-1}F \otimes S_q F$ in c) is the map in the Koszul complex, while $D_p F \otimes \wedge^{q-1}F \to D_{p-1}F \otimes \wedge^q F$ is "exterior differentiation."

If we let $R = Z$, then $D_p F$ and $S_p F$ are not isomorphic $\mathrm{GL}(F)$-representations. However, if we take $\mathbb{Q} \otimes_Z D_p F$ and $\mathbb{Q} \otimes_Z S_p F$ (where $\mathbb{Q}$ is the rational numbers), we do get isomorphic representations of $\mathrm{GL}(\mathbb{Q} \otimes F)$. The integral representations $D_p F$ and $S_p F$ are therefore inequivalent *Z-forms* of the rational representation $S_p F$.

When one takes the resolution of $S(F \otimes G)/I_p$ described by Lascoux (mentioned here in the Introduction), and attempts to imitate it over Z, one fails to get a resolution precisely because of the existence of inequivalent Z-forms. To be more explicit, in the case when $p = \min(m,n) - 1$ (we are here using the notation of our discussion in the Introduction), we were led in [**6**] to deal with Z-forms of the type $\mathrm{Ker}(D_\ell F \otimes \wedge^q F \to D_{\ell+1}F \otimes \wedge^{q-1}F)$ for various ℓ and q, which are Z-forms of the Schur modules described in c) above. As one decreases p, the Z-forms that enter the picture become ever more complicated and for this reason we turned our attention to the problem of studying Z-forms more systematically.

Let us leave, then, the context of resolutions of determinantal ideals, and look at some simple integral representations of $\mathrm{GL}(F)$. It is easy to see that we have an exact sequence

$$0 \to \wedge^{p+1} \to \wedge^p \otimes \wedge^1 \to L_{(p,1)} \to 0$$

of $\mathrm{GL}(F)$-representations (we have left out F in our notation, as that is fixed throughout). If we multiply $\wedge^{p+1}$ by an integer k, we obtain the commutative diagram

$$\begin{array}{ccccccccc}
0 & \longrightarrow & \wedge^{p+1} & \longrightarrow & \wedge^p \otimes \wedge^1 & \longrightarrow & L_{(p,1)} & \longrightarrow & 0 \\
 & & {\scriptstyle k}\downarrow & & \downarrow & & \| & & \\
0 & \longrightarrow & \wedge^{p+1} & \longrightarrow & H_k(p,1) & \longrightarrow & L_{(p,1)} & \longrightarrow & 0
\end{array}$$

where $H_k(p,1)$ is the push-out of the indicated maps. Each of the rows of the above diagram is an element of $\mathrm{Ext}^1_{\mathrm{GL}(F)}(L_{(p,1)}, \wedge^{p+1})$ which becomes the same element, namely zero, when tensored with $\mathbb{Q}$. Thus $H_k(p,1)$ is a Z-form of $\wedge^p \otimes \wedge^1$. It can be shown that $H_k(p,1) \approx H_\ell(p,1)$ if and only if $k \equiv \ell(p+1)$ (see the last section). We see, therefore, that one way to produce and/or study Z-forms is by means of Ext.

These Z-forms also arise in a more classical context. Consider the skew-shape λ/μ where $\lambda = (p-1, k+1)$ and $\mu = (k-1)$:

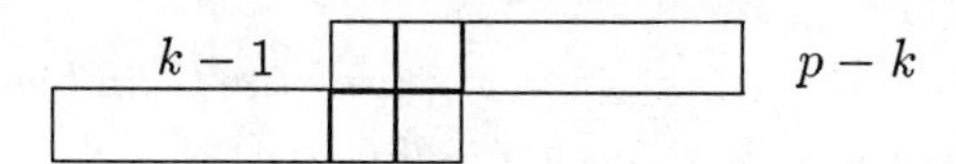

In characteristic zero, a classical formula of Giambelli tells us that the formal character of $L_{\lambda/\mu}$ is the symmetric function $e_{p-k}(x)\, e_{k+1}(x) - e_p(x)\, e_1(x)$, where $e_j(x)$ is the j-th elementary symmetric function. Since $e_j(x)$ is the character of the fundamental representation $\wedge^j$, we see that in the Grothendieck ring of GL(F)-modules, $[L_{\lambda/\mu}] = [\wedge^{p-k} \otimes \wedge^{k+1}] - [\wedge^p \otimes \wedge^1]$, where $[M]$ denotes the class of the representation, M, in the Grothendieck ring. This suggests that there is an exact sequence

$$(S) \qquad 0 \to \wedge^p \otimes \wedge^1 \to \wedge^{p-k} \otimes \wedge^{k+1} \to L_{\lambda/\mu} \to 0;$$

in fact, in characteristic zero, this is true. However, over Z, the sequence (S) is not exact. What we do have is an exact sequence

$$(S') \qquad 0 \to \wedge^{p+1} \to \wedge^p \otimes \wedge^1 \oplus \wedge^{p+1} \to \wedge^{p-k} \otimes \wedge^{k+1} \to L_{\lambda/\mu} \to 0,$$

where the map on the left sends $\wedge^{p+1} \to \wedge^p \otimes \wedge^1$ by diagonalization, and sends $\wedge^{p+1} \to \wedge^{p+1}$ by multiplication by $k+1$. Thus, the kernel of the map $\wedge^{p-k} \otimes \wedge^{k+1} \to L_{\lambda/\mu}$ is our Z-form $H_{k+1}(p,1)$. Similarly, for Weyl modules, we obtain an exact sequence

$$(S'') \qquad 0 \to D_{p+1} \to D_p \otimes D_1 \oplus D_{p+1} \to D_{p-k} \otimes D_{k+1} \to K_{\lambda/\mu} \to 0,$$

which ties in with the classical Jacobi-Trudi identity (an identity, similar to that of Giambelli, which expresses the character of a Schur (or Weyl) module in terms of the complete symmetric polynomials $h_j(x)$, which are the characters of $S_j = D_j$ (in characteristic zero)). A complete description of the Giambelli and Jacobi-Trudi identities can be found in [**4**], and will not be repeated here. Suffice it to say that these classical identities, and their interpretation as Euler-Poincaré characteristics in the Grothendieck ring, lead us naturally to the following basic questions:

(G). *Given a skew-shape $\lambda/\mu = (\lambda_1, \ldots, \lambda_k)/(\mu_1, \ldots, \mu_k)$, is there a finite acyclic complex*

$$0 \to X_n \to X_{n-1} \to \cdots \to X_0 \to L_{\lambda/\mu} \to 0$$

such that

i) *each module X_i is the direct sum of a finite number of tensor products of exterior powers;*

ii) $\sum(-1)^i[X_i] =$ *the expression given by the Giambelli identity in characteristic zero for the Schur module $L_{\lambda/\mu}$;*

iii) *each module X_i contains as a summand the module Y_i predicted by the Giambelli identity:*

$$Y_i = \sum_{\ell(\pi)=i} A_{1\pi(1)} \otimes \cdots \otimes A_{k\pi(k)},$$

where $\ell(\pi) =$ length of the permutation π, and $A_{ij} = \wedge^{\lambda_i - \mu_j + j - i}$.

(J-T). *Given a skew-shape λ/μ as above, is there a finite acyclic complex*

$$0 \to X'_n \to X'_{n-1} \to \cdots \to X'_0 \to K_{\lambda/\mu} \to 0$$

such that

i) *each module X'_i is the direct sum of a finite number of tensor products of divided powers;*

ii) $\sum(-1)^i[X'_i] =$ *the expression given by the Jacobi-Trudi identity in characteristic zero for the Schur module $L_{\tilde{\lambda}/\tilde{\mu}}$;*

iii) *each module X'_i contains as a summand the module Y'_i predicted by the Jacobi-Trudi identity:*

$$Y'_i = \sum_{\ell(\pi)=i} A'_{1\pi(1)} \otimes \cdots \otimes A'_{k\pi(k)},$$

where $\ell(\pi)$ is as above, and $A'_{ij} = D_{\lambda_i - \mu_j + j - i}$.

2. Some answers and an application to global dimension.

To relieve the suspense, we will say right off that the answers to both questions above are, yes. A fairly complete treatment of these problems is given in [4], so we will only touch on the main ideas involved in that treatment.

The point of departure for us was the observation that if λ/μ is a skew-shape, with $\lambda = (\lambda_1, \lambda_2)$, $\mu = (\mu_1, \mu_2)$, then we have an exact sequence:

$$(1) \qquad 0 \to M_\gamma \to M_\beta \to M_\alpha \to 0$$

where $\alpha = \lambda/\mu$

$$\gamma = (\lambda_1, \lambda_2 - (\mu_1 - \mu_2 + 2))/(\mu_2 - 1, \mu_2 - 1),$$
$$\beta = (\lambda_1, \lambda_2 - 1)/(\mu_1, \mu_2 - 1)$$

and M is either always L or always K.

If $\lambda_2 - \mu_1 \leq 0$, then $L_{\lambda/\mu} = \wedge^{\lambda_1 - \mu_1} \otimes \wedge^{\lambda_2 - \mu_2}$, $K_{\lambda/\mu} = D_{\lambda_1 - \mu_1} \otimes D_{\lambda_2 - \mu_2}$, so there is nothing more to be done. If $\lambda_2 - \mu_1 > 0$, we use induction on $\lambda_2 - \mu_1$, and observe that for both β and γ, this difference is less than that for λ/μ. Hence M_γ and M_β have finite resolutions of the desired type, say M_γ and M_β, and if we have a map of $\mathsf{M}_\alpha \to \mathsf{M}_\beta$ over $M_\alpha \to M_\beta$, the mapping cone will be a complex $\mathsf{M}_{\lambda/\mu}$ of the desired type over $M_{\lambda/\mu}$. Canonical maps $\mathsf{M}_\gamma \to \mathsf{M}_\beta$ are easy to define, so for two-rowed shapes the problem was easily solved.

As soon as one tries to extend the above procedure to shapes of three or more rows, however, the problem becomes much more complex. We will illustrate pictorially what happens in the case of three rows; namely, we get an exact sequence of shapes:

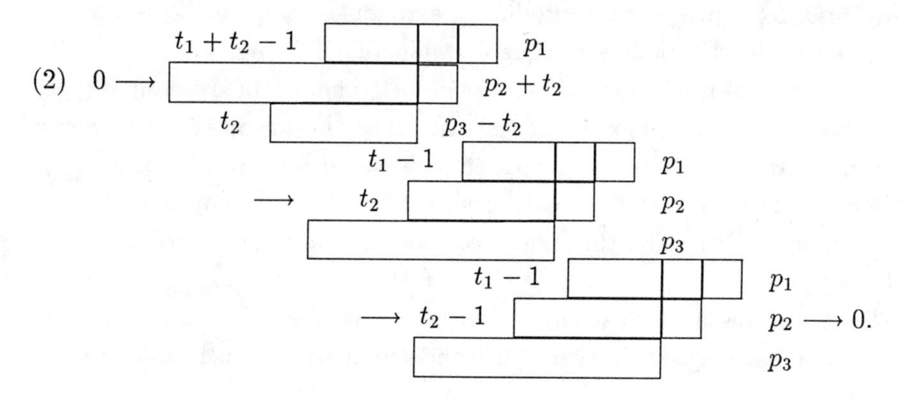

The shape on the left is not a skew-shape, but does have an associated shape matrix to which we can assign a Schur and Weyl module. We say that (2) is exact in the sense that the sequence of Schur and Weyl modules associated to those shapes is exact.

Because of the appearance of these new shapes, we had to consider the class J of shapes defined as follows. We let $\lambda = (\lambda_1, \ldots, \lambda_k)$ and $\mu = (\mu_1, \ldots, \mu_k)$ be sequences of non-negative integers such that

i) λ is a partition;
ii) $(\mu_1, \ldots, \mu_{k-1})$ is a partition and $\mu_i \leq \lambda_i$, $i = 1, \ldots, k$;
iii) $(\mu_{t+1} < \mu_k \leq \mu_t$ for some $t = 1, \ldots, k-2$, or $\mu_k \leq \mu_{k-1}$;
iv) $\lambda_{k-1} - \lambda_k \geq k - t - 1$.

To the pair of sequences $\alpha = (\lambda, \mu)$ we associate the shape matrix

$$A_\alpha = (a_{ij}) \text{ where } a_{ij} = \begin{cases} 1 & \text{for } \mu_i + 1 \leq j \leq \lambda_i \\ 0 & \text{otherwise.} \end{cases}$$

This class of shapes has associated Schur and Weyl modules denoted by L_α and K_α. It is clear that this class of shapes includes our usual skew shapes, as well as the shapes (for $k = 3$) that occur as kernels in our exact sequence (2).

To each shape α in the class J, we associate the non-negative integer $j(\alpha)$ which is the sum of the number of "overlaps" in adjacent rows of the shape α (i.e. $j(\alpha) = \sum_{i=2}^{k} \lambda_k - \sum_{i=1}^{k-2} \mu_i - \max(\mu_k, \mu_{k-1})$). We then established the existence of an exact sequence

$$(3) \qquad 0 \to M_\alpha \to M_\beta \to M_\alpha \to 0$$

similar to (1), with $j(\gamma) < j(\alpha)$ and $j(\beta) < j(\alpha)$, when $j(\alpha) > 0$. (When $j(\alpha) = 0$, L_α is just a tensor product of exterior powers, and K_α is a tensor product of divided powers.) Thus, by induction on $j(\alpha)$, we may assume the existence of resolutions M_α and M_β of M_α and M_β of the desired type. However, we could not explicitly construct maps between M_α and M_β over the map $M_\alpha \to M_\beta$ whose mapping cone would be a resolution of M_α. More to the point, it was not transparent that such a map from M_α to M_β existed. To handle this existence problem, we resorted to the Schur algebra.

Briefly, this was done as follows. Let F be a free R-module of rank n, and let s be a non-negative integer. Then the symmetric group Γ_s operates on

$\otimes^s F$, and we define the *Schur algebra of degree* s, A_s, to be the R-algebra of Γ_s-endomorphisms of $\otimes^s F$ [**12**]. The algebra A_s is the universal enveloping algebra for all polynomial representations of $\mathrm{GL}(F)$ of degree s. It turns out that

$$A_s = \sum D_{a_1}F \otimes \cdots \otimes D_{a_n}F$$

where the sum runs over all sequences of non-negative integers $(a_1, \ldots, a_n)$ with $\sum a_i = s$.

With the restriction that $s \leq n$, (which can later be removed), one can show that the resolutions $\mathbb{K}_\alpha$ and $\mathbb{K}_\beta$ of K_α and K_β are A_s-projective resolutions of K_α and K_β, so that for the Weyl modules, the desired map $\mathbb{K}_\alpha \to \mathbb{K}_\beta$ is known to exist. Thus, for Weyl modules, one can extend the methods of the two-rowed case to an arbitrary number of rows.

To take care of the resolutions of Schur modules, we proved that

$$(4) \quad \begin{aligned} &\mathrm{Hom}_{A_s}(D_{a_1}F \otimes \cdots \otimes D_{a_n}F, D_{b_1}F \otimes \cdots \otimes D_{b_n}F) \\ &\qquad \approx \mathrm{Hom}_{A_s}\left(\wedge^{a_1}F \otimes \cdots \otimes \wedge^{a_n}F, \wedge^{b_1}F \otimes \cdots \otimes \wedge^{b_n}F\right), \end{aligned}$$

and the proof of the existence of the desired resolutions is complete.

In fact, the isomorphism (4) is a special case of the general fact, proved in [**3**], that

$$(5) \qquad \mathrm{Ext}^i_{A_s}(K_\alpha, K_\beta) \approx \mathrm{Ext}^i_{A_s}(L_\alpha, L_\beta)$$

where α and β are any shapes in our class J.

In [**4,7**], we showed that when $\alpha = \lambda/\mu$ (i.e. α is a skew-shape), the resolutions $\mathbb{L}_\alpha$ and $\mathbb{K}_\alpha$ give affirmative answers to the questions (G) and (J-T) of Section 1.

Another consequence of the existence of these finite resolutions is that the R-algebras A_s have finite global dimension when R is a field, or when R is the ring of integers. The detailed proof can be found in [**3**]. The idea of the proof is this. When R is an infinite field (of any characteristic), the irreducible A_s-modules are of the form $K_\lambda/\operatorname{rad}(K_\lambda)$ where λ runs through all partitions $\lambda = (\lambda_1, \ldots, \lambda_k)$ with $\sum \lambda_i = s$ and $k \leq n$ ($n = \operatorname{rank} F$), and $\operatorname{rad}(K_\lambda)$ is the unique maximal submodule of K_λ (see [**5**]). The set of such partitions is totally ordered (lexicographically), and induction on this ordered set takes care of the case of infinite fields.

For finite fields R, one goes to the algebraic closure $\bar{R}$ and, by a simple change of rings argument, the result for $\bar{R}$ implies that for R.

When $R = Z$, the ring of integers, we first note that $Z/(p) \otimes A_s$ is the Schur algebra over $Z/(p)$, and this algebra is semi-simple for $s! < p$. Thus, for only finitely many primes do we have $\operatorname{gldim}(Z/(p) \otimes A_s) > 0$. Let $d = \max_p\{\operatorname{gldim} Z/(p) \otimes A_s\}$. Then, by a standard change of rings argument, it follows that $\operatorname{gldim} A_s \leq d + 2$.

3. Intertwining numbers.

The results of Section 2 are of some interest in that they answer certain natural questions that arise when one moves from the characteristic zero theory to the characteristic-free theory. It is reasonable to ask, though, whether this characteristic-free theory provides any significant new tools for the solution of some long-standing problems either in representation theory or elsewhere. In the Introduction, we discussed resolutions of determinantal ideals as the motivation for undertaking our investigations, and indicated that, at least in the case of the submaximal minors, we have had some success. In this section we will look at a fairly old question in modular representation theory and consider how our methods apply to it.

Let $\lambda = (\lambda_1, \ldots, \lambda_k)$ be a partition, d a positive integer, and $\mu(d; i, j)$ the sequence $\lambda_1, \ldots, \lambda_i + d, \lambda_{i+1}, \ldots, \lambda_j - d, \ldots, \lambda_k)$. Assume that $\mu(d; i, j)$ is a partition. For p a prime let $\bar{A}_s$ be the Schur algebra over $Z/(p)$ of degree $s = \sum \lambda_i$, and let $\bar{K}_\lambda$, $\bar{K}_{\mu(d;i,j)}$ be the appropriate Weyl modules over $Z/(p)$. There has been a great deal of interest for quite some time in determining the vanishing behavior of $\operatorname{Hom}_{\bar{A}_s}(\bar{K}_\lambda, K_{\mu(d;i,j)})$ and computing its dimension as a vector space over $Z/(p)$ when it does not vanish (see **[9**, §4], **[10]**). The dimensions of these homomorphism spaces are known as intertwining numbers; we will denote them by $\varepsilon_p^0(\lambda, d; i, j)$. As we are in a modular setting, the full intertwining behavior of the representations $\bar{K}_\lambda$ and $\bar{K}_{\mu(d;i,j)}$ is measured by the Poincaré polynomial $\sum_q \varepsilon_p^q(\lambda, d; i, j)t^q$, where $\varepsilon_p^q(\lambda, d; i, j)$ denotes the dimension of $\operatorname{Ext}^q_{\bar{A}_s}(\bar{K}_\lambda, \bar{K}_{\mu(d;i,j)})$ as a vector space over $Z/(p)$.

Our point of departure for the study of these numbers $\varepsilon_p^q(\lambda, d; i, j)$ is the following.

First of all, if we let K_λ be the Weyl module over Z of shape λ (where λ is any partition), and A_s the Schur algebra over Z of degree s, then for any prime p, $Z/(p) \otimes K_\lambda = \bar{K}_\lambda$ and $Z/(p) \otimes A_s = \bar{A}_s$. (This is just a special case of the universality of the Schur algebra and the Weyl and Schur modules.)

Our next observation is that we have, for each $i \geq 0$, an exact sequence

$$(1)\quad 0 \to Z/(p) \otimes \mathrm{Ext}^q_{A_s}(K_\lambda, K_{\mu(d;i,j)}) \to \mathrm{Ext}^q_{\bar{A}_s}(\bar{K}_\lambda, \bar{K}_{\mu(d;i,j)}) \to \mathrm{Tor}_1^Z\left(Z/(p), \mathrm{Ext}^{q+1}_{A_s}(K_\lambda, K_{\mu(d;i,j)})\right) \to 0.$$

We see immediately from (1) that the knowledge of the integral $\mathrm{Ext}^q_{A_s}$ groups gives us the integers $\varepsilon^q_p(\lambda, d; i, j)$ for all p. Since the resolutions $\mathbb{K}_\lambda$ of K_λ that we discussed in Section 2 are A_s-projective, an explicit description of these resolutions would be a first step in computing the groups $\mathrm{Ext}^q_{A_s}$.

Recall that the terms of the resolution $\mathbb{K}_\lambda$ are direct sums of $D_{a_1} \otimes \cdots \otimes D_{a_n}$ with $(a_1, \ldots, a_n)$ a weight of degree s (and $n = \mathrm{rank}\, F$, where F is our underlying free module). Thus $\mathrm{Hom}_{A_s}(\mathbb{K}_\lambda, K_{\mu(d;i,j)})$ is made up of terms $\mathrm{Hom}_{A_s}(D_{a_1} \otimes \cdots \otimes D_{a_n}, K_{\mu(d;i,j)})$, which are free abelian groups. Now for any A_s-module X, $\mathrm{Hom}_{A_s}(D_{a_1} \otimes \cdots \otimes D_{a_n}, X)$ is the weight submodule of X corresponding to the weight $(a_1, \ldots, a_n)$ (see [**3**]). If $X = K_{\mu(d;i,j)}$, then X has a basis consisting of standard tableaux, and it is easy to see that a basis for $\mathrm{Hom}_{A_s}(D_{a_1} \otimes \cdots \otimes D_{a_n}, X)$ is in 1–1 correspondence with those standard tableaux of content $(a_1, \ldots, a_n)$.

Although all of the above seems perfectly straightforward, it is by no means simple; even when we have the explicit resolution $\mathbb{K}_\lambda$, the computations are not transparent. However, we do have some fragmentary results which we hope will serve to illustrate the method as well as the problems.

From now on, we will be interested only in the intertwining numbers $\varepsilon^0_p(\lambda, d; 1, k)$, and we will let μ denote $\mu(d; 1, k)$. Since $\mathbb{Q} \otimes K_\lambda$ and $\mathbb{Q} \otimes K_\mu$ are irreducible representations, $\mathrm{Hom}_{\mathbb{Q}\otimes A_s}(\mathbb{Q} \otimes K_\lambda, \mathbb{Q} \otimes K_\mu) = 0$, so $\mathrm{Hom}_{A_s}(K_\lambda, K_\mu) = 0$. Therefore, we see that $\mathrm{Tor}_1^Z(Z/(p), \mathrm{Ext}^1_{A_s}(K_\lambda, K_\mu)) \approx \mathrm{Ext}^1_{\bar{A}_s}(\bar{K}_\lambda, \bar{K}_\mu)$; hence $\mathrm{Hom}_{\bar{A}_s}(\bar{K}_\lambda, \bar{K}_\mu)$ is the p-torsion subgroup of $\mathrm{Ext}^1_{A_s}(K_\lambda, K_\mu)$.

If $\ldots X_2 \to X_1 \to X_0 \to K_\lambda \to 0$ is the beginning of the A_s-projective resolution of K_λ, then (since $\mathrm{Hom}_{A_s}(K_\lambda, K_\mu) = 0$)

$$0 \to \mathrm{Hom}_{A_s}(X_0, K_\mu) \xrightarrow{u} \mathrm{Hom}_{A_s}(X_1, K_\mu) \xrightarrow{v} \mathrm{Hom}_{A_s}(X_2, K_\mu) \to \ldots$$

is a complex whose 0-dimensional cohomology is zero, and whose 1-dimensional cohomology is $\mathrm{Ext}^1_{A_s}(K_\lambda, K_\mu)$. Because all the Hom groups are free, the image of v is free, so $\mathrm{Ker}(v)$ is a summand of $\mathrm{Hom}_{A_s}(X_1, K_\mu)$. Therefore, if $\varepsilon_1, \ldots, \varepsilon_t$ are the invariant factors of the map u, we have

$\mathrm{Ext}^1_{A_s}(K_\lambda, K_\mu) \simeq Z/(\varepsilon_1) \oplus \ldots \oplus Z/(\varepsilon_t)$. (Recall that if M is a $t \times (t+\ell)$ integral matrix the invariant factors of M may be computed as follows: Let δ_r be the gcd of the minors of M of order r. Then $\varepsilon_1 = \delta_1$ and $\varepsilon_i = \frac{\delta_i}{\delta_{i-1}}$ for $i = 2, \ldots, t$. In this notation, we let $\frac{0}{0} = 0$. Of course, if M is an injection, $\delta_r \neq 0$ for $r = 1, \ldots, t$, so we don't run into the case of $\frac{0}{0}$.) As a result, in order to compute $\mathrm{Ext}^1_{A_s}(K_\lambda, K_\mu)$, we only need the first two terms, X_0 and X_1, of the resolution of K_λ, and these we know for any partition λ (see [**5**]). But as we said earlier, knowing the relevant terms of the resolution doesn't make the problem altogether trivial. We will therefore specialize even further and, for the rest of the paper, consider only partitions of length two or three, i.e. $\lambda = (\lambda_1, \lambda_2)$ or $\lambda = (\lambda_1, \lambda_2, \lambda_3)$.

A) THE TWO-ROWED CASE: $\lambda = (\lambda_1, \lambda_2)$:
In this case, we have the exact sequence

$$\sum_{\ell \geq 1} D_{\lambda_1+\ell} \otimes D_{\lambda_2-\ell} \xrightarrow{\square} D_{\lambda_1} \otimes D_{\lambda_2} \to K_\lambda \to 0$$

with the map $\square$ defined, for $x \otimes y \in D_{\lambda_1+\ell} \otimes D_{\lambda_2-\ell}$, by

$$\square(x \otimes y) = \sum x(\lambda_1) \otimes x'(\ell) y$$

where $\Delta(x) = \sum x(\lambda_1) \otimes x'(\ell) \in D_{\lambda_1} \otimes D_\ell$.

The partition μ that we are interested in is $\mu = (\lambda_1 + d, \lambda_2 - d)$, and we have to compute the matrix

$$\mathrm{Hom}_A(D_{\lambda_1} \otimes D_{\lambda_2}, K_\mu) \xrightarrow{\mathrm{Hom}_A(\square, K_\mu)} \mathrm{Hom}_A(D_{\lambda_1+\ell} \otimes D_{\lambda_2-\ell}, K_\mu).$$

To compute $\mathrm{Hom}_A(D_{\lambda_1} \otimes D_{\lambda_2}, K_\mu)$ we must determine how many standard tableaux K_μ has of content $1^{\lambda_1} \otimes 2^{\lambda_2}$. (The standard tableaux for a Weyl module, in terms of a given basis of F, are ones that are weakly increasing in the rows, and strictly increasing in the columns.) It's fairly clear that the only standard tableau is

$$\alpha = \begin{array}{l} \overbrace{\boxed{1}\;\cdots\;\boxed{1}}^{\lambda_1}\boxed{2}\;\cdots\;\boxed{2} \\ \underbrace{\boxed{2}\;\cdots\;\boxed{2}}_{\lambda_2-d} \end{array}$$

(From now on, we will denote this tableau by $1^{(\lambda_1)}2^{(d)}/2^{(\lambda_2-d)}$, and use similar notation for the tableaux that are to follow.)

The map $\phi_\alpha : D_{\lambda_1} \otimes D_{\lambda_2} \to K_\mu$ that corresponds to the generator α of the (λ_1, λ_2)-weight submodule of K_μ is:

$$\phi_\alpha(x \otimes y) = \sum xy(d) \otimes y'(\lambda_2 - d)$$

where, as before $\Delta(y) = \sum y(d) \otimes y'(\lambda_2 - d) \in D_d \otimes D_{\lambda_2 - d}$.

It is easy to see that $\mathrm{Hom}_A(D_{\lambda_1+\ell} \otimes D_{\lambda_2-\ell}, K_\mu)$ is zero for $\ell > d$, and that for $1 \leq \ell \leq d$, this group is generated by the single tableau

$$\beta_\ell = 1^{(\lambda_1+\ell)}2^{(d-\ell)}/2^{(\lambda_2-d)}$$

whose corresponding map $\psi_\ell \in \mathrm{Hom}_A(D_{\lambda_1+\ell} \otimes D_{\lambda_2-\ell}, K_\mu)$ is $\psi_\ell(x \otimes y) = \sum xy(d-\ell) \otimes y'(\lambda_2 - d)$.

To compute the matrix of $\mathrm{Hom}_A(\Box, K_\mu)$ with respect to the bases $\{\alpha\}$, $\{\beta_1, \ldots, \beta_d\}$, we have to compute $\mathrm{Hom}_A(\Box, K_\mu)(\phi_\alpha)\left(1^{(\lambda_1+\ell)} \otimes 2^{(\lambda_2-\ell)}\right)$ for $\ell = 1, \ldots, d$. But $\mathrm{Hom}_A(\Box, K_\mu)(\phi_\alpha) = \phi_\alpha\Box$, and

$$\begin{aligned}
\phi_\alpha\Box\left(1^{(\lambda_1+\ell)} \otimes 2^{(\lambda_2-\ell)}\right) &= \phi_\alpha\left(1^{(\lambda_1)} \otimes 1^{(\ell)}2^{(\lambda_2-\ell)}\right) \\
&= \sum_{u=0}^{\ell} 1^{(\lambda_1)}1^{(u)}2^{(d-u)} \otimes 1^{(\ell-u)}2^{(\lambda_2-d-(\ell-u))} \\
&= \sum_{u=0}^{\ell} \binom{\lambda_1+u}{u} 1^{(\lambda_1+u)}2^{(d-u)} \otimes 1^{(\ell-u)}2^{(\lambda_2-d-(\ell-u))}.
\end{aligned}$$

Unless $u = \ell$, the above terms do not give rise to standard tableaux in K_μ, so they must be straightened. Now

$$\begin{aligned}
&\Box\left(1^{(\lambda_1+\ell)}2^{(d-u)} \otimes 2^{(\lambda_2-d-(\ell-u))}\right) \\
&= \sum_{v=0}^{\ell-u-1} \binom{\lambda_2-d-\ell+u+v}{v} 1^{(\lambda_1+u+v)}2^{(d-u-v)} \otimes 1^{(\ell-(u+v))}2^{(\lambda_2-d-\ell+u+v)} \\
&\qquad + \binom{\lambda_2-d}{\ell-u} 1^{(\lambda_1+\ell)}2^{(d-k)} \otimes 2^{(\lambda_2-d)}.
\end{aligned}$$

Therefore,

$$\begin{aligned}
&\Box\left(\sum \binom{(\lambda_1-\lambda_2+d+\ell)}{u} 1^{(\lambda_1+\ell)}2^{(d-u)} \otimes 2^{(\lambda_2-d-\ell-u)}\right) \\
&= \sum_{w=0}^{\ell-1} \binom{\lambda_1+w}{w} 1^{(\lambda_1+w)}2^{(d-w)} \otimes 1^{(\ell-w)}2^{(\lambda_2-d-\ell+w)} \\
&\quad + \left[\binom{\lambda_1+\ell}{\ell} - \binom{\lambda_1-\lambda_2+d+\ell}{\ell}\right] 1^{(\lambda_1+\ell)}2^{(d-\ell)} \otimes 2^{(\lambda_2-d)}.
\end{aligned}$$

Thus, $\phi_\alpha(1^{(\lambda_1)} \otimes 1^{(\ell)}2^{(\lambda_2-k)}) = \binom{\lambda_1-\lambda_2+d+\ell}{\ell}\beta_\ell$, and the matrix we are looking for is the $1 \times d$ matrix:

$$\left(s_1+d+1, \binom{s_1+d+2}{2}, \dots, \binom{s_1+d+d}{d}\right)$$

where $s_1 = \lambda_1 - \lambda_2$. The group $\mathrm{Ext}^1_A(K_\lambda, K_\mu)$ is $Z/(a)$ where $a = \gcd\{s_1 + d+1, \dots, \binom{s_1+d+d}{d}\}$. Now it is fairly elementary to prove that if b and d are positive integers, then $\gcd\left\{b, \binom{b+1}{2}, \dots, \binom{b+d-1}{d}\right\} = \frac{b}{\delta}$ where $\delta = \gcd(b, \mathrm{lcm}\{1, 2, \dots, d\})$. Hence

$$\mathrm{Ext}^1_A(K_\lambda, K_\mu) \approx Z/(a)$$

where $a = \frac{s_1+d+1}{\gcd(s_1+d+1, \mathrm{lcm}\{1,\dots,d\})}$.

THE THREE-ROWED CASE: We now consider $\lambda = (\lambda_1, \lambda_2, \lambda_3)$, and $\mu = (\lambda_1 + d, \lambda_2, \lambda_3 - d)$. (If we add and subtract d to and from two adjacent rows, we immediately reduce to the two-rowed case so the case we are looking at is the only relevant one here.)

Our Weyl module K_λ is the cokernel of the map

$$\sum_{\ell>0} D_{\lambda_1+\ell} \otimes D_{\lambda_2-\ell} \otimes D_{\lambda_3} \oplus \sum_{\ell>0} D_{\lambda_1} \otimes D_{\lambda_2+\ell} \otimes D_{\lambda_3-\ell} \xrightarrow{\Box} D_{\lambda_1} \otimes D_{\lambda_2} \otimes D_{\lambda_3}$$

where, for $x \otimes y \otimes z \in D_{\lambda_1+\ell} \otimes D_{\lambda_2-\ell} \otimes D_{\lambda_3}$, $\Box(x \otimes y \otimes z) = \sum x(\lambda_1) \otimes x'(\ell) y \otimes z$ and, for $x \otimes y \otimes z \in D_{\lambda_1} \otimes D_{\lambda_2+\ell} \otimes D_{\lambda_3-\ell}$, $\Box(x \otimes y \otimes z) = \sum x \otimes y(\lambda_2) \otimes y'(\ell) z$. (The notation $\sum x(\lambda_1) \otimes x'(\ell)$, etc., is the one we have been using consistently.)

We find that the $(\lambda_1, \lambda_2, \lambda_3)$-weight submodule of K_μ has a basis consisting of $d+1$ standard tableaux:

$$\alpha_i = 1^{(\lambda_1)} 2^{(d-i)} 3^{(i)} / 2^{(\lambda_2-d+i)} 3^{(d-i)} / 3^{(\lambda_3-d)}, 0 \le i \le d,$$

corresponding to the maps (which we will also denote by α_i):

$$\alpha_1(u \otimes v \otimes w) = \sum uv(d-i)\, w(i) \otimes v'(\lambda_2 - d + i)\, w'(d-i) \otimes w''(\lambda_3 - d)$$
$$\text{where } \Delta(v) = \sum v(d-i) \otimes v'(\lambda_2 - d + i) \in D_{d-i} \otimes D_{\lambda_2-d+i}$$
$$\text{and } \Delta(w) = \sum w(i) \otimes w'(d-i) \otimes w''(\lambda_3 - d) \in D_i \otimes D_{d-i} \otimes D_{\lambda_3-d}.$$

The groups $\mathrm{Hom}_A(D_{\lambda_1+\ell}\otimes D_{\lambda_2-\ell}\otimes D_{\lambda_3}, K_\mu)$ are zero for $\ell > d$ and, for each ℓ with $1\le \ell\le d$, we have the basis

$$\beta_{\ell i} = 1^{(\lambda_1+\ell)}2^{(d-\ell-i)}3^{(i)}/2^{(\lambda_2-d+i)}3^{(d-i)}/3^{(\lambda_3-d)}$$

for $0\le i\le d-\ell$.

Similarly, for $\mathrm{Hom}_A(D_{\ell_1}\otimes D_{\lambda_2+\ell}\otimes D_{\lambda_3-\ell})$, we have, for each ℓ with $1\le\ell\le d$, the basis

$$\gamma_{\ell i} = 1^{(\lambda_1)}2^{(d-i)}3^{(i)}/2^{(\lambda_2-(d-i-\ell))}3^{(d-i-\ell)}/3^{(\lambda_3-d)} \text{ for } 0\le i\le d-\ell.$$

Thus our matrix for $\mathrm{Hom}_A(\square, K_\mu)$ is a $(d+1)\times 2\binom{d+1}{2}$ matrix, and we compute it by looking at $\alpha_i\square(1^{(\lambda_1+\ell)}\otimes 2^{(\lambda_2-\ell)}\otimes 3^{(\lambda_3)})$ and $\alpha_i\square(1^{(\lambda_1)}\otimes 2^{(\lambda_2+\ell)}\otimes 3^{(\lambda_3-\ell)})$.

$$\alpha_i\square\left(1^{(\lambda_1+\ell)}\otimes 2^{(\lambda_2-\ell)}\otimes 3^{(\lambda_3)}\right) = \alpha_i\left(1^{(\lambda_1)}\otimes 1^{(\ell)}2^{(\lambda_2-\ell)}\otimes 3^{(\lambda_3)}\right)$$
$$= \sum_{k=0}^{\ell}\binom{\lambda_1+k}{k}1^{(\lambda_1+k)}2^{(d-k-i)}3^{(i)}\otimes 1^{(\ell-k)}2^{(\lambda_2-\ell-(d-k-i))}3^{(d-i)}\otimes 3^{(\lambda_3-d)}$$

and, as in the two-rowed case, most of these terms require straightening. A simple induction on $\ell-k$ proves that

$$1^{(\lambda_1+\ell-k)}2^{(d-i-\ell+k)}3^{(i)}/1^{(k)}2^{(\lambda_2-(d-i)-k)}3^{(d-i)}/3^{(\lambda_3-d)}$$
$$= (-1)^k\sum_{u=0}^{k}\binom{\lambda_2-(d-i)-u}{k-u}\binom{d-i+u}{u}1^{(\lambda_1+\ell)}2^{(d-\ell-i+u)}3^{(i-u)}$$
$$/2^{(\lambda_2-(d-i)-u)}3^{(d-i+u)}/3^{(\lambda_3-d)}.$$

Therefore,

$$\alpha_i\square(1^{(\lambda_1+\ell)}\otimes 2^{(\lambda_2-\ell)}\otimes 3^{(\lambda_3)})$$
$$= \sum_u(-1)^u\binom{d-i+u}{u}\binom{s_1+d-i+\ell}{\ell-u}\beta_{\ell i-u},$$

where $s_1 = \lambda_1-\lambda_2$.

A similar calculation gives

$$\alpha_i\square(1^{(\lambda_1)}\otimes 2^{(\lambda_2+\ell)}\otimes 3^{(\lambda_3-\ell)}) = \sum_u\binom{d-i+u}{u}\binom{s_2+\ell+(i+u)}{\ell-u}\gamma_{\ell i-u},$$

where $s_2 = \lambda_2 - \lambda_3$.

We see that our matrix just depends on s_1, s_2 and d. Let us denote this matrix by $M(s_1, s_2; d)$. It is a $(d+1) \times 2\binom{d+1}{2}$-matrix defined recursively on d by:

$$M(s_1, s_2; 1) = \begin{pmatrix} s_1 + 2 & s_2 + 1 \\ -1 & 1 \end{pmatrix} \text{ for all } s_1, s_2 \geq 0;$$

$M(s_1, s_2; d) = (A, B, C)$ where $A = (a_{ij})$ is the $(d+1) \times d$ matrix defined by:

$$a_{ij} = (-1)^{i-1} \binom{d}{i-1} \binom{s_1 + d + j - i + 1}{j - i + 1};$$

$B = (b_{ij})$ is the $(d+1) \times d$ matrix defined by:

$$b_{ij} = \binom{d}{i-1} \binom{s_2 + j}{j - i + 1};$$

and $C = (c_{ij})$ is the $(d+1) \times 2\binom{2}{d}$ matrix defined by:

$$c_{ij} = 0 \text{ for } j = 1, \ldots, 2\binom{2}{d}$$

$$c_{ij} = M(s_1, s_2 + 1; d - 1)_{i-1j} \text{ for } i = 2, \ldots, d + 1,$$

where $M(s_1, s_2 + 1; d - 1)$ is, by our recursive assumption, already defined.

For example,

$$M(s_1, s_2; 2) = \begin{pmatrix} s_1 + 3 & \binom{s_1+4}{2} & s_2 + 1 & \binom{s_2+2}{2} & 0 & 0 \\ -2 & -2(s_1 + 3) & 2 & 2(s_2 + 2) & s_1 + 2 & s_2 + 2 \\ 0 & 1 & 0 & 1 & -1 & 1 \end{pmatrix}$$

It is easy to see that the minors of order 1 and 2 in $M(s_1, s_2; 2)$ generate the unit ideal. Thus $\text{Ext}^1 \approx Z/(a)$ where a is the gcd of the minors of order 3. Some elementary transformations and direct computation show that

$$a = \frac{s_1 + s_2 + 4}{\gcd(s_1 + 3, s_2 + 1, 2)}.$$

Thus, when $s_1 = s_2 = 0$, we get $a = 4$. A very tentative conjecture in the general three-rowed case is that $\text{Ext}^1 \approx Z/(a)$ where

$$a = \frac{s_1 + s_2 + d + 2}{\gcd(s_1 + d + 1, s_2 + 1, \text{lcm}\{1, \ldots, d\})}.$$

REFERENCES

1. K. Akin, *On complexes relating the Jacobi-Trudi identity with the Bernstein-Gelfand-Gelfand resolution*, J. Algebra (to appear).
2. K. Akin and D. A. Buchsbaum, *Characteristic-free representation theory of the general linear group*, Adv. in Math. **58** (1985).
3. K. Akin and D. A. Buchsbaum, *Characteristic-free representation theory of the general linear group II: Homological considerations*, Adv. in Math. (to appear).
4. K. Akin and D. A. Buchsbaum, *Characteristic-free realizations of the Giambelli and Jacobi-Trudi determinantal identities*, in "Proc. of K. I. T. Workshop on Algebra and Topology," 1987.
5. K. Akin, D. A. Buchsbaum and J. Weyman, *Schur functors and Schur complexes*, Adv. in Math. **44** (1982).
6. K. Akin, D. A. Buchsbaum and J. Weyman, *Resolutions of determinantal ideals: the submaximal minors*, Adv. in Math. **39** (1981).
7. D. A. Buchsbaum, *Jacobi-Trudi and Giambelli identities in characteristic-free form*, in "Proc. Special Session on Invariant Theory," AMS Meeting, Denton, Texas, 1986 (to appear).
8. D. A. Buchsabum and D. S. Rim, *A generalized Koszul complex II: Depth and multiplicity*, Trans. AMS **111** (1964).
9. R. W. Carter and G. Lusztig, *On the modular representations of the general linear and symmetric groups*, Math. Zeit. **136** (1974).
10. R. W. Carter and J. T. Payne, *On homomorphisms between Weyl modules and Specht modules*, Math. Proc. Camb. Phil. Soc. **87** (1980).
11. J. Eagon and D. G. Northcott, *Ideals defined by matrices and a certain complex associated to them*, Proc. Royal Soc. **A269** (1962).
12. J. A. Green, *Polynomial representations of* GL_n, Lecture Notes in Math., vol. 830, Springer-Verlag (1980).
13. D. Hilbert, *Über die Theorie der algebraischen Formen*, Math. Ann. **36** (1890).
14. A. Lascoux, *Thèse*, Paris (1977).
15. H. A. Nielsen, *Tensor functors of complexes*, Aarhus University Preprint Series, no. 15 (1978).
16. P. Pragacz and J. Weyman, *Complexes associated with trace and evaluation. Another approach to Lascoux's resolution*, Adv. in Math. **57** (1985).

Department of Mathematics, University of Oklahoma, Norman OK 73019

Department of Mathematics, Brandeis University, Waltham MA 02254

Prof. Buchsbaum's work supported in part by an NSF Grant.

Cohen-Macaulay Modules for Graded Cohen-Macaulay Rings and their Completions

MAURICE AUSLANDER AND IDUN REITEN

Introduction.

Let $\mathbb{Z}$ denote the integers and $R = \coprod_{i \in \mathbb{Z}} R_i$ a $\mathbb{Z}$-graded commutative ring with $R_0 = k$ a field, $R_i = (0)$ when $i < 0$, and where R is generated as a k-algebra by a finite number of homogenous elements. Denote by $\hat{R}$ the completion at the unique graded maximal ideal $m = \coprod_{i>0} R_i$. When R is Cohen-Macaulay, we shall study the relationship between R and $\hat{R}$, especially with respect to almost split sequences for Cohen-Macaulay modules. Existence theorems for almost split sequences were first proved in the complete case [**1,2,6**], and then in the graded case by imitating the method [**7**]. We here show that there is a more direct connection between the complete case and graded case, by deducing the existence theorem for R from that of $\hat{R}$. An interesting feature of this approach is that it shows at the same time that almost split sequences for R stay almost split under completion.

R is said to be of finite (graded) Cohen-Macaulay type if there is only a finite number of indecomposables up to shift, in the category $\mathrm{CM}(\mathrm{gr}\, R)_0$ of graded Cohen-Macaulay modules with morphisms of degree 0. Similarly $\hat{R}$ is said to be of finite Cohen-Macaulay type if there is only a finite number of indecomposables in the category $\mathrm{CM}(\hat{R})$ of Cohen-Macaulay modules over $\hat{R}$. The known examples indicated that R is of finite Cohen-Macaulay type if and only if $\hat{R}$ is, and that when $\hat{R}$ is of finite Cohen-Macaulay type, then each indecomposable in $\mathrm{CM}(\hat{R})$ is obtained by completion. The motivation for this paper was to show that this is the case, and this is done in Section 1. More generally, we show that if R_p is Gorenstein for each nonmaximal graded prime ideal p in R, then in a connected component of the AR-quiver of $\hat{R}$, if one module is the completion of an R-module, then all are. A corresponding result was proved by Gordon-Green for artin algebras [**15**]. In Section 3 we also show that the AR-quivers for R and $\hat{R}$ are closely related.

A crucial step in our proofs is to investigate the connection between indecomposability for an R-module C and its completion $\hat{C}$, and what it means that $\hat{C} \simeq \hat{C}'$. The results we need are given for artin algebras by Gordon-Green [**14**], and we can use Maranda type reduction to the artin case in the context where we need the result. As suggested to us by M. Van den Bergh, these results are however true more generally, and in Section 2 we give a proof using the method suggested by him.

We would like to thank S. Goto, J. Herzog and M. Van den Bergh for helpful conversations.

1. Finite Cohen-Macaulay type for R and $\hat{R}$.

Throughout this paper $R = \coprod_{i\in\mathbb{Z}} R_i$ is a $\mathbb{Z}$-graded commutative Cohen-Macaulay ring, such that $R_i = (0)$ for $i < 0$, $R_0 = k$ is a field and R is generated as a k-algebra by a finite number of homogenous elements. Then R is noetherian and $m = \coprod_{i>0} R_i$ is the unique graded maximal ideal in R. We denote by $\hat{R}$ the completion of R with respect to the m-adic topology. Then $\hat{R}$ is a complete local Cohen-Macaulay ring with maximal ideal $\hat{m}$. Denote by mod$\hat{R}$ the category of finitely generated $\hat{R}$-modules, by CM($\hat{R}$) the full subcategory whose objects are the Cohen-Macaulay $\hat{R}$-modules and by $L_p(\hat{R})$ the full subcategory of CM($\hat{R}$) whose objects C have the property that C_p is R_p-projective for all nonmaximal prime ideals p in $\hat{R}$. Similarly mod$(\operatorname{gr} R)_0$ denotes the category of finitely generated graded R-modules with degree zero maps, and CM($(\operatorname{gr} R)_0$ and $L_p(\operatorname{gr} R)_0$ subcategories defined similarly to the above. Note that a graded module C has the property that C_p is R_p-projective for every nonmaximal graded prime ideal p if and only if C_p is R_p-projective for every nonmaximal prime ideal p in R. In [**6**] we showed that if C is indecomposable in $L_p(\hat{R})$, there is an almost split sequence $0 \to A \to B \to C \to 0$ in CM($\hat{R}$), and A was expressed in terms of C. Using the same methods, an analogous result was proved for R in [**7**]. In this section we show that the existence theorem in the graded case can be deduced from our results in the complete case. At the same time we prove that the almost split sequences for $\hat{R}$ are obtained by completing those for R. We use this to show that R is of finite Cohen-Macaulay type if and only if $\hat{R}$ is.

We can choose a regular homogenous R-sequence $x_1, \ldots, x_d$ in m such that $T = k[x_1, \ldots, x_d]$ is a graded polynomial ring with an inclusion map $T \to R$ of degree zero and R is a finitely generated free T-module. We denote the dualizing module $\operatorname{Hom}_T(R, T)$ by ω_R. Then $\omega_{\hat{R}} = \hat{\omega}_R$ is the dualizing module for $\hat{R}$. If C is an R-module, $\hat{C}$ denotes the completion of C. We have the following crucial results, where the assumptions and notation are as above. In addition, for A in $\operatorname{mod}(\operatorname{gr} R)_0$, $A(n)$ is defined by $A(n)_i = A_{n+i}$.

LEMMA 1. *If A is indecomposable in $L_p(\operatorname{gr} R)_0$, then $\hat{A}$ is indecomposable in $L_p(\hat{R})$.*

LEMMA 2. *If A and B are indecomposable in $L_p(\operatorname{gr} R)_0$ and $\hat{A} \simeq \hat{B}$, then there is some n such that $A \simeq B(n)$.*

PROOF: If R is artin, so that $\hat{R} = R$, the lemmas are known by [**14**, Th. 2.4, Th. 3.6]. We can then use a graded version of Maranda type reduction to reduce to the artin case. Namely, we have the following. (See [**13,23**] for details in the complete case.) Since A and B are in $L_p(\operatorname{gr} R)_0$, the annihilator of $\operatorname{Ext}^1_R(A,A) \oplus \operatorname{Ext}^1_R(A,B) \oplus \operatorname{Ext}^1_R(B,B)$ contains a homogenous regular R-sequence $y_1, \ldots, y_d$ such that $A_1 = A/(y_1,\ldots,y_d)A$ and $B_1 = B/(y_1,\ldots,y_d)B$ are indecomposable in $\operatorname{mod}(\operatorname{gr} R)_0$, where $R_1 = R/(y_1,\ldots,y_d)$. And if A_1 and B_1 are isomorphic in $\operatorname{mod}(\operatorname{gr} R)_0$, then A and B are isomorphic in $L_p(\operatorname{gr} R)_0$. Since R_1 is artin, both lemmas follow easily from this.

In Section 2 we prove these lemmas for A and B in $\operatorname{mod}(\operatorname{gr} R)_0$, as suggested by M. Van der Bergh.

We can now prove the following, on the basis of [**6**].

THEOREM 3. *Let C be a nonfree indecomposable in $L_p(\operatorname{gr} R)_0$. Then there exists an almost split sequence $0 \to A \to B \to C \to 0$ in $\operatorname{CM}(\operatorname{gr} R)_0$, and $0 \to \hat{A} \to \hat{B} \to \hat{C} \to 0$ is almost split in $\operatorname{CM}(\hat{R})$.*

PROOF: Since C is in $L_p(\operatorname{gr} R)_0$, the endomorphism ring $\underline{\operatorname{End}}_R(C)$ modulo projectives is of finite length, and hence equal to its completion $\underline{\operatorname{End}}_{\hat{R}}(\hat{C})$. This shows that $\hat{C}$ is in $L_p(\hat{R})$ since $\hat{C}$ is clearly in $\operatorname{CM}(\hat{R})$. Since $\hat{C}$ is indecomposable by Lemma 1 we know that there is an almost split sequence $0 \to D\operatorname{Tr}_L \hat{C} \to E \to \hat{C} \to 0$ in $\operatorname{CM}(\hat{R})$, which is a generator for the simple $\underline{\operatorname{End}}_{\hat{R}}(\hat{C})^{\text{op}}$-socle of $\operatorname{Ext}^1_{\hat{R}}(\hat{C}, D\operatorname{Tr}_L \hat{C})$. (The second statement is a formal consequence of the first one.) We here recall that D denotes the duality $D = \operatorname{Hom}_{\hat{R}}(\ ,\hat{\omega}_R)$ in $\operatorname{CM}(\hat{R})$, and the transpose $\operatorname{Tr}\hat{C}$ is defined as follows. Let $P_1 \to P_0 \to \hat{C} \to 0$ be a minimal projective presentation in $\operatorname{mod}\hat{R}$. Then $\operatorname{Tr}\hat{C}$ is defined by the exact sequence $\operatorname{Hom}_{\hat{R}}(P_0,\hat{R}) \to \operatorname{Hom}_{\hat{R}}(P_1,\hat{R}) \to \operatorname{Tr}\hat{C} \to 0$, and $\operatorname{Tr}_L C$ is defined to be the d-th syzygy module $\Omega^d \operatorname{Tr} C$. The analogous construction can be done in the category $\operatorname{mod}(\operatorname{gr} R)_0$, and we clearly have $\widehat{D\operatorname{Tr}_L C} = D\operatorname{Tr}_L \hat{C}$.

Since C is in $L_p(\operatorname{gr} R)_0$, $\operatorname{Ext}^1_R(C, D\operatorname{Tr}_L C)$ has finite length, and is hence equal to its completion $\operatorname{Ext}^1_{\hat{R}}(\hat{C}, D\operatorname{Tr}_L \hat{C})$. Since we also have $\underline{\operatorname{End}}_R(C) \simeq \underline{\operatorname{End}}_{\hat{R}}(\hat{C})$, $\operatorname{Ext}^1_R(C, D\operatorname{Tr}_L C)$ is a graded $\underline{\operatorname{End}}_R(C)^{\mathrm{op}}$-module of finite length with a simple socle. $\underline{\operatorname{End}}_R(C)^{\mathrm{op}} \simeq \underline{\operatorname{End}}_{\hat{R}}(\hat{C})^{\mathrm{op}}$ is a local ring, and it is easy to see that its radical is $\coprod_{i\neq 0} \underline{\operatorname{End}}_R(C)^{\mathrm{op}} \coprod \operatorname{rad}(\underline{\operatorname{End}}_R(C)_0^{\mathrm{op}})$. (See [**14**], or Prop. 8.) It then follows that the simple socle of $\operatorname{Ext}^1_R(C, D\operatorname{Tr}_L C)$ is in $\operatorname{Ext}^1_R(C, D\operatorname{Tr}_L C)_i$ for some i. Hence the socle of $\operatorname{Ext}^1_R(C, D\operatorname{Tr}_L C(i))$ is in $\operatorname{Ext}^1_R(C, D\operatorname{Tr}_L C(i))_0$. Let $0 \to D\operatorname{Tr}_L C(i) \to B \to C \to 0$ be a generator for the socle. This is then a nonsplit exact sequence in $\operatorname{CM}(\operatorname{gr} R)_0$, and the completion is almost split in $\operatorname{CM}(\hat{R})$. We want to show that the sequence is almost split in $\operatorname{CM}(\operatorname{gr} R)_0$.

Let $f : X \to C$ be a nonisomorphism in $\operatorname{CM}(\operatorname{gr} R)_0$, with X indecomposable. By considering the exact sequence $0 \to \operatorname{Ker} f \to X \to C \to \operatorname{Coker} f \to 0$, and using that the completion of a nonzero graded module is nonzero, we see that $\hat{f} : \hat{X} \to \hat{C}$ is not an isomorphism. The map $\operatorname{Ext}^1_R(f, \mathrm{id}) : \operatorname{Ext}^1_{\hat{R}}(\hat{C}, D\operatorname{Tr}_L \hat{C}) \to \operatorname{Ext}^1_{\hat{R}}(\hat{X}, D\operatorname{Tr}_L \hat{C})$ sends the almost split sequence $0 \to \widehat{D\operatorname{Tr}_L C(i)} \to \hat{B} \to \hat{C} \to 0$ to 0. $\operatorname{Tr}_L C$ is clearly in $L_p(\operatorname{gr} R)_0$, so that $(D\operatorname{Tr}_L C)_p$ is injective in $\operatorname{CM}(R_p)$ when p is a graded nonmaximal prime ideal in R. Hence $\operatorname{Ext}^1_R(X, D\operatorname{Tr}_L C(i))$ has finite length. The commutative diagram

$$\begin{array}{ccc} \operatorname{Ext}^1_R(C, D\operatorname{Tr}_L C(i)) & \longrightarrow & \operatorname{Ext}^1_R(X, D\operatorname{Tr}_L C(i)) \\ \wr\| & & \wr\| \\ \operatorname{Ext}^1_{\hat{R}}(\hat{C}, D\operatorname{Tr}_L \hat{C}) & \longrightarrow & \operatorname{Ext}^1_{\hat{R}}(\hat{X}, D\operatorname{Tr}_L \hat{C}) \end{array}$$

then shows that $\operatorname{Ext}^1_R(f, \mathrm{id})$ sends $0 \to D\operatorname{Tr}_L C(i) \to B \to C \to 0$ to zero, so that the sequence is almost split.

REMARK: Since we know from [**7**, Th. 1.1] that if $0 \to A \to B \to C \to 0$ is almost split in $\operatorname{CM}(\operatorname{gr} R)_0$, then C is in $L_p(\operatorname{gr} R)_0$, we actually have that any almost split sequence in $\operatorname{CM}(\operatorname{gr} R)_0$ remains almost split under completion.

Before we give our application to finite Cohen-Macaulay type, we give a proof of the following $\mathbb{Z}$-graded version of a result in [**2,10**]. (See also [**8**] for a related result.)

PROPOSITION 4. *If R is of finite Cohen-Macaulay type, then R is an isolated singularity.*

PROOF: Assume that R is of finite Cohen-Macaulay type, and let C be indecomposable nonprojective in $\mathrm{CM}(\mathrm{gr}\, R)_0$. Since for every indecomposable X in $\mathrm{CM}(\mathrm{gr}\, R)_0$, $\mathrm{Hom}_R(X,C)$ is a finitely generated graded R-module, it is easy to see that we get a right almost split map $g: B \to C$ in $\mathrm{CM}(\mathrm{gr}\, R)_0$. This means that any $h: Y \to C$ in $\mathrm{CM}(\mathrm{gr}\, R)_0$, where Y is indecomposable, can be lifted through g. Considering $0 \to \mathrm{Ker}\,\beta \to B \to C \to 0$ and projecting $\mathrm{Ker}\,\beta$ onto an appropriate indecomposable summand, we get an almost split sequence $0 \to A \to B' \to C \to 0$. This implies that C is in $L_p(\mathrm{gr}\, R)_0$ [**7**], so that $\mathrm{CM}(\mathrm{gr}\, R)_0 = L_p(\mathrm{gr}\, R)_0$.

Let now C be in $\mathrm{mod}(\mathrm{gr}\, R)_0$. Since if $d = \dim R$, $\Omega^d C$ is Cohen-Macaulay and hence is in $L_p(\mathrm{gr}\, R)_0$, we get $pd_{R_p} C_p < \infty$ for any nonmaximal graded prime ideal p in R. This shows that gl. $\dim R_p < \infty$ [**18**, II 8.2], so that R is an isolated singularity.

We shall call the $\hat{R}$-modules of the form $\hat{C}$ gradable $\hat{R}$-modules. We then have the following.

THEOREM 5. *a) R is of finite graded Cohen-Macaulay type if and only if $\hat{R}$ is of finite Cohen-Macaulay type.*

b) If $\hat{R}$ is of finite Cohen-Macaulay type, then every module in $\mathrm{CM}(\hat{R})$ *is gradable.*

PROOF: a) Assume first that R is of finite Cohen-Macaulay type. By Proposition 4, R is an isolated singularity, hence $\hat{R}$ is the same. Then $\mathrm{CM}(\hat{R})$ has almost split sequences. The set $\mathcal{S}$ of $\hat{R}$-modules of the form $\hat{C}$ where C is indecomposable in $\mathrm{CM}(\mathrm{gr}\, R)_0$ is a finite set of modules in $\mathrm{CM}(\hat{R})$, which are indecomposable by Lemma 1. For each $\hat{C}$ in $\mathcal{S}$ we have an almost split sequence of the form $0 \to \hat{A} \to \hat{B} \to \hat{C} \to 0$ in $\mathrm{CM}(\hat{R})$, so that $\hat{A}$ and the indecomposable summands of $\hat{B}$ are in $\mathcal{S}$, using Lemma 1. Since $\mathcal{S}$ is finite and contains $\hat{R}$, we know that when $\hat{R}$ is not Gorenstein, $\mathcal{S}$ consists of all indecomposables in $\mathrm{CM}(\hat{R})$ by [**5**, Th. 1.1]. If $\hat{R}$ is Gorenstein, let $g: E \to \hat{R}$ be minimal right almost split. We know that the nonprojective part E' of E has the property that $\Omega^d \hat{m} \simeq \Omega^d E'$ [**5**, Prop. 1.2], and hence E' is clearly gradable. From [**5**, Th. 1.1] we can now conclude that $\mathcal{S}$ consists of all indecomposables also in this case.

Assume conversely that $\hat{R}$ is of finite Cohen-Macaulay type. If for B and C indecomposable in $\mathrm{CM}(\mathrm{gr}\, R)_0$, $B \not\simeq C(i)$ for all i, we know by Lemmas 1 and 2 that $\hat{B}$ and $\hat{C}$ are nonisomorphic indecomposables in $\mathrm{CM}(\hat{R})$. This shows our claim.

Assume now in addition to our previous assumptions that R_p is Gorenstein whenever p is a nonmaximal graded prime ideal in R. We will then say that R is isolated Gorenstein. Then $\hat{R}$ is also isolated Gorenstein. For $\underline{\mathrm{End}}_R(\omega)$ is a graded R-module which by the assumption on R has finite length, so that also $\underline{\mathrm{End}}_{\hat{R}}(\hat{\omega}_R) \simeq \underline{\mathrm{End}}_R(\omega)$ has finite length. In this case $D\,\mathrm{Tr}_L\, C$ is clearly in $L_p(\mathrm{gr}\, R)_0$ when C is, and similarly for $\hat{R}$, so that $L_p(\hat{R})$ and $L_p(\mathrm{gr}\, R)_0$ have almost split sequences. (This follows from the previous results in this section, or see also [**1**,**6**,**7**].) Also there is a minimal right almost split map $E \to \hat{R}$ (and a minimal left almost split map $\omega_{\hat{R}} \to E'$) in $L_p(\hat{R})$ [**9**, Th. 3.9] [**2**, §6]. And the corresponding result for $L_p(\mathrm{gr}\, R)_0$ follows similarly. (Alternatively one can use results from [**3**], as discussed in Section 3.) We recall that saying that a map $g : E \to \hat{R}$ is right almost split in $L_p(\hat{R})$ means that any nonisomorphism $h : X \to \hat{R}$ with X indecomposable in $L_p(\hat{R})$ can be lifted through g. And g is minimal if the restriction to any nonzero summand is nonzero (see [**4**]). We also recall that a map $g : B \to C$ between indecomposables in $L_p(\hat{R})$ or $L_p(\mathrm{gr}\, R)_0$ is irreducible if g is not an isomorphism, and whenever there is a commutative diagram

$$\begin{array}{ccc} X & = & X \\ {\scriptstyle s}\uparrow & & \downarrow{\scriptstyle t} \\ B & \xrightarrow{g} & C \end{array}$$

then s is split mono or t is split epi [**4**]. The indecomposables in $L_p(\hat{R})$ and $L_p(\mathrm{gr}\, R)_0$ are split up into equivalence classes with respect to the equivalence relation generated by the existence of an irreducible map. The following result shows that if one indecomposable in $L_p(\hat{R})$ is gradable, then all in the same equivalence class are. An analogous result was given by Gordon-Green for artin algebras [**15**].

THEOREM 6. *Let R be isolated Gorenstein, and $g : X \to Y$ an irreducible map between indecomposables in $L_p(\hat{R})$. Then X is gradable if and only if Y is gradable.*

PROOF: Assume that Y is gradable. (The proof is similar when we assume X is gradable.) Choose C in $L_p(\mathrm{gr}\, R)_0$ with $\hat{C} \simeq Y$. If C is not projective, let $0 \to A \to B \to C \to 0$ be almost split in $L_p(\mathrm{gr}\, R)_0$. Then $0 \to \hat{A} \to \hat{B} \to \hat{C} \to 0$ is almost split in $L_p(\hat{R})$ by Proposition 3. Since $X \to \hat{C}$ is irreducible we know that X is a summand of $\hat{B}$ [4], and hence gradable, by Lemma 1.

If $Y = R$, assume first that R is not Gorenstein. Then R is not injective in $L_p(\operatorname{gr} R)_0$, so that we have an almost split sequence $0 \to R \to B \to C \to 0$, and hence an almost split sequence $0 \to \hat{R} \to \hat{B} \to \hat{C} \to 0$. Since $\omega_{\hat{R}} \simeq \hat{\omega}_R$ is gradable, we can assume $X \not\simeq \omega_R$. Then we know that $X \simeq D\operatorname{Tr}_L E$ for some indecomposable summand E of $\hat{B}$ (see [4]), and hence X is gradable. If R is Gorenstein, we can by the construction in [**5**, Prop. 1.2], in the graded version used in [**21**], find a map $g : F \coprod R^i \to R$ in $\mathrm{CM}(\operatorname{gr} R)_0$, where F has no nonzero projective summands, $\Omega^d F \simeq \Omega^d m$, and any nonzero map $h : Z \to R$ with Z indecomposable in $\mathrm{CM}(\operatorname{gr} R)_0$ can be lifted through g. We can choose $i \geq 0$ such that g is minimal with respect to this property. By the construction it is clear that $F \coprod R^i$ is in $L_p(\operatorname{gr} R)_0$, so that $g : F \coprod R^i \to R$ is minimal right almost split in $L_p(\operatorname{gr} R)_0$. Since we have a similar construction of the minimal right almost split map $E \coprod \hat{R}^j \to \hat{R}$, we see that $E \simeq \hat{F}$. Since X must be a summand of $E \coprod \hat{R}^j$, X is gradable.

Let $R_1 = k[x,y,z,u,v]/(xz - y^2, xv - yu, yv - zu)$ and $R_2 = k[x,y,z]^{\mathbb{Z}_2}$, where the generator of $\mathbb{Z}_2$ acts by sending each variable to its negative, and assume k is algebraically closed. (For R_2, assume that the characteristic is different from 2, or use the characteristic free description of R_2). Then it was known that R_1, R_2 and $\hat{R}_1, \hat{R}_2$ are of finite Cohen-Macaulay type [**5,7,22**]. It now follows from Theorem 5 that one result can be deduced from the other one.

Also when R is a hypersurface given by an element which is a quadratic form, it was known that R is of finite graded Cohen-Macaulay type if and only if $\hat{R}$ is of finite Cohen-Macaulay type ([**11**], see also [**12**]). For other quasihomogenous hypersurfaces we get the following application, by using known results in the complete case [**17,12,21**].

THEOREM 7. *Let $R = k[x_1, \ldots, x_n]/(f)$, where f is quasihomogenous and k is algebraically closed.*

a) *If char $k \neq 2$, then R is of finite graded Cohen-Macaulay type if and only if $f = g + z_1{}^2 + \cdots + z_t{}^2$, where g is a simple isolated curve singularity in the sense of [**10**].*
b) *If char $k = 2$, R is of finite graded Cohen-Macaulay type if $f = g + x_1 y_1 + \cdots + x_t y_t$, where g is a simple isolated curve singularity in the sense of [**10**] or a rational double point.*

2. Indecomposable modules for R and $\hat{R}$.

In this section we prove the following generalizations of Lemmas 1 and 2.

PROPOSITION 8. *Let A be in* $\operatorname{mod}(\operatorname{gr} R)_0$.

a) *If A is indecomposable in* $\operatorname{mod}(\operatorname{gr} R)_0$, *then* $\operatorname{End}_R(A)$ *is a graded local ring.*

b) *A is indecomposable in* $\operatorname{mod}(\operatorname{gr} R)_0$ *if and only if $\hat{A}$ is an indecomposable $\hat{R}$-module.*

PROPOSITION 9. *Let A and B be indecomposable in* $\operatorname{mod}(\operatorname{gr} R)_0$. *Then $\hat{A} \simeq \hat{B}$ if and only if $A \simeq B(n)$ for some $n \in \mathbb{Z}$.*

To prove Proposition 8 we need the following.

LEMMA 10. *Let $\Lambda = \coprod_{i \in \mathbb{Z}} \Lambda_i$ be a $\mathbb{Z}$-graded ring such that Λ_0 is a local ring with radical $\underline{r}$, and there is an integer n such that $\Lambda_i = (0)$ for $i < n$. Then Λ is a graded local ring with unique graded maximal ideal $\underline{n} = \coprod_{i \neq 0} \Lambda_i \coprod \underline{r}$, and this ideal is also a maximal ideal in Λ.*

PROOF: Let x and y be homogenous elements of degrees s and t, where $s \leq t$. It is enough to show that xy is in $\underline{n}$ if either x or y is in $\underline{n}$. This is clear if $s \neq -t$ or if $s = t = 0$. If $s = -t \neq 0$, and xy is not in $\underline{n}$, there is some z is Λ_0 such that $xyz = 1$ since Λ_0 is a local ring. Since $\Lambda_i = (0)$ for $i < n$, x is nilpotent. Choose m minimal with $x^m = 0$. Then $0 = x^{m-1}(xyz) = x^{m-1}$ gives a contradiction. The argument is similar if y has negative degree, and this finishes the proof.

PROOF OF PROPOSITION 8: When A is indecomposable in $\operatorname{mod}(\operatorname{gr} R)_0$, $\operatorname{End}_R(A)$ is a graded local ring whose unique graded maximal ideal is a maximal ideal. Then $\hat{R} \otimes_R \operatorname{End}_R(A) \simeq \operatorname{End}_{\hat{R}}(\hat{A})$ is a local ring, so that $\hat{A}$ is an indecomposable $\hat{R}$-module.

PROOF OF PROPOSITION 9: Let A and B be indecomposable in $\operatorname{mod}(\operatorname{gr} R)_0$. Hence $\hat{A}$ and $\hat{B}$ are indecomposable $\hat{R}$-modules by Proposition 8. If $\hat{A} \simeq \hat{B}$, the natural $\hat{R}$-map

$$(*) \qquad \operatorname{Hom}_{\hat{R}}(\hat{A}, \hat{B}) \otimes_{\hat{R}} \operatorname{Hom}_{\hat{R}}(\hat{B}, \hat{A}) \xrightarrow{\beta} \operatorname{End}_{\hat{R}}(\hat{A}) \text{ is surjective.}$$

Consider also the natural R-map

$$(**) \qquad \operatorname{Hom}_R(A, B) \otimes_R \operatorname{Hom}_R(B, A) \xrightarrow{\alpha} \operatorname{End}_R(A).$$

Since (*) is the completion of (**) it follows that $\hat{R}\otimes_R \operatorname{Coker}\alpha = \operatorname{Coker}\beta = (0)$, which means that $\operatorname{Coker}\alpha$ is (0) since it is a graded module. Since by Proposition 8 $\operatorname{End}_R(A)$ is a graded local ring, there are maps $g : A \to B$ and $h : B \to A$ of degree j and $-j$ such that $hg = \mathrm{id}_A$. This shows that A is isomorphic to a summand of $B(j)$, and hence isomorphic to $B(j)$.

3. AR-quivers.

In this section we shall refine the results from Section 1 and give relationships between the AR-quivers for $L_p(\operatorname{gr} R)_0$ and $L_p(\hat{R})$, when R is isolated Gorenstein. We recall that the vertices of the AR-quiver are in one-to-one correspondence with the indecomposables in $L_p(\operatorname{gr} R)_0$ or $L_p(\hat{R})$, and there is an arrow between two vertices if there is an irreducible map between the corresponding modules. Associated with each arrow $A \to B$ there is an ordered pair (r_{AB}, s_{AB}) of positive integers, where s_{AB} denotes the multiplicity of B in a minimal left almost split map $A \to E$ and r_{AB} the multiplicity of A in a minimal right almost split map $F \to B$. We say r_{AB} and s_{AB} are zero if there is no arrow $A \to B$.

To be able to compare the AR-quivers for R and $\hat{R}$, we need more information about irreducible maps to R (and dually irreducible maps from ω). We have the following:

PROPOSITION 11. *If $f : E \to R$ is minimal right almost split in $L_p(\operatorname{gr} R)_0$, then $\hat{f} : \hat{E} \to \hat{R}$ is minimal right almost split in $L_p(\hat{R})$.*

PROOF: We use results on Cohen-Macaulay approximations by Auslander-Buchweitz to show that $\hat{f} : \hat{E} \to \hat{R}$ is right almost split. We first show that $f : E \to m \subset R$ is in fact a Cohen-Macaulay approximation, that is, any map $h : X \to m$ with X in $\operatorname{CM}(\operatorname{gr} R)_0$ can be factored through f. For we know by [**7,2**] that there is a minimal Cohen-Macaulay approximation $g : F \to m$, and that $\mathrm{id}_{\operatorname{gr} R} K < \infty$, where $K = \operatorname{Ker} f$ and $\mathrm{id}_{\operatorname{gr} R} K$ denotes the injective dimension in $\operatorname{mod}(\operatorname{gr} R)_0$ [**3**]. If p is a graded nonmaximal prime ideal, we have $m_p \simeq R_p$, and hence $F_p \simeq R_p \coprod K_p$. Since R_p is Gorenstein, $\operatorname{pd} K_p < \infty$, hence K_p is free since it is Cohen-Macaulay. This shows that F is in $L_p(\operatorname{gr} R)_0$, and hence $g : F \to R$ must be isomorphic to $f : E \to R$.

Consider now $0 \to L \to E \xrightarrow{f} m \to 0$ and its completion $0 \to \hat{L} \to \hat{E} \xrightarrow{\hat{f}} \hat{m} \to 0$. Since $\mathrm{id}_{\operatorname{gr} R} L < \infty$, it follows by a graded version of [**19**] that there is an exact sequence $0 \to \omega_t \to \cdots \to \omega_0 \to L \to 0$ in $\operatorname{mod}(\operatorname{gr} R)_0$, where each ω_i is isomorphic to a finite sum of shifts of ω. Completing, we see that $\operatorname{id}_{\hat{R}} \hat{L} < \infty$, and hence $\hat{f} : \hat{E} \to \hat{m}$ is a Cohen-Macaulay approximation by [**3**]. This means that $\hat{f} : \hat{E} \to \hat{R}$ is right almost split.

We next want to show that the map $\hat{f} : \hat{E} \to \hat{R}$ is minimal.

LEMMA 12. *Let $h : A \to B$ be an indecomposable map in* $\operatorname{mod}(\operatorname{gr} R)_0$*. Then $\hat{h} : \hat{A} \to \hat{B}$ is an indecomposable map in* $\operatorname{mod} \hat{R}$.

PROOF: Denote $h : A \to B$ by X as a module over the lower triangular matrix ring $T = \begin{pmatrix} R & 0 \\ R & R \end{pmatrix}$. Then the completion $\hat{T}$ is clearly $\begin{pmatrix} \hat{R} & 0 \\ \hat{R} & \hat{R} \end{pmatrix}$. $\operatorname{End}(X)$ is a $\mathbb{Z}$-graded ring, where the elements of degree i are the pairs (f_i, g_i) of maps of degree i in $\operatorname{mod}(\operatorname{gr} R)_0$, such that $g_i h = h f_i$. $\operatorname{End}(X)_0$ is an artin ring, which must be local since $h : A \to B$ is assumed to be indecomposable in $\operatorname{mod}(\operatorname{gr} R)_0$. Since there is some integer n such that $\operatorname{End}(X)_i = (0)$ if $i < n$, it follows from Lemma 10 that $\operatorname{End}(X)$ is a local ring. Since $\widehat{\operatorname{End}(X)} \simeq \operatorname{End}(\hat{X})$, $\hat{X}$ is an indecomposable $\hat{T}$-module. It is easy to see that $\hat{X}$ is isomorphic to the T-module $\hat{h} : \hat{A} \to \hat{B}$.

Combining Proposition 11 with our results in Section 1, we get the following.

THEOREM 13. *Assume R is isolated Gorenstein, and let A and B be indecomposable in $L_p(\operatorname{gr} R)_0$. Then there is an irreducible map $f : \hat{A} \to \hat{B}$ in $L_p(\hat{R})$ if and only if there is an irreducible map $g : A \to B(i)$ in $L_p(\operatorname{gr} R)_0$ for some i.*

And if there is an irreducible map $f : \hat{A} \to \hat{B}$, then $s_{AB} = \sum_j s_{AB(j)}$ and $r_{AB} = \sum_j r_{AB(j)}$.

This result shows that if $\mathcal{C}$ is a connected component of the AR-quiver for R, then the full subquiver $\hat{\mathcal{C}}$ of the AR-quiver for $\hat{R}$, whose vertices correspond to completions of the modules given by $\mathcal{C}$, is a connected component for $\hat{R}$, and it can be constructed from $\mathcal{C}$. Note however that $\mathcal{C}$ cannot be constructed from $\hat{\mathcal{C}}$, and that there may be more components for R with the same image. $\mathcal{C}$ and $\hat{\mathcal{C}}$ may be isomorphic, for example this is true for all $\mathcal{C}$ when R is the graded ring of an elliptic curve singularity (see [**19**]). But if R is of finite Cohen-Macaulay type, $\mathcal{C}$ and $\hat{\mathcal{C}}$ are clearly not isomorphic. Note that in this case the AR-quiver for R contains more information than the AR-quiver for $\hat{R}$.

References

1. M. Auslander, *Functors and morphisms determined by objects*, in "Proc. Conf. on Representation Theory, Philadelphia 1976," Marcel Dekker, 1978, pp. 1–327.
2. M. Auslander, *Isolated singularities and almost split sequences*, Proc. ICRA IV, Springer Lecture Notes in Math. **1178** (1986), 194–241.
3. M. Auslander and R.-O. Buchweitz, *The homological theory of maximal Cohen-Macaulay approximations.*
4. M. Auslander and I. Reiten, *Representation theory of artin algebras IV: Invariants given by almost split sequences*, Comm. in Algebra **5** (1977), 519–554.
5. M. Auslander and I. Reiten, *The Cohen-Macaulay type of Cohen-Macaulay rings*, Adv. in Math. (to appear).
6. M. Auslander and I. Reiten, *Almost split sequences for Cohen-Macaulay modules*, Math. Annalen **277** (1987), 345–350.
7. M. Auslander and I. Reiten, *Almost split sequences for graded rings*, Proc. Lambrecht Conf. on Vector Bundles, Sing. Theory and Representation Theory, Springer Lecture Notes in Math. **1273** (1987), 232–243.
8. M. Auslander and I. Reiten, *Almost split sequences for abelian group graded rings*, J. Algebra (to appear).
9. M. Auslander and S. Smalø, *Lattices over orders: Finitely presented functors and preprojective partitions*, Trans. AMS **273** (1982), 433–446.
10. W. Barth, C. Peters and A. Van den Ven, "Compact Complex Surfaces," Erg. d. Math. Band 4, Springer, Berlin, 1984.
11. R. O. Buchweitz, D. Eisenbud and J. Herzog, *Cohen-Macaulay modules on quadrics*, Proc. Lambrecht Conf. on Vector Bundles, Sing. Theory and Representation Theory, Springer Lecture Notes in Math. **1273** (1987), 58–133.
12. R. O. Buchweitz, G. M. Greuel and F. O. Schreyer, *Cohen-Macaulay modules on hypersurface singularities II*, Invent. in Math., Fasc 1 **88** (1987), 165–182.
13. E. Dieterich, *Reduction of isolated singularities*, Comm. Math. Helv.
14. R. Gordon and E. L. Green, *Graded artin algebras*, J. Algebra **76** (1982), 111–137.
15. R. Gordon and E. L. Green, *Representation theory of graded artin algebras*, J. Algebra **76** (1982), 138–152.
16. J. Herzog, *The representation type of a graded Cohen-Macaulay ring and its completion.*
17. H. Knörrer, *Cohen-Macaulay modules on hypersurface singularities II*, Invent. Math. **88** (1987), 153–164.
18. C. Natasescu and F. Van Oystayen, "Graded Ring Theory," North Holland, 1982.
19. I. Reiten, *Finite dimensional algebras and singularities*, Proc. Lambrecht Conf. on Vector Bundles, Sing. Theory and Representation Theory, Springer Lecture Notes in Math. **1273** (1987), 35–57.
20. R. Sharp, *Finitely generated modules of finite injective dimension over certain Cohen-Macaulay rings*, Proc. London Math. Soc. (3) **25** (1972), 303–328.
21. Ø. Solberg, *Hypersurface singularities of finite Cohen-Macaulay type*, Proc. Londong Math. Soc. (to appear).
22. Ø. Solberg, *A graded ring of finite Cohen-Macaulay type*, Comm. in Algebra (to appear).
23. Y. Yoshino, *Brauer-Thrall theorems for maximal Cohen-Macaulay modules*, J. Math. Soc. Japan **39, No. 4** (1987), 719–739.

Department of Mathematics, Brandeis University, Waltham MA 02254
Institutt for matematikk og statistikk, Universitetet i Trondheim, AVH, 7055 Dragvoll, NORWAY

Homological Asymptotics of Modules over Local Rings

LUCHEZAR L. AVRAMOV

1. Introduction.

In this paper we consider the asymptotic behavior of the ranks of the free modules in resolutions of finitely generated modules, M, over a commutative noetherian ring, R. We address two basic questions:

— *What* are the possible types of asymptotics for these ranks?

— *How* is any particular type related to the structure of M?

In order to be able to make precise statements, it is convenient to restrict the attention to the case where R is a commutative noetherian local ring, with maximal ideal $\mathbf{m}$, and residue field $k = R/\mathbf{m}$. This will be our usual assumption for the main part of the paper, and we shall describe this situation by writing: "$(R, \mathbf{m}, k)$ is a local ring." The *n-th Betti number*, $b_n^R(M)$, is then defined as the rank of the n-th module in a minimal R-free resolution, F, of the R-module M. Equivalently, this is the minimal number of generators of the *n-th syzygy* of M, $\mathrm{Syz}_n^R(M) = \mathrm{Coker}(d_{n+1} : F_{n+1} \to F_n)$, or the dimension of either of the two dual vector spaces, $\mathrm{Tor}_n^R(M, k)$ and $\mathrm{Ext}_R^n(M, k)$. It should be noted, that everything which follows also applies to the situation in which $R = \oplus_{i \geq 0} R_i$ is a finitely generated graded algebra over a field $k = R_0$, M is a finitely generated graded R-module, and F is the minimal resolution of M by graded free R-modules, with differentials d_n of degree zero. In this setup, $\mathbf{m}$ denotes the irrelevant maximal ideal $\oplus_{i \geq 1} R_i$, and I is given the structure of a graded R-module via the isomorphism $R/\mathbf{m} \cong k$.

Our problem now becomes to study the properties of the Betti sequence $\{b_n^R(M) \mid n \geq 0\}$, as n tends to infinity. Asymptotic properties are best described for eventually non-decreasing sequences, and it has been asked in [**Av4**] whether the Betti sequences have this property. Over any R, this is indeed the case for $M = k$, cf. [**Ta**], but the rings over which *all* M produce eventually non-decreasing sequences form a short list: beside the Golod rings [**Le2**] and those with $\mathbf{m}^3 = 0$ [**Le1**], it includes only scattered examples, described in terms of some specific presentation of M [**GR, Le1,**

Ch]. If one considers the weaker question of Ramras [**Ra**] — is it true that $\lim_{n\to\infty} b_n^R(M) = \infty$, or else the Betti sequence is eventually constant? — then an answer is available in one more case: when R is a *complete intersection*, that is, its $\mathbf{m}$-adic completion $\hat{R}$ has a Cohen presentation $\hat{R} = Q/I$ with Q a regular local ring, and I generated by a Q-regular sequence, then a positive answer follows from [**Gu2**].

Thus, there is at present no general result to guarantee that the Betti sequences have any regular behavior at infinity. This leads us, in the analysis of their asymptotic properties, to replace them by their first sum transforms $\{\beta_n^R(M) \mid n \geq 0\}$:

$$\beta_n^R(M) = \sum_{i=0}^{n} b_n^R(M).$$

(1.1) DEFINITIONS: The *complexity* of M, $\operatorname{cx}_R M$, is equal to d, if $d-1$ is the smallest degree of a polynomial in n, which bounds $b_n^R(M)$ from above (the zero polynomial is assigned degree -1); if no such polynomial exists, then we set $\operatorname{cx}_R M = \infty$. (Complexity is discussed in more detail in the Appendix.)

The Betti numbers are said to have *polynomial growth* of degree d, if there exist polynomials $p(X)$ and $q(X)$ with real coefficients, which are both of degree d and have the same leading term, such that the inequalities:

$$p(n) \leq \beta_n^R(M) \leq q(n)$$

hold for sufficiently large n.

The Betti numbers of M are said to have *exponential growth*, if there exist real numbers $1 < \alpha \leq \beta$, such that the inequalities

$$\alpha^n \leq \beta_n^R(M) \leq \beta^n$$

hold for all sufficiently large n.

In case one can replace $\beta_n^R(M)$ by $b_n^R(M)$ in the sandwiches above, the Betti sequence of M is said to have *strong polynomial growth* of degree $d+1$, resp. *strong exponential growth.* It should be noted that an exponential upper bound on the Betti numbers is known to exist for every finitely generated R-module, cf. (2.5) below.

In order to set the first question above in concrete terms, recall that the *embedding dimension* of R, $\operatorname{edim} R$, is the minimal number of generators of $\mathbf{m}$ ($= \dim_k \mathbf{m}/\mathbf{m}^2$), and that $\operatorname{depth} R$ denotes the maximum length of an R-regular sequence contained in $\mathbf{m}$.

(1.2) PROBLEM: Does there exist a finitely generated R-module M, such that $\operatorname{edim} R - \operatorname{depth} R < \operatorname{cx}_R M < \infty$?

Do the Betti numbers of M grow (strongly) polynomially when $\operatorname{cx}_R M < \infty$?

Do the Betti numbers of M grow (strongly) exponentially when $\operatorname{cx}_R M = \infty$?

We review the main cases in which an answer is known. If $\mathbf{m}^3 = 0$, then either the sequence $\{b_n^R(M) \mid n \geq 1\}$ is constant (hence $\operatorname{cx}_R M \leq 1$) or else it has strong exponential growth [**Le1**]. When R is a Golod ring, a result of [**Le2**] is easily sharpened to give an equally exhaustive answer, cf. (2.7) below. If $\operatorname{cx}_R k < \infty$, then R is a complete intersection [**Gu3**]. If R is not a complete intersection, then the Betti numbers of k grow strongly exponentially [**Av4**]. Finally, over a complete intersection R, every module has complexity $\leq \operatorname{edim} R - \operatorname{depth} R$ [**Gu2**].

The last result is contained in (1.4) below, which also extends it to a larger class of modules: those having finite virtual projective dimension. To define the concept, we need some more notation. A surjection of local rings $\rho : (Q, \mathbf{n}, k) \to (R, \mathbf{m}, k)$ will be called in this paper a (*codimension* c) *deformation* of R, if $\operatorname{Ker} \rho$ is generated by a Q-regular sequence (of length c); by abuse of language, we shall also refer to Q as being a deformation of R. A deformation is called *embedded*, resp. *constant*, if $\operatorname{Ker} \rho \subset \mathbf{n}^2$, resp. $\operatorname{Ker} \rho = 0$. When k is infinite, we set $\tilde{R} = \hat{R}$; when k is finite, $\tilde{R}$ stands for the maximal-ideal-adic completion of $R[Y]_{\mathbf{m}R[Y]}$, where Y is an indeterminate over R. In either case, set $\tilde{M} = M \otimes_R \tilde{R}$. Note that the equalities $b_n^{\tilde{R}}(\tilde{M}) = b_n^R(M)$, which hold for all n, reduce most of the questions dealing with the Betti numbers of M to a situation where the module is finitely generated over a complete local ring with infinite residue field.

(1.3) DEFINITION [**Av5**, (3.3)]: The *virtual projective dimension* of M, $\operatorname{vpd}_R M$, is defined by the formulas:

$$\begin{aligned} \operatorname{vpd}_R M &= \min\{\operatorname{pd}_{Q'} \tilde{M} \mid Q' \text{ is a deformation of } \tilde{R}\} \\ &= \min\{\operatorname{pd}_Q \tilde{M} \mid Q \text{ is an embedded deformation of } \tilde{R}\}. \end{aligned}$$

(That both expressions provide the same result is shown in [**Av5**, (3.4)], where it is also noted that $\operatorname{pd}_R M < \infty$ implies $\operatorname{vpd}_R M = \operatorname{pd}_R M$.)

The work in [**Av5**] now gives complete answers, for modules of finite virtual projective dimension, to the questions raised above.

(1.4) THEOREM. *If* $\operatorname{vpd}_R M < \infty$, *then there is an equality:*

$$\operatorname{vpd}_R M = \operatorname{depth} R - \operatorname{depth} M + \operatorname{cx}_R M,$$

and the Betti numbers of M *grow strongly polynomially of degree* $\operatorname{cx}_R M$.

When R *is not a complete intersection, and* $\operatorname{vpd}_R M < \infty$, *then* $\operatorname{cx}_R M \leq \operatorname{edim} R - \operatorname{depth} R - 2$, *and this can be improved to* $\operatorname{cx}_R M \leq \operatorname{edim} R - \operatorname{depth} R - 3$ *if furthermore* R *is Gorenstein.*

When R *is a complete intersection, then* $\operatorname{vpd}_R M < \infty$ *for every* M, *and in this case* $\operatorname{cx}_R M \leq \operatorname{edim} R - \operatorname{depth} R$.

PROOF: The equality is provided by [**Av5**, (3.5)]. The remaining assertions are easy consequences of [**Av5**, (4.1) and (3.7)] and their proofs. □

Having the complete story over complete intersections, we try to link away from them: R is m links from a complete intersection, if for some Cohen presentation $\hat{R} = Q/I$ there is a collection of ideals $I_0, I_1, \ldots, I_m = I$, and a collection of Q-regular sequences $\underline{x}_0, \underline{x}_1, \ldots, \underline{x}_m$ such that $(\underline{x}_0) = I_0$, and for $1 \leq j \leq m$ one has $\underline{x}_j \subset I_{j-1}$ and $I_j = (\underline{x}_j : I_{j-1})$.

We are ready to state the main result of this paper:

(1.5) THEOREM. *Let* R *be a local ring, which satisfies one of the following conditions:*

(a) $\operatorname{edim} R - \operatorname{depth} R \leq 3$;
(b) $\operatorname{edim} R - \operatorname{depth} R \leq 4$, *and* R *is Gorenstein;*
(c) R *is one link from a complete intersection;*
(d) R *is two links from a complete intersection, and* R *is Gorenstein.*

If M *is a finitely generated* R*-module with* $\operatorname{vpd}_R M = \infty$, *then the Betti numbers of* M *grow exponentially.*

Combined with results of [**Av5**], the theorem yields the following precise classification statement:

(1.6) THEOREM. *Let* M *be a finitely generated module over a local ring,* R, *which satisfies one of the conditions of the preceding theorem.*

Then M *belongs to one of four types, described below by sets of equivalent conditions:*

(I) $\operatorname{cx}_R M = 0$;
$b_n^R(M) = 0$ *for* $n \geq \operatorname{depth} R - \operatorname{depth} M + 1$;
$\operatorname{vpd}_R M = \operatorname{depth} R - \operatorname{depth} M$;

$\operatorname{pd}_R M < \infty$.

(II) $\operatorname{cx}_R M = 1$;
$b_n^R(M) = b > 0$ *for* $n \geq \operatorname{depth} R - \operatorname{depth} M + 1$;
$\operatorname{vpd}_R M = \operatorname{depth} R - \operatorname{depth} M + 1$;
$\operatorname{Syz}^R_{\operatorname{depth} R - \operatorname{depth} M+1}(M)$ *is nonzero and periodic of period 2.*

(III) *There is an integer* d, $2 \leq d \leq \operatorname{edim} R - \operatorname{depth} R$, *such that:*
$\operatorname{cx}_R M = d$;
$\{b_n^R(M)\}$ *grows strongly polynomially of degree* d;
$\operatorname{vpd}_R M = \operatorname{depth} R - \operatorname{depth} M + d$.

(IV) $\operatorname{cx}_R M = \infty$;
$\{b_n^R(M)\}$ *grows exponentially;*
$\operatorname{vpd}_R M = \infty$.

PROOF: Under our hypotheses on R, (1.4) and (1.5) show that the equality $\operatorname{vpd}_R M = \operatorname{depth} R - \operatorname{depth} M + \operatorname{cx}_R M$ holds for all M, hence the equivalence of the first and third condition in each of the four cases. In the first three cases the third condition implies the second one by (1.4), and the same holds in the last case by (1.5). Next, note that $\operatorname{cx}_R M = 0$ is equivalent to $\operatorname{pd}_R M < \infty$ by definition. Finally, since $\operatorname{cx}_R M = 1$ implies $\operatorname{vpd}_R M < \infty$ by (1.5), we conclude that the $(\operatorname{depth} R - \operatorname{depth} M + 1)$-st syzygy of M is periodic of period 2 by invoking [**Av5**, (4.4)]. □

The fact that in (II) the last condition is implied by the first one establishes, for the rings of the theorem, a conjecture of Eisenbud [**Ei**], which he himself proved over complete intersections. At the Berkeley Conference, I conjectured that over any local ring, finite complexity of an R-module M would imply its finite virtual projective dimension: in view of [**Av5**, (4.4)] this represents a strengthening of Eisenbud's conjecture.

Subsequent development has shown that, quite to the contrary, the results of (1.5) and (1.6) are best possible, at least with respect to increasing the value of $\operatorname{edim} R - \operatorname{depth} R$. In fact, in [**AGP**] a module M is constructed over an artinian ring R of embedding dimension 4, such that M is periodic of period 2 (in particular, $\operatorname{cx}_R M = 1$), but $\operatorname{vpd}_R M = \infty$. Following this, Gasharov and Peeva [**GP**] produced, over rings of the same kind, periodic modules with (minimal) periods different from two, and non-periodic modules with constant Betti numbers. They also constructed similar examples over Gorenstein artinian rings of embedding dimension 5. Thus, none of the properties of M, which have been proposed so far as reasons for the module to have finite complexity, is broad enough to account for all cases.

We now move in a different direction. If $\tilde{R}$ has a non-constant embedded deformation, it has a Cohen presentation $\tilde{R} = P/I$ with $(P, \mathbf{p}, k)$ a regular local ring, $I \subset \mathbf{p}^2$, and $I = J + (x)$ with x a nonzero divisor modulo J. But then $\operatorname{pd}_{P/(x)} \tilde{R} < \infty$ (cf. (3.3) below). Thus, there is a local homomorphism $S \to \tilde{R}$ from the singular hypersurface ring $S = P/(x)$, such that the flat dimension of R as an S-module, $\operatorname{fd}_S \tilde{R}$, is finite. That Theorem (1.5) implies a partial converse was pointed out to me by David Eisenbud.

(1.7) Corollary. *If R is as in (1.5), then $\tilde{R}$ has a non-constant embedded deformation if and only if there is a local homomorphism $S \to \tilde{R}$ with S a singular hypersurface ring and $\operatorname{fd}_S \tilde{R} < \infty$.*

Proof: If F is the minimal resolution of the residue field of S, then $\mathsf{F} \otimes_S \tilde{R}$ is a minimal complex of $\tilde{R}$-modules, with $H_i(\mathsf{F} \otimes_S \tilde{R}) = 0$ for $i > \operatorname{fd}_S \tilde{R} = m$. Thus $M = \operatorname{Coker}(F_{m+1} \otimes_S \tilde{R} \to F_m \otimes_S \tilde{R})$ is an $\tilde{R}$-module with $\operatorname{cx}_{\tilde{R}} M = 1$, hence by (1.5) $\operatorname{pd}_Q M < \infty$ for some non-constant embedded deformation Q of $\tilde{R}$. □

Finally, we take a look at the *Bass numbers* $\mu_R^n(M) = \dim_k \operatorname{Ext}_R^n(k, M)$, which count the quantity of copies of the injective envelope of k over R, contained in the n-th module of a minimal injective resolution of M. If the Bass sequence is bounded by a polynomial of degree $d-1$, and by no polynomial of lower degree, we say M has *plexity* d, and write $\operatorname{px}_R M = d$; in a similar vein, we denote by $\operatorname{vid}_R M$ the *virtual injective dimension* of M, i.e. $\min\{\operatorname{id}_Q \tilde{M} \mid Q$ is an embedded deformation of $\tilde{R}\}$ (cf. [**Av5**, (5.1)]). In this notation we have:

(1.8) Corollary. *If $\operatorname{vid}_R M < \infty$, then $\operatorname{vid}_R M = \operatorname{depth} R + \operatorname{px}_R M$, and the Bass number of M grow strongly polynomially of degree $\operatorname{px}_R M$.*

Assume furthermore that R satisfies one of the conditions of (1.5). Then $\operatorname{vid}_R M = \infty$ implies the Bass numbers of M grow exponentially.

Proof: The expression for $\operatorname{vid}_R M$ is taken from [**Av5**, (5.2)]. For any R-module M, Foxby [**Fo**, (3.10)] has constructed an $\tilde{R}$-module N, such that $b_n^{\tilde{R}}(N) = \mu_R^{n-\dim M}(M)$ for all $n \geq \dim M$. Thus, our claims will follow from (1.4) and (1.5) once we show $\operatorname{vid}_R M < \infty$ is equivalent to

$\mathrm{px}_R M < \infty$. One way holds always, cf. [**Gu2**, (4.2)] or [**Av5**, (2.6)], so assume the plexity is finite $\neq 0$, and let Q be an embedded deformation of $\tilde{R}$ over which $\mathrm{pd}_Q N$ is finite. If $\mathrm{id}_Q M < \infty$, we are done. Otherwise, iterate Foxby's construction with Q in place of R: the argument can be repeated, since Q satisfies the same condition as R, and only a finite number of iterations is possible, since $\operatorname{edim} Q - \operatorname{depth} Q < \operatorname{edim} R - \operatorname{depth} R$. □

Having outlined the problems treated in the paper and the main results obtained in it, we now briefly overview its contents.

Section 2 takes a close look at the significance of the rationality of the Poincaré series of a module for the growth of its Betti numbers.

All modules over the rings of Theorem (1.5) have rational Poincaré series, as shown by the main result of the joint work [**AKM**] with Andy Kustin and Matt Miller. This and other results from that paper are used in an essential way in Section 3. Theorem (1.5) is proved there, on the basis of homological criteria for a ring satisfying one of its conditions to be a hypersurface section: cf. Theorem (3.1) and Proposition (3.4).

In Section 4 we complete the investigation of the homotopy Lie algebra of local rings of small embedding codepth or small linking number, started in [**Av3**] and continued in [**JKM**] and [**AKM**]. In particular, it is shown here that all its central elements lie in degree 1 or 2, and that each nonzero central element of degree 2 corresponds to a non-trivial embedded deformation: cf. Theorem (4.4).

The notion of complexity of a finitely generated module over a noetherian local ring, as defined in [**Av5**] and recalled in (1.1) above, plays a crucial role in this paper. As noted in *loc. cit.*, this is an adaptation to our setup of a concept, introduced initially for finite dimensional modular representations by Alperin [**Al**] and Alperin with Evens [**AE**], and extensively used by a number of authors over the past ten years. In an Appendix to this paper, we place the notion of complexity in its natural setup — that of associative ring theory — and show it has satisfactory formal properties, which make it a meaningful refinement of the notion of projective dimension.

2. Modules with rational Poincaré series.

In this paper I shall use some specific properties of rational functions, whose Taylor coefficients are rational integers. The relevant facts, although possibly "well-known", do not seem to appear in the literature in the form required for the applications we have in mind. For this reason, proofs are included.

(2.1) NOTATION: Let $F(t) = \sum_{n\geq 0} b_n t^n \in \mathbb{Z}[[t]]$ represent a rational function of the complex variable t. Write $F(t)$ in the form $f(t)/g(t)$ with coprime complex polynomials $f(t)$ and $g(t)$, such that $g(0) = 1$; clearly, such an expression is unique. We shall also use the decomposition of $g(t)$ in the form:

$$g(t) = (1 - a_1 t)^{d_1} \dots (1 - a_s t)^{d_s}$$

with complex numbers a_i such that $a = |a_1| = \cdots = |a_t| > |a_{t+1}| \geq \cdots \geq |a_s| > 0$, and positive integers d_i, such that $d = d_1 = \cdots = d_u > d_{u+1} \geq \cdots \geq d_t$. The real number $1/a$, which is the radius of convergence of $F(t)$, will also be denoted by ρ.

The basic fact on rational functions with integer coefficients is the following theorem, published by Fatou in 1900. The original proof is reproduced in [**Bi**, pp. 124–126].

(2.2) FATOU'S THEOREM: In the preceding notation, $f(t)$ and $g(t)$ are polynomials with integer coefficient, and $a \geq 1$, with equality holding if and only if a_i is a root of unity for $1 \leq i \leq s$. □

In the next two results we use the fact that the notions of exponential growth and of complexity, defined in (1.1), make sense for arbitrary sequences of real numbers.

(2.3) PROPOSITION. *In the notation of (2.1), the following conditions are equivalent:*

(i) $\{|b_n| \mid n \geq 0\}$ *has infinite complexity;*

(ii) $\rho < 1$;

(iii) $\{|b_n| \mid n \geq 0\}$ *grows exponentially.*

If the sequence $\{b_n\}$ *is eventually non-negative and non-decreasing, they are also equivalent to:*

(iv) $\{|b_n| \mid n \geq 0\}$ *has strong exponential growth.*

PROOF: Expansion into prime fractions yields:

$$F(t) = \sum_{i=1}^{s} \sum_{j=1}^{d_i} \frac{c_{ij}}{(1 - a_i t)^j} + h(t)$$

for some $h(t) \in \mathbb{C}[t]$, and $c_{ij} \in \mathbb{C}$ such that $c_{id_i} \neq 0$. This leads to the expression:

$$(2.3.1) \qquad b_n = \sum_{i=1}^{s} \sum_{j=1}^{d_i} c_{ij}\, a_i^n \binom{j+n-1}{n-1} \quad \text{for } n \gg 0.$$

In particular, there is an $A > 0$, such that

$$(2.3.2) \qquad |b_n| \leq A n^{d-1}\, a^n \quad \text{for } n \gg 0.$$

On the other hand, since $a \geq 1$ by Fatou's Theorem, (2.3.1) shows there is a positive real B' such that

$$(2.3.3) \qquad |b_n| \geq B'\, n^{d-1}\, a^n |W_n| \quad \text{for } n \gg 0,$$

where we have set $W_n = \sum_{i=1}^{u} c_i\, \omega_i^n$, $c_i = c_{id_i}$ and $\omega_i = a_i/a$ for $i = 1, \ldots, u$.

We next use an argument of Ostrowski, cf. [**Bi**, Proof of (3.3.11)], to show there is a $B'' > 0$ such that for each integer $q \geq 0$, one has: $\max\{|W_n| \mid q \leq n \leq q+u-1\} \geq B''$. Denote by v_q the determinant of the $u \times u$ matrix $(\omega_j^{q+i-1})_{\substack{1 \leq i \leq u \\ 1 \leq j \leq u}}$, and by v_{qi} the determinant of its submatrix obtained by deleting its first column and i-th row. Since $|\omega_j| = 1$ for $j = 1, \ldots, u$, we have $|v_q| = |v_0|$ and $|v_{qi}| = |v_{0i}|$ for all $q \geq 0$ and $1 \leq i \leq u$. Furthermore, since the ω_j's are pairwise distinct, and v_0 is the Van der Monde determinant on $\omega_1, \ldots, \omega_u$, we have $v_0 \neq 0$. Setting $v = \max\{|v_{0i}| \mid 1 \leq i \leq u\}$, and expressing c_1 from the system of linear equations $\sum_{j=1}^{u} c_j\, \omega_j^{q+i-1} = W_{q+i-1}$, $i = 1, \ldots, u$, we now get:

$$\begin{aligned} |c_1| &= \left| \frac{1}{v_q} \sum_{i=1}^{u} W_{q+i-1} v_{qi} \right| = \frac{1}{|v_0|} \left| \sum_{i=1}^{u} W_{q+i-1} v_{qi} \right| \\ &\leq \frac{1}{|v_0|} \sum_{i=1}^{u} |W_{q+i-1}| |v_{qi}| \leq \frac{uv}{|v_0|} \max\{|W_n| \mid q \leq n \leq u+q-1\}. \end{aligned}$$

Since $c_1 = c_{1d_1} \neq 0$, we can take $B'' = |c_1|\, |v_0|/uv$.

Thus, we obtain from (2.3.3):

(2.3.4). *There are an integer $h \geq 0$ and a real number $B > 0$, such that the inequality $|b_n| \geq B n^{d-1}\, a^n$ holds for at least one of any u consecutive values of n, when $n \geq h$.*

Now we turn to the proof that the various conditions of the proposition are equivalent.

Noting the triviality of the implications (iii) $\Rightarrow$ (ii) $\Rightarrow$ (i), we assume (i) holds and prove (iii). From (2.3.2) we see that $a > 1$. Setting $\beta_n = \sum_{i=0}^{n} |b_i|$, we obtain from (2.3.4) that

$$\beta_{um+h-1} \geq \sum_{i=0}^{m-1} |b_{h+ui}| \geq \sum_{i=0}^{m-1} (C')^{h+ui} = \frac{(C')^h}{(C')^u - 1}((C')^{um} - 1)$$

for an appropriate $C' > 1$. It is straightforward to see that this implies the desired inequality: $\beta_n \geq C^n$ for $n > 0$ and some $C > 1$.

In case the b_n's are eventually non-negative and non-decreasing, a similar computation establishes the existence of a $C_1 > 1$ such that $b_n > C_1^n$ for $n \gg 0$, so that (i) implies (iv) in this case. Since obviously (iv) contains (iii), we are done. □

(2.4) Proposition. *In the notation of (2.1), the following conditions are equivalent:*

(i) $\{|b_n| \mid n \geq 0\}$ *has complexity* $d' < \infty$;

(ii) $\rho \geq 1$, *and* d' *is the highest order of a pole of* $F(t)$ *on the unit circle;*

(iii) *the decomposition of* $g(t)$ *into prime factors over the rationals is of the form* $\Phi_{m_1}(t)^{e_1} \dots \Phi_{m_\ell}(t)^{e_\ell}$, *where* Φ_{m_i} *denotes the* m_i*-th cyclotomic polynomial,* $m_i \neq m_j$ *for* $i \neq j$, $e_i > 0$ *for* $1 \leq i \leq \ell$, *and* $d' = \max\{e_i \mid 1 \leq i \leq \ell\}$ *(when* $\ell = 0$ *this is taken to mean* $d' = 0$*);*

(iv) *there are integer-valued polynomials* $p_1(z), \dots, p_r(z)$, *such that* $b_{mr+i} = p_i(mr+i)$ *for* $i = 1, \dots, r$ *and* $m \gg 0$, *and with* $\max\{\deg p_i(z) \mid 1 \leq i \leq r\} = d' - 1$.

If the sequence b_n *is eventually non-negative, then they are also equivalent to:*

(v) $\rho \geq 1$, *and* d' *is the order of the pole of* $F(t)$ *at* $t = 1$.

Proof: (i) $\Rightarrow$ (ii). Applying (2.3.4) one sees $F(t)$ has no poles inside the unit disk, and the complexity of $\{|b_n|\}$ is larger or equal than the highest order of a pole of $F(t)$ lying on the unit circle. The converse inequality follows from (2.3.2).

(ii) $\Rightarrow$ (iii): Apply Fatou's Theorem (2.2) to obtain the decomposition, and note that the roots of Φ_{m_i} and Φ_{m_j} are distinct when $m_i \neq m_j$.

(iii) $\Rightarrow$ (iv): Since a_i is the root of unity for $i = 1, \dots, s$, this is immediate from (2.3.1).

(iv) $\Rightarrow$ (i): Obvious.

In order to establish the last statement, we may assume the sequence $\{b_n\}$ has all its terms non-negative, and its complexity is $d' < \infty$. We have to prove that $g \geq d'$, where g denotes the order of the pole at $t = 1$, since the inequality $g \leq d'$ follows from the preceding discussion. Arguing as in the proof of (2.3), we obtain the sequence of inequalities:

$$\beta_n = \beta_{um+h-1} \geq B \sum_{j=0}^{m-1} (uj+h)^{d'-1} \geq Bu^{d'-1} \sum_{j=0}^{m-1} j^{d'-1}$$

$$\geq D''(m-1)^{d'} = D''(n-h-u+1)^{d'}/u^{d'} \geq D'n^{d'} \geq D\binom{n+d'}{d'}$$

for appropriate positive real numbers D'', D' and D, and for $n \gg 0$. Since the β_n are coefficients of $F(t)/(1-t)$, this implies the coefficient-wise inequality $F(t)/(1-t)+E(t) \geq D/(1-t)^{d'+1}$ holds with a suitable polynomial $E(t)$. It follows that $\lim_{t\to 1^-}(1-t)^{g+1}/(1-t)^{d'+1}$ is finite, i.e. $g \geq d'$. (The final twist of the argument is a substantial shortcut, indicated by one of the referees.)

(2.5) REMARK: Recall that the formal power series

$$P_M^R(t) = \sum_{n \geq 0} b_n^R(M) t^n$$

is called the Poincaré series of the R-module M. An argument due to Serre, which we now recall, shows that it represents the Taylor development around the origin of some analytic function: this is equivalent to the existence of an exponential upper bound on the Betti numbers, referred to in (1.1).

Indeed, let $\hat{R} = P/I$ be a Cohen presentation, with P regular local. The associated standard change of rings spectral sequence,

$${}_2E^{p,q} = \operatorname{Ext}^p_{\hat{R}}(\hat{M}, \operatorname{Ext}^q_P(\hat{R}, k)) \Rightarrow \operatorname{Ext}^{p+q}_P(\hat{M}, k),$$

provides by means of estimates of the ranks over k of its consecutive differentials, the coefficient-wise inequality of formal power series:

$$P_M^R(t) = P_{\hat{M}}^{\hat{R}}(t) \leq P_{\hat{M}}^P(t)/(1+t-tP_{\hat{R}}^P(t)).$$

Since the right-hand side is a rational function in t, this proves the claim.

$\square$

When $P_M^R(t)$, viewed as a function of a complex variable, represents a rational function, we say M has rational Poincaré series. That this is not always so, even when $M = k$, was shown by Anick [**An**]. We record next some consequences of rationality.

(2.6) COROLLARY. *If M has rational Poincaré series and bounded Betti numbers, then the sequence $\{b_n^R(M)\}$ is eventually periodic.*

PROOF: By (2.4) $F(t) = f(t)/g(t)$, with $f(t) \in \mathbb{Z}[t]$, and $g(t) = \prod_{i=1}^{\ell} \Phi_{m_i}(t)$ with different m_i's. Multiplying both numerator and denominator with the product of all $\Phi_{m'}(t)$, when m' divides $m = m_1 m_2 \ldots$, and $m' \neq m_i$ for $i = 1, 2, \ldots, \ell$, one sees that $F(t)$ can also be written as a fraction with numerator in $\mathbb{Z}[t]$ and with denominator equal to $1 - t^m$. □

The preceding may be viewed as a weak form of Eisenbud's conjecture. The following result adds a finishing touch to previous work of Ghione with Gulliksen and of Lescot.

(2.7) COROLLARY. *Let M be a finitely generated module over a Golod local ring, R. Then M belongs to one of the following three types:*

(I) $\operatorname{cx}_R M = 0$ *if and only if* $\operatorname{pd}_R M < \infty$.

(II) $\operatorname{cx}_R M = 1$ *if and only if* $\operatorname{pd}_R M = \infty$ *and R is a hypersurface ring, if and only if* $\operatorname{Syz}^R_{\operatorname{depth} R - \operatorname{depth} M + 1}(M)$ *is non-zero and periodic of period 2, if and only if the Betti numbers of M are eventually constant and nonzero.*

(III) $\operatorname{cx}_R M = \infty$ *if and only if* $\operatorname{vpd}_R M = \infty$, *if and only if the Betti numbers of M grow strongly exponentially.*

PROOF: Modules over hypersurface rings are treated in [**Ei**, (6.1)] (cf. also (1.4) above), so we assume R is not one. Then, in the notation of (2.5), $\sum_{i \geq 1} \dim_k \operatorname{Tor}_i^P(\hat{R}, k) > 2$, hence 1 is not a root of the polynomial $1 + t - tP_{\hat{R}}^P(t) = \operatorname{Den}^R(t)$. By [**GG**] $P_M^R(t)$ is a rational function with a divisor of $\operatorname{Den}^R(t)$ as denominator, while by [**Le2**, (6.5)], if $\operatorname{pd}_R M = \infty$, then the Betti numbers of M strictly increase starting from $n = \operatorname{edim} R + 1$. Now apply (2.3) and (2.4). □

For this paper, the primary supply of rational Poincaré series comes from:

(2.8) [**AKM**, (6.4)]. Let $(R, \mathbf{m}, k)$ be a local ring, which satisfies one of the conditions of (1.5), and has embedding dimension n. There exists then a polynomial $\operatorname{Den}^R(t) \in \mathbb{Z}[t]$, such that

$$P_k^R(t) = (1+t)^n / \operatorname{Den}^R(t),$$

and furthermore $\mathrm{Den}^R(t) \cdot P_M^R(t) \in \mathbb{Z}[t]$ for every finitely generated R-module M. □

It may be noted that the formulation in [**AKM**] is less precise, but the proof of the theorem gives the sharper statement recorded above. Due to a result of Levin, cf. [**AKM**, (5.18)], it is an immediate consequence of the following theorem, some particular cases of which have been obtained by a number of authors over a long period of time: cf. [**AKM**, Introduction] for precise references.

(2.9) [**AKM**, (6.1), (6.2) and (6.3)] Let $(R, \mathfrak{m}, k)$ be a local ring, which satisfies one of the conditions of (1.5), and has embedding dimension n; in case R is Gorenstein and $\dim R = n - 4$, assume furthermore that k is of characteristic 2, or k has square roots. Let $\hat{R} = P/I$ be a Cohen presentation, with P regular local of dimension n.

There exists then a factorization $\hat{R} \leftarrow C \leftarrow P$ of the canonical projection, in which the first map is an embedded deformation (i.e., C is a local complete intersection), and the second one is a Golod homomorphism.

□

3. Deformations of local rings of small embedding codepth or small linking number.

The last result quoted in the preceding section gives a "structure theorem" for Cohen presentations, by factoring them on the right through an embedded deformation. The main result of this section achieves a similar factorization on the left.

(3.1) THEOREM. *Let $(R, \mathfrak{m}, k)$ be a local ring of embedding dimension n, which satisfies one of the conditions of (1.5). Let $\hat{R} = P/I$ be a Cohen presentation, with P regular local of dimension n, and let c denote the order of the pole of the Poincaré series $P_k^R(t)$ at $t = 1$ (this makes sense in view of (2.8)).*

There is then a factorization $\hat{R} \leftarrow Q \leftarrow P$ of the canonical projection, in which the second map is a codimension c embedded deformation, and the first one is a surjection onto a ring Q, satisfying the same condition as R. Furthermore, Q has no non-constant embedded deformation, and $P_k^Q(t)$ is regular at $t = 1$.

Before embarking upon its proof, which takes up most of this section, let us show how it enters the

PROOF OF THEOREM (1.5): Once we establish that infinite virtual projective dimension implies infinite complexity, the exponential growth of the Betti numbers will come from (2.8) and (2.3).

So let M be a module with $\operatorname{cx}_R M < \infty$. In the notation of (3.1), $\hat{M}$ has finite complexity over Q by [**Av5**, (3.2.3)], cf. also (A.11) below. By (2.4) $\operatorname{cx}_Q \hat{M}$ equals the order of the pole of $P^Q_{\hat{M}}(t)$ at $t = 1$. By (2.8) this order is at most equal to that of $P^Q_k(t)$, which is zero by (3.1). Thus $\operatorname{cx}_Q \hat{M} = 0$, i.e. $\operatorname{pd}_Q \hat{M} < \infty$, so that $\operatorname{vpd}_R M < \infty$. □

For (3.1) we start with the easier part:

PROOF OF THEOREM (3.1) IN CASES (c) AND (d): It hinges on the structure theorems for R, obtained in [**AKM**, (3.1) and (3.2)]. Namely, it is shown there that for R as in (c) (resp. (d)), $\hat{R}$ has an embedded deformation, Q, of the form P/J, where J is a Northcott (resp. Herzog) ideal of grade m, whose defining parameters are specialized in the maximal ideal $\mathbf{p}$ of P. We assume, as we may, that R is not a complete intersection.

In case (c) then $m \geq 2$, and J is generated by the determinant of an $m \times m$ matrix u and by the entries of the vector vu, where $v \in P^m$, and both u and v have all their entries in $\mathbf{p}$. From [**Av1**, (6.10)] one get the expression:

$$P^Q_k(t) = (1+t)^{n-m}/((1-t)^m - t^2).$$

In case (d) $m \geq 3$, and J is generated by the entries of the vector vu, where both $v \in P^m$ and the $m \times (m-1)$ matrix u have their entries in $\mathbf{p}$, and by the m elements $(-1)^{i+1}u_i + wv_i$, where $w \in \mathbf{p}$, and u_i denotes the determinant of the matrix, obtained from u by deleting the i-th row. From [**HS**, Theorem 2], cf. also [**JKM**, (2.4)], one has:

$$P^Q_k(t) = (1+t)^{n-m}/((1-t)^{m-1} - t).$$

In case (c) (resp. (d)) J is a perfect (resp. Gorenstein) ideal, linked in 1 step (resp. 2 steps) to a complete intersection ideal in P, hence Q is of type (c) (resp. (d)). In either case, $P^Q_k(t)$ has no pole at $t = 1$. To finish the proof it remains to show that the order of the pole of $P^R_k(t)$ at $t = 1$ is precisely $\dim Q - \dim R$, and that Q has no non-constant embedded deformation. The first statement is immediate from the following well-known result of Tate; the second one will be established in (4.4). □

(3.2) [**Ta**, Theorem 5] If Q is a codimension c embedded deformation of an arbitrary local ring R, the equality

$$P^R_k(t) = P^Q_k(t)/(1-t^2)^c$$

holds. □

The proof of (3.1) in the remaining cases requires more effort. We start with some general remarks. For an ideal I of a local ring $(P, \mathbf{p}, k)$ we denote by T the graded k-algebra $\operatorname{Tor}^P(P/I, k)$. The composition of the natural projection $I \to I/\mathbf{p}I$ with the isomorphism $I/\mathbf{p}I \cong T_1$, induced by the exact sequence $0 \to I \to P \to P/I \to 0$ is denoted by σ. In this notation we have:

(3.3) LEMMA. *Let $(P, \mathbf{p}, k)$ be a local ring, and let $I \subset \mathbf{p}$ be an ideal of P, which has finite projective dimension as a P-module. Consider the following conditions on an element $x \in I$:*

(i) *there is an ideal $J \subset P$, such that $(J : x) = J$, and $J + (x) = I$;*
(ii) *x is a non-zero-divisor on P, $\sigma(x) \neq 0$ and T is a free module over its subalgebra generated by $\sigma(x)$;*
(iii) *$\operatorname{pd}_{P/(x)} R/I = \operatorname{pd}_P I - 1$; if furthermore I is a perfect (respectively: Gorenstein) ideal of P, then $I/(x)$ is a perfect (respectively: Gorenstein) ideal of $P/(x)$.*

The implications (i) ⇒ (ii) ⇒ (iii) then hold.

PROOF: (i) ⇒ (ii). Condition (i) asserts precisely that the sequence $0 \to P/J \xrightarrow{x} P/J \to P/I \to 0$ is exact. This implies $\operatorname{pd}_P P/J$ is finite. Thus, x is a non-zero-divisor on a P-module of finite projective dimension, hence is a non-zero-divisor of P itself, due to results of Peskine-Szpiro, Hochster, and P. Roberts.

The second fact to be proved does not require $\operatorname{pd}_P P/I$ to be finite. Indeed, if L is any free P-algebra resolution of P/J, cf. [**Ta**], and K is the Koszul complex on x, then the standard homology exact sequence shows $K \otimes_P L$ is exact in positive degrees, hence is a P-free algebra resolution of P/I. Thus, one has

$$\begin{aligned} T &= H(K \otimes_P L \otimes_P k) = H((K \otimes_P k) \otimes_k (L \otimes_P k)) \\ &\cong H(K \otimes_P k) \otimes_k H(L \otimes_P k) = k\langle \sigma(x) \rangle \otimes_k \operatorname{Tor}^P(P/J, k) \end{aligned}$$

as graded k-algebras.

(ii) ⇒ (iii). Consider the spectral sequence

$${}^2E_{pq} = \operatorname{Tor}_p^{\operatorname{Tor}^P(P/(x),k)}(\operatorname{Tor}^P(P/I,k),k)_q \Rightarrow \operatorname{Tor}_{p+q}^{P/(x)}(P/I,k)$$

where in the ${}^2E$-term p refers to the homological degree, and q refers to the internal degree in the Tor of graded modules over the graded ring $\mathrm{Tor}^P(P/(x), k)$: this is a special case of the sequence constructed in [**Av2**, (3.1)]. Since x is a non-zero-divisor on P, the Koszul complex K on x is a minimal algebra resolution of $P/(x)$ over P, hence $\mathrm{Tor}^P(P/(x), k) = H(K \otimes_P k) = K \otimes_P k$ is the exterior algebra over k on a generator h' of degree 1. Since $\sigma(x) \neq 0$, x is contained in a minimal set of generators of I, hence the map $\mathrm{Tor}^P(P/(x), k) \to \mathrm{Tor}^P(P/I, k) = T$ induced by the natural projection $P/(x) \to P/I$ can be identified with the unique homomorphism of k-algebras, which maps h' onto $\sigma(x)$. Since T is free over $k\langle\sigma(x)\rangle$, this shows the sequence above is concentrated in the line $p = 0$, hence degenerates to an isomorphism $T/\sigma(x)T \cong \mathrm{Tor}^{P/(x)}(P/I, k)$ of graded k-algebras.

Comparing the top degrees of these algebras, one obtains the claim in (iii) concerning the projective dimensions. The claim on perfection follows immediately, since $\mathrm{grade}_{P/(x)} I/(x) = \mathrm{grade}_P I - 1$ by the definition of grade. Finally, for the Gorenstein property use the result of [**AF**], which says that I is a Gorenstein ideal if and only if the k-algebra $\mathrm{Tor}^P(P/I, k)$ has Poincaré duality. Thus, one has to show that if T is a Poincaré duality algebra, which is a free module over its subalgebra generated by $h \in T_1 \setminus \{0\}$, then T/hT is a Poincaré duality algebra as well. This is an easy exercise. □

The next result represents a sharpening of the lemma for ideals of small projective dimension. I know of no counterexample to the proposition in higher dimensions.

(3.4) PROPOSITION. *Let $(P, \mathbf{p}, k)$ be a local ring, and let $I \subset \mathbf{p}$ be an ideal, such that either $\mathrm{pd}_P P/I \leq 3$, or $\mathrm{pd}_P P/I = 4$, and I is Gorenstein. Set $T = \mathrm{Tor}^P(P/I, k)$.*

The following conditions are then equivalent:

(i) *there is an ideal $J \subset P$ and an element $x \in P$ such that $(J : x) = J$ and $J + (x) = I$;*

(ii) *there is a nonzero element $h \in T_1$ such that T is free as a module over its subalgebras $k\langle h\rangle$.*

PROOF: Because of the lemma, we have to show that (ii) implies (i). Clearly, (ii) guarantees $\mathrm{pd}_P P/I \geq 1$. If equality holds, then I is free, that is $I = (x)$ for some non-zero-divisor x; in this case (ii) obtains with $J = 0$. Assume $\mathrm{pd}_P P/I \geq 2$, and let $x \in I$ be an element with $\sigma(x) = h$. Modifying it if necessary by something in $\mathbf{p}I$, we can furthermore assume x is a non-zero-divisor on P. By the lemma we conclude that $1 \leq \operatorname{pd}_{P/(x)} P/I = \operatorname{pd}_P P/I - 1 \leq 3$. Three cases arise.

$\operatorname{pd}_{P/(x)} P/I = 1$. Then $I/(x)$ is a principal ideal of $P/(x)$, generated by a non-zero-divisor $y + (x)$, so that x, y form a P-regular sequence; our claim follows by setting $J = (y)$.

$\operatorname{pd}_{P/(x)} P/I = 2$. By the Buchsbaum-Eisenbud lifting theorem [**BE1**, (9.2)] there is an ideal $J \subset P$, such that $\operatorname{pd}_P J = 1$, x is a non-zero-divisor on P/J, and $(P/J)/x(P/J) = P/I$; put in other words, this read $(J : x) = J$ and $J + (x) = I$, as claimed.

$\operatorname{pd}_{P/(x)} P/I = 3$. By the lemma in this case $I/(x)$ is a Gorenstein ideal of $P/(x)$, hence the preceding argument can be repeated, by using the lifting theorem for Gorenstein ideals of grade 3, cf. [**BE2**, Remark p. 478]. □

(3.5) THEOREM. *Let I be an ideal of the n-dimensional regular local ring $(P, \mathbf{p}, k)$, such that $I \subset \mathbf{p}^2$, and set:*

$$
\begin{aligned}
&R = P/I;\\
&T = \operatorname{Tor}^P(R, k);\\
&p = \dim_k (T_1)^2, \quad q = \dim_k (T_1\, T_2);\\
&\operatorname{Den}^R(t) = (1+t)^n / P_k^R(t).
\end{aligned}
$$

If either $\operatorname{pd}_P R \leq 3$, *or* $\operatorname{pd}_P R = 4$ *and R is Gorenstein, then one of several disjoint cases occurs* (see Table 1).

PROOF: Most claims are already in the literature.

When $R = P$ is regular, $\operatorname{Den}^R(t) = 1$ by the exactness of the Koszul complex on a minimal set of generators of $\mathbf{p}$. The rings of type CI are complete intersections of codimension $\operatorname{pd}_P R$, hence $\operatorname{Den}^R(t)$ is obtained from Tate's theorem (3.2). For $\operatorname{pd}_P = 2$ the table is filled in by Scheja's paper [**Sc**]. When $\operatorname{pd}_P R = 4$ and k has $\frac{1}{2}$ the classification is due to Kustin and Miller [**KM3**], who also computed the Poincaré series in joint work with Jacobsson [**JKM**]; the assumption on the characteristic was removed by Kustin [**Ku**]. Finally, the five types of Tor-algebras when $\operatorname{pd}_P R = 3$ were obtained by Weyman [**We**] (apparently, under characteristic zero assumptions), and then (in a characteristic-free way) in [**AKM**]; both papers note the rationality of $P_k^R(t)$, but do not provide a catalogue. This is what we do next, in order to complete the proof.

$\mathrm{pd}_P R$	Type	$P_R^P(t)$	$\frac{p}{q}$	$\mathrm{Den}^R(t)$
0	REG	1		1
1	CI	$1+t$		$(1+t)(1-t)$
2	CI	$(1+t)^2$		$(1+t)^2\,(1-t)^2$
	GO	$1+(\ell+1)t+\ell t^2$, $\ell+1\geq 2$	$\frac{p=0}{q=0}$	$(1+t)(1-t-\ell t^2)$
3	CI	$(1+t)^3$		$(1+t)^3(1-t)^3$
	TE		$\frac{p=3}{q=0}$	$(1+t)(1-t-\ell t^2-(m-\ell-3)t^3-t^5)$
	B	$1+(\ell+1)t+mt^2+(m-\ell)t^3$	$\frac{p=1}{q=1}$	$(1+t)(1-t-\ell t^2-(m-\ell-1)t^3+t^4)$
	$\mathrm{G}(r), (\ell+1\geq r\geq 2)$	$m\geq \ell+1\geq 3$	$\frac{p=0}{q=1}$	$(1+t)(1-t-\ell t^2-(m-\ell)t^3+t^4)$
	$\mathrm{H}(p,q)$		$\frac{\ell\geq p\geq 0}{m-\ell\geq q\geq 0}$	$(1+t)(1-t-\ell t^2-(m-\ell-p)t^3+qt^4)$
4	CI	$(1+t)^4$		$(1+t)^4(1-t)^4$
	GTE	$1+(\ell+1)t+2\ell t^2+(\ell+1)t^3+t^4$	$\frac{p=3}{q=3}$	$(1+t)^2(1-2t-(\ell-2)t^2+t^3+t^4-t^5)$
	GGO		$\frac{p=0}{q=0}$	$(1+t)^2(1-2t-(\ell-2)t^2-2t^3+t^4)$
	$\mathrm{GH}(p)$	$\ell+1\geq 6$	$\frac{\ell\geq p\geq 1}{q=p}$	$(1+t)^2(1-2t-(\ell-2)t^2+(p-2)t^3+2t^4-t^5)$

Table 1

We use the technique of [**Av1**]. First of all, since by [**BE2**, (1.3)] the minimal P-free resolution of R carries a structure of associative differential graded algebra, one has the equality $P_k^R(t) = (1+t)^n \cdot P_k^T(t)$ by [**Av1**, (3.3)], hence $\mathrm{Den}^R(t) = 1/P_k^T(t)$. Next we find in T a graded subalgebra, A, and a graded vector subspace, W, such that $T = A \oplus W$, $A_+W = 0$, and $W^2 = 0$. In other words, T is isomorphic to the trivial extension $A \ltimes W$, hence the graded version of a theorem of Gulliksen [**Gu1**, Theorem 2], given in [**Av1**, (9.1)], provides the expression

$$1/P_k^T(t) = (1/P_k^A(t)) - t\left(\sum_{i\geq 0} \dim_k W_i\, t^i\right).$$

Clearly, it is advantageous to take A as small as possible, and we use for this purpose the list established in [**AKM**, (2.1)]. For the rest of the proof, we adhere to the following conventions: $e_1, \ldots, e_s$ denotes a basis of A_1; $f_1, \ldots, f_t$ — one of A_2; $g_1, \ldots, g_u$ — one of A_3; only nonzero products of basis elements are indicated, and the multiplication table of A is completed by using skew-commutativity, and setting all other products equal to zero.

There are four cases to deal with:

Type TE: $s = t = 3$, $u = 0$; $e_1\, e_2 = f_3$, $e_2\, e_3 = f_1$, $e_3\, e_1 = f_2$. By [**Av1**, (9.2)] we have:

$$1/P_k^A(t) = (1-t^2)^3 - t^5.$$

Type B: $s = 2$, $t = 3$, $u = 1$; $e_1\, e_2 = f_3$, $e_1\, f_1 = g_1 = e_2\, f_2$. It is shown in the proof of [**AKM**, (6.1)] that the DG algebra $B = A\langle X; dX = e_1\rangle$ has a trivial Massey operation, hence by (a version of) Golod's theorem, $P_k^B(t) = 1/(1 - t\sum_{i\geq 1} \dim_k H_i(B)t^i)$, which gives:

$$\begin{aligned} 1/P_k^A(t) &= (1-t^2)/P_k^B(t) = (1-t^2)(1-t^2-t^3-t^3(1-t^2)^{-1}),\\ &= 1-2t^2-2t^3+t^4+t^5. \end{aligned}$$

Type G(r): $s = t = r \geq 2$, $u = 1$; $e_i\, f_i = g_1$, $1 \leq i \leq r$. The Poincaré series of such an algebra is computed in [**Av1**, (9.6)], the result being

$$1/P_k^A(g) = 1 - rt^2 - rt^3 + t^5.$$

Type H(p,q): $s = p+1$, $t = p+q$, $u = q$; $e_{p+1}\, e_i = f_i$ for $1 \leq i \leq p$, $e_{p+1}\, f_{p+j} = g_j$ for $1 \leq j \leq q$. Denote by C (resp. D) the subspace of A

spanned by 1 and e_{p+1} (resp. $1, e_1, \ldots, e_p, f_{p+1}, \ldots, f_{p+q}$). Then C and D are subalgebras of A, $(C_+)^2 = 0$, $(D_+)^2 = 0$, and $A = C \otimes_k D$. Thus:

$$\begin{aligned} 1/P_k^A(t) &= 1/P_k^C(t)\, P_k^D(t) = (1-t^2)(1-pt^2-qt^3) \\ &= 1-(p+1)t^2 - qt^3 + pt^4 + qt^5. \end{aligned}$$

□

We are now ready for the

PROOF OF THEOREM (3.1) IN CASES (a) AND (b): Replacing $\hat{R}$ by R we may assume we are under the hypotheses of (3.5). By this theorem, $P_k^R(t)$ is a rational function of the form $(1+t)^{\operatorname{edim} R} / \operatorname{Den}^R(t)$. Thus, Tate's theorem (3.2) shows that if R has a codimension $c > 0$ embedded deformation, then $\operatorname{Den}^R(t)$ has at $t = 1$ a zero of order at least c. Using the expressions for $\operatorname{Den}^R(t)$ listed in (3.5), one sees three cases appear: R is of type CI; R is of type H(p, q) with $p + q = m$; R is of type GH(p), with $p = \ell$.

In the first case R is a singular complete intersection of codimension $\operatorname{edim} R - \operatorname{depth} R$, and this integer is also the order of the pole of $P_k^R(t)$ at $t = 1$. In this situation, take $Q \leftarrow P$ to be the identity map.

In the remaining two cases, $P_k^R(t)$ has a simple pole at $t = 1$. In order to finish the proof, we shall show that R has then an embedded deformation Q. By the notes at the beginning of this proof, Q will have no non-constant embedded deformation.

Let R be of type H(p, q) with $p + q = m$. The inequalities $p \leq \ell$ and $q \leq m - \ell$ show that $p = \ell$ and $q = m - \ell$. In the notation of the proof of (3.5) this means $T = A$, hence we can take $h = e_{p+1}$.

Let finally R be of type GH(ℓ). When $\operatorname{char} k = 2$, a decomposition $T = k\langle e_{\ell+1}\rangle \otimes_k D$ (as k-algebras) is provided by [**Ku**, Corollary]. When $\operatorname{char} k \neq 2$ and k has square roots, the same conclusion is established in [**KM3**, (2.2)]. However, a careful reading of the argument of that paper shows that the assumption on k is not used in the choice of the element $e_{\ell+1}$ in T_1. Letting $P \to P'$ denote an inflation, which on the residue fields level induces an inclusion of k into its algebraic closure k', and setting $R' = P' \otimes_P R$, one has a canonical isomorphism of k'-algebras $T' = \operatorname{Tor}^{P'}(R', k') \cong k' \otimes_k T$. By the case already treated, T' is free as a $k'\langle e_{\ell+1}\rangle$-module, and the freeness of T over $k\langle e_{\ell+1}\rangle$ follows. Here again, set $h = e_{\ell+1}$. Now apply (3.4). □

(3.6) REMARK: Due to the structure theorems of Burch [**Bu**] and Buchsbaum-Eisenbud [**BE2**], part of Theorem (3.1) can be recast as a description of all local rings of small embedding codepth, which have embedded deformations. To do this we use the notation of (3.5). In particular, $\ell + 1 = \nu(I) = \dim_k T_1 = \dim_k H_1(K^R)$, where K^R denotes the Koszul complex on a minimal set of generators of $\mathbf{m}$. Thus we have:

When $\operatorname{edim} R - \operatorname{depth} R \leq 3$, the only local rings which are not completely intersections and have a non-constant embedded deformation, are those of type $H(\ell, \ell - 1)$. In this case $\ell \geq 2$, and $\hat{R} \cong P/(aJ' + (x))$, where J' is a height 2 ideal generated by the maximal minors of an $\ell \times (\ell - 1)$ matrix with entries in $\mathbf{p}$, $a \in R$ ($a \in \mathbf{p}$ when $\ell = 2$), and x is a non-zero-divisor modulo aJ'. Furthermore, R is Cohen-Macaulay if and only if $a \notin \mathbf{p}$, and $\ell \geq 3$.

When $\operatorname{edim} R - \operatorname{depth} R \leq 4$, the only local Gorenstein rings, which are not complete intersections and have a non-constant embedded deformation, are those of type $\mathrm{GH}(\ell)$. In this case ℓ is odd, $\ell \geq 5$, and $\hat{R} \cong P/(J + (x))$, where J is a height 3 ideal generated by the $(\ell - 1) \times (\ell - 1)$ pfaffians of an $\ell \times \ell$ alternating matrix with coefficients in $\mathbf{p}$, and x is a non-zero-divisor modulo J.

4. The representation of the homotopy Lie algebra on the cohomology of the module.

Yoneda products, cf. [**Ma**], give $\operatorname{Ext}_R^*(k, k)$ a structure of graded k-algebra, and $\operatorname{Ext}_R^*(M, k)$ one of a graded left module over it. Since $\operatorname{Ext}_R^*(k, k)$ is in a canonical way the universal enveloping algebra of a graded Lie algebra, $\pi^*(R)$, called the homotopy Lie algebra of R, cf. [**Av4**], one obtains a representation of $\pi^*(R)$ on the graded vector space $\operatorname{Ext}_R^*(M, k)$. Here we inspect this structure for its influence of the growth of the Betti sequence.

(4.1) PROBLEM: Let M be a finitely generated R-module with $\operatorname{cx}_R M < \infty$. Does there exist a finitely generated graded k-subalgebra $\mathcal{Q}$ of $\operatorname{Ext}_R^*(k, k)$, such that the sequence $\{\dim \mathcal{Q}_n \mid n \geq 0\}$ has finite complexity, and over which $\operatorname{Ext}_R^*(M, k)$ is a finitely generated module?

The main reason for expecting strong finiteness conditions should hold is that for non-complete intersections R the Betti numbers of k grow strongly exponentially, cf. [**Av4**, (6.2)]. When $\operatorname{vpd}_R M < \infty$, a positive solution is available, in a stronger form:

(4.2) If $\rho : Q \to R$ is a codimension c embedded deformation, then the natural homomorphism of graded Lie algebras

$$\pi^*(\rho) : \pi^*(R) \to \pi^*(Q)$$

is surjective, and its kernel is a c-dimensional central ideal, concentrated in degree 2.

If the finitely generated R-module M has $\operatorname{pd}_R M < \infty$, then $\operatorname{Ext}^*_R(M,k)$ is a finitely generated graded module over the universal enveloping algebra of $\operatorname{Ker} \pi^*(\rho)$, which is a central c-dimensional polynomial subalgebra of $\operatorname{Ext}^*_R(k,k)$. □

This is essentially a compilation of results of different authors, cf. [**Av5**, (6.1)] for references. In particular, it focuses attention on the center, $\zeta^*(R)$, of the homotopy Lie algebra. Jacobsson [**Ja**, p. 230] conjectured, that it is finite dimensional and concentrated in degrees 1 and 2. In the recent joint work [**FHJLT**] the first part was proved, and it was also shown that $\dim \zeta^{\text{even}} \leq \operatorname{edim} R - \operatorname{depth} R$. A positive answer to the next problem would give on the center the most detailed information possible. We use the ring $\tilde{R}$, defined in the Introduction, and identify $\pi^*(\tilde{R})$ with $\tilde{k} \otimes_k \pi^*(R)$.

(4.3) PROBLEM: Does every element $z \in \zeta^{\geq 2}(R)$ arise from a deformation, i.e. does there exist a codimension one deformation ρ of $\tilde{R}$, such that $1 \otimes z$ generates $\operatorname{Ker} \pi^*(\rho)$?

Clearly, the answer is positive when R is a complete intersection, since then $\pi^{\geq 3}(R) = 0$, and $\pi^2(R) = \operatorname{Ker} \pi^*(\rho)$, where ρ is a deformation of R to a regular local ring of the same embedding dimension. The main result of this section contains a positive solution for the rings of the Main Theorem.

(4.4) THEOREM. *Let R be a local ring, and let $\zeta^2(R)$ be the subspace of central elements of degree 2 in its homotopy Lie algebra, $\pi^*(R)$. Denote by $\bar{\pi}^*(R)$ the graded Lie algebra $\pi^{\geq 2}(R)/\zeta^2(R)$, and write $\bar{y}$ for the image $\bar{\pi}^*(R)$ of $y \in \pi^*(R)$.*

If R satisfies one of the conditions of (1.5), then the following hold:

1. *$\dim_k \zeta^2(R) = c$, where c is the order of the pole of $P^R_k(t)$ at $t = 1$;*
2. *every element of $\zeta^2(R)$ arises from a deformation;*
3. *for each nonzero $\bar{y}_1 \in \bar{\pi}^*(R)$ there is a $\bar{y}_2 \in \bar{\pi}^i(R)$ with $i = 2$ or 3, such that y_1 and y_2 generate a non-commutative free Lie subalgebra of $\pi^*(R)$.*

In particular, $\bar{\pi}^*(R)$ contains no non-trivial solvable ideals.

(4) *$\pi^{\geq 3}(R)$ is a free Lie algebra, which is zero if R is a complete intersection, and is non-commutative otherwise.*

PROOF: The case of complete intersections is clear from the preceding remarks, so for the rest of the proof we assume R is not a complete intersection. Also, one may assume R is complete, so we adopt the hypotheses and notation of (3.5).

We deal with the last claim first. Since it is invariant under field extensions, we may assume k has square roots, so let $C \to R$ be the Golod map from a complete intersection, provided by (2.9). By [**Av1**, (3.5)] there is an exact sequence of graded Lie algebras:

$$0 \to \mathsf{L}(\mathrm{sExt}_C^+(R,k)) \to \pi^*(R) \to \pi^*(C) \to 0,$$

where L is the free Lie algebra functor, and $(sW)^i = W^{i-1}$ for any graded k-vector space W. If it happened that $\dim_k \mathrm{Ext}_C^+(R,k)$ were 0 or 1, $\mathrm{Ker}(C \to R)$ would be trivial, or generated by a nonzero divisor, hence R would be a complete intersection, contradicting our assumption. Thus $\mathsf{L}(\mathrm{sExt}_C^+(R,k))$ is a non-commutative free Lie algebra. This property is inherited by its subalgebra consisting of the elements of degree ≥ 3, which — by the exact sequence above — is isomorphic to $\pi^{\geq 3}(R)$. This also establishes the claim of (3) when $\deg y_1 \geq 3$.

Let now $\rho : Q \to R$ be the codimension c deformation, supplied by (3.1). Because of (4.2), it gives rise to a c-dimensional subspace of $\zeta^2(R)$, all of whose elements arise from deformations. If z is a central element of $\pi^{\geq 2}(R)$, which is not contained in this subspace, the surjectivity of $\pi^*(\rho)$ implies its image in $\pi^*(Q)$ is a nonzero central element. Hence in order to prove (1) and (2) it suffices to show $\zeta^{\geq 2}(Q) = 0$, which in turn follows from the validity, for Q, of property (3). Thus, the theorem will be proved completely, once we establish the validity of (3) under the additional assumption that $P_k^R(t)$ has no pole at $t = 1$. We work under this assumption from now on, and show it implies the following:

CLAIM. *For each nonzero element $h_1 \in T_1$, there exists a $h_2 \in T$, such that $h_1 h_2 = 0$ and $(h_2)^2 = 0$, and furthermore:*

(1) *either $h_2 \in T_1$ and $h_2 \notin kh_1$, or*

(2) *$h_2 \in T_2$ and $h_2 \notin (T_1)^2$.*

To this end, we use decompositions $T = A \ltimes W$ with $A_+ W = 0$, provided by previous work: the claim will follow if we show that either W_1 or W_2 is not zero. Under condition (a) the relevant decompositions, obtained in [**We**] and [**AKM**], were already used in the proof of (3.5). They show $W_1 \neq 0$, unless R is of type $H(\ell, q)$, or of type TE with $\ell = 2$. In the first case $W_2 \neq 0$, since otherwise one obtains type $\mathrm{H}(\ell, m - \ell)$, and the Poincaré series has then a pole at $t = 1$. In the second case, $W_2 = 0$ means $T_2 = (T_1)^2$, which is a well-known criterion for complete intersections. Under condition (b), with R Gorenstein of codimension 4, the multiplication tables of [**KM3**, (2.2)] show that $W_1 \neq 0$, except for rings of type $\mathrm{GH}(\ell)$, but these have 1 as a singular point for their Poincaré series. Finally, assume condition (c) or (d) holds, and $P_k^R(t)$ is regular at $t = 1$. As recalled in the proof of Theorem (1.5), then I is a Northcott or Herzog ideal of P, whose defining parameters are specialized in $\mathbf{p}$. Using the explicit DG algebra structure on the minimal P-free resolution of R, constructed respectively in [**AKM**, (4.1)] and in [**KM2**, (1.6)], one sees that in both cases $W_1 \neq 0$ (cf. also [**JKM**, Proof of (2.4)], and [**AKM**, Proof of (6.3)]). The Claim is proved.

Let K^P denote the Koszul complex on a minimal system of generators of $\mathbf{p}$, and set $K^R = R \otimes_P K^P$. From [**Av4**, (3.1)] we know that the homomorphism of homotopy Lie algebras, induced by the inclusion $R \to K^R$, identifies $\pi^*(K^R)$ with $\pi^{\geq 2}(R)$. Furthermore, let F be a minimal P-free resolution of R, equipped with a DG algebra structure: its existence in case (c) and (d) has just been recalled, in case (a) it is due to Buchsbaum and Eisenbud (cf. the proof of (3.5)), and in case (b) and codimension 4 to Kustin and Miller [**KM1**, (4.3)] and Kustin [**K**, Theorem]. The homomorphisms of DG algebras

$$K^R = R \otimes_R K^P \leftarrow F \otimes_P K^P \to F \otimes_P k = T$$

induce isomorphisms in homology, hence $\pi^*(K^R) \cong \pi^*(T)$ by [**Av4**, (2.1)]. Hence, it suffices to show that $\pi^*(T)$ has property (3) of the Theorem.

Since $\pi^2(T) = \mathrm{Hom}_k(T_1, k)$, choose $h_1 \in T_1$ such that $y_1(h_1) \neq 0$, then pick the $h_2 \in T$ of the Claim, finally choose y_2 in $\pi^2(T)$ or in $\pi^3(T) = \mathrm{Hom}_k(T_2/(T_1)^2, k)$ in such a way, that $y_2(h_2) \neq 0$. Denote by U the k-linear span in T of 1, h_1, and h_2: this is a graded k-subalgebra of T, with $(U_+)^2 = 0$. It maps isomorphically into $T/(T_+)^2$ by the canonical projection from T. Any homogeneous subspace of $T/(T_+)^2$, which is complementary to the image of U, is a graded ideal: composing the canonical projection from T with the factorization of this ideal yields a homomorphism of graded k-algebras $T \to U$, which exhibits U as a retract of T. In particular, it maps the free Lie algebra $\pi^*(U) = \mathsf{L}(\mathrm{sHom}_k(U_+, k))$ isomorphically onto the subalgebra of $\pi^*(T)$, generated by y_1 and y_2. □

(4.5) REMARKS: (1) After establishing the Claim in the preceding proof, we could have finished the argument by a direct reference to [**Av3**, (1.4)]: the last part of the proof of (4.4) is, in fact, a concretization of the general construction used in the proof of *loc. cit.*

(2) With y_1 as in (4.4.3), it is not always possible to find y_2 of degree 2, such that $[y_1, y_2] \neq 0$. Indeed, let $Q = k[[X, Y, Z]]$ and $R = Q/(XZ, Y^2, X^2 - YZ)$. Then $T_1^2 \cong \wedge^2 T_1$ (cf. [**Av1**, (7.2) and (7.7)]), and it is easily seen from [**Av4**, (4.2)] that in this case $[\pi^2(T), \pi^2(T)] = 0$.

Appendix. Complexity.

Consider a sequence $\mathbf{b} = \{b_n \in \mathbb{R}_+ \cup \{\infty\} \mid n \in \mathbb{N}\}$. If d is an integer ≥ 0, write $\gamma(\mathbf{b}) \leq d$ if there is a positive real number A, such that the inequality $b_n \leq An^{d-1}$ holds for all sufficiently large n. In case the condition $\gamma(\mathbf{b}) \leq d$ holds, but $\gamma(\mathbf{b}) \leq d - 1$ does not, we set $\gamma(\mathbf{b}) = d$; in case there is no d such that $\gamma(\mathbf{b}) \leq d$, we write $\gamma(\mathbf{b}) = \infty$.

From now on R denotes an associative ring. Unless the contrary is stated explicitly, "module" will mean "left-R-module". The minimal number of generators of the R-module M is denoted $\nu_R(M)$: this is a non-negative integer or ∞, and by convention $\nu_R(0) = 0$.

(A.1) DEFINITION: The *complexity* of the R-module M, $\mathrm{cx}_R(M)$, is defined by the formula:

$$\mathrm{cx}_R(M) = \inf_{\mathsf{P}} \gamma(\{\nu_R(P_n)\}),$$

where P ranges over the R-projective resolutions

$$(0 \leftarrow M \leftarrow) P_0 \overset{d_0}{\leftarrow} P_1 \leftarrow \cdots \leftarrow P_{n-1} \overset{d_n}{\leftarrow} P_n \leftarrow \ldots$$

of M.

The first properties of complexity do not involve any finiteness hypotheses.

(A.2) *$\operatorname{cx}_R M = 0$ if and only if $\operatorname{pd}_R M < \infty$.* □

(A.3) *If $0 \to M_1 \to M_2 \to M_3 \to 0$ is an exact sequence of R-modules, then*

$$\operatorname{cx}_R M_i \leq \max\{\operatorname{cx}_R M_j, \operatorname{cx}_R M_k\}$$

where $\{i, j, k\} = \{1, 2, 3\}$.

PROOF: Use mapping cone constructions. □

Let $\varphi : R \to S$ be a homomorphism of rings, and let ${}_\varphi S$ (resp. S_φ) denote the left (resp. right) R-module structure it produces on S.

(A.4) *If S_φ is flat, then $\operatorname{cx}_S(S \otimes_R M) \leq \operatorname{cx}_R M$.* □

(A.5) *Let ${}_\varphi S$ be projective and finitely generated. For an R-module M, one then has $\operatorname{cx}_S \operatorname{Hom}_R(S, M) \leq \operatorname{cx}_R M$. For an S-module N, one then has $\operatorname{cx}_R N \leq \operatorname{cx}_S N$.* □

(A.6) *If U is a multiplicatively closed subset of the center of R, then $\operatorname{cx}_{U^{-1}R}(U^{-1}M) \leq \operatorname{cx}_R M$.* □

(A.7) *If x is a central non-zero-divisor in a ring Q and $R = Q/(x)$, then $\operatorname{cx}_R M \leq \operatorname{cx}_Q M + 1$.*

PROOF: Let P' be a Q-projective resolution of M, such that $\gamma(\{\nu_R(P_n)\}) = \operatorname{cx}_Q M$. A construction of Shamash [**Sh**, Section 3], cf. also [**Ei**, Section 7] and [**Av5**, (2.4)], shows there is an R-projective resolution P of M, with $P_n = P'_n/xP'_n \oplus P'_{n-2}/xP'_{n-2} \oplus \dots$ (all the references quoted assume R is commutative and P is free, but the arguments use only the centrality of x and the projectivity of P). □

Next we note several properties of complexity, which require finiteness conditions.

(A.8) *If R is left noetherian and M is finitely generated, then $\operatorname{cx}_R M \leq d$ is equivalent to the existence of a resolution P with $\nu_R(P_n) \leq An^{d-1}$ for all $n \geq 1$.* □

(A.9) *Let R be left noetherian, and semiperfect in the sense of Bass* [**Ba**], i.e. the factor-ring $k = R/\mathbf{m}$ modulo the Jacobsson radical $\mathbf{m}$ is semi-simple, and idempotents lift from k to R. This happens if and only if the following condition holds: every finitely generated R-module M has a resolution P^M, which has one of the equivalent properties: (a) if P is a resolution of M, and $\alpha : \mathsf{P}^M \to \mathsf{P}$ is a comparison map over id_M, then α is a split monomorphism of complexes; (b) $d(\mathsf{P}^M) \subset \mathbf{m}\,\mathsf{P}^M$.

In this case, one has $\operatorname{cx}_R M = \gamma(\{\nu_R(P_n^M)\})$.

Furthermore, Nakayama's lemma yields the equalities $\nu_R(P_n^M) = \nu_k(P_n^M/\mathbf{m}P_n^M)$. $\square$

(A.10) We assume M is finitely generated over the semi-perfect left noetherian ring R and introduce its *Betti number* $b_n^R(M)$, by slightly stretching the arguments of Jans [**Jn**]. Let $C_1, \ldots, C_s$ be representatives of all the different isomorphism classes of simple R-modules, and set $C = \bigoplus_{i=1}^s C_i$. Denote by D_i the division ring $\operatorname{Hom}_R(C_i, C_i)$, and consider C as a left R-D-bimodule, where $D = \operatorname{Hom}_R(C, C) = \prod_{i=1}^s D_i$. For any left R-module M, this endows $\operatorname{Ext}_R^n(M, C)$ with a natural structure of left D-module. Now set:

$$b_n^R(M) = \operatorname{length}_D \operatorname{Ext}_R^n(M, C).$$

Note that $b_n^R(M)$ is finite when M is finitely generated: if $\bigoplus_{i=1}^s C_i^{n_i}$ is the decomposition of $P_n^M/\mathbf{m}P_n^M$ into simple k-modules, then by the minimality of $\mathbf{P}^M$ one has that

$$\operatorname{Ext}_R^n(M, C) \cong \operatorname{Hom}_R(P_n^M/\mathbf{m}P_n^M, C) \cong \bigoplus_{i=1}^s D_i^{n_i}.$$

In particular, this provides equalities $b_n^R(M) = n_1 + \cdots + n_s = \operatorname{length}_k P_n^M/\mathbf{m}P_n^M$. Consider now the epimorphism $k^{n_1+\cdots+n_s} \to C_1^{n_1} \oplus \cdots \oplus C_s^{n_s}$, obtained from the canonical projections $k \to C_i$. Comparing numbers of generators one obtains the right-hand side of the next double inequality:

$$\frac{b_n^R(M)}{\operatorname{length}_k(k)} \leq \nu_k(P_n^M/\mathbf{m}P_n^M) \leq b_n^R(M),$$

whose left-hand side is obtained by comparing lengths in the epimorphism $k^\nu \to P_n^M/\mathbf{m}P_n^M$, with $\nu = \nu_k(P_n^M/\mathbf{m}P_n^M)$.

Summing up, we have established:

For finitely generated modules over semi-perfect left noetherian rings, one has:

$$\operatorname{cx}_R M = \gamma(\{b_n^R(M)\}).$$

$\square$

In particular, this shows that the definition of complexity, given in (A.1), agrees with (1.1) on finitely generated modules over local commutative noetherian rings.

(A.11) *Let Q be a semiperfect left noetherian ring and let x be a central non-zero-divisor in Q. If $R = Q/(x)$, and M is a finitely generated R-module, then*

$$\operatorname{cx}_Q M \leq \operatorname{cx}_R M \leq \operatorname{cx}_Q M + 1.$$

In particular, $\operatorname{cx}_R M$ and $\operatorname{cx}_Q M$ are finite or infinite simultaneously.

PROOF: The right-hand side has been established in (A.7) without restrictions on Q. For the other side, consider the exact sequence

$$\operatorname{Ext}_R^n(M,C) \to \operatorname{Ext}_Q^n(M,C) \to \operatorname{Ext}_R^{n-1}(M,C)$$

arising from the standard change of rings spectral sequence ${}_2E^{p,q} = \operatorname{Ext}_R^p(M, \operatorname{Ext}_Q^q(R,C)) \Rightarrow \operatorname{Ext}_Q^{p+q}(M,C)$, in view of the fact that $0 \to Q \xrightarrow{x} Q (\to R \to 0)$ is a projective resolution of R over Q. In the notation of (A.10) this yields $b_n^Q(M) \leq b_n^R(M) + b_{n-1}^R(M)$, and the conclusion follows by the last formula in (A.10). $\square$

(A.12) *If the ring R is left artinian, then*

$$\operatorname{cx}_R M = \gamma(\{\operatorname{length}_R(P_n^M)\}),$$

where $\mathbf{P}^M$ is a minimal projective resolution of the finitely generated R-module M.

PROOF: The results of (A.9) and (A.10) apply to this case, and the equality to be proved is obtained by a variation of the argument of (A.10). $\square$

(A.13) *If R is a finite dimensional algebra over a field k', then $\operatorname{cx}_R M = \gamma(\{\dim_{k'} P_n^M\})$, where $\mathbf{P}^M$ is a minimal R-projective resolution of M.*

$\square$

This is a special case of the preceding statement. If one specializes further to the group algebra $R = k'G$ of a finite group G, the formula in (A.13) is Alperin and Evens's definition of complexity: cf. [**AE**, p. 1], where it is denoted $c_G(M)$.

ACKNOWLEDGEMENT: I should like to thank the organizers of the Microprogram on Commutative Algebra and Professor Irving Kaplansky for providing me with the possibility of participating in this exciting meeting.

References

[Al] J. Alperin, *Periodicity in groups*, Illinois J. Math. **21** (1977), 776–783.

[AE] J. Alperin and L. Evens, *Representations, resolutions, and Quillen's dimension theorem*, J. Pure Appl. Algebra **22** (1981), 1-9.

[An] D. J. Anick, *A counterexample to a conjecture of Serre*, Ann. of Math. **115** (1982), 1–33.

[Av1] L. L. Avramov, *Small homomorphisms of local rings*, J. Algebra **50** (1978), 400–453.

[Av2] L. L. Avramov, *Obstructions to the existence of multiplicative structures on minimal free resolutions*, Amer. J. Math. **103** (1981), 1–31.

[Av3] L. L. Avramov, *Free Lie subalgebras of the cohomology of local rings*, Trans. AMS **270** (1982), 589–608.

[Av4] L. L. Avramov, *Local algebra and rational homotopy*, Astérisque **113–114** (1984), 15–43.

[Av5] L. L. Avramov, *Modules of finite virtual projective dimension*, Invent. Math. (to appear).

[AF] L. L. Avramov and H.-B. Foxby, *Locally Gorenstein homomorphisms*, Københavns Universitets Matematiske Instituts Preprint Series 1988, no. 16.

[AGP] L. L. Avramov, V. N. Gasharov and I. V. Peeva, *A periodic module of infinite virtual projective dimension*, J. Pure Appl. Algebra (to appear).

[AKM] L. L. Avramov, A. R. Kustin and M. Miller, *Poincaré series of modules over local rings of small embedding codepth or small linking number*, J. Algebra (to appear).

[Ba] H. Bass, *Finitistic dimension and a homological generalization of semiprimary rings*, Trans. AMS **95** (1960), 466–488.

[Bi] L. Bieberbach, "Analytische Fortsetzung," Springer-Verlag, Berlin, 1955.

[BE1] D. A. Buchsbaum and D. Eisenbud, *Some structure theorems for finite free resolutions*, Advances in Math. **12** (1974), 84–139.

[BE2] D. A. Buchsbaum and D. Eisenbud, *Algebra structures for finite free resolutions, and some structure theorems for ideals of codimension 3*, Amer. J. Math. **99** (1977), 447–485.

[Bu] L. Burch, *On ideals of finite homological dimension in local rings*, Proc. Camb. Phil. Soc. **64** (1968), 941–952.

[Ch] S. Choi, *Betti numbers and the integral closure of ideals*, preprint (1988).

[Ei] D. Eisenbud, *Homological algebra on a complete intersection, with an application to group representations*, Trans. AMS **260** (1980), 35–64.

[FHJLT] Y. Felix, S. Halperin, C. Jacobsson, C. Löfwall and J.-C. Thomas, *The radical of the homotopy Lie algebra*, Amer. J. Math. **110** (1988), 301-322.

[Fo] H.-B. Foxby, *On the μ^i in a minimal injective resolution, II*, Math. Scand. **41** (1977), 19–44.

[GG] F. Ghione and T. H. Gulliksen, *Some reduction formulas for the Poincaré series of modules*, Atti Acad. Naz. Lincei Rend. Cl. Sci. Fis. Natur. (8) **58** (1975), 82–91.

[GP] V. N. Gasharov and I. V. Peeva, *Boundedness versus periodicity over commutative local rings*, Trans. AMS (to appear).

[GR] E. H. Gover and M. Ramras, *Increasing sequences of Betti numbers*, Pacific J. Math. **87** (1980), 65–68.

[Gu1] T. H. Gulliksen, *Massey operations and the Poincaré series of certain local rings*, J. Algebra **22** (1972), 223–232.

[Gu2] T. H. Gulliksen, *A change of rings theorem with applications to Poincaré series and intersection multiplicity*, Math. Scand. **34** (1974), 167–183.

[Gu3] T. H. Gulliksen, *On the deviations of a local rings*, Math. Scand. **47** (1980), 5–20.

[HS] J. Herzog and M. Steurich, *Two applications of change of rings theorems for Poincaré series*, Proc. AMS **73** (1979), 163–168.

[Ja] C. Jacobsson, *On local flat homomorphisms and the Yoneda Ext-algebra of the fibre*, Astérisque **113-114** (1984), 227–233.

[JKM] C. Jacobsson, A. R. Kustin and M. Miller, *The Poincaré series of a codimension four Gorenstein ring is rational*, J. Pure Appl. Algebra **38** (1985), 255–275.

[Jn] J. P. Jans, *Some generalizations of finite projective dimension*, Illinois J. Math. **5** (1961), 334-343.

[Ku] A. R. Kustin, *Gorenstein algebras of codimension four and characteristic two*, Communications in Algebra **15** (1987), 2417–2429.

[KM1] A. R. Kustin and M. Miller, *Algebra structures on minimal resolutions of Gorenstein rings of codimension four*, Math. Z. **173** (1980), 171–184.

[KM2] A. R. Kustin and M. Miller, *Multiplicative structure on resolutions of algebras defined by Herzog ideals*, J. London Math. Soc. (2) **28** (1983), 247–260.

[KM3] A. R. Kustin and M. Miller, *Classification of the Tor-algebras of codimension four Gorenstein local rings*, Math. Z. **190** (1985), 341–355.

[Le1] J. Lescot, *Asymptotic properties of Betti numbers of modules over certain rings*, J. Pure Appl. Algebra **38** (1985), 287–298.

[Le2] J. Lescot, *Séries de Poincaré et modules inertes*, preprint (1986).

[Ma] S. MacLane, "Homology," Springer-Verlag, Berlin, 1963.

[Ra] M. Ramras, *Sequences of Betti numbers*, J. Algebra **66** (1980), 193–204.

[Sc] G. Scheja, *Über die Bettizahlen lokaler Ringe*, Math. Ann. **155** (1964), 155–172.

[Sh] J. Shamash, *The Poincaré series of a local ring*, J. Algebra **12** (1969), 453–470.

[Ta] J. Tate, *Homology of Noetherian rings and of local rings*, Illinois J. Math. **1** (1957), 14–27.

[We] J. Weyman, *On the structure of free resolutions of length 3*, preprint (1985).

Institute of Mathematics, ul. "Akad. G. Bončev" bl. 8, 1113 Sofia, BULGARIA

On Ulrich-Modules over Hypersurface Rings

JÖRGEN BACKELIN AND JÜRGEN HERZOG

Let $(R, \mathfrak{m}, k)$ be a local Cohen-Macaulay ring, and let M be a maximal Cohen-Macaulay module (MCM-module) over R. Then the minimal number of generators $\nu(M)$ of M is bounded by the multiplicity $e(M)$ of M.

Ulrich asks in his paper [U] whether there always exists an MCM-module such that $\nu(M) = e(M)$. We call a module satisfying this equality an *Ulrich-module* (or a *linear MCM-module*).

The Ulrich-modules have the nice property [**BHU**, 1.6] that the associated graded module $\mathrm{gr}_{\mathfrak{m}}(M) = \oplus_{i\geq 0} M/\mathfrak{m}^i M$ is again a (graded) MCM-module over the associated graded ring $\mathrm{gr}_{\mathfrak{m}}(R)$ (which need not be Cohen-Macaulay!).

Thus if one could prove the existence of Ulrich-modules in general, one could conclude the following interesting statement:

Let S be a homogeneous k-algebra. If there exists a local Cohen-Macaulay ring $(R, \mathfrak{m}, k)$ such that $\mathrm{gr}_{\mathfrak{m}}(R) \simeq S$, then S admits a graded MCM-module.

Unfortunately not so many existence theorems for Ulrich-modules are known so far. Most of them can be found in [**BHU**]. For the convenience of the reader we give here the complete list of known results in this direction.

Ulrich-modules exist for the following rings R:

(1) [**BHU**, 1.2, 2.1] $\dim R \leq 1$.
(2) [**BHU**, 2.5] R has minimal multiplicity.
(3) [**BHU**, 4.8] R is a homogeneous 2-dimensional CM-ring with infinite residue class field.
(4) [**BHS**] R is a homogeneous hypersurface ring whose residue class field has characteristic 0.

In this paper we will generalize the result (4).

THEOREM 1. *Let k be an arbitrary field, $f \in k[X_1, \ldots, X_N]$ a polynomial in n variables over k without constant, and let R be either the ring $k[X_1, \ldots, X_N]_{(X_1,\ldots,X_n)}$ or its completion. Then the hypersurface ring $S = R/fR$ has an Ulrich-module.*

The question remains open whether a complete hypersurface ring defined by a formal power series f also admits an Ulrich-module.

The proof of the theorem is based on results of [**BHS**] which need to be slightly generalized in order to be applicable to our situation.

Let $f \in k[X_1, \ldots, X_N]$ be a polynomial whose initial form has degree $d \geq 1$. We call an equation

$$f \cdot E_m = \alpha_1 \cdot \alpha_2 \cdot \ldots \cdot \alpha_d$$

a *matrix factorization* of f (of size m). Here E_m denotes the unit matrix of size m, and the α_1 are square matrices of size m whose entries all belong to the maximal ideal $(X_1, \ldots, X_n)$.

This notion generalizes the original definition of Eisenbud. Eisenbud proves in his paper [**E**, Corollary 3.6] the following result, from which we quote here only as much as is needed to prove Theorem 1.

THEOREM (Eisenbud). *If f has a matrix factorization*

$$f \cdot E_m = \alpha \cdot \beta$$

as above (but only with two factors), and if $L(\alpha)$ denotes the linear map $R^m \to R^m$ induced by α, then $\operatorname{coker} L(\alpha)$ *is an MCM-module over S.*

Applying Eisenbud's theorem we see that if we have a matrix factorization $f \cdot E_m = \alpha_1 \cdot \ldots \cdot \alpha_d$, then $\operatorname{coker} L(\alpha_1)$ is an MCM-module over S. We will see below that $\operatorname{coker} L(\alpha_1)$ is in fact an Ulrich-module.

Thus for the proof of Theorem 1 we also need the following:

THEOREM 2. *Let $f \in k[X_1, \ldots, X_N]$ be a polynomial with initial degree $d \geq 1$. Then for a suitable integer $m \geq 1$ there exists a matrix factorization*

$$f \cdot E_m = \alpha_1 \cdot \ldots \cdot \alpha_d.$$

We would like to thank Bokut who informed us that L'vov and Nesterenko have reported on the homogeneous version of this theorem at the 17th All Union Algebra Conference in Minsk, 1983. They announced this result (without proofs) in the Proceedings of this conference, Minsk 1983, pages 118 and 137.

The proof of Theorem 2 depends on the theory of generalized Clifford algebras. Let $f \in k[X_1, \ldots, X_N]$ be a homogeneous polynomial of degree $d \geq 1$, and let $V = \oplus_{i=1}^{n} k e_i$ be an n-dimensional k-vectorspace with basis $e_1, \ldots, e_n$. Then f defines a map $f : V \to k$

$$f(x_1 e_1 + \cdots + x_n e_n) = f(x_1, \ldots, x_n)$$

A $\mathbb{Z}/d\mathbb{Z}$-graded algebra C together with a monomorphism $V \to C_1$ is called a *generalized Clifford algebra*, if C is generated by V, and if $f(x) = x^d$ for all $x \in V$.

A matrix factorization $f \cdot E_m = \alpha_1 \cdot \ldots \cdot \alpha_d$ of the homogeneous polynomial f is called *linear* if all the matrices α_i have linear entries.

In [**BHS**, 1.11] the following is shown:

(i) Let f be a homogeneous polynomial in n variables for which there exists a finite dimensional generalized Clifford algebra, then f has a linear matrix factorization. Conversely, if f has a linear matrix factorization and f is non-degenerated, that is, cannot be transformed by a linear transformation into a polynomial in less than n variables, then f has a finite dimensional generalized Clifford algebra.

We will use this to prove the following generalization of a result of Childs [**C**, Theorem 2]:

(ii) Let f and g be homogeneous polynomials of degree d in disjoint sets of variables. If both f and g have linear matrix factorizations then $f + g$ has a linear matrix factorization.

Before we indicate a proof of (ii) we show how Theorem 2 follows from (ii):

Let $f = \sum_i a_i X_{i1} \cdot \ldots \cdot X_{id}$ be a polynomial with $X_{ij} \neq X_{kl}$ for $(i,j) \neq (k,l)$. It follows from (ii) that f has a linear matrix factorization since all the summands of f have such a factorization.

Now let $f = \sum_\nu a_\nu X^\nu \in k[X_1, \ldots, X_N]$ be an arbitrary (inhomogeneous) polynomial, whose initial degree is $d \geq 1$. Then each of its summands $a_\nu X^\nu$ can be factored into d factors, which are all of degree ≥ 1. Hence if we let g be the polynomial

$$g = \sum_\nu Y_{\nu 1} \ldots Y_{\nu d}$$

in the variables $Y_{\nu i}$, then there exists an epimorphism $\varphi : k[\ldots, Y_{\nu i}, \ldots] \to k[X_1, \ldots, X_N]$, mapping the irrelevant maximal ideal $(\ldots, Y_{\nu i}, \ldots)$ of $k[\ldots, Y_{\nu i}, \ldots]$ into $(X_1, \ldots, X_n)$, and such that $\varphi(g) = f$.

Let $g \cdot E_m = \alpha_i \cdot \ldots \cdot \alpha_d$ be a linear matrix factorization of g. Then it is clear that $f \cdot E_m = \beta_1 \cdot \ldots \cdot \beta_d$ is a matrix factorization of f, where β_i is obtained from α_i by applying φ to the entries of α_i.

Now we indicated the proof of (ii) which follows quite exactly the line of arguments in [**BHS**]:

First of all, by remark [**BHS**, 1.10], we may assume that k is algebraically closed, and secondly we may assume that f and g are non-degenerated. Because if, say, f is degenerated we may apply a linear transformation to it in order to obtain a non-degenerated polynomial in a fewer number of generators. Thus if we can prove (ii) for the new f and g, then applying the inverse linear transformation we get the result in the general case. Using (i) it then follows that f and g have finite dimensional generalized Clifford algebras, say C_1 and C_2 respectively. As in [**BHS**, 3.1] we will construct a finite dimensional generalized Clifford algebra for $f+g$. Then again using (i) it will follow that $f+g$ has a linear matrix factorization.

Let m_d denote the d'th cyclotomic polynomial, i.e., the minimal polynomial of a primitive d'th root of unity ω over $\mathbb{Q}$. m_d is a polynomial with integer coefficients. We write again m_d for the corresponding polynomial over the prime field of k, and let ξ be a root of m_d in the algebraically closed field k.

We now define the $\mathbb{Z}/d\mathbb{Z}$-graded tensor product $C_1\hat{\otimes}C_2$ as the ordinary tensor product of $\mathbb{Z}/d\mathbb{Z}$-graded k-vectorspaces with the multiplication given by

$$(a\otimes b)\cdot(c\otimes d)=\xi^{(\deg b)(\deg c)}(ac\otimes bd)$$

for all homogeneous $a,c\in C_1$, $b,d\in C_2$.

We claim that $C=C_1\hat{\otimes}C_2$ is a generalized Clifford algebra for $f+g$, which of course is again finite dimensional.

As in [**BHS**,3.1], the proof of this claim can be reduced to showing that

$$h_i(\xi):=\sum\left(\xi^{\sum_{j=0}^{i} j\,u_j}\right)=0 \text{ for } i=1,\ldots,d-1,$$

where the sum is taken over all integers u_j such that $u_j\geq 0$ and $\sum_{j=0}^{i}u_j=d-i$.

In [**BHS**, 3.1] it is shown for the primitive d-th root of unity ω that $h_i(\omega)=0$ for $i=1,\ldots,d-1$, which simply means that $m_d(X)$ divides $h_i(X)$ in $\mathbb{Z}[X]$. That is, $h_i(X)=a_i(X)\cdot m_d(X)$ with $a_i(X)\in\mathbb{Z}[X]$ for $i=1,\ldots,d-1$. But as $m_d(\xi)=0$, we get $h_i(\xi)=0$ for $i=1,\ldots,d-1$ as well, and this concludes the proof of (ii).

PROOF OF THEOREM 1: We choose a matrix factorization $f\cdot E_m=\alpha_1\cdot\ldots\cdot\alpha_d$ of f, and claim that the MCM-module $M:=\operatorname{coker}L(\alpha_1)$ is an Ulrich-module.

To prove this we will use a criterion that is derived in [**BHU**]: let N be an arbitrary finitely generated R-module, where $(R, \mathfrak{m}, k)$ is a noetherian local ring, and let

$$F_* \cdots \to F_2 \to F_1 \to F_0 \to 0$$

be a minimal free R-resolution of N. We define the following natural filtration on the complex F_*:

$$\mathcal{F}_i\, F_j := \begin{cases} F_j & \text{for } i < j \\ \mathfrak{m}^{i-j} F_j & \text{for } i \geq j \end{cases}$$

N is said to have a *linear resolution* if the associated graded complex $\operatorname{gr}_{\mathcal{F}} F_*$ is acyclic. If this is the case then $\operatorname{gr}_{\mathcal{F}} F_*$ is a minimal free homogeneous resolution of $\operatorname{gr}_{\mathfrak{m}}(N)$.

It follows from [**BHU**, 1.5] that if M is an MCM-module over the hypersurface ring $S = R/fR$, then M is an Ulrich-module if an only if it has a linear R-resolution. (This explain the alternative name "linear MCM-module".)

Using Eisenbud's Theorem we see that $M = \operatorname{coker} L(\alpha_1)$ is an MCM-module whose free R-resolution is

$$0 \to R^m \xrightarrow{L(\alpha_1)} R^m \to 0.$$

The associated graded complex of this resolution is

$$0 \to A^m \xrightarrow{L(\beta_1)} A^m \to 0$$

where $A = k[X_1, \ldots, X_N]$, and where β_1 is obtained from α_1 by taking the homogeneous components of degree 1 of the entries of α_1.

If we define the matrices β_i, $i = 2, \ldots, d$ similarly, then the equation $f \cdot E_m = \alpha_1 \cdot \ldots \cdot \alpha_d$ yields the equation $f_d \cdot E_m = \beta_1 \cdot \ldots \cdot \beta_d$, where f_d denotes the d-th homogeneous component of f. But this equation implies that β_1 has maximal rank, and hence the associated complex is acyclic. Thus we have shown that M is an Ulrich-module.

References

[**BHS**] J. Backelin, J. Herzog and H. Sanders, *Matrix factorizations of homogeneous polynomials*, to appear in the Proceedings of the Fifth National School in Algebra, Varna 1986, (SLN).

[**BHU**] J. Brennan, J. Herzog and B. Ulrich, *Maximally generated Cohen-Macaulay modules*, Math. Scand. **61** (1987), 181–203.

[C] L. N. Childs, *Linearizing n-ic forms and generalized Clifford algebras*, Linear and Multilinear Algebra **5** (1978), 267–278.

[E] D. Eisenbud, *Homological algebra on a complete intersection with an application to group representations*, Trans. AMS **260** (1980), 35–64.

[U] B. Ulrich, *Gorenstein rings and modules with high number of generators*, Math. Z. **188** (1984), 23–32.

Mr. Backelin supported in part by the NFR.

Matematiska Institutionen, Stockholms Universitet, Box 6701
S-13385 Stockholm, SWEDEN

Prof. Herzog supported in part by the DFG.

Universität-Gesamthochschule Essen, FB6 Mathematik, Universitätsstr. 3,
D-4300 Essen 1, FEDERAL REPUBLIC OF GERMANY

Differential Structure of Étale Extensions of Polynomial Algebras

HYMAN BASS

0. Introduction.

The Jacobian Conjecture (see [**BCW**]) asserts that an étale polynomial map $F : \mathbb{C}^n \to \mathbb{C}^n$ is an isomorphism. In the course of some work with Haboush on group actions [**BH**] we observed that this is easily proved if F is equivariant for a reductive group acting on the two $\mathbb{C}^n$'s and having a dense orbit and a fixed point. One is thus tempted to try to prove the Jacobian conjecture by "feeding" such group actions into the picture. For example, let $G = SL_n(\mathbb{C})$ act linearly on the target $\mathbb{C}^n$. We may assume that $F(0) = 0$. In order to make F equivariant for a G-action, we must make G act on the source $\mathbb{C}^n$ by "$F \circ G \circ F^{-1}$". Of course this begs the issue, since we don't know that F^{-1} exists. However F^{-1} exists analytically near 0. In particular we can pull back the action of the Lie algebra $\underline{sl}_n(\mathbb{C})$ as vector fields on $\mathbb{C}^n$. Moreover the fact that the Jacobian matrix of F has a polynomial inverse implies that $\underline{sl}_n(\mathbb{C})$ pulls back to *polynomial* vector fields on the source $\mathbb{C}^n$.

This motivates trying to exploit this action of $\underline{sl}_n(\mathbb{C})$ as polynomial vector fields in approaching the Jacobian Conjecture. After many fruitless inquiries among representation theorists about what tools might be useful, I decided to see directly where this idea might lead. The present paper is a report on that exploration. While it has not solved the Jacobian Conjecture, the calculations for $n = 2$ lead to some unexpected and (to me) very interesting contacts with diophantine geometry and classical complex function theory which seem to invite further study.

The general algebraic setting we consider is as follows: $A = k[x_1, \ldots, x_n]$, a polynomial algebra in n variables over an algebraically closed field k of characteristic zero; $A \subset B$ is an *étale* extension, where B is an affine k-domain. (The Jacobian Conjecture asserts that if B is also a polynomial algebra then $B = A$.) The derivations $\partial_i = \partial/\partial x_i$ extend to derivations of B, so B is a module over the Weyl algebra $W = k[x_1, \ldots, x_n, \partial_1, \ldots, \partial_n]$ of polynomial differential operators. In §2 we show that B is a "holonomic" W-module, hence cyclic and of finite length.

The derivations $\epsilon_{ij} = x_i\partial_j$ span $\underline{gl}_n(k) \subset W$, which acts linearly on A. In §3 we show that B/A has no nonzero $\underline{sl}_n(k)$-submodules of finite length. Moreover if $\underline{g}$ is a Lie subalgebra of $\underline{gl}_n(k)$ of dimension $> n$ then B is a *torsion* module over the enveloping algebra $U(\underline{g})$. Thus one could approach the Jacobian Conjecture by choosing such a $\underline{g}$ and trying to show that B/A is a *torsion free* $U(g)$-module. This is the approach we adopt, for $n = 2$, in §6.

Sections 4 and 5 record various elementary facts about the action of the derivations ∂_i on B. It is perhaps worth noting that none of our results use more than the following assumptions on B: (i) B is étale over A; (ii) $B^\times = k^\times$ (only constant units); and (iii) the field of fractions Frac(B) is unirational over k. Thus our results support the "Generalized Jacobian Conjecture" that (i), (ii), and (iii) imply that $B = A$.

The most interesting results here are the calculations for $n = 2$ in §6. Simplifying notation, consider $A = \mathbb{C}[x, y] \subset B$, satisfying (i), (ii), and (iii) above. Put $\partial_x = \partial/\partial x$, $\partial_y = \partial/\partial y$. Then $\underline{gl}_2(\mathbb{C}) \subset W$ is spanned by $\epsilon_x = x\partial_x$, $\epsilon_y = y\partial_y$, $\delta = x\partial_y$, and $\delta' = y\partial_x$. To prove that $B = A$ it would suffice, in view of the results of §3 quoted above, to show that B/A is a torsion free module over $U(\underline{b}) = \mathbb{C}[\epsilon_x, \epsilon_y, \delta]$ (or over $U(\underline{sl}_2(\mathbb{C})) = \mathbb{C}[\delta', \epsilon_x - \epsilon_y, \delta]$). Our calculations show that B/A is torsion free over $\mathbb{C}[\epsilon_x + \epsilon_y, \delta]$, over $\mathbb{C}[\partial]$ for all $\partial \in \underline{gl}_2(\mathbb{C})$, and over $\mathbb{C}[\epsilon_x, \epsilon_y]$. It is in the proof of the last result that extraordinary things occur. The proof invokes Siegel's theorem on algebraic curves with infinitely many integer points, as well as Fabry's Theorem (1896) on Gap Series. It was R. Narasimhan who, after hearing a weaker version of this result, pointed our Fabry's Theorem to me and how it could be applied to the issue at hand; I am most grateful to him.

While these results remain indecisive for the Jacobian Conjecture, the methods developed here for $n = 2$ constitute what appears to be a new approach, which merits further study.

There are several appendices which present background material in a form convenient for present applications, and for which I did not find a convenient reference.

Finally, I wish to thank especially the following people, among many others, for helpful discussions of various aspects of this work: Mel Hochster, Barry Mazur, M. P. Murthy, R. Narasimhan, Martha Smith, Harold Stark, Robert Steinberg, and David Wright.

Contents

Notation

The group of units of a ring R is denoted $R^{\times}$. If R is an integral domain then $\mathrm{Frac}(R)$ denotes its field of fractions. If D is a derivation of R we put $R_0^D = \mathrm{Ker}(D)$ and $R_{(0)}^D = \bigcup_{m \geq 0} \mathrm{Ker}(D^m)$.

1. The étale extension $A = k[x_1, \ldots, x_n] \subset B$.

We fix a polynomial algebra

$$A = k[x] = k[x_1, \ldots, x_n]$$

in n variables over an algebraically closed field k of characteristic zero. Putting $\partial_i = \partial/\partial x_i$ ($i = 1, \ldots, n$) we have the *Weyl algebra*

$$W = k[x_1, \ldots, x_n, \partial_1, \ldots, \partial_n]$$

of polynomial differential operators acting on A. It is a noncommutative integral domain with commutation relations $[x_i, x_j] = 0$, $[\partial_i, \partial_j] = 0$, and $[\partial_i, x_j] = \delta_{ij}$ $(1 \le i, j \le n)$. A k-basis of W is furnished by the monomials $x^\alpha \partial^\beta = x_1^{\alpha_1} \ldots x_n^{\alpha_n} \partial_1^{\beta_1} \ldots \partial_n^{\beta_n}$ (cf. [**Bj**,Ch. 1]).

Let $A \subset B$ be an unramified (= separable) extension of affine k-algebras (cf. [**W**, Section 1] or [**AK**, VI.3]). The derivations of A extend uniquely to B. Hence the ∂_i above are defined on B, so B is also a W-module. It is our purpose to study the structure of B as a module over W and various subalgebras of W. We have in mind especially the following.

FUNDAMENTAL EXAMPLE: Suppose that $A = k[x] \subset B$ where B itself is a polynomial algebra, $B = k[y] = k[y_1, \ldots, y_n]$ and we have the Jacobian condition

$$\frac{\partial(x_1, \ldots, x_n)}{\partial(y_1, \ldots, y_n)} = 1.$$

In this case the *Jacobian Conjecture* (cf. [**BCW**]) asserts that $A = B$. The extension $A \subset B$ is unramified, and the extension of ∂_i to B is given by

$$\partial_i(b) = \frac{\partial(x_1, \ldots, x_{i-1}, b, x_{i+1}, \ldots, x_n)}{\partial(y_1, \ldots \qquad \ldots, y_n)}.$$

Throughout this paper we fix the following notations.

$A = k[x] = k[x_1, \ldots, x_n]$

B = an affine integral domain over k which is an unramified extension of A; $A \subset B$

$E = \mathrm{Frac}(B)$, the field of fractions of B, which is a finite extension of $k(x) = \mathrm{Frac}(A)$.

E is a module over the Weyl algebra

$$W = k[x_1, \ldots, x_n, \partial_1, \ldots, \partial_n] \qquad (\partial_i = \partial/\partial x_i)$$

and A and B are submodules, as is any intermediate field L, $k(x) \subset L \subset E$, as well as $L \cap B$.

We start with a slight generalization of a recent result of Formanek [**F**, Theorem 1].

(1.1) PROPOSITION. *Assume that* $B = A[t]$ *for some* $t \in B$. *Then* $B = A[T]/f \cdot A[T]$ *for some* $f = f(T) \in A[T]$, *and* $f'(t) \in B^\times$. *If* $B^\times = k^\times$ *then* $B = A$.

PROOF: Since B is étale over A we have $0 = \Omega_{B/A} = B \otimes_{A[T]} (\Omega_{A[T]/A}/A[T]\,df)$. Now $\Omega_{A[T]/A}$ is $A[T]$-free with basis dT, and $df = f'(T)\,dT$, whence the first assertion. If $B^\times = k^\times$ then $f'(t) = c \in k$ so $f'(T) - c$ is divisible by f. But $\deg_T(f'(T) - c) < \deg_T(f)$. Hence $f'(T) = c$. It follows that $f = cT - a$ $(a \in A)$ so $t = c^{-1}a \in A$, and $B = A$ as claimed.

(1.2) PROPOSITION. (a) *Let* $\underline{m}$ *be a maximal ideal of* B *and* $\underline{m}_A = \underline{m} \cap A$. *The map of completions* $\hat{A}_{\underline{m}_A} \to \hat{B}_{\underline{m}}$ *is an isomorphism.*

(b) *B is regular, and B is étale (= unramified and flat) over A.*

PROOF: Since B is unramified over A, $\underline{m}B_{\underline{m}} = \underline{m}_A B_{\underline{m}}$. Since A is regular, $\underline{m}_A A_{\underline{m}_A}$ has n $(= \dim(B))$ generators, whence B is regular. Since k is algebraically closed both $A_{\underline{m}_A}$ and $B_{\underline{m}}$ have residue field k. If $y_1, \ldots, y_n \in \underline{m}_A$ generate $\underline{m}_A$ mod $\underline{m}_A^2$ then both $\hat{A}_{\underline{m}_A}$ and $\hat{B}_{\underline{m}}$ can be identified with $k[[y_1, \ldots, y_n]]$, whence (a). Since $\hat{A}_{\underline{m}_A} \to \hat{B}_{\underline{m}}$, being an isomorphism, is flat for all $\underline{m}$ it follows that B is flat over A, whence (b).

REMARK: The above argument did not require that B be an integral domain, but only that B have pure dimension n (i.e. $\dim(B_{\underline{m}}) = n$ for all maximal ideals $\underline{m}$ of B). One still concludes that B is regular, hence a finite product of regular integral domains, each étale over A.

(1.3) COROLLARY. *If* $B^\times = k^\times$ *then*

$$k(x) \cap B = A,$$

i.e. B/A *is a torsion free* A*-module.*

PROOF: Say $p/q \in B$ where p and q are relatively prime elements of A. Then $A/Aq \xrightarrow{p} A/Aq$ is injective. Applying $B \otimes_A -$, which is exact because

B is A-flat, we see that $B/Bq \xrightarrow{p} B/Bq$ is injective. But $p/q \in B$, i.e. $p \in Bq$. Thus the 0-map in B/Bq is injective, so $q \in B^\times = k^\times$, and hence $p/q \in A$.

REMARK: The proof shows, more generally, that if $C \subset D$ are integral domains with C factorial, $D^\times = C^\times$, and D flat over C, then $D \cap \text{Frac}(C) = C$.

Recall that a field extension $K \subset L$ is called *unirational* if L is a subfield of rational functions (in several variables) over K.

(1.4) PROPOSITION. *Assume that E is unirational over k. (e.g. that B is a subalgebra of a polynomial algebra.) Let $L \subset E$ be a field extension of k of transcendence degree 1.*

(a) *$L = k(t)$ for some $t \in L$.*

Assume further that $L \cap B \neq k$ and that $B^\times = k^\times$.

(b) *We can choose t in (a) so that $L \cap B = k[t]$.*

(c) *Let $u \in A$ be a homogenous polynomial of degree $d > 0$ which is not a proper power in A ($u = v^m$, $v \in A \Rightarrow m = 1$). If $u \in L$ then $L = k(u)$ and $L \cap B = k[u]$. In particular, $k(u)$ is algebraically closed in E.*

PROOF: (a) is just Luroth's Theorem [**N**, p. 137]. To prove (b), put $R = L \cap B \neq k$. Then R is integrally closed in L so $L = \text{Frac}(R)$. Since $R^\times \subset B^\times = k^\times$, R must be a polynomial algebra, whence (b).

To prove (c) consider the "Euler derivation" $\epsilon = x_1\partial_1 + \cdots + x_n\partial_n$. Then $\epsilon(u) = du$ since u is homogeneous of degree d. Since ϵ stabilizes $k(u)$ and L is algebraic over $k(u)$, ϵ stabilizes L, hence also $R = L \cap B = k[t]$. From $du = \epsilon(u) = \frac{du}{dt} \cdot \epsilon(t)$ we see that $\deg_t(\epsilon(t)) = 1$, say $\epsilon(t) = at + b$ ($a \in k^\times$, $b \in k$). Putting $s = t + a^{-1}b$ we have $\epsilon(s) = as$, hence $\epsilon(s^m) = ams^m$ for $m \geq 0$. Thus the s^m ($m \geq 0$) are eigenvectors of ϵ, with distinct eigenvalues, spanning $R = k[s]$. Since $\epsilon(u) = du$ we must have $u = cs^m$ for some $c \in k^\times$ and $am = d$. Since (by hypothesis) uA is not a proper power, and since B is unramified over A, $uB = s^mB$ is likewise not a proper power in B. It follows that $m = 1$, hence $R = k[u]$, whence (c).

(1.5) COROLLARY. *Assume that E is unirational and $B^\times = k^\times$. Then $F = k(x_1)$ is algebraically closed in E, $F = \bigcap_{j=2}^n E_0^{\partial_j}$, and $k[x_1] = \bigcap_{j=2}^n B_0^{\partial_j}$.*

This follows from (1.4) with $L = \bigcap_{j=2}^n E_0^{\partial_j}$ and $u = x_1$. (Note that $\text{tr deg}_k(L) \leq 1$ since $\partial_2, \ldots, \partial_n$ are linearly independent L-derivations of E.)

2. B is a holonomic module over the Weyl algebra W.

On $W = k[x_1, \ldots, x_n, \partial_1, \ldots, \partial_n]$ we have the Bernstein filtration $\Gamma = (\Gamma_r)_{r \geq 0}$, where Γ_r is the k-linear span of all $x^\alpha \partial^\beta = x_1^{\alpha_1} \ldots x_n^{\alpha_n} \partial_1^{\beta_1} \ldots \partial_n^{\beta_n}$ with $|\alpha| + |\beta| \leq r$ ($|\alpha| = \alpha_1 + \cdots + \alpha_n$, $|\beta| = \beta_1 + \cdots + \beta_n$).

Let M be a left W-module. An ascending filtration $\Sigma = (\Sigma_s)_{s \geq 0}$ of M (by k-submodules Σ_s) will be called Γ-compatible if $\dim_k(\Sigma_s) < \infty$, $\Gamma_r \cdot \Sigma_s \subset \Sigma_{r+s}$, and $\bigcup_s \Sigma_s = M$. Then the associated graded $\mathrm{gr}^\Sigma(M) = \bigoplus_{s \geq 0} (\Sigma_s / \Sigma_{s-1})$ (with $\Sigma_{-1} = 0$) is a $\mathrm{gr}^\Gamma(W)$-module. If $\mathrm{gr}^\Sigma(M)$ is a finitely generated $\mathrm{gr}^\Gamma(W)$-module we call Σ a *good* filtration. In this case (cf. [**BoD**, V 1.8 and 1.10]) M is a finitely generated W-module, and the Hilbert function

$$\chi(M, \Sigma, r) = \dim_k(\Sigma_r)$$

is, for $r \gg 0$, a polynomial in r of degree $d \leq 2n$, with leading term written $\frac{m}{d!} \cdot r^d$. The degree $d = d(M)$ and coefficent $m = m(M)$ are independent of the good filtration Σ. (Any finitely generated W-module possesses a good filtration.) Bernstein's Inequality [**BoD**, V, 1.12] says that, for $M \neq 0$, $d(M) \geq n$. One calls M *holonomic* if $d(M) \leq n$, i.e. if $M = 0$ or $M \neq 0$ and $d(M) = n$.

We shall use the following result.

(2.1) PROPOSITION [**BoD**, V, 1.15]. *Let M be a left W-module with a Γ-compatible filtration Σ. Suppose that $P(t) \in \mathbb{R}[t]$ is a polynomial with leading term $\frac{c}{n!} \cdot t^n$ such that*

$$\dim_k(\Sigma_r) \leq P(r)$$

for all $r \gg 0$. Then M is holonomic and $m(M) \leq c$. In particular M is a finitely generated W-module.

(2.2) THEOREM. *Let B be an étale extension of $A = k[x_1, \ldots, x_n]$, as in Section 1. Then B is a holonomic W-module.*

PROOF: It suffices to exhibit a filtration Σ of B satisfying the conditions of (2.1). Let $u_1, \ldots, u_m$ generate B as a k-algebra, and let B_r denote the k-linear span of all $u^\gamma = u_1^{\gamma_1} \ldots u_m^{\gamma_m}$ with $|\gamma| = \gamma_1 + \cdots + \gamma_m \leq r$. Choose q large enough so that,

$$x_i \in B_q \text{ and } \partial_i(u_j) \in B_{q+1} \qquad (i = 1, \ldots, n; j = 1, \ldots, m).$$

We filter B by setting

$$\Sigma_s = B_{sq}.$$

Let $u^\gamma \in B_s$, $|\gamma| \leq s$. Then $x_i u^\gamma \in B_{s+q}$, and $\partial_i(u^\gamma) = \sum_{j=1}^m \gamma_j(u^\gamma/u_j)\cdot \partial_i(u_j) \in B_{s+q}$. It follows that if $|\alpha| + |\beta| \leq r$ then $x^\alpha \partial^\beta u^\gamma \in B_{s+rq}$, and hence $\Gamma_r B_s \subset B_{s+rq}$, so $\Gamma_r \Sigma_s = \Gamma_r B_{sq} \subset B_{(s+r)q} = \Sigma_{s+r}$, as required.

According to (A.1) (Appendix A below) there is a polynomial $P(t)$ of degree $\dim(B) = n$ such that $P(r) = \dim_k(B_r)$ for all $r \gg 0$. Then $\dim_k(\Sigma_r) = \dim_k(B_{rq}) = P(rq) = Q(r)$. Since $Q(t) = P(qt)$ is a polynomial of degree n it follows that we have the holonomic estimate of (2.1) needed to conclude that B is holonomic.

(2.3) COROLLARY. *As a W-module, B is cyclic (i.e. has a single generator) and of finite length.*

These are general properties of holonomic modules [**BoD**, V, 1.16.1].

(2.4) COROLLARY. *If L is an intermediate field between $k(x) = \mathrm{Frac}(A)$ and $E = \mathrm{Frac}(B)$ then $L \cap B$ is a holonomic W-module.*

PROOF: It suffices to show that $L \cap B$ is a sub W-module of B. Since L is a finite (separable) extension of $k(x)$, the ∂_i stabilize L, so L is a W-module, hence so also is $L \cap B$.

(2.5) REMARKS: (1) Theorem (2.2) generalizes the special case $B = A[1/p]$, $p \in A$ [**BoD**, V, 1.16.2].

(2) It is natural to seek a similar generalization of [**BoD**, V, 1.16.3]. Namely, if M is a holonomic W-module then $B \otimes_A M$ is a W-module, where ∂_i acts by $\partial_i(b \otimes m) = \partial_i(b) \otimes m + b \otimes \partial_i(m)$. It seems plausible that $B \otimes_A M$ is holonomic, but I was not able to confirm this.

3. B as a $U(\underline{g})$-module.

The Lie algebra $\underline{gl}_n(k)$ acts as linear derivations on $A = k[x]$, the basic derivations being

$$\epsilon_{ij} = x_i \partial_j \qquad (1 \leq i, j \leq n).$$

We have $A = \oplus_{d \geq 0} A_{(d)}$, where $A_{(d)}$ denotes the k-module of homogeneous polynomials of degree d.

Fix a maximal ideal $\underline{m}$ of B, and put $\underline{m}_A = \underline{m} \cap A$. After translating variables, if necessary, we can assume that $\underline{m}_A = \sum_{i=1}^n A x_i$.

(3.1) PROPOSITION. *Let $\underline{g}$ denote one of $\underline{gl}_n(k)$ or $\underline{sl}_n(k)$.*

(a) *For $d \geq 0$, $A_{(d)}$ is a simple $\underline{g}$-module of dimension $\binom{d+n-1}{n-1}$. Hence A is a semi-simple multiplicity free $\underline{g}$-module.*

(b) *For $r \geq 0$, $A/\underline{m}_A^r \to B/\underline{m}^r$ is an isomorphism of $\underline{g}$-modules.*

(c) *The $\underline{g}$-module B/A contains no non-zero submodule of finite length.*

PROOF: Part (a) is classical (and easily verified directly).

(b). By (1.2)(a), $A/\underline{m}_A^r \to B/\underline{m}^r$ is an isomorphism. It remains only to observe that $\underline{m}$, and hence also all $\underline{m}^r$, are $\underline{g}$-invariant. If $b \in B$ then $\epsilon_{ij}(b) = x_i\partial_j(b) \in \underline{m}_A \cdot B \subset \underline{m}$, whence (b).

For (c), let $\bar{L} \subset B/A$ be a $\underline{g}$-module of finite length. Let $L \subset B$ be a finitely genrated module over the enveloping algebra $U = U(\underline{g})$ that projects mod A onto $\bar{L} \cong L/L\cap A$. Since U is noetherian, $L \cap A$ is finitely generated, and also semi-simple, by (a), hence of finite length. Thus L itself is of finite length. Since $\bigcap_{r\geq 0}\underline{m}^r = 0$ it follows that $\underline{m}^r \cap L = 0$ for some r. Thus L embeds in $B/\underline{m}^r \cong A/\underline{m}_A^r$, by (b), so, by (a) again, L is semi-simple. To show, as required, that $L \subset A$, we may therefore assume that L is simple. Then as above we see that $L \cong A_{(d)}$ for some d. If $L \neq A_{(d)}$ then $L + A_{(d)} = L \oplus A_{(d)}$. Choose r large enough so that $(L + A_{(d)}) \cap \underline{m}^r = 0$. Then $L \oplus A_{(d)}$ embeds in $B/\underline{m}^r \cong A/\underline{m}_A^r$, contadicting the fact that A is multiplicity free.

Let $\underline{g}$ be a Lie subalgebra of $\underline{gl}_n(k)$. Then the universal enveloping algebra $U = U(\underline{g})$ maps into W with the image the subalgebra generated by $\underline{g}$. U is a non-commutative noetherian integral domain. Let M be a U-module, $m \in M$, and $\mathrm{ann}_U(m) = \{u \in U \mid um = 0\}$. We call M *torsion* (resp. *torsion free*) if $\mathrm{ann}_U(m) \neq 0$ for all $m \in M$ (resp. $\mathrm{ann}_U(m) = 0$ for all $m \neq 0$ in M).

(3.2) PROPOSITION. *Let $\underline{g}$ be a Lie subalgebra of $\underline{gl}_n(k)$. If $\dim_k(\underline{g}) > n$ then B is a torsion $U(\underline{g})$-module.*

PROOF: Suppose the contrary. Then for some $b \in B$, $\mathrm{ann}_U(b) = 0$, where $U = U(\underline{g})$. Let $(U_r)_{r\geq 0}$ be the ascending filtration of U defined by the generating module $\underline{g}$ of U : $U_0 = k \subset U_1 = k + \underline{g}$, and $U_{r+1} = U_1 \cdot U_r$ for $r \geq 0$. By the Poincaré-Birkhoff-Witt Theorem (cf. [**5**, Ch. 3, Theorem 4.3]), $\mathrm{gr}(U) \cong S_k(\underline{g})$, the symmetric algebra of $\underline{g}$. If $\dim_k(\underline{g}) = N$ it follows that $\dim_k(U_r \cdot b) = \dim_k(U_r) = \binom{r+N}{N}$.

Let $u_1, \ldots, u_m$ generate the affine algebra B and let B_r denote the k-linear span of all u^γ ($|\gamma| \leq r$) (cf. proof of (2.2)). Choose $q > 0$ so that

$x_i \in B_q$ and $\partial_i(u_j) \in B_{q+1}$ $(i = 1, \ldots, n; j = 1, \ldots, m)$. If $b \in B_t$ then, as in the proof of (2.2), we see that $\epsilon_{ij}(b) = x_i \partial_j(b) \in B_{t+2q}$, and so $U_r b \subset B_{t+2rq}$. It follows that

$$\binom{N+r}{N} \leq \dim_k(B_{t+2rq})$$

for all $r \geq 0$. Let $P(s) = \dim_k(B_s)$. For $s \gg 0$, $P(s)$ is a polynomial in s of degree $\dim(B) = n$, by (A.1). Since $\binom{N+r}{N} \leq P(t+2qr)$ and the left side is a polynomial in r with leading term $r^N/N!$, it follows that $N \leq n$, whence the proposition.

(3.3) REMARKS: (1) Proposition 3.2 furnishes an approach to the Jacobian Conjecture. Namely, given $A \subset B$ as in the Jacobian Conjecture, choose a $\underline{g} \subset \underline{gl}_n(k)$ with $\dim_k(\underline{g}) > n$, and try to show that B/A is a *torsion free* $U(\underline{g})$-module, hence zero by (3.2). We shall pursue this idea in §6, for $n = 2$.

(2) Part of the difficulty of this method stems from the non-commutativity of $U(\underline{g})$. However for $n \geq 4$ one can choose $\underline{g}$ to be commutative. For example if $n = 2m$ let $\underline{g}$ be spanned by $\left[\begin{smallmatrix} 0 & * \\ 0 & 0 \end{smallmatrix}\right]$ and the scalars, of dimension $m^2 + 1 > 2m$.

The next lemma is helpful in focusing some of our calculations.

(3.4) LEMMA. *Let $\underline{g} \subset \underline{gl}_n(k)$ be a Lie subalgebra and $U = U(\underline{g})$.*

(a) *If $u \in U - k$ then U is a product of irreducible elements. (An element $v \in U$ is irreducible if $v \notin U^\times$ and whenever $v = v_1 v_2$ in U either, $v_1 \in U^\times$ or $v_2 \in U^\times$.)*

(b) *Suppose that there exist $u \in U$, $u \neq 0$ and $b \in B - A$ such that $ub \in A$. Then there exists an irreducible $v \in U$ and a $c \in B - A$ such that $vc = 0$.*

PROOF: For $u \neq 0$ in U put $|u| = r$ if $u \in U_r - U_{r-1}$, where $(U_r)_{r \geq 0}$ is the Poincaré-Birkhoff-Witt filtration of U (cf. proof of (3.2)). Then $U_0 = k$ and $|uv| = |u| + |v|$ for $uv \neq 0$, so (a) follows immediately by induction on $|u|$.

Let M be a U-module, $x \neq 0$ in M, and $u \neq 0$ in U such that $ux = 0$. If $u = vw$ then either $wx = 0$, or $y = wx \neq 0$, and $vy = 0$. In this way we see by induction on $|u|$ that some non-zero element of M is killed by an irreducible element of U. Taking $M = B/A$ then in (b) we find a $b' \in B - A$ and an irreducible $v \in U$ such that $vb' \in A$. We have $A = \oplus_{d \geq 0} A_{(d)} \subset$

$B \subset \hat{A} = \prod_{d\geq 0} A_{(d)}$, and each $A_{(d)}$ is U-invariant. Writing $b' = \sum_{d\geq 0} b_d$ with $b_d \in A_{(d)}$, we have $vb' = \sum_{d\geq 0} vb_j$ with $vb_d \in A_{(d)}$. Since $vb' \in A$ we have $vb_d = 0$ for all $d >$ some N. Put $c = b' - (\sum_{d\leq N} b_d) \in B - A$. Then clearly $vc = 0$.

(3.5) REMARK: Remark (3.3)(1) and Lemma (3.4) lead us to comtemplate the following situation. Let u be an irreducible element of $U(\underline{g})(\underline{g} \subset \underline{gl}_n(k))$. We would like to know that $B_0^u = \mathrm{Ker}(u \text{ on } B)$ is contained in A. As a first step consider $\hat{A}_0^u = \mathrm{Ker}(u \text{ on } \hat{A})$. We have $\hat{A} = \prod_{d\geq 0} A_{(d)}$, and so $\hat{A}_0^u = \prod_{d\geq 0} (A_{(d)})_0^u$. Thus $\hat{A}_0^u \subset A$ iff u acts invertibly on $A_{(d)}$ for all but finitely many d. Here $A_{(d)} = S^d(V)$, the dth symmetric power of the natural representation $V = k^n$ of $gl_n(k)$.

Thus we are led to seek special properties of irreducible elements $u \in U(\underline{g})$ which annihilate non-zero elements of $S^d(V)$ for infinitely many $d \geq 0$. Obvious examples are the nilpotent elements of $\underline{g}$, and certain semi-simple elements of $\underline{g} + k$. More elaborate examples stem from the fact that $U(\underline{gl}_n(k))$ does not act faithfully on $\hat{A}$. For example (see the Remark following (6.2)) if $n = 2$ then $\epsilon(\epsilon + 2) - c$ acts trivially on $\hat{A}$, where $\epsilon = \epsilon_{11} + \epsilon_{22}$ ($\epsilon_{ij} = x_i \partial_j$) is the Euler derivation, and c is the Casimir element in $U(\underline{sl}_2(k))$. Robert Steinberg showed me further examples not accounted for by the above observations. Let $u = 2 + (\epsilon_{11} - \epsilon_{22}) - \epsilon_{21}\epsilon_{12} \in U(\underline{sl}_2(k))$. Though u is irreducible in $U(\underline{sl}_2(k))$ we have $u = (2 + \epsilon_{11})(1 - \epsilon_{22})$, so $u(x_1^d x_2) = 0$ for all $d \geq 0$. Moreover, relative to the basis of $A_{(d)}$ formed by monomials, u is diagonal and ϵ_{21} is nilpotent triangular, hence $v = u + \epsilon_{21}$ has a 0 eigenvalue. Moreover v is even irreducible in $U(\underline{gl}_2(k))$.

(3.6) PROPOSITION. *Let $n = 2$ and let $\underline{g} = \underline{sl}_2(k)$ or $\underline{b} = k\epsilon_{11} + k\epsilon_{22} + k\epsilon_{12}$, the triangular subalgebra of $\underline{gl}_2(k)$. If $B \neq A$ then B/A is a faithful module over $U(\underline{g})$.*

This follows from (C.7) (Appendix C).

4. The action of ∂_i on B.

We use the following notation as in Appendix B. If V is a k-module, $D \in \mathrm{End}_k(V)$, and $\lambda \in k$ put $V_\lambda^D = \mathrm{Ker}(D - \lambda)$ and $V_{(\lambda)}^D = \bigcup_{m\geq 0} \mathrm{Ker}(D - \lambda)^m$. We put $V_{\mathrm{tors}}^D = \oplus_{\lambda\in k} V_{(\lambda)}^D$.

As always, we have

$$\begin{array}{ccc} k(x) & \subset & E = \mathrm{Frac}(B) \\ \cup & & \cup \\ A = k[x] & \subset & B \end{array}$$

with B étale over A, and $\partial_i = \partial/\partial x_i$ $(i = 1, \dots, n)$.

(4.1) PROPOSITION. $\bigcap_{j=1}^n E_{(0)}^{\partial_j} = A$, and $\bigcap_{j=1}^n E_0^{\partial_j} = k$.

PROOF: From (B.2) we have $\bigcap_{j=1}^n E_{(0)}^{\partial_j} = K[x_1, \dots, x_n]$, where $K = \bigcap_{j=1}^n E_0^{\partial_j}$ is clearly algebraic over k, hence $K = k$.

We put $\bar{A}$ = the integral closure of A in B.

(4.2) COROLLARY. *If $\partial_j(\bar{A}) \subset \bar{A}$ $(j = 1, \dots, n)$ then $\bar{A} = A$ and so $E = k(x)$. If further $B^\times = k^\times$ then $B = A$.*

PROOF: Since $A \subset B_{(0)}^{\partial_j}$ it follows from Vasconcelos's Theorem ((B.7)(b)) that if $\partial_j(\bar{A}) \subset \bar{A}$ then $\bar{A} = \bar{A}_{(0)}^{\partial_j}$. Hence $\bar{A} \subset \bigcap_j E_{(0)}^{\partial_j} = A$, by (4.1). The last assertion follows from the first one plus (1.3).

(4.3) PROPOSITION. *Consider an intermediate ring $A \subset C \subset B$ invariant under each ∂_j, e.g. $C = L \cap B$ with $k(x) \subset L \subset E$ an intermediate field.*

(a) *C is étale over A.*
(b) *If $C \subset \bar{A}$ then $C = A$ and $L = k(x)$.*

PROOF: Note first that the ∂_j stabilize L and hence also C. If $C \subset \bar{A}$ then one shows, as in the proof of (4.2), that $C = A$, and so $L = \mathrm{Frac}(C) = \mathrm{Frac}(A) = k(x)$.

To prove (a) we want to show that $f : C \otimes_A \Omega_A \to \Omega_C$ is an isomorphism. Applying $\mathrm{Hom}_C(-, C)$ yields $f^* : \mathrm{Der}(C, C) \xrightarrow{\text{restr.}} \mathrm{Der}(A, C) = \oplus_j C\partial_j$. Invariance of C under the ∂_j implies that f^* is surjective. Since $C \otimes_A \Omega_A$ is a finitely generated free C-module it follows then from [**B**, Ch. IV, Prop. (1.1)] that f is a split monomorphism. Thus $\Omega_C \cong (C \otimes_A \Omega_A) \oplus T$. Since $L/k(x)$ is separable ($\mathrm{char}(k) = 0$) T must be a torsion module. Hence $\Omega_C \to \Omega_C^{**}$ is surjective. Now it follows from results of Lipman [**Li**, proof of Theorem 1] that C is smooth, so Ω_C is a projective C-module, hence $T = 0$ and f is an isomorphism, as claimed.

REMARK: Consider the following *"Generalized Jacobian Conjecture"*. Suppose that B, in addition to being étale over A, satisfies $B^\times = k^\times$ and $E = \mathrm{Frac}(B)$ is unirational over k. Then $B = A$.

A convenient feature of this formulation is that if $k(x) \subset L \subset E$ is an intermediate field then, in view of (4.3), $C = L \cap B$ inherits all of the above assumptions on B. Hence one can reduce the above conjecture to the case where $k(x) \subset E$ has no proper intermediate fields.

(4.4) PROPOSITION. *Let* $\partial = f\partial_1$ *where* $f \neq 0$, $f \in k[x_2, \ldots, x_n]$. *Put* $L = E_0^\partial$ *and* $R = B_0^\partial = L \cap B$ *which contains* $k[x_2, \ldots, x_n]$.

(a) $L = \mathrm{Frac}(R) =$ *the algebraic closure of* $k(x_2, \ldots, x_n)$ *in* E. *(Both* L *and* R *are independent of* f.*)*

(b) $B_{(0)}^\partial = B_{\mathrm{tors}}^\partial = R[x_1]$, *and* R *is étale over* $k[x_2, \ldots, x_n]$.

(c) $E_{(0)}^\partial = E_{\mathrm{tors}}^\partial = L[x_1]$.

PROOF: Since $\mathrm{char}(k) = 0$, L is algebraically closed in E. Since $E_0^\partial = E_0^{\partial_1}$ and $E_{(0)}^\partial = E_{(0)}^{\partial_1}$ ($\partial_1(f) = 0$ so $\partial^m = f^m \partial_1^m$) it follows from (B.1) that $E_{(0)}^\partial = L[x_1]$ and $B_{(0)}^\partial = R[x_1]$, which have transcendence degree n over k. Hence $\mathrm{tr}\,\deg_k(L) = n - 1$, so L is the algebraic closure of $k(x_2, \ldots, x_n)$ in E. If $\lambda \in k^\times$ and $y \in E_{(\lambda)}^\partial$, $y \neq 0$, then from the graded algebra structure of $E_{\mathrm{tors}}^\partial = \oplus_{\mu \in k} E_{(\mu)}^\partial$ (cf. (B.4)(b)) we see that y is transcendental over $E_{(0)}^\partial = L[x_1]$, which is impossible. Thus $E_{\mathrm{tors}}^\partial = E_{(0)}^\partial = L[x_1]$, and $B_{\mathrm{tors}}^\partial = B_{(0)}^\partial = R[x_1]$. Since $R = B \cap L$ is integrally closed in L, and L is algebraic over $k(x_2, \ldots, x_n)$, with $k[x_2, \ldots, x_n] \subset R$, it follows finally that $L = \mathrm{Frac}(R)$. That R is étale over $k[x_2, \ldots, x_n]$ follows from (4.3)(a).

(4.5) PROPOSITION. *Let* $F = k(x_1)$. *Assume* $B^\times = k^\times$ *and that* E *is unirational.*

(a) F *is algebraically closed in* E.

(b) $F = \bigcap_{j=2}^n E_0^{\partial_j}$, *and* $k[x_1] = \bigcap_{j=2}^n B_0^{\partial_j}$.

(c) $F[x_2, \ldots, x_n] = \bigcap_{j=2}^n E_{(0)}^{\partial_j} = \bigcap_{j=2}^n E_{\mathrm{tors}}^{\partial_j}$.

(d) $A = k[x_1, x_2, \ldots, x_n] = \bigcap_{j=2}^n B_{(0)}^{\partial_j} = \bigcap_{j=2}^n B_{\mathrm{tors}}^{\partial_j}$.

PROOF: Corollary (1.5) supplies (a) and (b). By (B.2) and (b) we have $\bigcap_{j=2}^n E_{(0)}^{\partial_j} = F[x_2, \ldots, x_n]$ and $\bigcap_{j=2}^n B_{(0)}^{\partial_j} = A$. Moreover $E_{(0)}^{\partial_j} = E_{\mathrm{tors}}^{\partial_j}$ and $B_{(0)}^{\partial_j} = B_{\mathrm{tors}}^{\partial_j}$ by (4.4)(b) and (c).

(4.6) COROLLARY. *If* $n = 2$ *(i.e.* $A = k[x_1, x_2]$*) in (4.5) then* $B_0^{\partial_2} = k[x_1]$, $B_{(0)}^{\partial_2} = B_{\mathrm{tors}}^{\partial_2} = A$, $E_0^{\partial_2} = k(x_1)$, *and* $E_{(0)}^{\partial_2} = E_{\mathrm{tors}}^{\partial_2} = k(x_1)[x_2]$.

(4.7) PROPOSITION. *Let* M *be a* k*-submodule of* $\bar{A}$, *the integral closure of* A *in* B.

(a) *If* $\partial_j(M) \subset M$ *then* $M \subset B_{(0)}^{\partial_j} = B_0^{\partial_j}[x_j]$.

(b) *Assume that $B^\times = k^\times$ and that E is unirational. If $\partial_j(M) \subset M$ for $j = 2, \ldots, n$ then $M \subset A$.*

PROOF: Let $C = A[M] \subset \bar{A}$. If $\partial_j(M) \subset M$ then $\partial_j(C) \subset C$ so, by (B.7)(b), $C = C^{\partial_j}_{(0)} \subset B^{\partial_j}_{(0)} = B^{\partial_j}_0[x_j]$, the last by (4.4)(b). If $\partial_j(M) \subset M$ for $j = 2, \ldots, n$ then $C \subset \bigcap_{j=2}^n B^{\partial_j}_{(0)} = A$ by (4.5)(d).

(4.8) PROPOSITION. *Let $\epsilon = \sum_{i=1}^n x_i\partial_i$, the "Euler derivation".*

(a) $B^\epsilon_{\text{tors}} = A$.

(b) $E^\epsilon_{\text{tors}} = A_S = F[x_1, x_1^{-1}]$, *where* $F = k(x_2/x_1, \ldots, x_n/x_1)$

and S is the multiplicative set of non-zero homogeneous polynomials in A. If $d \in \mathbb{Z}$ then $E^\epsilon_{(d)} = E^\epsilon_d = Fx_1^d$, and if $\lambda \notin \mathbb{Z}$ then $E^\epsilon_{(\lambda)} = 0$.

PROOF: We have $A = \oplus_{d \geq 0} A_{(d)} \subset B \subset \hat{A} = \prod_{d \geq 0} A_{(d)} = k[[x]]$, and $\epsilon|_{A_{(d)}}$ is the multiplication by d. Hence $(\hat{A})^\epsilon_{(d)} = (\hat{A})^\epsilon_d = A_{(d)}$ for $d \in \mathbb{N}$, and $(\hat{A})^\epsilon_\lambda = 0$ if $\lambda \notin \mathbb{N}$. Thus $\hat{A}^\epsilon_{\text{tors}} = A$, whence (a).

To prove (b) put $L = E^\epsilon_0$. Then $L \supset F$ and $\operatorname{tr} \deg_k(L) < n$, so L must be the algebraic closure of F in E. If $L \neq E^\epsilon_{(0)}$ there must be a $y \in E^\epsilon_{(0)}$ with $\epsilon(y) = 1$. Then by (B.1) we have $E^\epsilon_{(0)} = L[y]$. Since $\epsilon(x_1) = x_1$, $L[y, x_1]$ must then be a polynomial L-algebra in 2 variables (cf. (B.4)), of transcendence degree $n + 1$ over k, which is impossible. Thus $L = E^\epsilon_{(0)}$.

Set $\Lambda = \{\lambda \in k \mid E^\epsilon_\lambda \neq 0\}$. We have $x_1 \in E^\epsilon_1$, $E^\epsilon_\lambda E^\epsilon_\mu \subset E^\epsilon_{\lambda+\mu}$, and if $b \in E^\epsilon_\lambda$, $b \neq 0$, then $b^{-1} \in E^\epsilon_{-\lambda}$. Thus Λ is an additive subgroup of k containing $\mathbb{Z}$. For each $\lambda \in \Lambda$ choose $y_\lambda \neq 0$ in E^ϵ_λ. If $z \in E^\epsilon_{(\lambda)}$ then $y_\lambda^{-1} z \in E^\epsilon_{(0)} = L$, and so $E^\epsilon_{(\lambda)} = Ly_\lambda = E^\epsilon_\lambda$. If $\lambda \in \Lambda$ and $\mathbb{Z}\lambda \cap \mathbb{Z} = 0$ then $L[x_1, y_\lambda]$ is a polynomial L-algebra in 2 variables which, as we saw above, is impossible. Thus the abelian group Λ has rank 1. To show that $\Lambda = \mathbb{Z}$ it suffices to show that if $d \in \mathbb{Z}$, $d > 0$, $1/d \in \Lambda$, then $d = 1$. Put $y = y_{1/d}$. Then $y^d \in Lx_1$. We have $F[x_1] \subset F[x_1, x_1^{-1}] = A_S$. Since B is flat over A, we have

$$B_1 = F[x_1] \bigotimes_A B \subset F[x_1, x_1^{-1}] \bigotimes_A B = B_S \subset E,$$

and B_1 is étale over $F[x_1]$, hence normal (cf. (1.2)). Since L is algebraic over $F \subset B_1$ we have $F[x_1] \subset L[x_1] \subset B_1$. Since $y^d \in Lx_1$, y is integral over B_1, hence $y \in B_1$. We have $x_1 B_1 = (yB_1)^d$. But $x_1 F[x_1]$ is not a proper power in $F[x_1]$, and B_1 is unramified over $F[x_1]$, hence $x_1 B_1$ is not a proper power in B_1. Hence $d = 1$, as claimed, and so $\Lambda = \mathbb{Z}$. It follows that $E^\epsilon_{(\lambda)} = 0$ for $\lambda \notin \mathbb{Z}$, while for $d \in \mathbb{Z}$ we have $E^\epsilon_{(d)} = E^\epsilon_d = Lx_1^d$; thus $E^\epsilon_{\text{tors}} = L[x_1, x_1^{-1}]$.

It remains to prove that $L = F$, i.e. that $e = [L : F]$ equals 1. Clearly $e = [L(x_1) : F(x_1)] = [L(x_1) : k(x)]$. Let A' denote the integral closure of A in $L(x_1)$. Since $L(x_1)$ is algebraic over $k(x) = \text{Frac}(A)$ we have $L(x_1) = \text{Frac}(A')$. Since $A \subset A_S = F[x_1, x_1^{-1}] \subset L[x_1, x_1^{-1}] = E^{\epsilon}_{\text{tors}}$, and $L[x_1, x_1^{-1}]$ is integrally closed, we have $A' \subset (E^{\epsilon}_{\text{tors}} \cap B) = B^{\epsilon}_{\text{tors}} = A$, by (a). Thus $L(x_1) = \text{Frac}(A') = \text{Frac}(A) = k(x)$, so $e = [L(x_1) : k(x)] = 1$, and $L = F$, as claimed. This completes the proof of (4.8).

5. The extension $k[x_1] \subset B$.

(5.1) PROPOSITION. (a) *For each* $\lambda \in k$,

$$(x_1 - \lambda)B = P_{\lambda 1} \cap \cdots \cap P_{\lambda m_\lambda},$$

where the P_{λ_j} *are pairwise comaximal height 1 primes, and* B/P_{λ_j} *is regular (of dimension* $n-1$*) and étale over* $A/(x_1 - \lambda)A = k[x_2, \ldots, x_n]$.

(b) *Suppose that* $B^\times = k^\times$ *and* E *is unirational. Then* $F = k(x_1)$ *is algebraically closed in* E*, and, if* $B_F = F \otimes_{k[x_1]} B$*, then* $B_F^\times / F^\times$ *is a finitely generated abelian group.*

(c) *Suppose further that* B *is factorial, so that each* P_{λ_j} *is principal,* $P_{\lambda_j} = Bp_{\lambda_j}$*. Then* $B_F^\times$ *is generated by* $k^\times$ *and all* p_{λ_j} $(\lambda \in k,\ j = 1, \ldots, m_\lambda)$*. The abelian group* $B_F^\times / F^\times$ *is free of rank* $\sum_{\lambda \in k}(m_\lambda - 1)$*. Hence* $m_\lambda = 1$*, i.e.* $x_1 - \lambda$ *is prime in* B*, for all but finitely many* λ*. For all* $\lambda \in k$, $j = 1, \ldots, m_\lambda$*, we have* $\partial_i(P_{\lambda_j}) \subset P_{\lambda_j}$ *for* $i = 2, \ldots, n$*, whereas* $\partial_1(p_{\lambda_j})$ *is invertible* $\bmod P_{\lambda_j}$.

PROOF: Since B is a normal affine domain $B/(x_1 - \lambda)B$ is of pure dimension $n-1$. Moreover it is étale over $A/(x_1 - \lambda)A = k[x_2, \ldots, x_n]$. Hence, by (1.2) and the subsequent remark, $B/(x_1 - \lambda)B$ is a finite product of regular domains, each étale over $A/(x_1 - \lambda)A$; whence (a).

Assume that $B^\times = k^\times$ and E is unirational. By (1.5) $F = k(x_1)$ is algebraically closed in E. Hence, by a theorem of Rosenlicht (cf. [**L**, Ch. 2, Cor. 7.3], $B_F^\times / F^\times$ is a finitely generated group (isomorphic to a subgroup of the group of divisors supported at infinity in a normal projective completion of $\text{Spec}(B_F)$ over F). Suppose further that B is factorial. Then $B_F^\times$ is generated by $k^\times$ and the p_{λ_j} $(\lambda \in k,\ j = 1, \ldots, m_\lambda)$, while $F^\times$ is generated by $k^\times$ and the $(x_1 - \lambda)$ $(\lambda \in k)$. Clearly then $B_F^\times / F^\times$ is free abelian of rank $\sum_{\lambda \in k}(m_\lambda - 1)$. Hence $m_\lambda = 1$ for almost all λ.

Writing $x_1 - \lambda = p_{\lambda 1} \dots p_{\lambda m_\lambda}$, and putting $p'_{\lambda_j} = (x_1 - \lambda)/p_{\lambda_j} \in B - P_{\lambda_j}$, we have

$$\partial_i(x_1 - \lambda) = \sum_{h=1}^{m_\lambda} \partial_i(p_{\lambda_h}) \cdot p'_{\lambda_h}$$

so

$$\partial_i(x_1 - \lambda) \equiv \partial_i(p_{\lambda_j}) \cdot p'_{\lambda_j} \bmod P_{\lambda_j}.$$

For $i = 1$, $\partial_1(x_1 - \lambda) = 1$, so $\partial_1(p_{\lambda_j})$ is invertible $\bmod P_{\lambda_j}$, whereas for $i = 2, \dots, n$, $\partial_i(x_1 - \lambda) = 0$, so $\partial_i(p_{\lambda_j}) \in P_{\lambda_j}$. This establishes all the assertions of (5.1).

(5.2) REMARKS: (1) Consider the morphism $Y = \mathrm{Spec}(B) \to \mathbb{A}^1 = \mathrm{Spec}(k[x_1])$. The fiber Y_λ over $\lambda \in k$ is a disjoint union of smooth hypersurfaces in Y, each étale over $\mathbb{A}^{n-1} = \mathrm{Spec}(k[x_2, \dots, x_n])$. If $B^\times = k^\times$, B is factorial, and E is unirational then all but finitely many Y_λ are irreducible.

(2) With $F = k(x_1)$ it is natural to try to apply induction arguments to the étale extension $A_F = F[x_2, \dots, x_n] \subset B_F = F \otimes_A B$. If we assume that $B^\times = k^\times$ and E is unirational over k then we must seek analogous conditions not just for B_F over F, but for $B_{\bar{F}} = \bar{F} \otimes_A B$ over $\bar{F}$, the algebraic closure of F. This does not seem to be a simple matter.

(3) The following observations, pointed out to me by M. P. Murthy, illustrate phenomena related to (5.1) and Remark (2). Let $B = k[u, v]$, a polynomial algebra in 2 variables, and let $x \in B - k$. Consider the map $\varphi : Y = \mathbb{A}^2 \to \mathbb{A}^1$ defined by $k[x] \subset B$, and put $F = k(x)$. (For $f \in B$ write $f_u = \partial f/\partial u$ and $f_v = \partial f/\partial v$.)

(a) If B/xB were a polynomial algebra (in one variable) then, by the Epimorphism Theorem of Abhyankar and Moh [**AM**] one would have $B = k[x, y]$ for some $y \in B$.

(b) Take $x = (uv + 1)u$. Since $x_u = 2uv + 1$ and $x_v = u^2$ are comaximal ($(1 - 2uv)x_u + 4v^2 x_v = 1$) all fibers Y_λ ($u^2 v + u = \lambda$) of φ are smooth. For $\lambda \neq 0$, Y_λ is a once-punctured affine line, while Y_0 is the disjoint union of the v-axis ($u = 0$) and the hyperbola $uv = -1$. Thus $B_{\bar{F}}^\times / F^\times$ is cyclic, generated by the image of u. On the other hand it can be shown that there is no $y \in B$ such that $x_u y_v - x_v y_u = 1$ (which would contradict the Jacobian Conjecture).

(c) Next take $x = u^3 - v^2$. For $\lambda \neq 0$, Y_λ ($v^2 = u^3 - \lambda$) is an elliptic curve, while Y_0 is a rational curve with a cusp; hence $B_{\bar{F}}^\times = F^\times$. Moreover B_F, being a localization of B, is factorial. But $B_F =$

$F[u,v] = F[U,V]/(V^2 - U^3 - x)$ defines an elliptic curve over $F = k(x)$. Therefore over the algebraic closure $\bar{F}$, $\mathrm{Pic}(B_{\bar{F}})$ is huge.

6. Calculations for $n = 2$.

As usual we have

$$A = \bigoplus_{d \geq 0} A_{(d)} \subset B \subset \hat{A} = \prod_{d \geq 0} A_{(d)},$$

where $A_{(d)}$ denotes the space of homogeneous polynomials of degree d, and B is étale over A. For convenience, and without loss of generality, we take $k = \mathbb{C}$. We study here the case $n = 2$. Writing x, y for x_1, x_2, we then have

$$A = \mathbb{C}[x,y] \subset B \subset \hat{A} = \mathbb{C}[[x,y]].$$

We make only the following assumptions on B:

(i) B is étale over A.

(ii) $B^\times = \mathbb{C}^\times$.

(iii) $E = \mathrm{Frac}(B)$ is unirational over $\mathbb{C}$.

Note that we do not assume B to be factorial; this, together with (i), (ii), and (iii) would imply that B is a polynomial ring.

Put $\partial_x = \partial/\partial x$ and $\partial_y = \partial/\partial y$. The standard basis of $\underline{gl}_2(\mathbb{C})$ is:

$$\begin{aligned} \epsilon_x &= x\partial_x, & \delta &= x\partial_y \\ \delta' &= y\partial_x, & \epsilon_y &= y\partial_y. \end{aligned}$$

We have the following 3-dimensional subalgebras with bases:

$$\underline{sl}_2(\mathbb{C}) : \delta, \epsilon_x - \epsilon_y, \delta'$$
$$\underline{b} : \epsilon_x, \epsilon_y, \delta.$$

To prove the Jacobian Conjecture (that $A = B$) it would suffice, according to (3.3), to show that B/A is a torsion free module over say

$$U = U(\underline{b}) = \mathbb{C}[\epsilon_x, \epsilon_y, \delta].$$

We shall prove several partial results in this direction.

If $\varphi \in U$ and M is a U-module we write $M_0^\varphi = \mathrm{Ker}(\varphi \text{ on } M) = \{m \in M \mid \varphi m = 0\}$.

(6.1) PROPOSITION. *Let $\epsilon = \epsilon_x + \epsilon_y$, the "Euler derivation". Then B/A is a torsion free $\mathbb{C}[\epsilon, \delta]$-module.*

PROOF: By (3.4)(b) it suffices to show that if $b \in B$ and $\varphi b = 0$ for some irreducible $\varphi = \varphi(\epsilon, \delta) \in \mathbb{C}[\epsilon, \delta]$ then $b \in A$.
Case 1: $\varphi = \delta$. Then $b \in B_0^\delta = B_0^{\partial_y} = \mathbb{C}[x] \subset A$, by (4.6).
It suffices now to treat
Case 2: $\varphi(\epsilon, 0) \neq 0$. Write $\varphi = \sum_{r \geq 0} \varphi_r(\epsilon)\delta^r$, with $\varphi_0(\epsilon) \neq 0$. Write $b = \sum_{d \geq 0} b_d \in \hat{A}$ with $b_d \in A_{(d)}$. Now $\varphi \mid_{A_{(d)}} = \varphi(d, \delta) = \sum_r \varphi_r(d)\delta^r$, and $\delta \mid_{A_{(d)}}$ is nilpotent. Hence $\varphi \mid_{A_{(d)}}$ is the sum of the homothety $\varphi_0(d) \cdot Id$ and a nilpotent endomorphism, so it is invertible if $\varphi_0(d) \neq 0$. Since φ_0 has only finitely many roots, $\hat{A}_0^\varphi = \mathrm{Ker}(\varphi \text{ on } \hat{A}) \subset A$, so $b \in A$.

(6.2) COROLLARY. *In $U(\underline{sl}_2(\mathbb{C}))$ let c be the (central) Casimir element, $c = 2\delta'\delta + (\epsilon_x - \epsilon_y)^2 + (\epsilon_x - \epsilon_y)$. Then B/A is a torsion free $\mathbb{C}[c, \delta]$-module.*

PROOF: We have $cx^py^q = rx^py^q$, where $r = 4(p+1)q + (p-q)^2 + (p-q) = (p+q)(p+q+2)$. Hence c acts on $A_{(d)}$ as multiplication by $d(d+2)$. Thus the action of c on $B \subset \hat{A} = \prod_d A_{(d)}$ is the same as that of $\epsilon(\epsilon+2)$, where $\epsilon = \epsilon_x + \epsilon_y$ as above. Hence $\mathbb{C}[c, \delta]$ acts on B as does $\mathbb{C}[\epsilon(\epsilon+2), \delta] \subset \mathbb{C}[\epsilon, \delta]$. Thus the corollary follows from (6.1).

REMARK: Since $c \neq \epsilon(\epsilon + 2)$ in $U(\underline{gl}_2(\mathbb{C}))$ we see that $U(\underline{gl}_2(\mathbb{C}))$ does not act faithfully on $\hat{A}$, and the natural homomorphism $U(\underline{gl}_2(\mathbb{C})) \to W = \mathbb{C}[x, y, \partial_x, \partial_y]$ is not injective. This contrasts with the situation for $\underline{sl}_2(\mathbb{C})$ (see (C.7)).

(6.3) THEOREM. *Let $\partial \in \underline{gl}_2(\mathbb{C})$. Then B/A is a torsion free $\mathbb{C}[\partial]$-module.*

PROOF: We can assume $\partial \neq 0$. After a linear homogenous change of coordinates in $\mathbb{C}[x, y]$ we can assume that ∂ is in Jordan canonical form. Then either $\partial = a\epsilon + \delta$ $(a \in \mathbb{C})$ or $\partial = a\epsilon_x - b\epsilon_y$ $(a, b \in \mathbb{C})$. In the first case $\partial \in \mathbb{C}[\epsilon, \delta]$ and the result follows from (6.1). Assume now that $\partial = a\epsilon_x - b\epsilon_y$; then $\partial x^py^q = (ap - bq)x^py^q$. Let $L_\lambda = \{(p, q) \in \mathbb{C}^2 \mid ap - bq = \lambda\}$, for $\lambda \in \mathbb{C}$. Then the kernel of $\partial - \lambda$ on $\hat{A}$ is clearly

$$\hat{A}_\lambda^\partial = \hat{A}_{(\lambda)}^\partial = \prod_{(p,q) \in L_\lambda \cap \mathbb{N}^2} \mathbb{C}x^py^q.$$

If $L_\lambda \cap \mathbb{N}^2$ is finite, then $\hat{A}_\lambda^\partial \subset A$, hence $B_\lambda^\partial \subset A$, as desired. To show that $B_\lambda^\partial (= \mathrm{Ker}(\partial - \lambda \text{ on } B)) \subset A$ in general, we need only consider the case when $L_\lambda \cap \mathbb{N}^2$ is infinite. In this case L_λ is a rational line with positive slope. Multiplying ∂ by a scalar, if necessary, we may therefore assume that $a, b \in \mathbb{N}$ and $\gcd(a,b) = 1$. This case is covered by the following proposition.

(6.4) PROPOSITION. *Let $a, b \geq 0$ be relatively prime integers. Put $\partial = a\epsilon_x - b\epsilon_y$ and $u = x^b y^a$. The eigenspaces of ∂ in $\hat{A} = \mathbb{C}[[x,y]]$ are of the form $x^r y^s \mathbb{C}[[u]]$ with $r, s \geq 0$ and either $r < b$ or $s < a$. We have $x^r y^s \mathbb{C}[[u]] \cap B = x^r y^s \mathbb{C}[u] \subset A$; hence $B^\partial_{\text{tors}} \subset A$. Further, $\mathbb{C}(u)$ is algebraically closed in $E = \mathrm{Frac}(B)$, and $\mathbb{C}(u) = E^\partial_0$.*

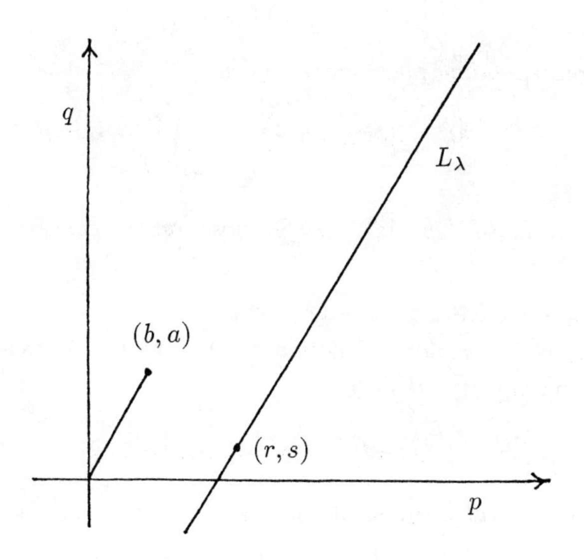

Figure 1.

PROOF: With $L_\lambda = \{(p,q) \mid ap - bq = \lambda\}$ as above, $L_\lambda \cap \mathbb{N}^2 = \emptyset$ unless $\lambda \in \mathbb{Z}$. If $L_\lambda \cap \mathbb{N}^2 \neq \emptyset$ choose $(r,s) \in L_\lambda \cap \mathbb{N}^2$ nearest the origin. Since $(r-b, s-a) \in L_\lambda$, either $r < b$ or $s < a$. Moreover it is easy to see that

$$L_\lambda \cap \mathbb{N}^2 = \{(r,s) + n(b,a) \mid n \in \mathbb{N}\},$$

whence $\hat{A}^\partial_\lambda = x^r y^s \mathbb{C}[[u]]$.

Let $L = E^\partial_0 = \mathrm{Ker}(\partial \text{ on } E = \mathrm{Frac}(B))$. Then $\mathbb{C}(u) \subset L$ and $\operatorname{tr\,deg}_{\mathbb{C}}(L) = 1$. Moreover $u = x^b y^a$ is not a proper power in A since $\gcd(b,a) = 1$. It follows therefore from (1.4) that $L = \mathbb{C}(u)$, i.e. $E^\partial_0 = \mathbb{C}(u)$, and $B^\partial_0 = B \cap \mathbb{C}[[u]] = \mathbb{C}[u] \subset A$. Moreover $\mathbb{C}(u)$ is algebraically closed in E.

It remains to show that $x^r y^s \mathbb{C}[[u]] \cap B = x^r y^s \mathbb{C}[u]$. Let $b = x^r y^s f(u) \in B$ with $f(u) \in \mathbb{C}[[u]]$. Then $f(u) \in E^\partial_0 = \mathbb{C}(u)$, as observed above. Since

$\mathbb{C}(u) \subset \mathbb{C}(x,y)$ we have $b \in B \cap \mathbb{C}(x,y) = A$, by (1.3). Thus $f(u) \in \mathbb{C}[u]$, as claimed.

We next study B as a module over $\mathbb{C}[\epsilon_x, \epsilon_y]$. We shall make essential use of Siegel's Theorem about algebraic curves with infinitely many integral points. Its consequences for the matters at hand are formulated in the next theorem, whose proof is developed in Appendix E.

(6.5) THEOREM. *Let $\varphi = \varphi(\epsilon_x, \epsilon_y) \in \mathbb{C}[\epsilon_x, \epsilon_y]$ be an irreducible polynomial, and let*

$$C = \{(p,q) \mid \varphi(p,q) = 0\}$$

be the corresponding plane curve. Then

$$\hat{A}_0^{\varphi} = \mathrm{Ker}(\varphi \text{ on } \hat{A}) = \prod_{(p,q)\in C\cap\mathbb{N}^2} \mathbb{C}x^p y^q.$$

If $C \cap \mathbb{N}^2$ is finite then $\hat{A}_0^{\varphi} \subset A$. Suppose, on the contrary, that $C \cap \mathbb{N}^2$ is infinite.

(a) *For some $c \in \mathbb{C}^{\times}$, $c\varphi \in \mathbb{Q}[\epsilon_x, \epsilon_y]$.*
(b) *There are rational functions $p(t)$, $q(t) \in \mathbb{Q}(t)$ such that $\mathbb{Q}(t) = \mathbb{Q}(p(t), q(t))$ and such that*

$$C \cap \mathbb{Q}^2 = \{(p(s), q(s)) \mid s \in \mathbb{Q}\}.$$

(c) *For a suitable choice of t we have one of the following two cases:*

(1) *$p(t), q(t) \in \mathbb{Q}[t]$ and have leading coefficients ≥ 0. Moreover if $s \in \mathbb{Q}$ and $p(s), q(s) \in \mathbb{Z}$ then $s \in \mathbb{Z}$.*
(2) *$p(t) = P(t)/(t^2 - d)^e$ and $q(t) = Q(t)/(t^2 - d)^e$, where $e > 0$, $P(t)$, $Q(t) \in \mathbb{Q}[t]$ and have degrees $\leq 2e$, and where d is a square free integer > 1. Put $r(t) = p(t) + q(t)$ and $X_N = \{s \in \mathbb{Q} \mid r(s) \in \mathbb{Z}$ and $|r(s)| \leq N\}$. Then $\mathrm{Card}(X_N) \leq K \cdot \log(N)$ for some constant $K > 0$.*

(6.6) THEOREM. *B/A is a torsion free module over $\mathbb{C}[\epsilon_x, \epsilon_y]$.*

PROOF: By (3.4) it suffices to show that if $\varphi = \varphi(\epsilon_x, \epsilon_y) \in \mathbb{C}[\epsilon_x, \epsilon_y]$ is irreducible then

$$B_0^{\varphi} = \mathrm{Ker}(\varphi \text{ on } B) = B \cap \hat{A}_0^{\varphi}$$

is contained in A. This is evident if $\hat{A}_0^{\varphi} \subset A$, so assume that $\hat{A}_0^{\varphi} \not\subset A$. Let then $C = \{(p,q) \mid \varphi(p,q) = 0\}$ and $p(t), q(t) \in \mathbb{Q}(t)$ be as above in

Theorem (6.5). If C is a straight line then φ is linear, so $\varphi = \partial - \lambda$ for some $\partial \in \underline{gl}_2(C)$, in which case $B_0^\varphi \subset A$ by (6.3). Suppose therefore that C is not a straight line. Assuming there is an $h \in B_0^\varphi$, $h \notin A$, we shall reach a contradiction. By (6.5) we can write

$$h = \sum_{s\in\mathbb{Q}} a_s\, x^{p(s)} y^{q(s)},$$

where $a_s = 0$ unless $(p(s), q(s)) \in \mathbb{N}^2$. Since $h \in B$, h is algebraic over $\mathbb{C}(x,y)$. By (D.4) (Appendix D) we can choose $b \in C^\times$ so that $h_b = h(x, bx) \in \mathbb{C}[[x]]$ is not in $\mathbb{C}[x]$, and h_b is algebraic over $\mathbb{C}(x)$. It follows then from (D.1) (Appendix D) that h_b has finite radius of convergence R. We can write

$$h_b = \sum_{s\in\mathbb{Q}} c_s x^{r(s)}$$

where $c_s = a_s b^{q(s)}$ and $r(t) = p(t)+q(t)$. Moreover $c_s = 0$ unless $p(s), q(s) \in \mathbb{N}$, in particular $c_s = 0$ unless $r(s) \in \mathbb{N}$. Let us list

$$(*) \qquad r(\mathbb{Q}) \cap \mathbb{N} = \{n_1 < n_2 < n_3 < \ldots\}.$$

Then, collecting coefficients, we can write

$$h_b = \sum_{m=1}^{\infty} d_m x^{n_m}.$$

We claim that

$$(**) \qquad n_m/m \to \infty \text{ as } m \to \infty.$$

Assuming (**), we conclude the proof as follows. By Fabry's Theorem (D.3) (Appendix D) every point on the circle of convergence $|x| = R$ is singular for h_b. But h_b, being an algebraic function, can have only finitely many singularities. This contradiction concludes the proof of (6.6), modulo the verification of (**). For this we distinguish the two cases of (6.5)(c) above.

Case 1: $p(t), q(t) \in \mathbb{Q}[t]$ have leading coefficients ≥ 0, and if $p(s), q(s) \in \mathbb{Z}$ then $s \in \mathbb{Z}$.

In this case $h_b = \sum_{s\in\mathbb{Z}} c_s x^{r(s)}$. Since C is not a straight line one of $p(t)$ and $q(t)$, and hence also $r(t) = p(t) + q(t)$ has degree ≥ 2. Hence $|r(s)/s| \to \infty$ as $|s| \to \infty$, and (**) follows immediately from this.

Case 2: $p(t) = P(t)/(t^2 - d)^e$ and $q(t) = Q(t)/(t^2 - d)^e$ as in (6.5)(c). Putting $X_N = \{s \in \mathbb{Q} \mid r(s) \in \mathbb{Z} \text{ and } |r(s)| \leq N\}$ we further have from (6.5)(c) that $\mathrm{Card}(X_N) \leq K \cdot \log(N)$ for some $K > 0$. In the notation of (*) we then have, for all $m \geq 1$, that $m \leq \mathrm{Card}(X_{n_m}) \leq K \cdot \log(n_m)$. It follows that $e^{(m/K)} \leq n_m$ so $e^{(m/K)}/m \leq n_m/m$, whence (**) in this case as well.

To prove the Jacobian Conjecture by these methods it remains to show that B/A is a torsion free module over the full algebra

$$U = \mathbb{C}[\epsilon_x, \epsilon_y, \delta].$$

Let $\varphi \in U$ be irreducible. As above it suffices to show that $B_0^\varphi \subset A$. Write

$$\varphi = \sum_{i=0}^{r} \varphi_i(\epsilon_x, \epsilon_y)\delta^i.$$

Since φ is irreducible and we have already treated the case $\varphi = \delta$, we may assume that $\varphi_0(\epsilon_x, \epsilon_y) \neq 0$. Consider the action of φ on $A_{(d)}$, relative to the monomial basis $x^d, x^{d-1}y, \ldots, xy^{d-1}, y^d$. the matrix of δ is

$$\delta \mid_{A_{(d)}} = \begin{bmatrix} 0 & 1 & 0 & \cdots & \cdots \\ 0 & 0 & 2 & \cdots & \cdots \\ \cdots & \cdots & \cdots & \cdots & \cdots \\ 0 & 0 & 0 & \cdots & d \\ 0 & 0 & 0 & \cdots & 0 \end{bmatrix}$$

and

$$\varphi_0 \mid_{A_{(d)}} = \mathrm{diag}\,[\varphi_0(d,0), \varphi_0(d-1,1), \ldots, \varphi_0(0,d)]$$

and so

$$\varphi \mid_{A_{(d)}} = \varphi_0 + \varphi_1\delta + \cdots + \varphi_r\delta^r \mid_{A_{(d)}}$$

is upper triangular with diagonal $\varphi_0 \mid_{A_{(d)}}$.

Let $C = \{(p,q) \mid \varphi_0(p,q) = 0\}$, a (possibly reducible) plane curve, and $N_d = \{(p,q) \in \mathbb{N}^2 \mid p + q = d\}$. Then the zero diagonal entries of $\varphi \mid_{A_{(d)}}$ correspond to the points of $C \cap N_d$. Thus

$$\dim_{\mathbb{C}}(A_{(d)})_0^\varphi \leq \mathrm{Card}(C \cap N_d).$$

Let $C_1, \ldots, C_m$ be the irreducible components of C, and $C_1, \ldots, C_\ell$ those for which $C_i \cap \mathbb{N}^2$ is infinite. Outside a large disk in $\mathbb{R}^2$, $C_1, \ldots, C_\ell$ are pairwise disjoint and they meet each N_d at most once for $d \gg 0$.

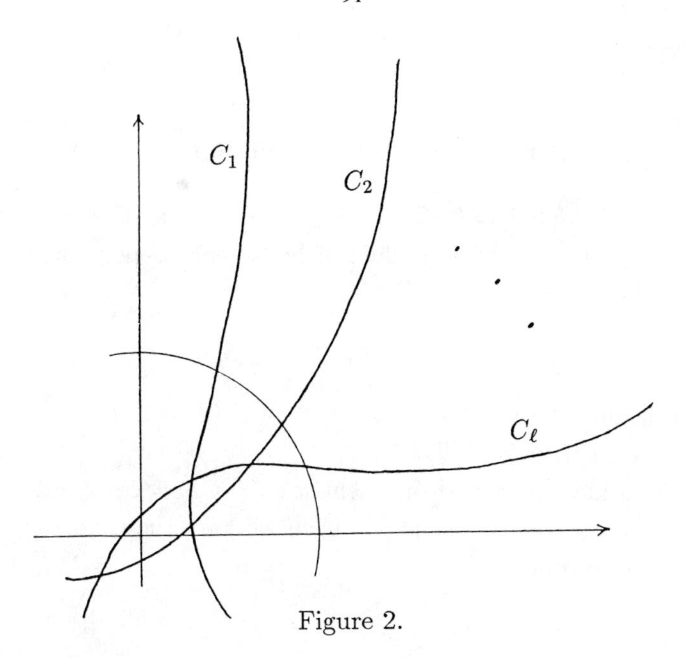

Figure 2.

It follows that, for $d \gg 0$,

$$\dim_{\mathbb{C}}((A_{(d)})_0^{\varphi}) \leq \ell.$$

Suppose that none of $C_1, \ldots, C_\ell$ is a straight line. Then, as in the proof of (6.6) we see that the integral points in $C \cap \mathbb{N}^2$ tend rapidly to infinity, so if $h \in \hat{A}_0^{\varphi}$, $h \notin A$, then, for suitable $b \in \mathbb{C}$, $h_b = h(x, bx)$ is a lacunary series to which we can apply Fabry's Theorem (D.3) to reach a contradiction. Thus if we have an $h \in B_0^{\varphi}$, $h \notin A$, then at least one of $C_1, \ldots, C_\ell$ must be a straight line. This is the case we have not yet been able to eliminate.

It might be useful to observe that, by (3.2) again, h must also be annihilated by some non-zero operator $\varphi' \in U' = \mathbb{C}[\epsilon_x, \epsilon_y, \delta']$, where $\delta' = y\partial_x$. Moreover h is annihilated by a non-zero $\psi \in U(\underline{sl}_2(\mathbb{C})) = \mathbb{C}[\delta, \epsilon_x - \epsilon_y, \delta']$.

APPENDICES.

Appendix A. Generator filtrations of affine k-algebras.

Let k be a field and $S = k[X_1, \ldots, X_m]$ a polynomial algebra in m variables. Let $S_{(d)}$ denote the k-module of homogeneous polynomials of degree d, and filter S by

$$S_d = S_{(0)} \oplus S_{(1)} \oplus \cdots \oplus S_{(d)}.$$

Define the leading form map $\ell : S \to S$ by $\ell(0) = 0$ and $\ell(f) = f_d$ if $f = f_0 + \cdots + f_d$, $f_i \in S_{(i)}$, $f_d \neq 0$.

Let J be an ideal of S with quotient $\pi : S \to B = S/J$. Filter J and B by $J_d = J \cap S_d$ and $B_d = \pi(S_d)$. Then we have the exact sequence of associated graded objects

$$0 \longrightarrow \mathrm{gr}(J) \longrightarrow \mathrm{gr}(S) \longrightarrow \mathrm{gr}(B) \longrightarrow 0.$$

Identifying $\mathrm{gr}(S)$ with S, $\mathrm{gr}(J)$ is the ideal generated by $\ell(J)$, the set of leading forms of elements of J.

Put $\chi(r) = \dim_k(B_r) = \sum_{d \leq r}(\mathrm{gr}_d(B))$, the Hilbert function of the graded affine algebra $\mathrm{gr}(B)$. For $r \gg 0$ it is a polynomial in r of degree $\dim(\mathrm{gr}(B))$ (cf. [**NB**, AC VIII, § 4.2] or [**M**, 10.F]).

(A.1) PROPOSITION. $\dim(\mathrm{gr}(B)) = \dim(B)$. *Hence* $\dim_k(B_r)$ *is, for* $r \gg 0$, *a polynomial in* r *of degree* $\dim(B)$.

We are grateful to M. Hochster for suggesting the following proof. Let $S[T]$ be a polynomial extension in one variable T. Define the homogenization operator $h : S \to S[T]$ by $h(0) = 0$ and, for $f = f_0 + f_1 + \cdots + f_d \in S$ of degree $d \geq 0$ ($f_i \in S_{(i)}$), put $h(f) = T^d f_0 + T^{d-1} f + 1 + \cdots + f_d$. Let H denote the (homogenous) ideal of $S[T]$ generated by $h(J)$.

Define retractions $\epsilon_i : S[T] \to S$ by $\epsilon_i(T) = i$ $(i = 0, 1)$. Then for $f \in S$ we have

$$\epsilon_1(h(f)) = f \text{ and } \epsilon_0(h(f)) = \ell(f).$$

It follows that

$$\epsilon_1(H) = J \text{ and } \epsilon_0(H) = \mathrm{gr}(J).$$

Let $C = S[T]/H$, a graded k-algebra, and $t = (T \bmod H) \in C$. Then

$$B = S/J \cong S[T]/(H, T-1) \cong C/(t-1)$$

and

$$\mathrm{gr}(B) = S/\mathrm{gr}(J) \cong S[T]/(H,T) \cong C/(t).$$

It therefore suffices to show that $\dim(C/(t)) = \dim(C) - 1 = \dim(C/(t-1))$.

(A.2) LEMMA. *Let D be an affine domain over k. Let $s \in D$ be non-zero and non-invertible. Then* $\dim(D/sD) = \dim(D) - 1$.

This follows from the Principal Ideal Theorem and the fact that, since D is affine, $\dim(D/P) = \dim(D) - 1$ for every height 1 prime P (cf. [**NB**,AC VIII, § 3.1, Cor. 2, and § 2.4,Th. 3]).

In order to conclude the proof of (A.1) it suffices, in view of (A.2), to show that if P is any minimal prime of C, $p : C \to C/P$ the quotient map, then $p(t)$ and $p(1-t)$ are non-zero and non-invertible in C/P. Note first that P is a homogeneous ideal, so C/P is graded and $\deg(p(t)) = 1$. Hence $p(t)$ is not invertible and $1 - p(t) \neq 0$. Moreover $1 - p(t)$ is invertible only if $p(t)$ is nilpotent, hence 0 since C/P is a domain. Hence it suffices to show that $p(t) \neq 0$. We show this by showing that t is not a zero divisor in $C = S[T]/H$. It suffices for this to show that if $f \in S[T]$ is homogeneous and $Tf \in H$ then $f \in H$. We may assume that $f \neq 0$ and write $f = T^r h(g)$ where $g = \epsilon_1(f) \in S$, and $r \geq 0$. Note that $g = \epsilon_1(Tf) = \epsilon_1(f) \in \epsilon_1(H) = J$. Hence $f = T^r \cdot h(g) \in H$ as claimed.

Appendix B. Some differential algebra.

Let k be an algebraically closed field of characteristic zero. Let B be a k-module. Let $D \in \mathrm{End}_k(B)$. For $\lambda \in k$ we put

$$B^D_\lambda = \mathrm{Ker}(D - \lambda)$$
$$B^D_{(\lambda)} = \bigcup_{n \geq 0} \mathrm{Ker}(D - \lambda)^n$$

and

$$B^D_{\mathrm{tors}} = \bigoplus_{\lambda \in k} B^D_{(\lambda)} \subset B.$$

Now assume that B is a *k-algebra* and D is a *k-derivation*, $D \in \mathrm{Der}_k(B)$, i.e. $D(ab) = D(a)b + aD(b)$.

(B.1) PROPOSITION ("TAYLOR EXPANSION"). *Suppose that $x \in B$ is central and $Dx = 1$. Then $B^D_{(0)} = B^D_0[x]$, a polynomial algebra over B^D_0, on which D acts like $\frac{d}{dx}$.*

PROOF: (cf. [**W**, Prop. 2.2]) Let $C = B^D_{(0)}$, $C_0 = B^D_0$, $\bar{C} = C/xC$, and $p : C \to \bar{C}$, $p(c) = \bar{c}$, the natural projection. Define $\varphi : C \to \bar{C}[X]$, X an indeterminate, by $\varphi(c) = \sum_{n\geq 0}(\overline{D^n c}/n!)X^n$. (Recall that $D^n c = 0$ for $n \gg 0$.) Then φ is (easily checked to be) a ring homomorphism, $\varphi(Dc) = \frac{d}{dx}\varphi(c)$, and $\varphi(x) = X$. To show that φ is surjective it suffices to show that $\bar{C} \subset \varphi(C)$. Let $c \in C$, $\varphi(c) = \bar{c} + \overline{D(c)}X + \cdots + \frac{1}{n!}\overline{D^n(c)}X^n$ and $\overline{D^m c} = 0$ for $m > n$. Put $c_1 = c - \frac{1}{n!}D^n(c)x^n$. Then $\bar{c}_1 = \bar{c}$ and $\deg_X(\varphi(c_1)) \leq n-1$. Continuing in this way we reach a $c_n \in C$ such that $\bar{c} = \bar{c}_n = \varphi(c_n)$.

Suppose that $c \in \mathrm{Ker}(\varphi)$. Then $D^n(c) \in Cx$ for all $n \geq 0$. Write $c = c_1 x$. Then $0 = \varphi(c) = \varphi(c_1)X$ so $\varphi(c_1) = 0$. It follows inductively that $c \in \bigcap_{n\geq 0} Cx^n$. To show finally that $\bigcap_{n\geq 0} Cx^n = 0$ consider the homomorphism $\psi : C \to C[X]$, $\psi(c) = \sum_{n\geq 0}(D^n(c)/n!)X^n$. Then $\psi(x) = X$ so $\psi(\bigcap_{n\geq 0} Cx^n) \subset \bigcap_{n\geq 0} C[X]X^n = 0$. But ψ is injective, since $\psi(c) = c + D(c)X + \ldots$.

(B.2) COROLLARY. *Suppose that $D_1, \ldots, D_n \in \mathrm{Der}_k(B)$ commute, $x_1, \ldots, x_n \in B$ are central, and $D_i(x_j) = \delta_{ij}$ $(1 \leq i, j \leq n)$. Let $C = \bigcap_{i=1}^n B^{D_i}_{(0)}$ and $C_0 = \bigcap_{i=1}^n B^{D_i}_0$. Then $C = C_0[x_1, \ldots, x_n]$, a polynomial algebra over C_0 in n variables, and $D_i(c) = \partial c/\partial x_i$ for $c \in C$.*

PROOF: By (B.1), $C = C_0^{D_n}[x_n]$, a polynomial algebra. By induction, $C_0^{D_n} = C_0[x_1, \ldots, x_{n-1}]$. (Note that $C_0^{D_n}$ is invariant under $D_1, \ldots, D_{n-1}$ since the latter commute with D_n.)

(B.3) PROPOSITION. *For $\lambda, \mu \in K$, $a, b \in B$, $N \geq 0$, we have*

$$(D + (\lambda + \mu))^N(ab) = \sum_{p=0}^{N} \binom{N}{p} (D + \lambda)^p(a) \cdot (D + \mu)^{N-p}(b).$$

PROOF:

$$
\begin{aligned}
(D+(\lambda+\mu))^N(ab) &= \sum_{p+q=N} \binom{N}{p}(\lambda+\mu)^p D^q(ab) \\
&= \sum_{p+q=N} \binom{N}{p} \left(\sum_{r+r'=p} \binom{p}{r} \lambda^r \mu^{r'} \right) \left(\sum_{s+s'=q} \binom{q}{s} D^s(a) D^{s'}(b) \right) \\
&= \sum_{\substack{p+q=N \\ r+r'=p \\ s+s'=q}} \binom{N}{p}\binom{p}{r}\binom{q}{s} \lambda^r D^s(a) \cdot \mu^{r'} D^{s'}(b)
\end{aligned}
$$

(Note that $\binom{N}{p}\binom{p}{r}\binom{q}{s} = \frac{N!}{p!q!} \frac{p!}{r!r'!} \frac{q!}{s!s'!} = \frac{N!}{r!r'!s!s'!}$.)

$$
\begin{aligned}
&= \sum_{\substack{t+t'=N \\ r+s=t \\ r'+s'=t'}} \binom{N}{t}\binom{t}{r}\binom{t'}{r'} \lambda^r D^s(a) \cdot \mu^{r'} D^{s'}(b) \\
&= \sum_{t+t'=N} \binom{N}{t} \left(\sum_{r+s=t} \binom{t}{r} \lambda^r D^s(a) \right) \left(\sum_{r'+s'=t'} \binom{t'}{r'} \mu^{r'} D^{s'}(b) \right) \\
&= \sum_{t+t'=N} \binom{N}{t} (D+\lambda)^t(a)(D+\mu)^{t'}(b).
\end{aligned}
$$

(B.4) COROLLARY. *(a) If $(D-\lambda)^n(a) = 0$ and $(D-\mu)^m(b) = 0$ then $(D-(\lambda+\mu))^N(ab) = 0$ for $N \geq n+m-1$.*

(b) The decomposition $B^D_{\text{tors}} = \oplus_\lambda B^D_{(\lambda)}$ is a k-graded k-algebra structure ($B^D_{(\lambda)} \cdot B^D_{(\mu)} \subset B^D_{(\lambda+\mu)}$) and $\oplus_\lambda B^D_\lambda$ is a graded subalgebra.

(a) follows from (B.3), and (b) follows from (a).

(B.5) PROPOSITION. *Let $D, E \in \operatorname{Der}_k(B)$ and assume that $[E, D]$ $(= ED - DE) = \alpha D$ $(\alpha \in k)$. Let $\lambda \in k$.*

(a) $(D+\lambda)^n E = E(D+\lambda)^n - n\alpha D(D+\lambda)^{n-1}$.

(b) $(E+\lambda)^n D = D(E+\lambda+\alpha)^n$.

PROOF OF (a): $X \to [E, X]$ is a derivation, $[E, D+\lambda] = [E, D] = \alpha D$, and so $[E, (D+\lambda)^n] = n(D+\lambda)^{n-1}\alpha D$, whence (a).

PROOF OF (b): $(E+\lambda)D = D(E+\lambda) + [E+\lambda, D] = D(E+\lambda) + \alpha D = D(E+\lambda+\alpha)$. If $(E+\lambda)^n D = D(E+\lambda+\alpha)^n$ then $(E+\lambda)^{n+1} D = (E+\lambda)D(E+\lambda+\alpha)^n = D(E+\lambda+\alpha)^{n+1}$.

(B.6) COROLLARY. *With* $[E, D] = \alpha D$ *as above, and* $\lambda \in k$*:*

(a) $EB^D_{(\lambda)} \subset B^D_{(\lambda)}$; *and*

(b) $DB^E_{(\lambda)} \subset B^E_{(\lambda+\alpha)}$ *and* $DB^E_\lambda \subset B^E_{\lambda+\alpha}$.

PROOF: It follows from (B.5)(a) that

$$E \cdot \operatorname{Ker}(D-\lambda)^{n-1} \subset \operatorname{Ker}(D-\lambda)^n$$

and from (B.5)(b) that

$$D \cdot \operatorname{Ker}(E-\lambda)^n \subset \operatorname{Ker}(E-\lambda-\alpha)^n$$

for all $n \geq 1$.

(B.7) PROPOSITION. *Let* B *be an integral domain containing* $\mathbb{Q}$*,* $E = \operatorname{Frac}(B)$*, and* $D \in \operatorname{Der}_{\mathbb{Q}}(B)$*.*

(a) *(Seidenberg) The complete integral closure of* B *(in* E*) is* D*-invariant.*

(b) *(Vasconcelos) If* B *is integral over* $B^D_{(0)}$ *then* $B = B^D_{(0)}$*.*

Proofs of these results may be found in [**W**, Prop. (2.3) and (2.5)].

(B.8) REMARK: Let $\mathbb{Q} \subset A \subset B$ be noetherian integral domains with B algebraic over A, and let $\bar{A}$ be the integral closure of A in B. Let $D \in \operatorname{Der}_{\mathbb{Q}}(B)$ and assume that $D(A) \subset A \subset B^D_{(0)}$. This situation presents itself in the setting of the Jacobian Conjecture, in which case the conjecture would follow if we could conclude that $D(\bar{A}) \subset \bar{A}$ (see Corollary (4.2) above).

One approach is to try to use the automorphism generated by D. Explicitly, let D act on $B[[t]]$ (t an indeterminate) by $D(\sum b_i t^i) = \sum D(b_i)t^i$. Then $E = e^{Dt} = \sum_{n \geq 0} D^n t^n/n!$ defines an automorphism of $B[[t]]$, $E(f) = f + D(f)t + D^2(f)t^2/2 + \ldots$, with inverse e^{-Dt}. Since $D(A) \subset A \subset B^D_{(0)}$ we have $E(A) \subset A[t]$. Hence $E(\bar{A})$ is contained in

$$C = \text{ the integral closure of } A[t] \text{ in } B[[t]].$$

It is tempting to try to argue that $C \subset \bar{A}[[t]]$. Were this to be so then, for $a \in \bar{A}$, we would have $E(a) = a + D(a)t + \cdots \in \bar{A}[[t]]$, whence $D(a) \in \bar{A}$, as desired.

However, it is never the case that $C \subset \bar{A}[[t]]$ unless $B = \bar{A}$. Indeed let $b \in B$. Since b is algebraic over A, $ab \in \bar{A}$ for some $a \neq 0$ in A. Put $f = a(1+bt)^{1/2} = a\sum_{n \geq 0}\binom{1/2}{n} b^n t^n \in B[[t]]$. Then $f^2 = a^2(1+bt) = a(a+abt) \in \bar{A}[t] \subset C$. If $C \subset \bar{A}[[t]]$ we conclude that $ab^n \in \bar{A}$ for all $n \geq 0$, hence $b \in \bar{A}$ as claimed.

This observation was worked out in a conversation with Mel Hochster.

Appendix C. The universal enveloping algebra of $\underline{gl}_2(k)$.

Let k be a field of characteristic zero. The standard basis of $\underline{gl}_2(k)$ will be denoted ϵ_{ij} $(1 \le i,j \le 2)$, and we put

$$e = \epsilon_{12} \qquad f = \epsilon_{21}$$
$$s = \epsilon_{11} + \epsilon_{22} \qquad h = \epsilon_{11} - \epsilon_{22}.$$

Then ks is the center of $\underline{gl}_2(k)$, and we have the commutators $[e,f] = h$, $[h,e] = 2e$, and $[h,f] = -2f$. We shall consider the Lie subalgebras

$$\underline{s} = \underline{sl}_2(k) = kf \oplus kh \oplus ke$$
$$\underline{b} = ks \oplus kh \oplus ke$$
$$\underline{b}_0 = kh \oplus ke.$$

(C.1) LEMMA. *Put $d = h/2$, so that $[d,e] = e$. Let $f(d) \in k[d]$ $(= k[h])$. Then $[e, f(d)] = E(f(d)) \cdot e$ where $E : k[d] \to k[d]$ is the k-linear map such that $E(d^r) = (d-1)^r - d^r$.*

We argue by induction on r, the case $r = 0$ being trivial. We have

$$\begin{aligned}
[e, d^{r+1}] &= [e, d^r]d + d^r \cdot [e,d] \\
&= ((d-1)^r - d^r)ed + d^r \cdot (-e) \quad \text{(induction)} \\
&= ((d-1)^r - d^r)(de - e) - d^r e \\
&= (((d-1)^r - d^r)(d-1) - d^r)e \\
&= ((d-1)^{r+1} - d^r(d-1+1))e \\
&= ((d-1)^{r+1} - d^{r+1})e.
\end{aligned}$$

(C.2) PROPOSITION. (a) *In $k[h]$ let J_n $(n \ge 0)$ be a sequence of ideals such that $J_n + E(J_n) \subset J_{n+1}$ for all n. Then $J = \oplus_{n\ge 0} J_n e^n$ is a 2-sided ideal of $U\underline{b}_0 = k[h,e] = \oplus_{n\ge 0} k[h]e^n$.*

(b) *Every 2-sided ideal J of $U\underline{b}_0$ is of the above form. If $J \ne 0$ then $J_n = k[h]$, i.e. $e^n \in J$, for all $n \gg 0$.*

PROOF: Since $k[h,e] = \oplus_{n\ge 0} k[h]e^n$, and since $[d, e^n] = ne^n$ $(d = h/2)$ it follows that $k[d]e^n$ is the eigenspace for the eigenvalue n of d. Hence any k-module $J \subset k[h,e]$ invariant under $[d, \cdot]$ must be of the form $J = \oplus_{n\ge 0} J_n e^n$, where $J_n \subset k[d]$. If J is to be a 2-sided ideal of $k[d,e]$ then J_n must be an ideal of $k[d]$ and $J_n \subset J_{n+1}$. From (C.1) we have $[e, J_n e^n] = [e, J_n]e^n = E(J_n)e^{n+1}$, and so $E(J_n) \subset J_{n+1}$ as well.

Suppose, conversely, that J_n is an ideal of $k[d]$ and $J_n + E(J_n) \subset J_{n+1}$ for all n. We claim that $J = \oplus_{n \geq 0} J_n e^n$ is a 2-sided ideal. It suffices to show that each $J_n e^n$, left or right multiplied by d or e, stays in J. Now $d(J_n e^n) \subset J$ because J_n is an ideal, and $(J_n e^n)e \subset J$ because $J_n \subset J_{n+1}$. If $f \in J_n$ then $(fe^n)d = f([e^n, d] - de^n) = f \cdot ((-ne^n) - de^n) = -f \cdot (n + d)e^n \in J$ because J_n is an ideal. Finally $e(fe^n) = ([e, f] - fe)e^n = (E(f) - f)e^{n+1}$, by (C.1), and the latter is in J because $J_n + E(J_n) \subset J_{n+1}$.

Finally, suppose $f \in J_m$, $f \neq 0$; say $\deg_d(f) = s$. Since $E(d^r) = (d-1)^r - d^r$ we see that $\deg_d(E(f)) = s - 1$. Hence $E^s(f) \in J_{m+s}$ is a non-zero constant, so $J_n = k[d]$ for $n \geq m + s$.

(C.3) COROLLARY. *Let $M \neq 0$ be a $U\underline{b}_0$-module. If $M \xrightarrow{e} M$ is injective then M is a faithful $U\underline{b}_0$-module.*

PROOF: $J = \operatorname{ann}_{U\underline{b}_0}(M)$ is a two-sided ideal containing no power of e, hence $J = 0$ by (C.2)(b).

(C.4) COROLLARY. *Let $M \neq 0$ be a $U\underline{b}$-module. If M is a torsion free $k[s]$-module and if $M \xrightarrow{e} M$ is injective then M is a faithful $U\underline{b}$-module.*

PROOF: Put $F = k(s)$ and $X_F = F \otimes_{k[s]} X$ for any $k[s]$-module X. Then M embeds in M_F which is a $(U\underline{b})_F$-module on which multiplication by e is injective. Now $(U\underline{b})_F = F[h, e]$ is the enveloping algebra of $F \otimes_k \underline{b}_0$ over F, so (C.3) implies that M_F is a faithful $(U\underline{b})_F$-module. Since s is central in $U\underline{b}$ it follows that M is a faithful $U\underline{b}$-module.

We next consider

$$U\underline{s} = U\underline{sl}_2(k) = k[f, h, e].$$

Its center is generated by the Casimir element,

$$c = 4fe + h^2 + 2h = 4ef + h^2 - 2h = 2ef + 2fe + h^2.$$

(C.5) PROPOSITION. *Let $M \neq 0$ be a $U\underline{s}$-module. If M is a torsion free $k[c]$-module, and if $M \xrightarrow{e} M$ is injective then M is a faithful $U\underline{s}$-module.*

PROOF: Let $V = k[c, h, e] \subset U\underline{s}$. Clearly $V \cong U\underline{b}$ so, by (C.4), M is a faithful V-module. It suffices now to observe that the ideal $J = \operatorname{ann}_{U\underline{s}}(M)$, if non-zero, would have non-zero intersection with V; this follows from (C.6)(b) below.

Recall that $U\underline{s}$ has a division ring of fractions D. The inverses introduced below are taken in D.

(C.6) PROPOSITION. *Let* $V = k[c,h,e] \subset U\underline{s} = k[f,h,e]$.

(a)

$$(U\underline{s})[e^{-1}] = V[e^{-1}] = k[c, e^{\pm 1}, h] = k[c, e^{\pm 1}, g]$$
$$= \bigoplus_{n \geq 0} k[c, e^{\pm 1}]h^n,$$

where

$$g = \frac{e^{-1}h}{2}, \text{ and } [g,e] = 1.$$

(b) *If* $\varphi \in (U\underline{s})[e^{-1}]$ *then* $e^n\varphi \in V$ *for* $n \gg 0$.

PROOF: We have $4fe = c - h^2 - 2h$ so $f = \frac{1}{4}(c - h^2 - 2h)e^{-1} \in V[e^{-1}] = k[c, e^{\pm 1}, h] = k[c, e^{\pm 1}, g]$. Since $[h, \cdot]$ leaves $k[c, e^{\pm 1}]$ invariant ($[h, e^n] = 2ne^n$ for all $n \in \mathbb{Z}$) it follows that $k[c, e^{\pm 1}, h] = \oplus_{m \geq 0} k[c, e^{\pm 1}]h^m$. If $\varphi \in k[c, e^{\pm 1}, h]$ then for $n \gg 0$ we have $e^n\varphi \in \oplus_{m \geq 0} k[c,e]h^m = k[c,e,h] = V$, whence (b). The formula $[g,e] = 1$ follows directly from $[h,e] = 2e$.

(C.7) REMARK: Note that $W = k[c,e,g] \subset (U\underline{s})[e^{-1}]$ is just the Weyl algebra over $k[c]$ defined by the relation $[g,e] = 1$, and $k(c)[e,g]$ is therefore a simple ring [**Bj**,Ch. I, Prop. 1.3], as is $k(c)[e^{\pm 1}, g]$. Moreover $W \supset V = k[c, e, eg]$, so $W[e^{-1}] = V[e^{-1}] = (U\underline{s})[e^{-1}]$.

Let $A = k[x,y] = \oplus_{d \geq 0} A_{(d)}$, where $A_{(d)}$ denotes the space of homogeneous polynomials of degree d. Then $\hat{A} = k[[x,y]] = \prod_{d \geq 0} A_{(d)}$. Put $\partial_x = \frac{\partial}{\partial x}$, $\partial_y = \frac{\partial}{\partial y}$. Then $\underline{gl}_2(k)$ acts as linear derivations on A and $\hat{A}$ by setting $\epsilon_{11} = x\partial_x$, $\epsilon_{12} = x\partial_y$, $\epsilon_{21} = y\partial_x$, $\epsilon_{22} = y\partial_y$. Then $s = x\partial_x + y\partial_y$ acts on $A_{(d)}$ as multiplication by d, so it follows easily that

$$\hat{A}^s_{\text{tors}} = \bigoplus_{d \geq 0} A_{(d)} = A.$$

On the other hand it is clear that

$$\hat{A}^e_{(0)} = \hat{A}^{\partial_y}_{(0)} = k[[x]][y];$$

recall that $e = \epsilon_{12} = x\partial_y$. Put $\bar{A} = k[x][[y]] \subset \hat{A}$. Then clearly $\bar{A}$ is $\underline{gl}_2(k)$-invariant, $\bar{A}^s_{\text{tors}} = A$, and $\bar{A}^e_{(0)} = k[[x]][y] \cap k[x][[y]] = k[x,y] = A$. Thus $M = \bar{A}/A$ is a non-zero module over $U(\underline{gl}_2(k))$ which is a torsion free $k[s]$-module such that $M \xrightarrow{e} M$ is injective. It follows from (C.4) that M is a faithful $U\underline{b}$-module. The Casimir element $c = 4fe + h^2 + 2h$ acts on $A_{(d)}$ by multiplication by $d(d+2)$ (see (6.2)), hence c acts on $\hat{A}$ as does $s(s+2)$.

(Thus $\hat{A}$ has non-zero annihilator $s(s+2)-c \in U(\underline{gl}_2(k))$.) It follows that $\hat{A}/A$ is a torsion free $k[c]$-module. Now (C.5) implies that $\bar{A}/A$ is a faithful $U\underline{s}$-module.

Let $U = U\underline{b}$ or $U\underline{s}$. We have just seen that the action of U on $\hat{A}$ (in fact on $\bar{A}/A$) is faithful. But U acts via its homomorphism into the Weyl algebra $W = k[w, y, \partial_x, \partial_y]$. It follows that $U \to W$ is injective. Since W is a simple ring, A itself is a faithful W-module. Hence A is a faithful U-module.

We now record these conclusions.

(C.7) PROPOSITION. *Let $U = U\underline{b}$ or $U\underline{s}$. The natural homomorphism $U \to W = k[x, y, \partial_x, \partial_y]$ is injective. Hence any non-zero W-module (e.g. A) is a faithful U-module.*

(C.8) REMARK: As noted above, the kernel of $U(\underline{gl}_2(k)) \to W$ contains $s(s+2)-c \neq 0$.

Appendix D. Analytic and algebraic functions.

(D.1) PROPOSITION. *If $f \in \mathbb{C}[[x_1, \ldots, x_n]]$ is algebraic (over $\mathbb{C}(x_1, \ldots, x_n)$) and entire on $\mathbb{C}^n$ then $f \in \mathbb{C}[x_1, \ldots, x_n]$.*

PROOF: Let $P(x,y) = P(x_1, \ldots, x_n, y)$ be an irreducible polynomial such that $P(x, f) = 0$. Let $X = \{(x,y) \in \mathbb{C}^{n+1} \mid P(x,y) = 0\}$, an irreducible affine variety. Let $\bar{f} : \mathbb{C}^n \to \mathbb{C}$ be the entire function defined by f, and consider its graph, $Y = \{(x, \bar{f}(x)) \mid x \in \mathbb{C}^n\}$. Define $\pi : \mathbb{C}^{n+1} \to \mathbb{C}^n$ by $\pi(x,y) = x$ and $p : \mathbb{C}^{n+1} \to \mathbb{C}$ by $p(x,y) = y$. Then we evidently have

(1) $$Y \subset X$$

(2) $$\pi : Y \cong \mathbb{C}^n$$

(3) $$p(x,y) = \bar{f}(\pi(x,y)) \quad \forall (x,y) \in Y.$$

Since $\pi : X \to \mathbb{C}^n$ is quasi-finite it follows from (2) that Y contains a non-empty open set of X (euclidean topology). From (3) therefore, p and $\bar{f} \circ \pi$ define analytic functions $X \to \mathbb{C}$ which agree on a non-empty open set, and hence on all of X, since X is irreducible. Thus $Y = X$ and so $\pi : X \xrightarrow{\cong} \mathbb{C}^n$. It follows, as claimed, that $f \in \mathbb{C}[x_1, \ldots, x_n]$.

(D.2) PROPOSITION. *Let $h \in \mathbb{C}[[x]]$ be an algebraic function (over $\mathbb{C}(x)$) with radius of convergence $R < \infty$. Then h has only finitely many singularities on the circle of convergence $|x| = R$.*

PROOF: Let $X = \operatorname{Spec}(\mathbb{C}[x, h])$ and $p : X \to \mathbb{C}$ the ramified covering defined by $\mathbb{C}[x] \subset \mathbb{C}[x, h]$. In a neighborhood of any point on $|x| = R$ above which p is unramified, h may be analytically continued, whence the Proposition.

(D.3) THEOREM (FABRY, 1896). *Let $h = \sum_{m=1}^{\infty} a_m x^{n_m} \in \mathbb{C}[[x]]$ have radius of convergence $R < \infty$. Assume that*

$$n_m/m \longrightarrow \infty \text{ as } m \longrightarrow \infty$$

(i.e. h is a "lacunary series"). Then every point on the circle of convergence $|x| = R$ is a singular point of h.

Proofs of this can be found in Hille [**H**, Theorem 11.7.2] or Dienes [**D**, Ch. XI, 93.II].

(D.4) LEMMA. *Let $h(x, y) \in \mathbb{C}[[x, y]]$. For $b \in \mathbb{C}$ put $h_b = h(x, bx) \in \mathbb{C}[[x]]$.*

(a) *If $h \notin \mathbb{C}[x, y]$ then there is a countable subset $R \subset \mathbb{C}$ such that $h_b \notin \mathbb{C}[x]$ for all $b \notin R$.*

(b) *Suppose that h is algebraic of degree d over $\mathbb{C}(x, y)$. Then there is a finite subset $R_0 \subset \mathbb{C}$ such that, for all $b \notin R_0$, h_b is algebraic of degree $\leq d$ over $\mathbb{C}(x)$.*

PROOF: (a) Say $h = \sum_{p,q \geq 0} a_{p,q}\, x^p y^q$. Then $h_b = \sum_{d \geq 0} c_d\, x^d$, where $c_d = \sum_{q=0}^{d} a_{d-q,q}\, b^q$. Put $f_d(t) = \sum_{q=0}^{d} a_{d-q,q}\, t^q$. For each d such that $f_d \neq 0$, let R_d be the set of (complex) roots of $f_d(t)$, and let R denote the union of these R_d. Since $c_d = f_d(b)$ and $f_d \neq 0$ for infinitely many d, we have $h_b \notin \mathbb{C}[x]$ if $b \notin R$.

(b) Suppose that $P(x, y, h) = 0$ where $P(x, y, z) \in \mathbb{C}[x, y, z]$ is irreducible of degree d in z, say $P = P_0(x, y) + P_1(x, y)z + \cdots + P_d(x, y)z^d$. We have $P(x, bx, h_b) = 0$. Therefore h_b is algebraic of degree $\leq d$ over $\mathbb{C}(x)$ provided that $P(x, bx, z) \neq 0$, for example that $P_d(x, bx) \neq 0$. As in the proof of (a) above we see that this is true for all b outside some finite set $R_0 \subset \mathbb{C}$.

Appendix E. Curves with infinitely many integer points.

We present here the results (due to Siegel) from which Theorem (6.5) follows.

Let $\varphi(X,Y) \in \mathbb{C}[X,Y]$ be an irreducible polynomial in two variables (corresponding to the variables ϵ_x and ϵ_y of Theorem (6.5)), and let

$$C = \{(\xi,\eta) \in \mathbb{C}^2 \mid \varphi(\xi,\eta) = 0\}$$

be the corresponding plane curve, with affine algebra $A = \mathbb{C}[x,y] = \mathbb{C}[X,Y]/(\varphi)$ We make the *basic assumption*

$$\text{(0)} \qquad C \cap \mathbb{Z}^2 \textit{ is infinite.}$$

In particular, $C \cap \mathbb{Q}^2$ is Zariski dense in C so it follows that C is defined over Q (cf. [**Bo**, Ch. AG, Theorem (14.4)(3)], and so

$$\text{(1)} \qquad \varphi_0 = \alpha\varphi \in \mathbb{Q}[X,Y] \textit{ for some } \alpha \in \mathbb{C}^\times.$$

Further it follows from (0) and Siegel's Theorem [**L**, Ch. 8, Theorem 2.4] that C has genus 0, in fact that C is a rational curve over $\mathbb{Q}$. Explicitly, if we put $A_{\mathbb{Q}} = \mathbb{Q}[x,y] = \mathbb{Q}[X,Y]/(\varphi_0)$ then the function field of C over $\mathbb{Q}$ is

$$\text{(2)} \qquad \mathbb{Q}(C) = \mathbb{Q}(x,y) = \mathbb{Q}(t),$$

a rational field in one variable t. Moreover C is rationally parametrized over $\mathbb{Q}$:

$$\text{(3)} \qquad C \cap \mathbb{Q}^2 = \{(x(\tau), y(\tau)) \mid \tau \in \mathbb{Q}\}.$$

Assertions (a) and (b) of Theorem (6.5) follow from (1), (2), and (3).

Now consider any non-constant $z = z(t) \in \mathbb{Z}[x,y] \subset \mathbb{Q}(t)$. From (0) and (3) we see that

$$\text{(4)} \qquad z(\mathbb{Q}) \cap \mathbb{Z} \textit{ is infinite.}$$

A further result of Siegel [**L**, Ch. 8, Theorem 5.1] then tells us that

$$\text{(5)} \qquad z(t) \textit{ has at most two poles (on } \mathbb{P}^1\textit{).}$$

Suppose that z has a pole that is rational over $\mathbb{Q}$. Choose t so that this pole is at ∞ and the other, if another pole exists, is at 0. Then we have

$$(6) \qquad z(t) \in \mathbb{Q}[t, t^{-1}], \quad z \notin \mathbb{Q}[t^{-1}].$$

(E.1) LEMMA. *Conditions (4) and (6) imply that $z \in \mathbb{Q}[t]$, and that there is an integer $d > 0$ such that if $\tau \in \mathbb{Q}$ and $z(\tau) \in \mathbb{Z}$ then $d\tau \in \mathbb{Z}$.*

PROOF: Write $z(t) = a_m t^m + \cdots + a_n t^n$ with $a_m a_n \neq 0$ and $m \leq n$. By assumption (b), $n > 0$. Let p be a prime and $v = v_p$ the p-adic valuation. Put

$$(7) \qquad m_p = \min\left(-\frac{v(a_n)}{n}, \frac{v(a_i) - v(a_n)}{n-i} \quad (\text{all } i < n)\right).$$

Let $q \in \mathbb{Q}$. Suppose that $v(q) < m_p$. Then

$$v(a_n q^n) = v(a_n) + nv(q) < 0 \quad (\text{since } n > 0)$$

and, for $i < n$,

$$v(a_i q^i) = v(a_i) + iv(q) > v(a_n) + nv(q) = v(a_n q^n).$$

It follows that $v(z(q)) = v(a_n q^n) < 0$. Thus for $q \in \mathbb{Q}$,

$$(8) \qquad z(q) \in \mathbb{Z} \Longrightarrow v_p(q) \geq m_p \qquad \text{for all primes } p.$$

Note that, for almost all primes p, we have $v_p(a_n) = 0$, and for $i < n$, $v_p(a_i) = 0$ or ∞. Hence $m_p = 0$ for almost all primes p. Put $D = \prod_p p^{m_p} \in \mathbb{Q}^\times$. Then we can restate (8) as:

$$z(q) \in \mathbb{Z} \Longrightarrow q/D \in \mathbb{Z}.$$

Put $d = (\text{denominator of } D) = \prod_{m_p < 0} p^{-m_p} \in \mathbb{Z}$. Then from the above we see that, for $q \in \mathbb{Q}$,

$$(8^+) \qquad z(q) \in \mathbb{Z} \Longrightarrow dq \in \mathbb{Z}.$$

It remains only to show that $z(t) \in \mathbb{Q}[t]$, i.e. that $m \geq 0$. If, on the contrary, $m < 0$, then we can argue as above with t^{-1} in place of t, and conclude that, for some integer $d' > 0$,

$$(8^-) \qquad z(q) \in \mathbb{Z} \Longrightarrow d'q^{-1} \in \mathbb{Z}.$$

But clearly (8^+) and (8^-) together bound q to finitely many possibilities, contradicting (4). This proves (E.1).

The following corollary covers case (c)(1) of Theorem (6.5).

(E.2) COROLLARY. *Suppose that x or y has a pole rational over $\mathbb{Q}$. Then we can choose t so that $x, y \in \mathbb{Q}[t]$ and, for $\tau \in \mathbb{Q}$,*

$$(x(\tau), y(\tau)) \in \mathbb{Z}^2 \Longrightarrow \tau \in \mathbb{Z}.$$

If $C \cap \mathbb{N}^2$ is infinite we can further arrange that $x(t)$ and $y(t)$ have positive leading coefficients.

PROOF: Say x has a pole rational over $\mathbb{Q}$, which may be assumed to be at ∞. By (E.1), $x \in \mathbb{Q}[t]$ and for some integer $d > 0$, $x(\tau) \in \mathbb{Z} \Rightarrow d\tau \in \mathbb{Z}$. Replacing t by t/d, we may assume that $d = 1$. If y has a pole at ∞ then $y \in \mathbb{Q}[t]$ also by (E.1). If y has no pole at ∞ then $x + y$ does, hence $x + y \in \mathbb{Q}[t]$, so again $y \in \mathbb{Q}[t]$. Moreover for $\tau \in \mathbb{Q}$, $(x(\tau), y(\tau)) \in \mathbb{Z}^2 \Rightarrow x(\tau) \in \mathbb{Z} \Rightarrow \tau \in \mathbb{Z}$.

Finally, suppose that $C \cap \mathbb{N}^2$ is infinite. Replacing t by $-t$ if necessary, we can assume that $(x(\tau), y(\tau)) \in \mathbb{N}^2$ for infinitely many $\tau \in \mathbb{N}$. Then $x(t)$ and $y(t)$, being positive for large positive values of t, have positive leading coefficients.

Returning to the non-constant $z = z(t) \in \mathbb{Z}[x, y] \subset \mathbb{Q}(t)$ satisfying (4) and (5) above, there remains the case when z has no pole rational over $\mathbb{Q}$. In that case, since the set of poles of z is rational over $\mathbb{Q}$, z must have two poles which are conjugate quadratic irrationalities over $\mathbb{Q}$. Moreover these poles must be real, since otherwise z would be bounded on $\mathbb{Q}$, contrary to (4). Explicitly, we can write

$$(9) \quad \begin{cases} z(t) = f(t)/g(t); \\ g(t) = q(t)^e, \ \textit{where } q(t) \in \mathbb{Q}[t] \textit{ is monic irreducible} \\ \qquad \textit{quadratic}; \\ f(t) \in \mathbb{Q}[t], \deg(f) \leq \deg(g) = 2e, \ \textit{and } f \textit{ is} \\ \qquad \textit{relatively prime to } g \end{cases}$$

By completing squares and rescaling t we may further arrange that,

$$(10) \qquad q(t) = t^2 - d, \text{ where } d \text{ is a square free integer} > 1.$$

These observations together with part (b) of (E.3) below easily cover case (c)(2) of Theorem (6.5), and so complete the proof of Theorem (6.5).

(E.3) PROPOSITION. *Let $z(t) \in \mathbb{Q}(t)$ satisfy (4), (9), and (10) above.*

(a) *There is an integer $D > 0$ such that if $\tau = u/v$ with u, v relatively prime integers and $z(\tau) \in \mathbb{Z}$ then $(u^2 - dv^2)^e$ divides D.*

(b) *There is a constant $K > 0$ such that if $N \in \mathbb{N}$ and $X_N = \{\tau \in \mathbb{Q} \mid z(\tau) \in \mathbb{Z}$ and $|z(\tau)| \leq N\}$ then*

$$\operatorname{Card}(X_N) \leq K \cdot \log(N).$$

PROOF OF (E.3)(a): Let u, v be indeterminates and put $t = u/v$. If $\deg(f(t)) = r$ $(\leq 2e)$ then

$$z(t) = z(u/v) = \frac{f(u/v)}{g(u/v)} = \frac{f(u,v)}{g(u,v)} \cdot v^{2e-r}$$

where $f(u,v) = v^r f(u/v)$ and $g(u,v) = v^{2e} g(u/v) = (u^2 - dv^2)^e$; we have $f(u,v)$, $g(u,v) \in \mathbb{Q}[u,v]$.

Since $f(t)$ and $g(t)$ are relatively prime we can solve

$$A(t)\, f(t) + B(t)\, g(t) = 1 \tag{11}$$

for $A, B \in \mathbb{Q}[t]$. Putting $t = u/v$ and multiplying by v^N for suitably large N we obtain

$$A_1(u,v)\, f(u,v) + B_1(u,v)\, g(u,v) = v^N \tag{12}$$

with $A_1, B_1 \in \mathbb{Q}[u,v]$. Since $g(u,v) = (u^2 - dv^2)^e \equiv u^{2e} \bmod (v)$ it follows that we can also solve

$$A_2(u,v)\, f(u,v) + B_2(u,v)\, g(u,v) = u^M \tag{13}$$

for suitable $A_2, B_2 \in \mathbb{Q}[u,v]$ and $M > 0$. Let D be a common denominator for all coefficients (in $\mathbb{Q}$) of A_1, A_2, B_1, B_2. Multiply (12) and (13) by $\frac{D \cdot v^{2e-r}}{g(u,v)}$, and recall that $\frac{v^{2e-r}\, f(u,v)}{g(u,v)} = z(u/v)$; we obtain,

$$D \cdot A_1(u,v) \cdot z(u/v) + D \cdot B_1(u,v) \cdot v^{2e-r} = \frac{D \cdot v^{2e-r+N}}{g(u,v)} \tag{14}$$

and

$$D \cdot A_2(u,v) \cdot z(u/v) + D \cdot B_2(u,v) \cdot v^{2e-r} = \frac{D \cdot v^{2e-r} \cdot u^M}{g(u,v)} \tag{15}$$

Now suppose we take u and v to be relatively prime integers such that $z(u/v) \in \mathbb{Z}$. Then the left sides of (14) and (15), hence also the right sides, are integers. It follows that $g(u,v)$ divides $\gcd(D\cdot v^{2e-r}\cdot v^N, D\cdot v^{2e-r}\cdot u^M) = D\cdot v^{2e-r}$. Since $g(u,v) = (u^2-dv^2)^e \equiv u^{2e} \bmod (v)$, $g(u,v)$ is prime to v, hence $g(u,v)$ divides D. This proves (E.3)(a).

PROOF OF (E.3)(b): For $N > 0$ and $X_N = \{\tau \in \mathbb{Q} \mid z(\tau) \in \mathbb{Z}$ and $|z(\tau)| \leq N\}$ we must show that

$$\text{(16)} \qquad \operatorname{Card}(X_N) = O(\log(N)).$$

Consider $z(t) = f(t)/g(t)$ as a (finite to one) function on $\mathbb{R}$; it has poles precisely at $\pm\sqrt{d}$, and is bounded at ∞. For $\epsilon > 0$ put $U_\epsilon = (-\sqrt{d}-\epsilon, -\sqrt{d}+\epsilon) \cup (\sqrt{d}-\epsilon, \sqrt{d}+\epsilon)$. Choose ϵ so that $f(t)$ has no roots in $U_{2\epsilon}$. Then there are constants c, $N_0 > 0$ such that

$$\text{(17)} \qquad |f(t)| \geq c \text{ for all } t \in U_\epsilon$$

and

$$\text{(18)} \qquad |z(t)| \leq N_0 \text{ for all } t \notin U_\epsilon.$$

From (18) we see that

$$\text{(19)} \qquad X_N - X_{N_0} \subset U_\epsilon \text{ for all } N > N_0.$$

Put $R = \mathbb{Z}[\sqrt{d}]$ and define

$$q : R - \mathbb{Z} \longrightarrow \mathbb{Q}, q(\alpha) = u/v \text{ for } \alpha = u + v\sqrt{d}.$$

Then q is surjective, and

$$\text{(20)} \qquad z(q(\alpha)) = z(u/v) = \frac{f(u/v)}{q(u/v)} = v^{2e}\frac{f(u/v)}{N(\alpha)^e},$$

where

$$N(\alpha) = u^2 - dv^2 = \alpha\bar{\alpha}, \quad \bar{\alpha} = u - v\sqrt{d}.$$

If $u, v \in \mathbb{Z}$ are relatively prime and $z(u/v) \in \mathbb{Z}$ then it follows from (E.3)(a) that $N(\alpha)^e$ divides the integer D of (E.3)(a).

Put $\Delta = \{\delta \in \mathbb{Z} \mid \delta^e \text{ divides } D\}$, a finite set. For any $\delta \in \mathbb{Z}$ put

$$R_\delta = \{\alpha \in R \mid N(\alpha)(= u^2 - dv^2) = \delta\}.$$

For $N > N_0$ put

$$R_{\delta,N} = \{\alpha \in R_\delta \mid N_0 < z(q(\alpha)) \leq N\}.$$

The above observations show that, for $N > N_0$,

$$X_N - X_{N_0} \subset q\left(\bigcup_{\delta \in \Delta} R_{\delta,N}\right). \tag{21}$$

Since Δ is a finite set (16) will follow from (21) once we show that, for integers $\delta \neq 0$ and $N > N_0$,

$$\mathrm{Card}(R_{\delta,N}) = O(\log(N)) \tag{22}$$

(where the implicit constant depends on δ).

Let $\alpha = u + v\sqrt{d} \in R_{\delta,N}$. Then it follows from (20) and (17) that

$$N \geq z(q(\alpha)) = \left|\frac{v^{2e} f(u/v)}{\delta^e}\right| \geq \left(\frac{c}{\delta^e}\right) v^{2e},$$

and hence

$$|v| \leq c_1 \cdot N^{1/2e} \text{ for } \alpha = u + v\sqrt{d} \in R_{\delta,N}, \tag{23}$$

where $c_1 = \left(\frac{\delta^e}{c}\right)^{1/2e}$.

Let $\omega = a + b\sqrt{d}$ be a unit in R of infinite order and norm $N(\omega) = 1$. We may assume that $a, b > 0$, after replacing ω by one of $\pm\omega, \pm\bar{\omega}$. Since $R^\times$ has rank 1 and there are only finitely many ideals with given norm it follows that there is a finite sequence $\alpha_1, \ldots, \alpha_h \in R$ such that every $\alpha \in R_\delta$ has the form $\alpha = \alpha_i\,\omega^n$ $(i = 1, \ldots, h; n \in \mathbb{Z})$. We claim that there is an integer $n_0 > 0$ and a constant $c_2 > 0$ such that,

$$\text{If } \alpha = \alpha_i\,\omega^n \in R_\delta \text{ with } n \geq n_0 \text{ then } |\alpha - \bar{\alpha}| \geq c_2 |\omega|^{|n|}. \tag{24}$$

(Here $|\ |$ refers to real absolute value in $\mathbb{Q}(\sqrt{d}) \subset \mathbb{R}$.) To see (24) we can assume that $n \geq 0$, replacing α by $\bar{\alpha}$ if necessary. Write $\alpha - \bar{\alpha} = \omega^n (\alpha_i - \bar{\alpha}_i \omega^{-2n})$. Since $\omega^{-2n} \to 0$ as $n \to \infty$, (24) follows.

Now let $\alpha = \alpha_i\,\omega^n \in R_{\delta,N}$. If $n \geq n_0$ then, by (23) and (24), we have

$$c_1\, N^{1/2e} \geq |v| = \left|\frac{1}{2\sqrt{d}}(\alpha - \bar{\alpha})\right| \geq \frac{c_2}{2\sqrt{d}}\, |\omega|^{|n|},$$

and so

$$\begin{aligned} |n| &\le \frac{1}{\log|\omega|}\left(\frac{1}{2e}\log(N) + \log\left(\frac{2\sqrt{d}\,c_1}{c_2}\right)\right) \\ &\le c_3 \log(N) \end{aligned}$$

for suitable $c_3 > 0$. For N large enough that $n_0 \le c_3 \log(N)$ we conclude that $|n| \le c_3 \log(N)$ for all $\alpha = \alpha_i\,\omega^n \in R_{\delta,N}$. It follows that

$$\text{Card}(R_{\delta,N}) \le h \cdot (2c_3 \log(N) + 1) = O(\log(N)).$$

This proves (22) and so concludes the proof of (E.3)(b).

References

[**AM**] S. S. Abhyankar and T.-T. Moh, *Embeddings of the line in the plane*, J. Reine angew. Math. **276** (1975), 149–166.

[**AK**] A. Altman and S. Kleiman, *Introduction to Grothendieck Duality*, Springer Lecture Notes in Math. **146** (1970).

[**B**] H. Bass, "Algebraic K-theory," W. A. Benjamin, New York, 1968.

[**BCW**] H. Bass, E. H. Connell and D. Wright, *The Jacobian Conjecture: Reduction of degree and formal expansion of the inverse*, Bull. AMS **7** (1982), 287–330.

[**BH**] H. Bass and W. Haboush, *Linearizing certain reductive group actions*, Trans. AMS **292** (1985), 463–482.

[**Bj**] J. E. Bjork, "Rings of Differential Operators," North Holland Publ. Co., Amsterdam, 1979.

[**Bo**] A. Borel, "Linear Algebraic Groups," W. A. Benjamin, New York, 1969.

[**BoD**] A. Borel, et al., "Algebraic D-modules," Perspectives in Mathematics, Academic Press, Boston, 1987.

[**NB**] N. Bourbaki, "Algèbre Commutative," Chs. 8 and 9, Masson, Paris, 1983.

[**D**] P. Dienes, "The Taylor Series: An Introduction to the Theory of Functions of a Complex Variable," Clarendon Press, Oxford, 1931.

[**F**] E. Formanek, *Two notes on the Jacobian Conjecture*, preprint (1987).

[**H**] E. Hille, "Analytic Function Theory," Ginn and Co., Boston, 1959.

[**L**] S. Lang, "Fundamentals of Diophantine Geometry," Springer-Verlag, New York, 1983.

[**Li**] J. Lipman, *Free derivation modules on algebraic varieties*, Amer. Jour. Math. **87** (1965), 874–898.

[**M**] H. Matsumura, "Commutative Algebra," W. A. Benjamin, New York, 1970.

[**N**] M. Nagata, "Field Theory," M. Dekker, New York, 1977.

[**S**] J.-P. Serre, "Lie Algebras and Lie Groups," Lectures at Harvard, W. A. Benjamin, New York, 1964.

[**W**] D. Wright, *On the Jacobian Conjecture*, Illinois Jour. Math. **25** (1981), 423–440.

Department of Mathematics, Columbia University, New York NY 10027

Additions to the Theory of Algebras with Straightening Law

WINFRIED BRUNS

In this article we want to supplement the theory of algebras with straightening law, ASLs for short, by two additions. The first addition concerns the arithmetical rank of ideals generated by an ideal of the poset underlying the ASL. (The arithmetical rank is the least number of elements generating an ideal up to radical.) It turns out that there is a general upper bound only depending on the combinatorial data of the poset. In particular we discuss ideals generated by monomials and show that the ideas leading to the general bound can be used to derive sharper results in this special case. On the other hand there exists a class of ASLs, called symmetric, in which the general bound is always precise. This class includes the homogeneous coordinate rings of Grassmannians, and perhaps the most prominent result in this context is the determination of the least number of equations defining a Schubert subvariety of a Grassmannian.

As a second addition we introduce the notion of a module M with straightening law over an ASL A. Such a module has a partially ordered set of generators, a basis of standard elements each of which is a product of a standard monomial and a generator, and finally the multiplication $A \times M \to M$ satisfies a straightening law similar to the straightening law in A. We discuss some examples, among them the powers of certain ideals generated by poset ideals and the generic modules. Remarkable facts: (i) the first syzygy of a module with straightening law has itself a straightening law, and (ii) the existence of natural filtrations which, for example, lead to a lower bound on the depth of a module with straightening law. In the last part we introduce a natural strengthening of the axioms which under special conditions leads to a straightening law on the symmetric algebra of the module. The most interesting examples to which we will apply this result are the generic modules.

The theory of ASLs has been developed in [**Ei**] and [**DEP.2**]; the treatment in [**BV.1**] also satisfies our needs. For the readers convenience we have collected the definition, results relevant for us, and some significant examples in the first section. Matsumura's book [**Ma**] may serve as a source for the commutative ring theory needed in this article.

1. Algebras with Straightening Laws.

An algebra with straightening law is defined over a ring B of coefficients. In order to avoid problems of secondary importance in the following sections we will assume throughout that B is a noetherian ring.

DEFINITION. Let A be a B-algebra and $\Pi \subset A$ a finite subset with partial order $\leq$. A is an *algebra with straightening law on* Π (*over* B) if the following conditions are satisfied:
(ASL-0) $A = \bigoplus_{i\geq 0} A_i$ is a graded B-algebra such that $A_0 = B$, Π consists of homogeneous elements of positive degree and generates A as a B-algebra.
(ASL-1) The products $\xi_1 \cdots \xi_m$, $m \geq 0$, $\xi_1 \leq \cdots \leq \xi_m$ are a free basis of A as a B-module. They are called *standard monomials*.
(ASL-2) (*Straightening law*) For all incomparable $\xi, v \in \Pi$ the product ξv has a representation

$$\xi v = \sum a_\mu \mu, \qquad a_\mu \in B, a_\mu \neq 0, \quad \mu \quad \text{standard monomial},$$

satisfying the following condition: every μ contains a factor $\zeta \in \Pi$ such that $\zeta \leq \xi$, $\zeta \leq v$. (It is of course allowed that $\xi v = 0$, the sum $\sum a_\mu \mu$ being empty.)

In [**Ei**] and [**BV.1**] B-algebras satisfying the axioms above are called graded ASLs, whereas in [**DEP.2**] they figure as graded ordinal Hodge algebras.

In terms of generators and relations an ASL is defined by its poset and the straightening law:

(1.1) PROPOSITION. *Let A be an ASL on Π. Then the kernel of the natural epimorphism*

$$B[T_\pi\colon \pi \in \Pi] \longrightarrow A, \qquad T_\pi \longrightarrow \pi,$$

is generated by the relations required in (ASL-2), *i.e. the elements*

$$T_\xi T_v - \sum a_\mu T_\mu, \qquad T_\mu = T_{\xi_1} \cdots T_{\xi_m} \quad \textit{if} \quad \mu = \xi_1 \cdots \xi_m.$$

See [**DEP.2**], 1.1 or [**BV.1**], (4.2).

(1.2) PROPOSITION. *Let A be an ASL on Π, and $\Psi \subset \Pi$ an ideal, i.e. $\psi \in \Psi$, $\phi \leq \psi$ implies $\phi \in \Psi$. Then the ideal $A\Psi$ is generated as a B-module by all the standard monomials containing a factor $\psi \in \Psi$, and*

$A/A\Psi$ is an ASL on $\Pi \setminus \Psi$ ($\Pi \setminus \Psi$ being embedded into $A/A\Psi$ in a natural way.)

This is obvious, but nevertheless extremely important. First several proofs by induction on $|\Pi|$, say, can be based on (1.2), secondly the ASL structure of many important examples is established this way.

For an element $\xi \in \Pi$ we define its rank by

$$\operatorname{rk}\xi = k \quad\Longleftrightarrow\quad \begin{array}{l}\text{there is a chain } \xi = \xi_k > \xi_{k-1} > \cdots > \xi_1,\ \xi_i \in \Pi,\\ \text{and no such chain of greater length exists.}\end{array}$$

For a subset $\Omega \subset \Pi$ let

$$\operatorname{rk}\Omega = \max\{\operatorname{rk}\xi \colon \xi \in \Omega\}.$$

The preceding definition differs from the one in [**Ei**] and [**DEP.2**] which gives a result smaller by 1. In order to reconcile the two definitions the reader should imagine an element $-\infty$ added to Π, vaguely representing $0 \in A$.

(1.3) PROPOSITION. *Let A be an ASL on Π. Then*

$$\dim A = \dim B + \operatorname{rk}\Pi \quad \textit{and} \quad \operatorname{ht} A\Pi = \operatorname{rk}\Pi.$$

Here of course $\dim A$ denotes the Krull dimension of A and $\operatorname{ht} A\Pi$ the height of the ideal $A\Pi$. A quick proof of (1.3) may be found in [**BV.1**], (5.10).

In the context of ASLs A we denote the length of a maximal M-sequence in $A\Pi$, M a finitely generated A-module, by $\operatorname{depth} M$.

We list three important examples of ASLs to which we will pay special attention in the following sections.

(1.4) EXAMPLES. (a) In order to study ideals generated by square-free monomials in the indeterminates of the polynomial ring $B[X_1, \ldots, X_n]$ one chooses Π as the set of all square-free monomials ordered by:

$$\xi \le \upsilon \quad\Longleftrightarrow\quad \upsilon \text{ divides } \xi.$$

(ASL-0) is satisfied for trivial reasons, and (ASL-1) holds since the standard monomials correspond bijectively to the ordinary monomials in $X_1, \ldots, X_n$. The straightening law is given by

$$\xi\upsilon = (\xi \sqcap \upsilon)(\xi \sqcup \upsilon)$$

where $\xi \sqcap \upsilon$ is the greatest common divisor and $\xi \sqcup \upsilon$ the least common multiple of ξ and υ. If $\Omega \subset \Pi$ is an arbitrary subset, then Ω and the smallest ideal $\Psi \supset \Omega$, $\Psi = \{\psi\colon \psi \le \omega \text{ for some } \omega \in \Omega\}$, generate the same ideal in $B[X_1, \dots, X_n]$, so the ideals generated by square-free monomials belong to the class covered by (1.2).

For a given poset Σ the *discrete ASL on* Σ is constructed as follows: One makes the polynomial ring $B[T_\sigma\colon \sigma \in \Sigma]$ an ASL as just described and passes to the residue class modulo the ideal generated by all products $T_\sigma T_\tau$, σ, τ incomparable. Thus one obtains an ASL on Σ in which the straightening law takes the special form $\sigma\tau = 0$ for all incomparable $\sigma, \tau \in \Sigma$.

(b) Let X be an $m \times n$ matrix of indeterminates over B, and $\mathrm{I}_t(X)$ denote the ideal generated by the t-minors (i.e. the determinants of the $t \times t$ submatrices) of X. For the investigation of the ideals $\mathrm{I}_t(X)$ and the residue class rings $\mathrm{R}_t(X) = B[X]/\mathrm{I}_t(X)$ one makes $B[X]$ an ASL on the set $\Delta(X)$ of all minors of X. Denote by $[a_1, \dots, a_t | b_1, \dots, b_t]$ the minor with row indices $a_1, \dots, a_t$ and column indices $b_1, \dots, b_t$. The partial order on $\Delta(X)$ is given by

$$[a_1, \dots, a_u | b_1, \dots, b_u] \le [c_1, \dots, c_v | d_1, \dots, d_v] \iff u \ge v \quad \text{and} \quad a_i \le c_i,\ b_i \le d_i,\ i = 1, \dots, v.$$

Then $B[X]$ is an ASL on $\Delta(X)$; cf. [**BV.1**], Section 4 for a complete proof. Obviously $\mathrm{I}_t(X)$ is generated by an ideal in the poset $\Delta(X)$, so $\mathrm{R}_t(X)$ is an ASL on the poset $\Delta_{t-1}(X)$ consisting of all the i-minors, $i \le t-1$.

(c) In the situation of (b) assume that $m \le n$. Then the B-subalgebra $\mathrm{G}(X)$ generated by the m-minors of X is a sub-ASL in a natural way, its poset being given by the set $\Gamma(X)$ of m-minors. This result is essentially due to Hodge. Again we refer to [**BV.1**], Section 4 for a proof. In denoting an m-minor we omit the row indices.

If $B = K$ is a field, then $\mathrm{G}(X)$ is the homogeneous coordinate ring of the Grassmannian $\mathrm{G}_m(K^n)$ of m-dimensional subspaces of the vector space K^n. The (special) Schubert subvariety $\Omega(a_1, \dots, a_m)$ of the Grassmannian is defined by the ideal generated by

$$\{\, \delta \in \Gamma(X)\colon \delta \not\ge [n+1-a_m, \dots, n+1-a_1] \,\},$$

an ideal in $\Gamma(X)$.

(d) Another example needed below is given by "pfaffian" rings. Let X_{ij}, $1 \le i < j \le n$, be a family of indeterminates over B, $X_{ji} = -X_{ij}$, $X_{ii} = 0$. The pfaffian of the alternating matrix $(X_{i_u i_v}\colon 1 \le u, v \le t)$, t even, is denoted by $[i_1, \dots, i_t]$. The polynomial ring $B[X]$ is an ASL on the set $\Phi(X)$ of the pfaffians $[i_1, \dots, i_t]$, $i_1 < \dots < i_t$, $t \le n$. The pfaffians are partially ordered in the same way as the minors in (b). The residue class ring $\mathrm{P}_{r+2}(X) = B[X]/\operatorname{Pf}_{r+2}(X)$, $\operatorname{Pf}_{r+2}(X)$ being generated by the $(r+2)$-pfaffians, inherits its ASL structure from $B[X]$ according to (1.2). The poset underlying $\mathrm{P}_{r+2}(X)$ is denoted $\Phi_r(X)$. Note that the rings $\mathrm{P}_{r+2}(X)$ are Gorenstein rings over a Gorenstein B—in fact factorial over a factorial B, cf. [**Av.1**], [**KL**]. —

2. The Arithmetical Rank of a Poset Ideal.

Let V be an affine or projective algebraic variety, and W a closed subvariety of V. In general it is difficult to determine the smallest number w of hypersurfaces H_i in the ambient affine or projective space such that

$$W = V \cap H_1 \cap \dots \cap H_w.$$

In more general and algebraic terms the problem above amounts to the determination of the *arithmetical rank* $\operatorname{ara} I$ of an ideal I in a commutative (noetherian) ring R, the arithmetical rank being defined to be the smallest number t for which there are elements $x_1, \dots, x_t \in R$ such that

$$\operatorname{Rad} I = \operatorname{Rad} \sum_{i=1}^{t} R x_i.$$

(In the projective situation one of course requires the x_i to be homogeneous.) In this section we obtain an upper bound for the arithmetical rank of an ideal I generated by an ideal Ω of the poset Π underlying an ASL:

(2.1) Proposition. *Let A be an ASL on Π over B, $\Omega \subset \Pi$ an ideal, and $I = A\Omega$. Then there are homogeneous elements $x_1, \dots, x_r \in I$, $r = \operatorname{rk} \Omega$, such that* $\operatorname{Rad} I = \operatorname{Rad} \sum_{i=1}^{r} A x_i$. *In particular* $\operatorname{ara} I \le \operatorname{rk} \Omega$.

Proof: Let m be the least common multiple of the degrees of the elements $\xi \in \Omega$, and $e(\xi) = m/\deg \xi$. We put

$$x_i = \sum_{\substack{\xi \in \Omega \\ \operatorname{rk} \xi = i}} \xi^{e(\xi)}, \qquad i = 1, \dots, r.$$

Let υ be a minimal element of Ω (and, hence, of Π). Then $\upsilon\zeta = 0$ for every different minimal element $\zeta \in \Omega$, and one concludes

$$\upsilon x_1 = \upsilon^{e(\upsilon)+1} \in \operatorname{Rad} \sum_{i=1}^{r} A x_i.$$

Now an induction argument finishes the proof: Let $\widetilde{\Omega}$ be the set of minimal elements of Ω, $\overline{\Omega} = \Omega \setminus \widetilde{\Omega}$, $\overline{A} = A/A\widetilde{\Omega}$, $\overline{\Pi} = \Pi \setminus \widetilde{\Omega}$. The data $\overline{A}, \overline{\Pi}, \overline{\Omega}$ satisfy the hypotheses of the proposition, and it follows that

$$\operatorname{Rad} A\Omega \subset \operatorname{Rad}(A\widetilde{\Omega} + \sum_{i=2}^{r} A x_i) = \operatorname{Rad} \sum_{i=1}^{r} A x_i. \text{ —}$$

In the next section we shall see that the bound of (2.1) is sharp under special circumstances. A first specialization:

(2.2) COROLLARY. *Let X be an $m \times n$ matrix of indeterminates over B. Then*

$$\operatorname{ara} \mathrm{I}_t(X) \le mn - t^2 + 1.$$

Obviously $[m-t+1,\ldots,,m|n-t+1,\ldots,,n]$ is the only maximal element of the poset ideal generating $\mathrm{I}_t(X)$, and one easily computes its rank. Cf. (1.4),(b) for the ASL structure on $B[X]$.

(2.3) REMARKS. (a) In general the bound given by (2.1) is not sharp: Consider the ideal generated by X_1, say, under the hypotheses of (1.4),(a), or the ideal generated by $[1|1]$, $[2|1]$, and $[1\,2|1\,2]$ under the hypotheses of (2.2), $m = n = 2$. Admittedly, none of these counterexamples is completely convincing: If one first takes the ideals in their "natural" rings $B[X_1]$ and $B[X_{11}, X_{21}]$ resp. and then extends the ideal to $B[X_1, \ldots, X_n]$ or $B[X]$, the precise bounds are obtained. It is quite clear that $\operatorname{ara} I$ in general cannot be determined from the combinatorial data given by Π and Ω. In (2.5) below we will note an improvement of (2.1) for a specific ASL which depends on the form of the straightening relations.

(b) In [**Ne**], p. 180, Example (i),(a) Newstead showed that the bound in (2.2) is precise for $t = 2$, B a field of characteristic 0. As Cowsik told us, Newstead's argument goes through for every t and can be transferred to characteristic $p > 0$ via the use of étale cohomology. There is of course no restriction in assuming that B is a field; otherwise one factors by a maximal ideal first.

For the case $t = \min(m, n)$ Hochster has given an invariant-theoretic argument which shows that $\operatorname{ara} \mathrm{I}_t(X) = mn - t^2 + 1$ in characteristic 0. Suppose $m \le n$. Then (in all characteristics) $\mathrm{G}(X)$ is the ring of invariants of the $\mathrm{SL}(m, B)$-action induced by the substitutions $X \to TX$, $T \in \mathrm{SL}(m, B)$, on $B[X]$. In characteristic 0 the group $\mathrm{SL}(m, B)$ is linearly reductive. This implies that $\mathrm{G}(X)$ is a direct $\mathrm{G}(X)$-summand of $B[X]$, $B[X] = \mathrm{G}(X) \oplus C$. Let $I = \mathrm{I}_m(X) \subset B[X]$, and $J = I \cap \mathrm{G}(X)$. Then $I = JB[X]$, and

$$\mathrm{H}_I^d(B[X]) = \mathrm{H}_J^d(B[X]) = \mathrm{H}_J^d(\mathrm{G}(X)) \oplus \mathrm{H}_J^d(C);$$

here H_I denotes cohomology with support in I, cf. [**Ha**]. Taking $d = nm - m^2 + 1$, one concludes $\mathrm{H}_I^d(B[X]) \ne 0$ since $d = \dim \mathrm{G}(X)$. By [**Ha**], p. 414, Example 2, $\operatorname{ara} I \ge d$.

The preceding argument breaks down in positive characteristic since $\mathrm{H}_I^d(B[X]) = 0$ then, provided $n > m$: $\mathrm{H}_I^i(B[X]) = 0$ for all $i > \operatorname{ht} I = n - m + 1$ according to [**PS**], p. 110, Proposition (4.1). It likewise fails for $t < \min(m, n)$ since the subalgebra of $B[X]$ generated by the t-minors has the same dimension as $B[X]$, cf. [**CN**] or [**BV.1**], Section 10. —

Specializing (2.1) to the example (1.4),(a) we obtain the following result of Gräbe ([**Gr**], Theorem 1):

(2.4) COROLLARY. *Let $X_1, \dots, X_n$ be indeterminates over B, and I an ideal generated by square-free monomials $f_1, \dots, f_m$. Let p be the smallest number of factors occuring among the f_i. Then*

$$\operatorname{ara} I \le n - p + 1.$$

The ASL A considered in the preceding corollary has a very special property: If Ω is an ideal in its underlying poset, then the maximal elements of Ω generate $A\Omega$. This allows a slight improvement of (2.1) which we only give under the hypotheses of (2.4), a result almost obtained by Gräbe ([**Gr**], Theorem 2). The straightening relations in (2.4) are the equations

$$\xi v = (\xi \sqcup v)(\xi \sqcap v),$$

cf. (1.4),(a), and $(\xi \sqcup v) < \xi, v$. Therefore it is enough to consider the smallest subset of Ω which contains the generators of I and is closed under taking least common multiples, i.e. the subset $\widetilde{\Omega}$ formed by the least common multiples of the subsets of $\{f_1, \dots, f_m\}$.

(2.5) COROLLARY. *Under the hypotheses of* (2.4) *one has*

$$\operatorname{ara} I \leq \operatorname{rk} \widetilde{\Omega}$$

(where of course $\operatorname{rk}\widetilde{\Omega}$ *is measured by chains in* $\widetilde{\Omega}$*). In particular, if* q *is the smallest number of factors occuring in any of the least common multiples of two of the elements* $f_1, \ldots, f_m$*, then*

$$\operatorname{ara} I \leq n - q + 2.$$

PROOF: One takes

$$x_i = \sum_{\substack{\xi \in \widetilde{\Omega} \\ \operatorname{rk} \xi = i}} \xi^{e(\xi)}, \qquad i = 1, \ldots, \operatorname{rk} \widetilde{\Omega},$$

and argues as in the proof of (2.1). The second part follows since the submaximal elements of $\widetilde{\Omega}$ are given as $\xi \sqcup \upsilon$, $\xi, \upsilon \in \{f_1, \ldots, f_m\}$, and $\operatorname{rk}_{\widetilde{\Omega}} \xi \sqcup \upsilon \leq \operatorname{rk}_{\Omega} \xi \sqcup \upsilon \leq n - q + 1$. —

Let $n = 5$, $f_1 = X_2X_4$, $f_2 = X_1X_3X_4$, $f_3 = X_1X_3X_5$, $f_4 = X_2X_3X_5$. Then (2.4) gives the trivial bound $\operatorname{ara} I \leq 4$ whereas (2.5) yields $\operatorname{ara} I \leq 3$, and $\operatorname{ara} I = 3$, as shown in [**Gr**], 5., Beispiel 4. An example for which (2.5) fails to give the precise value: Take $n = 5$, $f_1 = X_2X_4$, $f_2 = X_2X_5$, $f_3 = X_1X_4$, $f_4 = X_1X_3X_5$ ([**Gr**], 5., Beispiel 2). It is easily seen that $x_1 = X_2X_4X_5$, $x_2 = X_1X_3X_5 + X_1X_2X_4$, $x_3 = X_1X_4 + X_2X_5 + X_2X_4$ generate the ideal up to radical. The algorithm by which Gräbe finds x_1, x_2, x_3 can be described in the following way: One chooses subsets $\Psi_1, \ldots, \Psi_r \subset \Omega$ such that

(i) Ψ_i consists of incomparable elements of Ω, $i = 1, \ldots, r$,
(ii) the least common multiple of every pair $\psi_1, \psi_2 \in \Psi_i$, $\psi_1 \neq \psi_2$, is in the ideal generated by $\Psi_1 \cup \cdots \cup \Psi_{i-1}$,
(iii) $\Psi_1 \cup \cdots \cup \Psi_r$ contains the maximal elements of Ω.

Then it follows that the elements $x_i = \sum_{\omega \in \Psi_i} \omega$, $i = 1, \ldots, r$ generate I up to radical. (2.4) and (2.5) reflect two special choices of $\Psi_1, \ldots, \Psi_r$ which work for all Ω.

3. Symmetric ASLs.

In this section we introduce a special class of ASLs which, though certainly very small, contains some important examples.

DEFINITION. Let A be an ASL on Π over B. A is called *symmetric* if it is also a graded ASL with respect to the reverse order on Π.

Note that this is only a condition concerning (ASL-2): the standard monomials in the reverse order are the same as those with respect to the given one.

(3.1) EXAMPLES. (a) If A is a symmetric ASL and $\Omega \subset \Pi$ an ideal or a coideal (i.e. the complement of an ideal), then $A/A\Omega$ is a symmetric ASL.

(b) The discrete ASLs are symmetric.

(c) $\mathrm{G}(X)$, the homogeneous coordinate ring of a Grassmann variety, is a symmetric ASL. This is stated in [**Ho**], Lemma 2.1 and [**BV.1**], (4.6), and can in fact be seen very easily: The automorphism of $B[X]$ which reverses the order of the columns of X, induces an automorphism of $\mathrm{G}(X)$ which (up to sign) permutes the maximal minors of X and reverses the order of $\Gamma(X)$. It follows from (a) that the homogeneous coordinate rings of the Schubert subvarieties are symmetric ASLs, too.

(d) More generally than (c), the multihomogeneous coordinate ring of a flag variety is a symmetric ASL, cf. [**Ei**], Example (5). It can be described in the following way. Let X be an $n \times n$ matrix of indeterminates over B, and $n_1 > n_2 > \cdots > n_k$ a sequence of integers, $n_1 \le n$, $n_k \ge 1$. Then one considers the B-subalgebra generated by the n_i-minors of the first n_i rows of X, $i = 1, \ldots, k$. It is a sub-ASL of $B[X]$ in a natural way since in a standard representation

$$[1,\ldots,i|a_1,\ldots,a_i][1,\ldots,j|b_1,\ldots,b_j] = \sum a_\mu \mu$$

every standard monomial μ is of the form $[1,\ldots,i|\ldots][1,\ldots,j|\ldots]$: first it has at most two factors, and secondly every row index appears in μ with the same multiplicity as on the left side, cf. [**DEP.1**], Theorem 2.1 or [**BV.1**], (11.3). Now one can again apply the automorphism argument from (c). In this case the automorphism does not completely reverse the order on the poset; nevertheless the argument goes through as the reader may check.

(e) Let $(L, \sqcap, \sqcup)$ be a finite lattice, K a field, and A the residue class ring of the polynomial ring $K[X_\alpha \colon \alpha \in L]$ modulo the ideal generated by all the polynomials

$$X_\alpha X_\beta - X_{\alpha\sqcap\beta} X_{\alpha\sqcup\beta}, \qquad \alpha, \beta \text{ incomparable.}$$

In [**Hi**], Theorem, Hibi has shown that the following conditions are equivalent: (i) A is a graded ASL on L (relative to the embedding $L \to A$, $\alpha \to \overline{X}_\alpha$), (ii) A is an integral domain, (iii) L is distributive. Of course A is a symmetric ASL if L is distributive.

Moreover Hibi shows on p. 103 of [**Hi**] that a lattice L must be distributive if there is a symmetric ASL on L which is a domain and in which the standard monomials in the straightening relations all have exactly two factors. —

In symmetric ASLs the arithmetical rank on an ideal $A\Omega$ is always given by the rank of Ω, and sometimes this holds under more general circumstances:

(3.2) PROPOSITION. *Let A be an ASL on Π over B, $\Omega \subset \Pi$ an ideal, and $I = A\Omega$. Let $C(\Omega)$ denote the B-submodule generated by all the standard monomials which have a factor in $\Pi \setminus \Omega$. Suppose that one of the following hypotheses is satisfied:*
(i) $B[\Omega] \cap C(\Omega) = 0$ and $C(\Omega)$ is a $B[\Omega]$-submodule of A,
(ii) $C(\Omega)$ is an ideal in A.
Then $\operatorname{ara} I = \operatorname{rk} \Omega$.

(3.3) COROLLARY. *Let A be a symmetric ASL on Π over B, $\Omega \subset \Pi$ an ideal, and $I = A\Omega$. Then* $\operatorname{ara} I = \operatorname{rk} \Omega$.

The corollary follows immediately from the proposition since $C(\Omega)$ is the ideal generated by $\Pi \setminus \Omega$ if A is symmetric. By the way, one easily finds examples which demonstrate that none of the hypotheses (i) or (ii) in (3.2) implies the other one.

PROOF OF (3.2): In view of (2.1) we may first factor out a maximal ideal of B and assume that B is a field.

If hypothesis (i) is satisfied, $B[\Omega]$ is the B-module generated by all the standard monomials consisting entirely of factors from Ω. Therefore $B[\Omega]$ is an ASL in a natural way, and $\dim B[\Omega] = \operatorname{rk} \Omega$ (cf. (1.3)). Furthermore $B[\Omega]$ is a direct $B[\Omega]$-summand, and now one applies the cohomological argument detailed in (2.3),(b).

If hypothesis (ii) is satisfied, one passes to $\overline{A} = A/C(\Omega)$ which is an ASL on Ω in a natural way. Let $\overline{I} = \overline{A}\Omega$. Then

$$\operatorname{ara} I \geq \operatorname{ara} \overline{I} \geq \operatorname{ht} \overline{I} = \operatorname{rk} \Omega. \text{ —}$$

(3.4) Corollary. *Let K be an algebraically closed field. The minimal number of equations defining the Schubert variety $\Omega(a_1, \dots, a_m)$ as a subvariety of $\mathrm{G}_m(V)$, $\dim V = n$, is given by*

$$\max_{a_k - k < n - m} m(n-m) - (m-k+1)(a_k - k) + 1.$$

Proof: Using the information presented in (1.4),(c) we let $b_i = n - a_{m-i+1} + 1$, $i = 1, \dots, m$, and $\gamma = [b_1, \dots, b_m]$. The maximal elements of $\Omega = \{\delta \in \Gamma(X) \colon \delta \not\geq \gamma\}$ are given by

$$\tau_i = [b_i - i, \dots, b_i - 1, n - (m-i) + 1, \dots, n], \qquad b_i > i.$$

An easy computation yields

$$\operatorname{rk} \tau_i = m(n-m) + 1 - i(n - m - b_i + i + 1).$$

Replacing i by $m - k + 1$ and b_i by $n - a_k + 1$, one obtains the desired result. —

(3.5) Remark. Let B be an integral domain, $A = B[X]$, $I = \mathrm{I}_t(X)$ as in (2.2). The "symbolic graded ring"

$$\widetilde{A} = \bigoplus_{i=0}^{\infty} I^{(i)}/I^{(i+1)}$$

is a graded ASL over B on the poset Δ^* given by the leading forms of the minors of X. The ideal $\Omega^* \subset \Delta^*$ consisting of the leading forms of the ideal $\Omega \subset \Delta$ generating I satisfies both of the hypotheses (i) and (ii) of (3.2) though $\widetilde{A}$ is not a symmetric ASL, cf. [**BV.1**], Section 10. It follows that $\operatorname{ara} \widetilde{A}\Omega^* = mn - t^2 + 1$, a result we cannot prove by ASL methods for $I = A\Omega$. —

4. Straightening Laws on Modules.

It occurs frequently that a module M over an ASL A has a structure closely related to that of A: the generators of M are partially ordered, a distinguished set of "standard elements" forms a B-basis of M, and the multiplication $A \times M \to A$ satisfies a straightening law similar to the straightening law in A itself. In this section we introduce the notion of a module with straightening law whereas the next section contains a strengthening of this notion.

DEFINITION. Let A be an ASL over B on Π. An A-module M is called a *module with straightening law* (MSL) on the finite poset $\mathcal{X} \subset M$ if the following conditions are satisfied:
(MSL-1) For every $x \in \mathcal{X}$ there exists an ideal $\mathcal{I}(x) \subset \Pi$ such that the elements

$$\xi_1 \cdots \xi_n x, \qquad x \in \mathcal{X}, \quad \xi_1 \notin \mathcal{I}(x), \quad \xi_1 \leq \cdots \leq \xi_n, \quad n \geq 0,$$

constitute a B-basis of M. These elements are called *standard elements.*
(MSL-2) For every $x \in \mathcal{X}$ and $\xi \in \mathcal{I}(x)$ one has

$$\xi x \in \sum_{y<x} Ay.$$

It follows immediately by induction on the rank of x that the element ξx as in (MSL-2) has a standard representation

$$\xi x = \sum_{y<x} (\sum b_{\xi x \mu y} \mu) y, \qquad b_{\xi x \mu y} \in B, \ b_{\xi x \mu y} \neq 0,$$

in which each μy is a standard element.

(4.1) REMARKS. (a) Suppose M is an MSL, and $\mathcal{T} \subset \mathcal{X}$ an ideal. Then the submodule of M generated by $\mathcal{T}$ is an MSL, too. This fact allows one to prove theorems on MSLs by noetherian induction on the set of ideals of $\mathcal{X}$.

(b) It would have been enough to require that the standard elements are linearly independent. If just (MSL-2) is satisfied then the induction principle in (a) proves that M is generated as a B-module by the standard elements. —

(4.2) EXAMPLES. (a) A itself is an MSL if one takes $\mathcal{X} = \{1\}$, $\mathcal{I}(1) = \emptyset$. Another choice is $\mathcal{X} = \Pi \cup \{1\}$, $\mathcal{I}(\xi) = \{\pi \in \Pi \colon \pi \not\geq \xi\}$, $\mathcal{I}(1) = \Pi$, $1 > \pi$

for each $\pi \in \Pi$. The relations necessary for (MSL-2) are then given by the identities $\pi 1 = \pi$, the straightening relations

$$\xi\upsilon = \sum b_\mu \mu, \qquad \xi, \upsilon \quad \text{incomparable},$$

and the Koszul relations

$$\xi\upsilon = \upsilon\xi, \qquad \xi < \upsilon.$$

By (4.1),(a) for every poset ideal $\Psi \subset \Pi$ the ideal $A\Psi$ is an MSL, too.

(b) Suppose that Ψ as in (a) additionally satisfies the following condition: Whenever $\phi, \psi \in \Psi$ are incomparable, then every standard monomial μ in the standard representation $\phi\psi = \sum a_\mu \mu$, $a_\mu \neq 0$, contains at least two factors from Ψ. This condition appears in [**Hu**], [**EH**], [**BST**], and in [**BV.1**], Section 9 where the ideal $I = A\Psi$ is called *straightening-closed*. As a consequence of (d) below the powers I^n of $I = A\Psi$ are MSLs. Observe in particular that the condition above is satisfied if every μ a priori contains at most two factors and Ψ consists of the elements in Π of highest degree.

(c) In order to prove and to generalize the statements in (b) let us consider an MSL M on $\mathcal{X}$ and an ideal $\Psi \subset \Pi$ such that $I = A\Psi$ is straightening-closed and the following condition holds:

(∗) The standard monomials in the standard representation of a product ψx, $\psi \in \Psi$, $x \in \mathcal{X}$, all contain a factor from Ψ.

Then it is easy to see that IM is again an MSL on the set $\{\psi x \colon x \in \mathcal{X},\ \psi \in \Psi \setminus \mathcal{I}(x)\}$ partially ordered by

$$\psi x \leq \phi y \qquad \Longleftrightarrow \qquad x < y \quad \text{or} \quad x = y,\ \psi \leq \phi,$$

if one takes

$$\mathcal{I}(\psi x) = \{\pi \in \Pi \colon \pi \not\geq \psi\}.$$

Furthermore (∗) holds again. Thus $I^n M$ is an MSL for all $n \geq 1$, and in particular one obtains (b) from the special case $M = A$.

The residue class module M/IM also carries the structure of an MSL on the set $\overline{\mathcal{X}}$ of residues of $\mathcal{X}$ if we let

$$\mathcal{I}(\overline{x}) = \mathcal{I}(x) \cup \Psi.$$

Combining the previous arguments we get that $I^n M / I^{n+1} M$ is an MSL for all $n \geq 0$.

In the situation just considered the associated graded ring $\operatorname{Gr}_I A$ is an ASL on the set Π^* of leading forms (ordered in the same way as Π), cf. [**BST**] or [**BV.1**], (9.8), and obviously $\operatorname{Gr}_I M$ is an MSL on $\mathcal{X}^*$.

(d) Let $A = B[X]/I_{r+1}(X)$ as in (1.4), (b), $0 \le r \le \min(m,n)$ (so $A = B[X]$ is included). The matrix $\overline{X}$ over A whose entries are the residue classes of the indeterminates defines a map $A^m \to A^n$, also denoted by $\overline{X}$. The modules $\operatorname{Im}\overline{X}$ and $\operatorname{Coker}\overline{X}$ have been investigated in [**Br.1**]. A simplified treatment has been given in [**BV.1**], Section 13, from where we draw some of the arguments below. Let $d_1, \dots, d_m$ and $e_1, \dots, e_n$ denote the canonical bases of A^m and A^n. Then we order the system $\overline{e}_1, \dots, \overline{e}_n$ of generators of $\operatorname{Coker}\overline{X}$ linearly by

$$\overline{e}_1 > \cdots > \overline{e}_n.$$

Furthermore we put

$$\mathcal{I}(\overline{e}_i) = \begin{cases} \{\, \delta \in \Delta_r(X) : \delta \not\geq [1,\dots,r|1,\dots,\widehat{i},\dots,r+1] \,\} & \text{for } i \le r, \\ \emptyset & \text{else,} \end{cases}$$

if $r < n$, and in the case in which $r = n$

$$\mathcal{I}(\overline{e}_i) = \{\, \delta \in \Delta_r(X) : \delta \not\geq [1,\dots,r-1|1,\dots,\widehat{i},\dots,r] \,\}.$$

(where $\widehat{i}$ denotes that i is to be omitted). We claim: $\operatorname{Coker}\overline{X}$ is an MSL with respect to these data.

Suppose that $\delta \in \mathcal{I}(\overline{e}_i)$. Then

$$\delta = [a_1, \dots, a_s|1, \dots, i, b_{i+1}, \dots, b_s], \qquad s \le r.$$

The element

$$\sum_{j=1}^{s} (-1)^{j+i} [a_1, \dots, \widehat{a_j}, \dots, a_s|1, \dots, i-1, b_{i+1}, \dots, b_s] \overline{X}(d_{a_j})$$

of $\operatorname{Im}\overline{X}$ is a suitable relation for (MSL-2):

$$\delta \overline{e}_i = \sum_{k=i+1}^{n} \pm [a_1, \dots, a_s|1, \dots, i-1, k, b_{i+1}, \dots, b_s] \overline{e}_k. \tag{1}$$

Rearranging the column indices $1,\dots,i-1,k,b_{i+1},\dots,b_s$ in ascending order one makes (1) the standard representation of $\delta\overline{e}_i$, and observes the following fact recorded for later purpose:

(2) $\delta \notin \mathcal{I}(\overline{e}_k)$ for all $k \geq i+1$ such that

$$[a_1,\dots,a_s|1,\dots,i-1,k,b_{i+1},\dots,b_s] \neq 0.$$

In order to prove the linear independence of the standard elements one may assume that $r < n$ since $\mathrm{I}_n(X)$ annihilates M. Let

$$\widetilde{M} = \sum_{i=r+1}^{n} A\overline{e}_i, \quad \Psi = \{\, \delta \in \Delta_r(X)\colon \delta \not\geq [1,\dots,r|1,\dots,r-1,r+1] \,\}$$

and $I = A\Psi$.

We claim:
(i) $\widetilde{M}$ is a free A-module.
(ii) $M/\widetilde{M}$ is (over A/I) isomorphic to the ideal generated by the minors $[1,\dots,r|1,\dots,\widehat{i},\dots,r+1]$, $1 \leq i \leq r$, in A/I.
In fact, the minors just specified form a linearly ordered ideal in the poset $\Delta_r(X) \setminus \Psi$ underlying the ASL A/I, and the linear independence of the standard elements follows immediately from (i) and (ii).

Statement (i) simply holds since $\operatorname{rank} \overline{X} = r$, and the $r \times r$-minor in the left upper corner of $\overline{X}$, being the minimal element of $\Delta_r(X)$, is not a zero-divisor in A. For (ii) one applies (4.5) below to show that $M/\widetilde{M}$ and the ideal in (ii) have the same representation given by the matrix

$$\begin{pmatrix} \overline{X}_{11} & \dots & \overline{X}_{1r} \\ \vdots & & \vdots \\ \overline{X}_{m1} & \dots & \overline{X}_{mr} \end{pmatrix},$$

the entries taken in A/I: The assignment $\overline{e}_i \to (-1)^{i+1}[1,\dots,r|1,\dots,\widehat{i},\dots,r+1]$ induces the isomorphism. The computations needed for the application of (4.5) are covered by (1).

By similar arguments one can show that $\operatorname{Im}\overline{X}$ is also an MSL, see [**BV.1**], proof of (13.6) where a filtration argument is given which shows the linear independence of the standard elements. Such a filtration argument could also have been applied to prove (MSL-1) for M.

(e) Another example is furnished by the modules defined by generic alternating maps. Recalling the notations of (1.4), (d) we let $A = B[X]/\operatorname{Pf}_{r+2}(X)$ and M be the cokernel of the linear map

$$\overline{X}\colon F \longrightarrow F^*, \qquad F = A^n.$$

In complete analogy with the preceding example M is an MSL on $\{\overline{e}_1,\dots,\overline{e}_n\}$, the canonical basis of F^*, $\overline{e}_1 > \cdots > \overline{e}_n$, if one puts

$$\mathcal{I}(\overline{e}_i) = \begin{cases} \{\pi \in \Phi_r(X) : \pi \not\geq [1,\dots,\widehat{i},\dots,r+1]\} & \text{for } i \leq r, \\ \emptyset & \text{else,} \end{cases}$$

if $r < n$, and in the case in which $r = n$

$$\mathcal{I}(\overline{e}_i) = \begin{cases} \{\pi \in \Phi(X) : \pi \not\geq [1,\dots,\widehat{i},\dots,r-1]\} & \text{for } i \leq n-1, \\ \{[1,\dots,n]\} & \text{for } i = n. \end{cases}$$

The straightening law (1) is replaced by the equation

$$(1') \qquad \pi\overline{e}_i = \sum_{k=i+1}^{n} \pm[1,\dots,i-1,k,b_{i+1},\dots,b_s]\,\overline{e}_k,$$

obtained from Laplace type expansion of pfaffians as (1) has been derived from Laplace expansion of minors. Observe that the analogue (2′) of (2) is satisfied. The linear independence of the standard elements is proved in entire analogy with (d).

A notable special case is n odd, $r = n-1$. In this case $\operatorname{Coker} X \cong \operatorname{Pf}_r(X)$ is an ideal of projective dimension 2 [**BE**] and generated by a linearly ordered poset ideal in $\Phi(X)$. —

The following proposition helps to detect further MSLs:

(4.3) PROPOSITION. *Let M, M_1, M_2 be modules over an ASL A, connected by an exact sequence*

$$0 \longrightarrow M_1 \longrightarrow M \longrightarrow M_2 \longrightarrow 0.$$

Let M_1 and M_2 be MSLs on $\mathcal{X}_1$ and $\mathcal{X}_2$, and choose a splitting f of the epimorphism $M \to M_2$ over B. Then M is an MSL on $\mathcal{X} = \mathcal{X}_1 \cup f(\mathcal{X}_2)$ ordered by $x_1 < f(x_2)$ for all $x_1 \in \mathcal{X}_1$, $x_2 \in \mathcal{X}_2$, and the given partial orders on $\mathcal{X}_1$ and the copy $f(\mathcal{X}_2)$ of $\mathcal{X}_2$. Moreover one chooses $\mathcal{I}(x)$, $x \in \mathcal{X}_1$, as in M_1 and $\mathcal{I}(f(x)) = \mathcal{I}(x)$ for all $x \in \mathcal{X}_2$.

The proof is straightforward and can be left to the reader.

(4.4) EXAMPLE. The preceding proposition helps to supplement (4.2),(c). Under the hypotheses there one has that M/I^nM is an MSL for all $n \geq 1$.

It has been stated in (4.2),(c) that all the quotients $I^nM/I^{n+1}M$ are MSLs, and therefore we may argue inductively by the exact sequence

$$0 \longrightarrow I^nM/I^{n+1}M \longrightarrow M/I^{n+1}M \longrightarrow M/I^nM \longrightarrow 0.$$

In particular one has that A/I^n is an MSL over A (though it is not an ASL for $n \geq 2$). —

In terms of generators and relations an ASL is defined by its generating poset and its straightening relations, cf. (1.1). This holds similarly for MSLs:

(4.5) PROPOSITION. *Let A be an ASL on Π over B, and M an MSL on $\mathcal{X}$ over A. Let e_x, $x \in \mathcal{X}$, denote the elements of the canonical basis of the free module $A^{\mathcal{X}}$. Then the kernel $K_{\mathcal{X}}$ of the natural epimorphism*

$$A^{\mathcal{X}} \longrightarrow M, \qquad e_x \longrightarrow x,$$

is generated by the relations required for (MSL-2):

$$\rho_{\xi x} = \xi e_x - \sum_{y<x} a_{\xi xy} e_y, \qquad x \in \mathcal{X},\ \xi \in \mathcal{I}(x).$$

PROOF: We use the induction principle indicated in (4.1), (a). Let $\widetilde{x} \in \mathcal{X}$ be a maximal element. Then $\mathcal{T} = \mathcal{X} \setminus \{\widetilde{x}\}$ is an ideal. By induction $A\mathcal{T}$ is defined by the relations $\rho_{\xi x}$, $x \in \mathcal{T}$, $\xi \in \mathcal{I}(x)$. Furthermore (MSL-1) and (MSL-2) imply

$$M/A\mathcal{T} \cong A/A\mathcal{I}(\widetilde{x}) \tag{3}$$

If $a_{\widetilde{x}}\widetilde{x} - \sum_{y\in\mathcal{T}} a_y y = 0$, one has $a_{\widetilde{x}} \in A\mathcal{I}(\widetilde{x})$ and subtracting a linear combination of the elements $\rho_{\xi\widetilde{x}}$ from $a_{\widetilde{x}}e_{\widetilde{x}} - \sum_{y\in\mathcal{T}} a_y e_y$ one obtains a relation of the elements $y \in \mathcal{T}$ as desired. —

The kernel of the epimorphism $A^{\mathcal{X}} \rightarrow M$ is again an MSL:

(4.6) PROPOSITION. *With the notations and hypotheses of (4.5) the kernel $K_{\mathcal{X}}$ of the epimorphism $A^{\mathcal{X}} \rightarrow M$ is an MSL if we let*

$$\mathcal{I}(\rho_{\xi x}) = \{\pi \in \Pi \colon \pi \not\geq \xi\}$$

and

$$\rho_{\xi x} \leq \rho_{vy} \quad \Longleftrightarrow \quad x < y \quad \text{or} \quad x = y,\ \xi \leq v.$$

PROOF: Choose $\widetilde{x}$ and $\mathcal{T}$ as in the proof of (4.5). By virtue of (4.5) the projection $A^{\mathcal{X}} \to Ae_{\widetilde{x}}$ with kernel $A^{\mathcal{T}}$ induces an exact sequence

$$0 \longrightarrow K_{\mathcal{T}} \longrightarrow K_{\mathcal{X}} \longrightarrow A\mathcal{I}(\widetilde{x}) \longrightarrow 0.$$

Now (4.3) and induction finish the argument. —

If a module M is given in terms of generators and relations, it is in general more difficult to establish (MSL-1) than (MSL-2). For (MSL-2) one "only" has to show that elements $\rho_{\xi x}$ as in the proof of (4.5) can be obtained as linear combinations of the given relations. In this connection the following proposition may be useful: it is enough that the module generated by the $\rho_{\xi x}$ satisfies (MSL-2) again.

(4.7) PROPOSITION. *Let the data $M, \mathcal{X}, \mathcal{I}(x), x \in \mathcal{X}$, be given as in the definition, and suppose that* (MSL-2) *is satisfied. Suppose that the kernel $K_{\mathcal{X}}$ of the natural epimorphism $A^{\mathcal{X}} \to M$ is generated by the elements $\rho_{\xi x} \in A^{\mathcal{X}}$ representing the relations in* (MSL-2). *Order the $\rho_{\xi x}$ and choose $\mathcal{I}(\rho_{\xi x})$ as in* (4.6). *If $K_{\mathcal{X}}$ satisfies* (MSL-2) *again, M is an MSL.*

PROOF: Let $\widetilde{x} \in \mathcal{X}$ be a maximal element, $\mathcal{T} = \mathcal{X} \setminus \{\widetilde{x}\}$. We consider the induced epimorphism

$$A^{\mathcal{T}} \longrightarrow A\mathcal{T}$$

with kernel $K_{\mathcal{T}}$. One has $K_{\mathcal{T}} = K_{\mathcal{X}} \cap A^{\mathcal{T}}$. Since the $\rho_{\xi x}$ satisfy (MSL-2), every element in $K_{\mathcal{X}}$ can be written as a B-linear combination of standard elements, and only the $\rho_{\xi \widetilde{x}}$ have a nonzero coefficient with respect to $e_{\widetilde{x}}$. The projection onto the component $Ae_{\widetilde{x}}$ with kernel $A^{\mathcal{T}}$ shows that $K_{\mathcal{T}}$ is generated by the $\rho_{\xi x}$, $x \in \mathcal{T}$. Now one can argue inductively, and the split-exact sequence

$$0 \longrightarrow A\mathcal{T} \longrightarrow M \longrightarrow M/A\mathcal{T} \cong A/A\mathcal{I}(\widetilde{x}) \longrightarrow 0$$

of B-modules finishes the proof. —

Modules with a straightening law have a distinguished filtration with cyclic quotients; by the usual induction this follows immediately from the isomorphism (3) above:

(4.8) PROPOSITION. *Let M be an MSL on $\mathcal{X}$ over A. Then M has a filtration $0 = M_0 \subset M_1 \subset \cdots \subset M_n = M$ such that each quotient M_{i+1}/M_i*

is isomorphic with one of the residue class rings $A/A\mathcal{I}(x)$, $x \in \mathcal{X}$, and conversely each such residue class ring appears as a quotient in the filtration.

As a consequence one can bound the depth of an MSL (cf. Section 1 for the definition of depth in this context).

(4.9) COROLLARY. *Let M be an MSL on $\mathcal{X}$ over A. Then*

$$\operatorname{depth} M \geq \min\{\operatorname{depth} A/A\mathcal{I}(x) \colon x \in \mathcal{X}\}.$$

We specialize to ASLs over wonderful posets (cf. [**Ei**], [**DEP.2**], or [**BV.1**] for this notion and the properties of ASLs over wonderful posets).

(4.10) COROLLARY. *Let A be an ASL on the wonderful poset Π.*
(a) *If M is an MSL on $\mathcal{X}$ over A, then*

$$\operatorname{depth} M \geq \min\{\operatorname{rk}\Pi - \operatorname{rk}\mathcal{I}(x) \colon x \in \mathcal{X}\}.$$

(b) *Let $\Psi \subset \Pi$ be an ideal. Then*

$$\operatorname{depth} A/A\Psi \geq \operatorname{rk}\Pi - \operatorname{rk}\Psi.$$

(c) *Suppose furthermore that $I = A\Psi$ is straightening-closed. Then*

$$\operatorname{depth} A/I^n \geq \operatorname{rk}\Pi - \operatorname{rk}\Psi \qquad \textit{for all} \quad n \geq 1.$$

PROOF: In (b) and (c) $A\Psi$ and I^n resp. are MSLs on a certain poset $\mathcal{X}$, cf. (4.2),(b) and (c) above. In both cases one has

$$\mathcal{I}(x) = \{\pi \in \Pi \colon \pi \not\geq \psi\} \qquad \text{for some} \quad \psi \in \Psi$$

for all $x \in \mathcal{X}$. $\Pi \setminus \mathcal{I}(x)$ is wonderful again (cf. [**DEP.2**], 8.2 or [**BV.1**], (5.13)) and therefore

$$\operatorname{depth} A/A\mathcal{I}(x) \geq \operatorname{rk}\Pi - \operatorname{rk}\psi + 1 \geq \operatorname{rk}\Pi - \operatorname{rk}\Psi + 1$$

by virtue of [**DEP.2**], 8.1. Now one applies (4.9) and switches from $A\Psi$ and I^n to the residue class rings. Part (a) finally follows from (4.9) and (b). —

Of course the inequalities (4.9) and (4.10) can be improved in many cases. For example, $A/A\Psi$ may be a Cohen-Macaulay ring. On the other hand there is a class of ideals I such that one has equality in (4.10),(c) for $n \gg 0$, cf. [**BV.1**], (9.22). The depth of the generic modules (4.2),(d) and (e) has been determined in [**BV.1**], Section 13 and [**BV.2**] resp. using the fact that, with the notations of (4.2),(d), the depth of $M/\widetilde{M}$ can be computed exactly.

Further consequences concern the annihilator, the localizations with respect to prime ideals $P \in \operatorname{Ass} A$, and the rank of an MSL.

(4.11) PROPOSITION. *Let M be an MSL on $\mathcal{X}$ over A, and*

$$J = A(\bigcap_{x \in \mathcal{X}} \mathcal{I}(x)).$$

Then

$$J \supset \operatorname{Ann} M \supset J^n, \qquad n = \operatorname{rk} \mathcal{X}.$$

PROOF: Note that $A(\bigcap \mathcal{I}(x)) = \bigcap A\mathcal{I}(x)$ (as a consequence of (1.2)). Since $\operatorname{Ann} M$ annihilates every subquotient of M, the inclusion $\operatorname{Ann} M \subset J$ follows from (4.8). Furthermore (MSL-2) implies inductively that

$$J^i M \subset \sum_{\operatorname{rk} x \leq \operatorname{rk} \Pi - i} Ax$$

for all i, in particular $J^n M = 0$. —

(4.12) PROPOSITION. *Let M be an MSL on $\mathcal{X}$ over A, and $P \in \operatorname{Ass} A$.*
(a) Then $\{\pi \in \Pi : \pi \notin P\}$ has a single minimal element σ, and σ is also a minimal element of Π.
(b) Let $\mathcal{Y} = \{x \in \mathcal{X} : \sigma \notin \mathcal{I}(x)\}$. Then $\mathcal{Y}$ is a basis of the free A_P-module M_P. Furthermore $(K_{\mathcal{X}})_P$ is generated by the elements $\varrho_{\sigma x}$, $x \notin \mathcal{Y}$.

PROOF: (a) If π_1, π_2, $\pi_1 \neq \pi_2$, are minimal elements of $\{\pi \in \Pi : \pi \notin P\}$, then, by (ASL-2), $\pi_1, \pi_2 \in P$. So there is a single minimal element σ. It has to be a single minimal element of Π, too, since otherwise P would contain all the minimal elements of Π whose sum, however, is not zero-divisor in A ([**BV.1**], (5.11)).

(b) Consider the exact sequence

$$0 \longrightarrow A\mathcal{T} \longrightarrow M \longrightarrow A/A\mathcal{I}(\widetilde{x}) \longrightarrow 0$$

introduced in the proof of (4.5). If $\widetilde{x} \notin \mathcal{Y}$, then $\widetilde{x} \in A_P\mathcal{T}$ by the relation $\varrho_{\sigma\widetilde{x}}$, and we are through by induction. If $\widetilde{x} \in \mathcal{Y}$, then σ and all the elements of $\mathcal{I}(\widetilde{x})$ are incomparable, so they are annihilated by σ (because of (ASL-2)). Consequently $(A/A\mathcal{I}(\widetilde{x}))_P \cong A_P$, $\widetilde{x}$ generates a free summand of M_P, and induction finishes the argument again. —

We say that a module M over A has rank r if $M \otimes L$ is free of rank r as an L-module, L denoting the total ring of fractions of A. Cf. [**BV.1**], 16.A for the properties of this notion.

(4.13) COROLLARY. *Let M be an MSL on $\mathcal{X}$ over the ASL A on Π. Suppose that Π has a single minimal element π, a condition satisfied if A is a domain. Then*

$$\operatorname{rank} M = |\{x \in \mathcal{X} : \mathcal{I}(x) = \emptyset\}|.$$

5. Modules with a Strict Straightening Law.

Some MSLs satisfy further natural axioms which strengthen (MSL-1) and (MSL-2). Let M be an MSL on $\mathcal{X}$ over A. The first additional axiom:

(MSL-3) For all $x, y \in \mathcal{X}$: $\quad x < y \Rightarrow \mathcal{I}(x) \subset \mathcal{I}(y)$.

The property (MSL-3) implies that $\Pi \cup \mathcal{X}$ is a partially ordered set if we order its subsets Π and $\mathcal{X}$ as given and all other relations are given by

$$x < \xi \qquad \Longleftrightarrow \qquad \xi \notin \mathcal{I}(x).$$

(MSL-3) simply guarantees transitivity. If it is satisfied, one can consider the following strengthening of (MSL-2):

(MSL-4) $\xi x = \sum_{y<x,\xi} a_{\xi xy} y \quad$ for all $x \in \mathcal{X}$, $\xi \in \mathcal{I}(x)$.

DEFINITION. We say that M has a *strict straightening law* if it is an MSL satisfying (MSL-3) and (MSL-4).

An ideal $I \subset A$ generated by an ideal $\Psi \subset \Pi$ is a trivial example of a module with a strict straightening law, and the generic modules (4.2),(d) and (e) may be considered significant examples. On the other hand not every MSL has a strict straightening law. The following proposition which strengthens (4.11) excludes all the modules $M/I^n M$, $n \geq 2$, as in (4.4), in particular the residue class rings $A/I^n A$, $n \geq 2$, $I = A\Psi$ straightening-closed.

(5.1) PROPOSITION. *Let M be a module with a strict straightening law on $\mathcal{X}$ over A. Then*

$$\operatorname{Ann} M = A(\bigcap_{x \in \mathcal{X}} \mathcal{I}(x)).$$

PROOF: In fact, if $\xi \in \bigcap \mathcal{I}(x)$, then $\xi x = 0$ for all $x \in \mathcal{X}$, since there is no element $y \in \mathcal{X}$, $y < \xi$. —

Suppose that $\mathcal{X}$ is linearly ordered. Then the straightening laws (MSL-4) and (ASL-2) constitute a set of straightening relations on $\Pi \cup \mathcal{X}$, and the following question suggests itself: Is the symmetric algebra $\mathrm{S}(M)$ an ASL over B? In general the answer is "no", as the following example demonstrates: $A = B[X_1, X_2, X_3]$, $X_1 < X_2 < X_3$,

$$M = A^3/(A(X_1, 0, 0) + A(X_2, 0, 0) + A(0, X_1, X_3)),$$

the residue classes of the canonical basis ordered by $\bar{e}_1 > \bar{e}_2 > \bar{e}_3$. On the other hand $\mathrm{S}(I)$ is an ASL if I is generated by a linearly ordered poset ideal, cf. [**BV.1**], (9.13) or [**BST**]; one uses that the Rees algebra $\mathcal{R}(I)$ of A with respect to I is an ASL, and concludes easily that the natural epimorphism $\mathrm{S}(I) \to \mathcal{R}(I)$ is an isomorphism. We will give a new proof of this fact below.

The following proposition may not be considered ultima ratio, but it covers the case just discussed and also the generic modules.

(5.2) PROPOSITION. *Let M be a graded module with strict straightening law on the linearly ordered set $\mathcal{X} = \{x_1, \dots, x_n\}$, $x_1 < \cdots < x_n$. Put $\mathcal{X}_i = \{x_1, \dots, x_i\}$, $M_i = A\mathcal{X}_i$, $\overline{M}_{i+1} = M/M_i$, $i = 0, \dots, n$. Suppose that for all $j > i$ and all prime ideals $P \in \operatorname{Ass}(A/A\mathcal{I}(x_j))$ the localization $(\overline{M}_i)_P$ is a free $(A/A\mathcal{I}(x_i))_P$-module, $i = 1, \dots, n$.*
(a) Then $\mathrm{S}(M)$ is an ASL on $\Pi \cup \mathcal{X}$.
(b) If $\mathcal{I}(x_1) = \emptyset$, then $\mathrm{S}(M)$ is a torsionfree A-module.

PROOF: Since $\Pi \cup \mathcal{X}$ generates $\mathrm{S}(M)$ as a B-algebra (and $\mathrm{S}(M)$ is a graded B-algebra in a natural way) and (ASL-2) is obviously satisfied, it remains to show that the standard monomials containing k factors from $\mathcal{X}$ are linearly independent for all $k \geq 0$. Since $\mathrm{S}^0(M) = A$ this is obviously true for $k = 0$, and it remains true if $\operatorname{Ann} M = A\mathcal{I}(x_1)$ is factored out; since this does not affect the symmetric powers $\mathrm{S}^k(M)$, $k > 0$, we may assume that $\operatorname{Ann} M = 0$. If $n = 1$, then M is now a free A-module and the contention holds for trivial reasons.

The hypotheses indicate that an inductive argument is in order. Independent of the special assumptions on M_i and $\mathcal{I}(x_i)$ there is an exact sequence

$$\mathrm{S}^k(M) \xrightarrow{\ g\ } \mathrm{S}^{k+1}(M) \xrightarrow{\ f\ } \mathrm{S}^{k+1}(M/Ax_1) \longrightarrow 0 \tag{5}$$

in which f is the natural epimorphism and g is the multiplication by x_1. Let $P \in \operatorname{Ass} A$. By (4.12) x_1 generates a free direct summand of M_P. Therefore (5) splits over A_P, and $g \otimes A_P$ is injective. It is now enough to show that $\mathrm{S}^k(M)$ is torsionfree; then g is injective itself and (5) splits as a sequence of B-modules as desired: By induction the standard elements in $\mathrm{S}^k(M)$ as well as in $\mathrm{S}^{k+1}(M/Ax_1)$ are linearly independent.

The linear independence of the standard elements in $\mathrm{S}^k(M)$ implies that $\mathrm{S}^k(M)$ is an MSL over A on the set of monomials of length k in $\mathcal{X}$ with respect to a suitable partial order and the choice

$$\mathcal{I}(x_{i_1}\cdots x_{i_k})=\mathcal{I}(x_{i_k}),\qquad i_1\le\cdots\le i_k.$$

Let $P\in\operatorname{Spec}A$, $P\notin\operatorname{Ass}A$. Then $P\notin\operatorname{Ass}(A/A\mathcal{I}(x_1))$, since $\mathcal{I}(x_1)=\emptyset$ by assumption. If $P\notin\operatorname{Ass}(A/A\mathcal{I}(x_j))$ for all $j=2,\dots,n$, then $P\notin\operatorname{Ass}\mathrm{S}^k(M)$ by virtue of (4.8); otherwise $\mathrm{S}^k(M)_P$ is a free A_P-module by hypothesis. Altogether: $\operatorname{Ass}\mathrm{S}^k(M)=\operatorname{Ass}A$, and $\mathrm{S}^k(M)$ is torsionfree. —

(5.3) COROLLARY. *With the notations and hypotheses of (5.2), the symmetric algebra* $\mathrm{S}(M_i)$ *is an ASL on* $\Pi\cup\mathcal{X}_i$ *for all* $i=1,\dots,n$. $\mathrm{S}(M_i)$ *is a sub-ASL of* $\mathrm{S}(M)$ *in a natural way.*

PROOF: There is a natural homomorphism $\mathrm{S}(M_i)\to\mathrm{S}(M)$ induced by the inclusion $M_i\to M$. Since $\mathrm{S}(M_i)$ satisfies (ASL-2), it is generated as a B-module by the standard monomials in $\Pi\cup\mathcal{X}_i$. Since these standard monomials are linearly independent in $\mathrm{S}(M)$, they are linearly independent in $\mathrm{S}(M_i)$, too, and $\mathrm{S}(M_i)$ is a subalgebra of $\mathrm{S}(M)$. —

The following corollary has already been mentioned:

(5.4) COROLLARY. *Let* A *be an ASL on* Π, *and* $\Psi\subset\Pi$ *a linearly ordered ideal. Then* $\mathrm{S}(A\Psi)$ *is an ASL on the disjoint union of* Π *and* Ψ.

PROOF: For each $\psi\in\Psi$ the poset $\Pi\setminus\mathcal{I}(\psi)$ has ψ as its single minimal element. Let $\Psi=\{\psi_1,\dots,\psi_n\}$, $\psi_1<\cdots<\psi_n$. If $P\in\operatorname{Ass}(A/A\mathcal{I}(\psi_j))$, then $\psi_j\notin P$ since ψ_j is not a zero-divisor of the ASL $A/A\mathcal{I}(\psi_j)$. Consequently $(A\Psi/(\sum_{k=1}^{i}A\psi_k))_P$ is isomorphic to $(A/\mathcal{I}(\psi_i))_P$ for all $i<j$. —

We want to apply (5.2) to the generic modules discussed in (4.2), (d), and recall the notations introduced there: $A=B[X]/\mathrm{I}_{r+1}(X)$ is an ASL on $\Delta_r(X)$, the set of all i-minors, $i\le r$, of X. M is the cokernel of the map $A^m\to A^n$ defined by the matrix $\overline{X}$, $\overline{e}_1,\dots,\overline{e}_n$ are the residue classes of the canonical basis $e_1,\dots,e_n$ of A^n. (Thus M_k is the submodule of M generated by $\overline{e}_{n-k+1},\dots,\overline{e}_n$.)

(5.5) COROLLARY. (a) *With the notations just recalled, the symmetric algebra of a generic module* M *is an ASL. If* $r+1\le n$, $\mathrm{S}(M)$ *is torsionsfree over* A.
(b) *Let* B *be a Cohen-Macaulay ring.* $\mathrm{S}(M)$ *is Cohen-Macaulay if and only if* $r+1\le n$ *or* $r=m=n$.

PROOF: (a) Factoring out the ideal generated by $\mathcal{I}(\overline{e}_n)$ we may suppose that $r<n$. Note that with the notations introduced in (4.2),(d) one has

$\overline{e}_n < \cdots < \overline{e}_1$. Because of statement (ii) in (4.2),(d) the validity of the hypothesis of (5.2) for $i \ge n-r+1$ follows from the proof of (5.4).

Let $i \le n-r$, $j > i$, $k = n-j+1$, $\delta = [1,\ldots,r|1,\ldots,r]$ for $k \ge r+1$ and $\delta = [1,\ldots,r|1,\ldots,\widehat{k},\ldots,r+1]$ for $k \le r$. Then δ is the minimal element of the poset underlying $A/\mathcal{I}(\overline{x}_j) = A/\mathcal{I}(\overline{e}_k)$, thus not contained in an associated prime ideal of the latter. On the other hand $(\overline{M}_i)_P$ is free for every prime P not containing δ.

(b) in order to form the poset $\Pi \cup \{\overline{e}_1,\ldots,\overline{e}_n\}$ one attaches $\{\overline{e}_1,\ldots,\overline{e}_n\}$ to Π as indicated by the following diagrams for the cases $r+1 \le n$ and $r = m = n$ resp. In the first case we let $\delta_i = [1,\ldots,r|1,\ldots,\widehat{i},\ldots,r+1]$, in the second $\delta_i = [1,\ldots,r-1|1,\ldots,\widehat{i},\ldots,r]$.

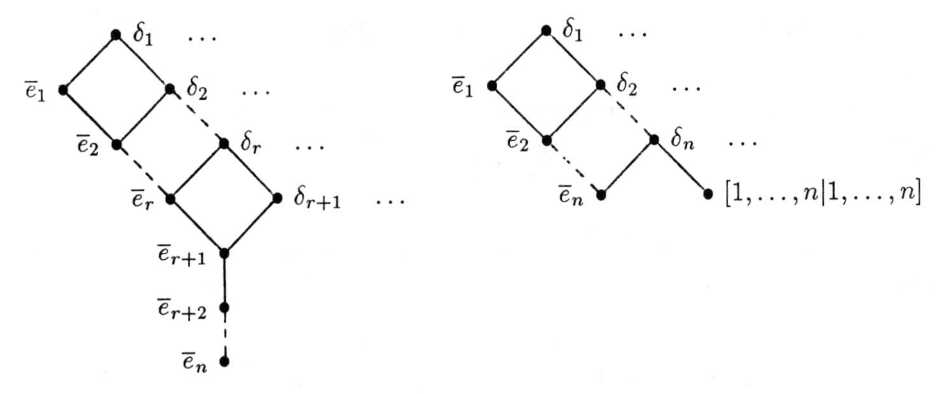

It is an easy exercise to show that $\Pi \cup \{\overline{e}_1,\ldots,\overline{e}_n\}$ and $\Pi \cup \{\overline{e}_{n-k+1},\ldots,\overline{e}_n\}$ are wonderful, implying the Cohen-Macaulay property for ASL's defined on the poset ([**BV.1**], Section 5 or [**DEP.2**]).

In the case in which $m > n = r$, the ideal $\mathrm{I}_n(X)\,\mathrm{S}(M)$ annihilates $\bigoplus_{i>0} \mathrm{S}^i(M)$, and $\dim \mathrm{S}(M)/\mathrm{I}_n(X) < \dim \mathrm{S}(M)$ by (1.3), excluding the Cohen-Macaulay property. —

Admittedly the preceding corollary is not a new result. In fact, let Y be an $n \times 1$ matrix of new indeterminates. Then

$$\mathrm{S}(M) \cong B[X,Y]/(\mathrm{I}_{r+1}(X) + \mathrm{I}_1(XY))$$

can be regarded as the coordinate ring of a variety of complexes, which has been shown to be a Hodge algebra in [**DS**]. The results of [**DS**] include part (b) of (5.5) as well as the fact that $\mathrm{S}(M)$ is a (normal) domain if $r+1 \le n$ and B is a (normal) domain. The divisor classgroup of $\mathrm{S}(M)$ in case $r+1 \le b$, B normal, has been computed in [**Br.2**]: $\mathrm{Cl}(\mathrm{S}(M)) = \mathrm{Cl}(B)$ if

$m = r < n-1$, $\mathrm{Cl}(\mathrm{S}(M)) = \mathrm{Cl}(B) \oplus \mathbb{Z}$ else. The algebras $\mathrm{S}(M)$, in particular for the cases $r+1 > \min(m, n)$, i.e. $A = B[X]$, and $r+1 = \min(m, n)$, have received much attention in the literature, cf. [**Av.2**], [**BE**], [**BKM**], and the references given there. Note that (5.5) also applies to the subalgebras $\mathrm{S}(M_k)$. In the case $A = B[X]$, $m \le n$, these rings have been analyzed in [**BS**].

The analogue (5.6) of (5.5) seems to be new however. We recall the notations of (4.2), (e): X is an alternating $n \times n$-matrix of indeterminates, $A = B[X]/\mathrm{Pf}_{r+2}(X)$, $F = A^n$, $\overline{X}\colon F \to F^*$ given by the residue class of X, and $M = \operatorname{Coker} \overline{X}$.

(5.6) COROLLARY. (a) *With the notations just recalled, the symmetric algebra of an "alternating" generic module M is an ASL. If $r < n$, $\mathrm{S}(M)$ is a torsionfree A-module.*
(b) *Let B be a Cohen-Macaulay ring. Then $\mathrm{S}(M)$ is Cohen-Macaulay if and only if $r < n$.*
(c) *Let B be a (normal) domain. Then $\mathrm{S}(M)$ is a (normal) domain if and only if $r < n$.*
(d) *Let B be normal and $r < n$. Then $\mathrm{Cl}(\mathrm{S}(M)) \cong \mathrm{Cl}(B) \oplus \mathbb{Z}$ if $r = n-1$, and $\mathrm{Cl}(\mathrm{S}(M)) \cong \mathrm{Cl}(B)$ if $r < n-1$. In particular $\mathrm{S}(M)$ is factorial if $r < n-1$ and B is factorial.*

PROOF: (a) and (b) are proved in the same way as (5.5).

Standard arguments involving flatness reduce (c) to the case in which B is a field (cf. [**BV.1**], Section 3 for example). Thus we may certainly suppose that B is a normal domain.

In the case in which $r = n-1$ the module M is just $I = \mathrm{Pf}_{n-1}(X)$ as remarked above, an ideal generated by a linearly ordered poset ideal. Then (i) $\mathrm{Gr}_I A$ is an ASL, in particular reduced, and (ii) $\mathrm{S}(M)$ is the Rees algebra of A with respect to I (cf. [**BST**] for example). Thus we can apply the main result of [**HV**] to conclude (c) and (d).

Let $r \le n-2$ now. In the spirit of this paper a "linear" argument seems to be most appropriate: By [**Fo**], Theorem 10.11 and [**Av.1**] it is sufficient that all the symmetric powers of M are reflexive. Since M_P, hence $\mathrm{S}^k(M_P)$ is free for prime ideals $P \not\supset \mathrm{Pf}_r(\overline{X})$ it is enough to show that $\mathrm{Pf}_r(\overline{X})$ contains an $\mathrm{S}^k(M)$-sequence of length 2 for every k. Each $\mathrm{S}^k(M)$ is an MSL whose data $\mathcal{I}(\dots)$ coincide with those of M itself. Therefore (4.8) can be applied and we can replace the $\mathrm{S}^k(M)$ by the residue class rings A/I_i, $I_i = A\{\pi \in \Phi_r(x)\colon \pi \not\geq [1, \dots, \widehat{i}, \dots, r+1]\}$, $i = 1, \dots, r$. One has

$\mathrm{Pf}_r(X) \supset I_i$.

The poset Π underlying A/I_i is wonderful (cf. [**DEP.2**], Lemma 8.2 or [**BV.1**], (5.13)). Therefore the elements

$$[1,\ldots,\widehat{i},\ldots,r+1] = \sum_{\substack{\pi\in\Pi \\ rk\,\pi=1}} \pi \qquad \text{and} \qquad \sum_{\substack{\pi\in\Pi \\ rk\,\pi=2}} \pi$$

form an A/I_i-sequence by [**DEP.2**], Theorem 8.1. Both these elements are contained in $\mathrm{Pf}_r(\overline{X})$.

References

[**Av.1**] L. L. Avramov, *A class of factorial domains*, Serdica **5** (1979), 378–379.

[**Av.2**] L. L. Avramov, *Complete intersections and symmetric algebras*, J. Algebra **73** (1981), 248–263.

[**Br.1**] W. Bruns, *Generic maps and modules*, Compos. Math. **47** (1982), 171–193.

[**Br.2**] W. Bruns, *Divisors on varieties of complexes*, Math. Ann. **264** (1983), 53–71.

[**BKM**] W. Bruns, A. Kustin and M. Miller, *The resolution of the generic residual intersection of a complete intersection*, (to appear).

[**BS**] W. Bruns and A. Simis, *Symmetric algebras of modules arising from a fixed submatrix of a generic matrix.*, J. Pure Appl. Algebra **49** (1987), 227-245.

[**BST**] W. Bruns, A. Simis and Ngô Việt Trung, *Blow-up of straightening closed ideals in ordinal Hodge algebras*, (to appear).

[**BV.1**] W. Bruns and U. Vetter, "Determinantal rings," Springer Lect. Notes Math. **1327**, 1988.

[**BV.2**] W. Bruns and U. Vetter, *Modules defined by generic symmetric and alternating maps*, (to appear).

[**BE**] D. Buchsbaum and D. Eisenbud, *Algebra structures for finite free resolutions, and some structure theorems for ideals of codimension 3*, Amer. J. Math. **99** (1977), 447-485.

[**CN**] R. Cowsik and M. V. Nori, *On the fibers of blowing up*, J. Indian Math. Soc. **40** (1976), 217–222.

[**DEP.1**] C. De Concini, D. Eisenbud and C. Procesi, *Young diagrams and determinantal varieties*, Invent. Math. **56** (1980), 129–165.

[**DEP.2**] C. De Concini, D. Eisenbud and C. Procesi, "Hodge algebras," Astérisque **91**, 1982.

[**DS**] C. De Concini and E. Strickland, *On the variety of complexes*, Adv. Math. **41** (1981), 57–77.

[**Ei**] D. Eisenbud, *Introduction to algebras with straightening laws*, in "Ring Theory and Algebra III," M. Dekker, New York and Basel, 1980, pp. 243–267.

[**EH**] D. Eisenbud and C. Huneke, *Cohen-Macaulay Rees algebras and their specializations*, J. Algebra **81** (1983), 202–224.

[**Gr**] H.-G. Gräbe, *Über den arithmetischen Rang quadratfreier Potenzproduktideale*, Math. Nachr. **120** (1985), 217–227.

[**Ha**] R. Hartshorne, *Cohomological dimension of algebraic varieties*, Ann. Math. **88** (1968), 403–450.

[**HV**] J. Herzog and V. Vasconcelos, *On the divisor class group of Rees algebras*, J. Algebra **93** (1985), 182–188.

[**Ho**] M. Hochster, *Grassmannians and their Schubert subvarieties are arithmetically Cohen-Macaulay*, J. Algebra **25** (1973), 40–57.

[**HE**] M. Hochster and J. A. Eagon, *Cohen-Macaulay rings, invariant theory, and the generic perfection of determinantal loci*, Amer. J. Math. **93** (1971), 1020–1058.

[**Hi**] T. Hibi, *Distributive lattices, affine semigroup rings and algebras with straightening laws*, in "Commutative algebra and combinatorics," Advanced Studies in Pure Mathematics **11**, 1987, pp. 93–109.

[**Hu**] C. Huneke, *Powers of ideals generated by weak d-sequences*, J. Algebra **68** (1981), 471–509.

[**KL**] H. Kleppe and D. Laksov, *The algebraic structure and deformation of Pfaffian schemes*, J. Algebra **64** (1980), 167–189.

[**Ma**] H. Matsumura, "Commutative Algebra," Second Ed., Benjamin/Cummings, Reading, 1980.

[Ne] P. E. Newstead, *Some subvarieties of Grassmannians of codimension 3*, Bull. London Math. Soc. **12** (1980), 176–182.

Universität Osnabrück, Abt. Vechta, Driverstr. 22, D-2848 Vechta

Partially supported by a DFG travel grant

Syzygies and Homological Conjectures

S. P. DUTTA

Throughout this work (A, m, k) denotes a commutative Noetherian local ring of dimension d, m is the maximal ideal of A and $k = A/m$. The main purpose of this paper is to study the canonical element conjecture of M. Hochster from the point of view of syzygies. The conjecture can be stated as follows:

CONJECTURE. *For every free resolution F:*

$$\to A^{s_i} \to A^{s_{i-1}} \to \cdots \to A^{s_0} \to k \to 0$$

of k and for every system of parameters $x_1, \ldots, x_d$ of A, if ϕ is any map from the Koszul complex $K_*(\underline{x}; A)$ ($\underline{x}$ stands for $x_1, \ldots, x_d$) to F which lifts the natural surjection $A/(x_1, \ldots, x_n) \to k$, then $\phi_n : K_n(\underline{x}; A) \to A^{s_n}$ is non-zero.

Hochster has proved the conjecture in the equicharacteristic case and has also shown that it occupies a central position among several homological conjectures [**H**]. The conjecture is open in mixed characteristic. Some special cases were proved in [**D1**].

Following [**D1**] let me first point out how the study of the above conjecture boils down to the study of a certain property of syzygies.

In [**D1**] we have shown that given any free complex $\cdots \to F_2 \to F_1 \to F_0 \to 0$ of finitely generated free modules with $H_0(F\bullet) = M$ say, and a non-zero submodule $N \subsetneq M$ there exists a minimal free complex $L\bullet$ and a map $\psi_0 : L\bullet \to F\bullet$ such that i) $\tilde{\psi}_i : H_i(L\bullet) \simeq H_i(F\bullet)$ is an isomorphism for $i > 0$, ii) $H_0(L\bullet) = N$ and $\tilde{\psi}_0$ induces the natural injection $N \hookrightarrow M$. We call such a complex $L\bullet$ an induced complex for N corresponding to $F\bullet$ and M (just an induced complex when there is no ambiguity). Now consider the truncated Koszul complex $\overset{t}{K\bullet}(\underline{x}; A)$:

$$0 \to A \xrightarrow{\beta_d} A^d \to \cdots \to A^{\binom{d}{2}} \xrightarrow{\beta_2} A^d \to 0.$$

Let $G_1(\underline{x}; A)$ denote coker β_2 then $H_1(\underline{x}; A) \hookrightarrow G_1(\underline{x}; A)$ (when there is no ambiguity we abbreviate these notations as $\overset{t}{K\bullet}$, G_1 and H_1 respectively).

Let $L\bullet$ denote an induced complex for H_1 corresponding to $\overset{t}{K\bullet}$ and G_1. We then have the following commutative diagram

$$\begin{array}{cccccccccccccc}
L\bullet : \longrightarrow & A^{s_d} & \xrightarrow[\alpha_d]{} & A^{s_{d-1}} & \xrightarrow[\alpha_{d-1}]{} & A^{s_{d-2}} & \longrightarrow & \dots & \longrightarrow & A^{s-1} & \xrightarrow[\alpha_1]{} & A^{s_0} & \longrightarrow & H_1 \\
(1) & \downarrow{\psi_d} & & \downarrow{\psi_{d-1}} & & \downarrow & & & & \downarrow & & \downarrow & & \downarrow \\
K\bullet : & 0 & \longrightarrow & A & \xrightarrow[\beta_d]{} & A^d & \longrightarrow & \dots & \longrightarrow & A^{\binom{d}{2}} & \xrightarrow[\beta_2]{} & A^d & \longrightarrow & G_1
\end{array}$$

We have proved in [**D1**] that the canonical element conjecture (henceforth C.E.C.) holds if and only if $\operatorname{Im}\psi_{d-1} \neq A$. It follows easily from the commutativity of (1) that $\operatorname{Im}\psi_{d-1} \neq A$ if and only if $\operatorname{coker}\alpha_d$ does not have a free summand (see (0.2) if necessary). Now when $0 < \operatorname{depth} A < d$ and $2 < d$ (it is enough to study this group of conjectures under these assumptions and one can even assume A to be complete normal) say $\operatorname{depth} A = d - r > 0$, then $\operatorname{coker}\alpha_d = \operatorname{Im}\alpha_{d-1} = \operatorname{syz}^{d-r}(\operatorname{coker}\alpha_r)$ and we are asking whether this particular syzygy has a free summand. We study this question here from several viewpoints.

Consider a minimal free complex (F_i, α_i), such that $F_j = 0$ for $j < 0$, α_i is the i-th boundary map, $\ell(H_i(F\bullet)) < \infty$, $H_0(F\bullet) \neq 0$ and $H_i(F\bullet) = 0$ for $i \geq r$ (recall $\operatorname{depth} A = d - r$); we ask whether $\operatorname{coker}\alpha_i$ can have a free summand. We show that for $i \neq d$, no $\operatorname{coker}\alpha_i$ can have a free summand (1.1). In particular if $F\bullet$ is a minimal free resolution of a module M of finite length then i) if $\operatorname{depth} A < d - 1$, no $\operatorname{syz}^i(M)$ (= i-th syzygy of M) can have a free summand and when ii) $\operatorname{depth} A = d - 1$, for $i \neq d - 1$, the same conclusion holds (1.2). Moreover when $M = k$, we establish in (1.3) that A is regular if and only if for some i, $\operatorname{syz}^i(k)$ has a free summand.

In Section 2, we study the monomial conjecture. The conjecture asserts that given any system of parameters (henceforth s.o.p.) $x_1, \dots, x_d$ of A, $(x_1 x_2 \dots x_d)^t \notin (x_1^{t+1}, \dots, x_d^{t+1}), \forall t > 0$. Hochster has proved this conjecture in the equicharacteristic case [**H1**] and has shown that it is equivalent to the C.E.C. [**H2**]. He has also shown that given $x_1, \dots, x_d$, $\exists s \gg 0$ such $x_1^s, \dots, x_d^s$ satisfy the monomial conjecture [**H1**]. Afterwards R. Y. Sharp and H. Zakeri have shown that given an s.o.p. $x_1, \dots, x_d$ $\exists s \gg 0$ such that $x_1^s, x_2, \dots, x_d$ satisfies the monomial conjecture provided one assumes that the conjecture already holds in lower dimensions [**SZ**]. Here we reduce this problem to the study of the above property of syzygies (directly) and annihilators of local cohomology modules $H_m^{d-i}(A)$. We point out that if $J = J_1 \dots J_r$, where $J_i = \operatorname{ann} H_m^{d-i}(A)$ and $J \not\subset (x_1, \dots, x_d)$, then $x_1, \dots, x_d$ satisfies the monomial conjecture (2.3). In particular this implies that when $\operatorname{depth} A = d-1$ and $x_1, \dots, x_d$ is an s.o.p. such that $\{x_1, \dots, x_{i-1}, x_{i+1}, \dots, x_d\}$ is an A-sequence, then for $s \gg 0$, $x_1, \dots, x_{i-1}, x_i^s, x_{i+1}, \dots, x_d$ satisfies the above conjecture. When A is a normal domain with $\operatorname{depth} A \geq 2$, given any s.o.p. $x_1, \dots, x_d$, there exists $s \gg 0$ such that $x_1, x_2, x_3^s, \dots, x_d^s$ satisfies the conjecture. (While proving (2.3) we first prove Proposition (2.1) which in turn provides another proof of the improved new intersection conjecture in the positive characteristics–the line of proof follows essentially the technique used by P. Roberts [**R1**]).

In Section 3, we study the C.E.C. for a s.o.p. of the type $p, x_2, \dots, x_d$ where $0 < p =$ characteristic of k, characteristic of $A = 0$, and k is perfect. Write $\bar{A} = A/pA$, $f^n : \bar{A} \to \bar{A}$ given by $f^n(x) = x^{p^n}$ and let ${}^{f^n}\bar{A}$ denote $\bar{A}$ as an $\bar{A}$-algebra via f^n. We raise the following question: For sufficiently large s is

$$\lim_{n\to\infty} \ell(\operatorname{Tor}_1^A(A/(p^s, x_2, \dots, x_d), {}^{f^n}\bar{A}))/p^{n(d-1)}$$
$$= \lim_{n\to\infty} \ell(\operatorname{Tor}_0^A(A/p^s, x_2, \dots, x_d), {}^{f^n}\bar{A}))/p^{n(d-1)} \;?$$

In (3.2) we prove that one way of studying the C.E.C. is by boiling it down to the study of asymptotic behavior of homologies of free complexes with finite length homologies in positive ch. and then using (3.2) (main theorem of [**D2**]) we derive that an affirmative answer to the above question implies the C.E.C. Since we are going to use the improved new intersection conjecture (henceforth I.N.I.C.), we will briefly describe it in the remaining part of this section.

(0.1) STATEMENT OF I.N.I.C. *Let $F\bullet$ be a finite complex with finitely generated free modules*

$$0 \to F_s \to \cdots \to F_0 \to 0$$

such that $H_i(F\bullet)$ has finite length for $i > 0$ and $H_0(F\bullet)$ has a (non-zero) minimal generator z such that Az has finite length. Then $s \geq d$.

E. G. Evans and P. Griffith proved it in the equicharacteristic case by using big Cohen-Macaulay modules [**EG**]. Recently several proofs have come out where big $C-M$ modules have not been used [**H2,D2**]. Moreover the equivalence of the C.E.C. and the I.N.I.C. has been pointed out in [**D1**]. The new intersection conjecture [**PS,R1**] has been proved in mixed characteristic: see [**R2**]; also see P. Roberts's paper in this volume.

Section 1.

(1.0) In this section first we study a certain fact which will be used many times in the work. Let $L\bullet$ be a complex of finitely generated free modules L_i $(= A^{s_i})$ with $L_j = 0$ for $j < 0$ and $\psi\bullet$ be a map from $L\bullet$ to $\overset{t}{K}\bullet(\underline{x}; A)$ where $\underline{x}$ stands for $x_1, \ldots, x_d$, a system of parameters for A; i.e. we have the following commutative diagram:

$$(1)\qquad \begin{array}{cccccccccc} L\bullet: & \longrightarrow & A^{s_d} & \xrightarrow{\alpha_d} & A^{s_{d-1}} & \xrightarrow{\alpha_{d-1}} & A^{s_{d-2}} & \longrightarrow & \cdots \\ & & \downarrow & & \downarrow{\scriptstyle \psi_{d-1}} & & \downarrow{\scriptstyle \psi_{d-2}} & & \\ \overset{t}{K}\bullet & & 0 & \longrightarrow & A & \underset{\beta_d}{\longrightarrow} & A^d & \longrightarrow & \cdots \end{array}$$

Assume $\operatorname{Im}\alpha_d = \operatorname{Ker}\alpha_{d-1}$. We want to explain what does it mean to have $\operatorname{Im}\psi_{d-1} = A$ in the following lines. Since $\operatorname{Im}\psi_{d-1} = A$, we can choose a basis $e_1, \ldots, e_{s_{d-1}}$ of $A^{s_{d-1}}$ such that $\psi_{d-1}(e_1) = 1$ and $\psi_{d-1}(e_i) = 0$ when $i > 1$. This implies, by the commutativity of (1), that $\operatorname{Im}\alpha_d \subset \langle e_2, \ldots, e_{s_{d-1}}\rangle =$ the submodule of $A^{s_{d-1}}$ generated by $e_2, \ldots, e_{s_{d-1}}$. Write $S_{d-1} = \operatorname{coker}\alpha_d = \operatorname{Im}\alpha_{d-1}$ and $S'_{d-1} = \langle e_2, \ldots, e_{s_{d-1}}\rangle / \operatorname{Im}(\alpha_d)$. Then $S_{d-1} = Ae_1 \oplus S'_{d-1} \simeq A \oplus S'_{d-1}$. If we write $\alpha_{d-1}(e_1) = (a_1, \ldots, a_{s_{d-2}}) \in A^{s_{d-2}}$, since $\beta_d(1) = (x_1, \ldots, \pm x_i, \ldots, \pm x_d) \in A^d$ by the commutativity of (1) it follows that $x_i \in (a_1, \ldots, a_{s_{d-2}})$ $\forall i$. Moreover denoting $\operatorname{Hom}_A(M, A)$ by M^* for any module M and f^* as dual of any A-linear map f, we see that $\alpha_d^*(e_1^*) = 0$ i.e. $H_{d-1}(\overset{*}{L\bullet}) = (Ae_1^* \oplus S'_{d-1}) / \operatorname{Im}(A^{\overset{*}{s_d}-2})$. Now projecting $H_{d-1}(\overset{*}{L\bullet})$ onto its e_1^* component we get a map $\Pi : H_{d-1}(L^*) \to A/(a_1, \ldots, a_{s_{d-2}})$ such that $\Pi(e_1^*) = \bar{1}$ in $A/\underline{a}$ (here $\underline{a} = (a_1, \ldots, a_{s_{d-2}})$). Applying $\operatorname{Hom}(-, A)$ to (1) we get the following commutative diagram:

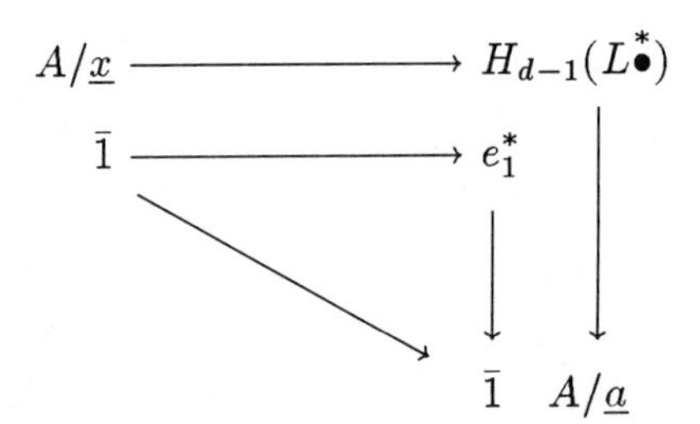

NOTATION: In the following sections, as we have already mentioned, depth $A > 0$, say depth $A = d - r$; all complexes consist of finitely generated modules and excepting the higher local cohomology modules and the injective modules all other modules are finitely generated. For any finitely generated module M, $\operatorname{syz}^i(M)$ usually denotes the i-th syzygy in a minimal free resolution of M (unless otherwise mentioned.)

THEOREM 1.1. *Let notation be as in the last paragraph of 1.0. Let* $(L\bullet, \alpha\bullet)$ *be a minimal free complex with finite length homologies and* $H_i(L\bullet) = 0$ *for* $i \geq r$, $H_0(L\bullet) \neq 0$. *Then* $\operatorname{coker} \alpha_i$ *cannot have a free summand for* $i > 1$ *and* $i \neq d$. Coker α_d *cannot have a free summand if and only if canonical element conjecture holds.*

PROOF: Notice that for $i > r$, $\operatorname{coker} \alpha_{i+1} = \operatorname{Im} \alpha_i = \operatorname{Ker} \alpha_{i-1}$ and $(L\bullet)_{i \geq r}$ is a minimal free resolution of $\operatorname{coker} \alpha_r$. Moreover, since $\ell(H_i(L\bullet)) < \infty$ for all i, $\ell(H_i(\overset{*}{L\bullet})) < \infty$ where $* = \operatorname{Hom}(-, A)$.

Case 1: $i \leq d$. Applying $\operatorname{Hom}(-, A)$ to $L\bullet$ we get a complex

$$0 \to L_0^* \to L_1^* \to \cdots \to L_{i-2}^* \to L_{i-1}^* \to \operatorname{coker}(\alpha_{i-1}^*) \to 0.$$

If $\operatorname{coker}(\alpha_i)$ has a free summand, applying $\operatorname{Hom}(-, A)$ to the exact sequence $L_i \to L_{i-1} \to \operatorname{coker} \alpha_i \to 0$ we see that there exists a free generator e^* of L_{i-1}^* such that $\alpha_i^*(e^*) = 0$. Now since $H_{i-1}(\overset{*}{L\bullet}) \subset \operatorname{coker}(\alpha_{i-1}^*)$, this implies that $\operatorname{coker}(\alpha_{i-1}^*)$ has a minimal generator killed by a power of m. If $i < d$, this cannot happen, since even in mixed characteristic, tensoring (1) with A/p where $p = \operatorname{ch} k > 0$, we get a complex which violates the I.N.I.C. in characteristic $p > 0$. When $i = d$, this cannot happen if and only if the I.N.I.C. holds. Since the C.E.C. and the I.N.I.C. are equivalent, we are done in this case [**D1**].

Case 2: $i > d$. In this case since $\operatorname{coker} \alpha_i = \operatorname{syz}^{i-r}(\operatorname{coker} \alpha_r)$, it has depth $d - r$. Write S_{i-r} for $\operatorname{coker} \alpha_i$. If possible let $S_{i-r} = A \oplus S'_{i-r}$, i.e. S_{i-r} has a free summand.

We consider the following exact sequences from $L\bullet$.

$$0 \to S_{i-r} \to L_{i-2} \to S_{i-r-1} \to 0$$
$$0 \to A \to L_{i-2} \to T \to 0$$
$$0 \to S'_{i-r} \to L_{i-2} \to W \to 0$$

Since $S_{i-r} = A \oplus S'_{i-r}$ we get an exact sequence

$$0 \to L_{i-2} \to T \oplus W \to S_{i-r-1} \to 0.$$

But as $i > d$, depth $S_{i-d-1} = d - r$ and depth $T = d - r - 1$ and thus we arrive at a contradiction. (Since the projective dimension of T is 1, depth $T = d - r - 1$).

COROLLARY 1.2. *Let notation be as in the last paragraph of 1.0. Let M be a module of finite length. Then*

i) *if $r = 0$, only $\operatorname{syz}^d(M)$ may have a free summand*
ii) *if $r > 1$, no $\operatorname{syz}^i(M)$ can have a free summand*
iii) *if $r = 1$, no $\operatorname{syz}^i(M)$ can have a free summand for $i \neq d-1$ and $\operatorname{syz}^{d-1}(M)$ cannot have a free summand if and only if the C.E.C. holds in this case.*

PROOF: Similar arguments as above cover all three cases. When $r = 0$ for $i < d$, arguments as in Case 1 and for $i > d$ arguments as in Case 2 prove the assertion for $\operatorname{syz}^i(M)$. When $r > 1$, for $i < d-1$, arguments as in Case 1 and for $i \geq d-1$ arguments as in Case 2 finish the proof. When $r = 1$, for $i \leq d-1$, arguments as in Case 1 and for $i > d-1$, arguments as in Case 2 prove the statement for $\operatorname{syz}^i(M)$.

REMARK: In the case when depth $A = d-1$, it is enough to prove that for any s.o.p. $x_1, \ldots, x_d$, $\operatorname{syz}^{d-1}(H_1(\underline{x}; A))$ cannot have a free summand. So the question reduces to showing that the map $\operatorname{Ext}^{d-1}(H_1, A) \to \operatorname{Ext}^{d-1}(H_1, k)$ is 0 (obtained from the surjection $A \to A/m = k$). Again, since the above question is equivalent to the canonical element conjecture [**D1**], by [**D1**] we know the above map is 0 in the following cases: (a) $H_m^{d-1}(A)$ is decomposable and (b) $\operatorname{Hom}(H_m^{d-1}(A), E(k))$ is cyclic, where $E(k)$ is the injective hull of k.

COROLLARY 1.3. *(A characterization of regular local rings). Let A be a local ring. A is regular if and only if for some i, $\operatorname{syz}^i(k)$ has a free summand.*

PROOF: If A is regular, $\operatorname{syz}^d(k)$ is free.

On the other hand, it follows from the theorem above that if A is Cohen-Macaulay, we have to consider only $\operatorname{syz}^d(k)$ and if depth $A = d-1$, we have to consider only $\operatorname{syz}^{d-1}(k)$. In other words, it is enough to prove if $i =$ depth $A > 0$, $\operatorname{syz}^i(k)$ cannot have a free summand unless it is regular. If possible set $S_i = \operatorname{syz}^i(k) = A \oplus S_i'$. Let $F\bullet$ be a minimal resolution of k and write $F_i = A^{t_i}$. We consider the exact sequences

$$0 \to S_i \to A^{t_{i-1}} \to S_{i-1} \to 0, \quad 0 \to A \to A^{t_{i-1}} \to T \to 0$$
$$0 \to S_i' \to A^{t_{i-1}} \to W \to 0 \qquad \left(1 \to (a_1, \ldots, a_{t_{a-1}}) \in A^{t_{i-1}}\right).$$

Since $S_i = A \oplus S'_i$ we obtain exact sequences

$$(2) \qquad 0 \to S'_i \to T \to S_{i-1} \to 0, \quad 0 \to A \to W \to S_{i-1} \to 0$$

and

$$(3) \qquad 0 \to A^{t_{i-1}} \to T \oplus W \to S_{i-1} \to 0.$$

Applying $\mathrm{Hom}(-, A)$ to (2) we get an exact sequence

$$(4) \qquad 0 \to A/I \to \mathrm{Ext}^1(S_{i-1}, A) \to \mathrm{Ext}^1(W, A) \to 0.$$

Since $\mathrm{Ext}^1(S_{i-1}, A) = \mathrm{Ext}^i(k, A)$ is a vector space, it follows that $I = m$ and $\mathrm{Ext}^1(W, A)$ is also a vector space. We write $\underline{a} = (a_1, \ldots, a_{t_{i-1}})$.

Now, applying $\mathrm{Hom}(-, A)$ to (3), we get an exact sequence

$$(5) \quad 0 \to S^*_{i-1} \to T^* \oplus W^* \to A^{t^*_{i-1}} \to \mathrm{Ext}^i(k, A) \to A/\underline{a} \oplus \mathrm{Ext}^1(W, A) \to 0.$$

This implies $\underline{a} = m$. From (4) we conclude that $0 \to S^*_{i-1} \to T^* \oplus W^* \to A^{t^*_{i-1}} \to 0$ is exact. Since $\operatorname{depth} A = i$, S^*_{i-1} is of finite projective dimension and this implies that T^* is of finite projective dimension. Now from the exact sequence

$$0 \to T^* \to A^{t^*_{i-1}} \to A \to A/\underline{a} \to 0$$

we conclude that $A/\underline{a} = A/m$ is also of finite projective dimension. Thus A is regular.

Section 2.

In this section we are going to study the monomial conjecture (already stated in the introduction). Without any loss of generality we can assume that A is complete. There exists a complete Gorenstein ring B with $\dim B = \dim A$ and a surjective ring homomorphism $B \to A$. Via this homomorphism we can use local duality for finitely generated modules M over A : $\mathrm{Ext}^i_B(M, B) \simeq \mathrm{Hom}(H^{d-i}_m(M), E(k))$ where $E(k)$ is the injective hull of k. We will use this fact in this section and in the next section. Let us remind the reader that for any module M, $H^0_m(M) = \{x \in M \mid m^t x = 0$ for some positive integer $t\}$.

First we prove a proposition which is a generalized version of a theorem of P. Roberts [**R1**], the idea of the proof being essentially the same. In this case because the proposition is more general, one has to dig a bit deeper into spectral sequences.

We write $J_i = \text{annihilator}_A H_m^{d-i}(A)$ and $J = J_1 J_2 \dots J_r$ (recall Depth $A = d - r > 0$).

PROPOSITION 2.1. *Let* $0 \to F_{d-1} \to F_{d-2} \to \cdots \to F_1 \to F_0 \to 0$ *be a finite free complex with* $\ell(H_i(F\bullet)) < \infty$ *for* $i > 0$ *and* $H_m^0(H_0(F\bullet)) \neq 0$. *Then* $JH_m^0(H_0(F\bullet)) = 0$.

PROOF: Let $0 \to I_0 \to I_1 \to \cdots \to I_d \to 0$ be a dualizing complex for A with $I_i = \bigoplus_{\dim \frac{A}{p} = d-i} E(A/p)$, where $E(A/p)$ is the injective hull of A/p, p a prime ideal of A. In the above complex I_d is $E(k)$, the injective hull of k.

We consider the double complex $\text{Hom}(F\bullet, I\bullet)$:

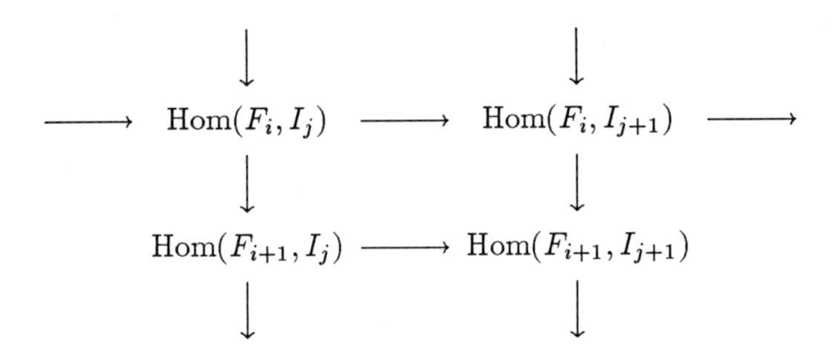

Since $\ell(H_i(F\bullet)) < \infty$ for $i > 0$, for $j \neq d$ the j-th column of the above double complex has zero homology everywhere except at $i = 0$. This in turn implies that the spectral sequence, obtained by considering the columns first, has $E_2^{i,j} = 0$ for $i > 0$ and $j < d$; $E_2^{0,j} = \text{Ext}_B^j(M, B)$ where $M = H_0(F\bullet)$ and $E_2^{i,d} = \text{Hom}(H_i(F\bullet), E(k))$ for $i > 0$. Moreover we have an edge homomorphism $\alpha : E_2^{0,d} = \text{Ext}_B^d(M, B) \to H_d(\text{Hom}(F\bullet, I\bullet))$. The crucial point is that the above fact about homologies on columns forces α to be injective; this can easily be checked by chasing diagrams.

On the other hand for the spectral sequence $E_{i,j}^2$ obtained by considering the horizontal rows first we have $E_{i,j}^2 = H_i(\text{Hom}(F\bullet, \text{Ext}^j(A, R)))$. This implies that J kills $H_d(\text{Hom}(F\bullet, I\bullet))$ and hence $\text{Ext}^d(M, B)$, which, in turn, by local duality over complete local rings, implies that J kills $H_m^0(H_0(F\bullet))$ (since $\text{Hom}(H_m^0(H_0(F)), E(k)) \cong \text{Ext}^d(M, B)$).

PROPOSITION 2.2. *Let $x_1, \dots, x_d$ be a s.o.p. for A and let $P\bullet$ be a free resolution of $A/\underline{x}$ (here $\underline{x}$ stands for the ideal $(x_1, \dots, x_d)$) and $\theta\bullet : K\bullet(\underline{x}; A) \to P\bullet$ denote a lift of the identity map $A/\underline{x} \to A/\underline{x}$. Then $\mathrm{Im}(\mathrm{Ext}^d(A/\underline{x}, A) \xrightarrow{\tilde{\theta}_d} A/\underline{x})$ contains $J + \underline{x}/\underline{x}$.*

PROOF: It is enough to prove the assertion for any particular free resolution of $A/\underline{x}$. Let $L\bullet$ be an induced complex for $H\bullet(\underline{x}; A)$ corresponding to $\overset{t}{K}\bullet(\underline{x}; A)$ and $G_1(\underline{x}; A)$. The exact sequence

$$0 \to H_1 \to G_1 \to \underline{x}A \to 0$$

implies that the mapping cone $\overset{t}{K}\bullet \wedge L\bullet$ of $\psi\bullet : L\bullet \to \overset{t}{K}\bullet$ gives rise to a free resolution of $\underline{x}A$. We have an exact sequence of complexes

$$0 \to \overset{t}{K}\bullet \xrightarrow{\theta\bullet} \overset{t}{K}\bullet \wedge L\bullet \to (L\bullet)_{-1} \to 0.$$

Applying $\mathrm{Hom}(-, A)$ to the above sequence we get an exact sequence

$$\cdots \to \mathrm{Ext}^d(A/\underline{x}, A) \xrightarrow{\tilde{\theta}_d} A/\underline{x} \to H_{d-1}(\overset{*}{L}\bullet) \to \dots$$

Since $H_{d-1}(\overset{*}{L}\bullet) \subset \mathrm{coker}(\alpha^*_{d-1})$ and $\ell(H_{d-1}(\overset{*}{L}\bullet)) < \infty$ for all i, it follows, by considering the complex

$$0 \to L^*_0 \to L^*_1 \to \cdots \to L^*_{d-1} \to 0,$$

by Proposition 2.1 that $JH_{d-1}(\overset{*}{L}\bullet) = 0$. This implies that $\mathrm{Im}\,\tilde{\theta}_d$ contains $J + \underline{x}/\underline{x}$.

REMARKS: (1) By considering $\mathrm{Ext}^d(E, \mathrm{Hom}_B(A, B))$ one can give a direct proof of Proposition 2.2 without using Proposition 2.1.

(2) Proposition 2.1 provides a proof of the improved new intersection conjecture in characteristic $p > 0$ (following arguments as in [**R1**].

THEOREM 2.3. *Let $x_1, \dots, x_d$ be a s.o.p. for A such that $J \not\subseteq (x_1, \dots, x_d)$. Then $x_1, \dots, x_d$ satisfies the monomial conjecture, i.e. for any $t \geq 0$, $(x_1 x_2 \dots x_d)^t \notin (x_1^{t+1}, \dots, x_d^{t+1})$.*

PROOF: If possible, let $(x_1 x_2 \dots x_d)^t \in (x_1^{t+1}, \dots, x_d^{t+1})$, i.e.

$$(1) \qquad (x_1 \dots x_d)^t = \sum_{i=1}^{d} \lambda_i x_i^{t+1}, \lambda_i \in A.$$

Consider the map $\delta\bullet : K\bullet(\underline{x}^{t+1}; A) \to K\bullet(\underline{x}, A)$ defined by $\delta_r(e_{i_1} \wedge \cdots \wedge e_{i_r}) = (x_{i_1} \ldots x_{i_r})^t e_{i_1} \wedge \cdots \wedge e_{i_r}$. We have the following commutative diagram:

$$\begin{array}{ccccccccccc} 0 & \longrightarrow & A & \xrightarrow{\beta_d^{[t+1]}} & A^d & \longrightarrow & \ldots & \longrightarrow & A^d & \longrightarrow & A & \longrightarrow & 0 \\ & & {\scriptstyle (x_1\ldots x_d)^t}\downarrow{\scriptstyle \delta_d} & \swarrow{\scriptstyle g} & \downarrow{\scriptstyle \delta_{d-1}} & & & & \downarrow & & \| & & \\ 0 & \longrightarrow & A & \xrightarrow[\beta_d]{} & A^d & \longrightarrow & \ldots & \longrightarrow & A^d & \longrightarrow & A & \longrightarrow & 0 \end{array}$$

By (1), we can define a map $g : A^d \to A$, $e_i \to \pm\lambda_i$ (choosing the sign appropriately) so that the triangle on the extreme left commutes; now change $\delta\bullet$ to $\gamma\bullet$ such that $\gamma_d = \delta_d - g \cdot \beta_d^{[t+1]}$, $\gamma_{d-1} = \delta_{d-1} - \beta_d \cdot g$, $\gamma_i = \delta_i$ for $i < d-1$. Because of (1), $\gamma_d = 0$ and we have the following commutative diagram:

$$(2)\quad \begin{array}{ccccccccccc} 0 & \longrightarrow & A & \xrightarrow{\beta_d^{[t+1]}} & A^d & \longrightarrow & \ldots & \longrightarrow & A^d & \xrightarrow{\beta_1^{[t+1]}} & A & \longrightarrow & 0 \\ & & \downarrow{\scriptstyle 0} & & \downarrow{\scriptstyle \gamma_{d-1}} & & & & \downarrow{\scriptstyle \gamma_1} & & \| & & \\ 0 & \longrightarrow & A & \xrightarrow[\beta_d]{} & A^d & \longrightarrow & \ldots & \longrightarrow & A^d & \xrightarrow[\beta_1]{} & A & \longrightarrow & 0 \end{array}$$

Applying $\operatorname{Hom}(-, A)$ to (2), the following commutative diagram is obtained:

$$(3)\quad \begin{array}{ccccccccccc} 0 & \longrightarrow & A & \xrightarrow{\beta_1^*} & A^d & \longrightarrow & \ldots & \longrightarrow & A^d & \longrightarrow & A & \longrightarrow & 0 \\ & & \| & & \downarrow{\scriptstyle \gamma_1^*} & & & & \downarrow{\scriptstyle \gamma_{d-1}^*} & & \downarrow{\scriptstyle 0} & & \\ 0 & \longrightarrow & A & \xrightarrow[\beta_1^{[t+1]*}]{} & A^d & \longrightarrow & \ldots & \longrightarrow & A^d & \longrightarrow & A & \longrightarrow & 0 \end{array}$$

Let $L\bullet$ be an induced complex for $H_1(\underline{x}^{t+1}; A)$ corresponding to $\overset{t}{K\bullet}(\underline{x}^{t+1}; A)$ and $G_1(\underline{x}^{t+1}; A)$. Then we have the following commutative diagram (writing $L_i = A^{s_i}$)

$$(4)\quad \begin{array}{lccccccccccc} L\bullet : & \longrightarrow A^{s_d} & \xrightarrow{\alpha_d} & A^{s_{d-1}} & \xrightarrow{\alpha_{d-1}} & A^{s_{d-2}} & \longrightarrow \ldots \longrightarrow & A^{s_1} & \longrightarrow & A^{s_0} & \longrightarrow & H_1(\underline{x}^{t\text{-}} \\ & \downarrow & & \downarrow{\scriptstyle \psi_{d-1}} & & \downarrow & & {\scriptstyle \psi_1}\downarrow & & \downarrow{\scriptstyle \psi_0} & & \downarrow \\ \overset{t}{K\bullet}(\underline{x}^{t+1}) : 0 & \longrightarrow & A & \longrightarrow & A^d & \longrightarrow & \ldots \longrightarrow & A^{\binom{k}{2}} & \longrightarrow & A^d & \longrightarrow & G_1(\underline{x}^{t\text{-}} \end{array}$$

Hence a free resolution of $A/\underline{x}^{t+1}$ is obtained by taking the mapping cone of (4):

$$\to A^{s_{d-1}} \to A \oplus A^{s_{d-2}} \to \cdots \to A^{\binom{d}{2}} \oplus A^{s_0} \to A^d \to A \to A/\underline{x} \to 0$$

Moreover we have the following commutative diagram:

$$
(5)\qquad
\begin{array}{ccccccccccc}
\longrightarrow & A^{s_{d-1}} & \longrightarrow & A \oplus A^{s_{d-2}} & \longrightarrow & \cdots & \longrightarrow & A^{\binom{d}{2}} \oplus A^{s_0} & \longrightarrow & A^d & \longrightarrow A \\
 & \downarrow{\scriptstyle \pi_{d-1}} & & {\scriptstyle \pi_{d-2}}\downarrow & & & & {\scriptstyle \pi_0}\downarrow & & & \\
\longrightarrow & A^{s_{d-1}} & \longrightarrow & A^{s_{d-2}} & \longrightarrow & \cdots & \longrightarrow & A^{s_0} & \longrightarrow & 0 &
\end{array}
$$

Notice that the map $A^{s_{d-1}} \to A \oplus A^{s_{d-2}}$ is $-\psi_{d-1} \oplus \alpha_{d-1}$.

Now from (3), we get the following commutative diagram:

$$
(6)\qquad
\begin{array}{ccccccccccc}
 & 0 & \longrightarrow & A & \xrightarrow{\beta_1^*} & A^d & \longrightarrow \cdots \longrightarrow & A^d & \longrightarrow & A \\
 & & & \| & & \downarrow & & \downarrow{\scriptstyle \gamma_{d-1}^*} & & \downarrow{\scriptstyle 0} \\
 & 0 & \longrightarrow & A & \xrightarrow{\beta_1^{[t+1]*}} & A^d & \longrightarrow \cdots \longrightarrow & A^d & \longrightarrow & A \\
 & \downarrow & & \downarrow{\scriptstyle \theta_d} & & \downarrow{\scriptstyle \theta_{d-1}} & & \downarrow{\scriptstyle \theta_1} & & \downarrow{\scriptstyle \mathrm{id}} \\
\longrightarrow & A^{s_{d-1}} & \longrightarrow & A \oplus A^{s_{d-2}} & \longrightarrow & A^d \oplus A^{s_{d-3}} & \longrightarrow \cdots \longrightarrow & A^d & \longrightarrow & A
\end{array}
$$

where the map $\theta_\bullet$ between the last two rows is the natural injection obtained from (4), and hence $\theta_d(1) = (1, 0)$. Since $\theta_\bullet \cdot \gamma_\bullet^*$ lifts the 0 map $A/\underline{x} \to A/\underline{x}^{t+1}$ and the bottom row is a resolution of $A/\underline{x}^{t+1}$, $\theta_\bullet \cdot \gamma_\bullet^*$ is homotopic to 0. This forces $\operatorname{Im} \psi_{d-1}$ to be equal to A in (4). $\theta_\bullet \cdot \gamma_\bullet^*$ being homotopic to 0, let $\{h'_i\}$ denote the homotopy maps between the top and the bottom rows of (6). Now together with (5) we get maps $h_i = \pi_i \cdot h'_i$ between the top row of (6) and $L_\bullet$. Moreover one can check that $\{\gamma_i^*\}$ is homotopic to $\{\psi_i \cdot h_i\}$. Now applying $\operatorname{Hom}(-, A)$ to (4) and using (1.0) we get an exact sequence and a commutative diagram as follows:

$$
\begin{array}{ccccc}
\operatorname{Ext}^d(A/\underline{x}^{t+1}, A) & \longrightarrow & A/\underline{x}^{t+1} & \longrightarrow & H_{d-1}(L_\bullet^*) \\
 & & \bar{1} & \longrightarrow & e_1^* \qquad \Big| \\
 & & & \searrow & \downarrow \qquad \downarrow \\
 & & & & \bar{1} \in A/\underline{x}
\end{array}
$$

This forces J to be contained in $(x_1, \ldots, x_d)$ which is a contradiction.

Corollary 2.4. *Let depth $A = d - 1$ and $x_1, \ldots, x_d$ be a s.o.p. for A such that for some i, $1 \leq i \leq d$, $\{x_1, \ldots, x_{i-1}, x_{i+1}, \ldots, x_d\}$ is a maxi-*

mal A-sequence. Then for $t \gg 0$, $x_1, \dots, x_{i-1}, x_i^t, x_{i+1}, \dots, x_d$ satisfies the monomial conjecture.

PROOF: Here $J = J_1 = \operatorname{ann} \operatorname{Ext}^1_B(A, B)$. Consider the ring $\bar{A} = A/(x_1, \dots, x_{i-1}, x_{i+1}, \dots, x_d)$. Then $\operatorname{Hom}(H^0_m(\bar{A}), E(k)) \xrightarrow{\sim} \operatorname{Ext}^1(A, B)/(x_1, \dots, x_{i-1}, x_{i+1}, \dots, x_d) \operatorname{Ext}^1(A, B) \neq 0$. This implies that $J + (x_1, \dots, x_{i-1}, x_{i+1}, \dots, x_d)$ is m-primary. Hence J cannot be contained in $(x_1, \dots, x_{i-1}, x_i^t, x_{i+1}, \dots, x_d)$ for $t \gg 0$. Now the assertion follows from the above theorem.

COROLLARY 2.5. *Let A be a complete normal domain. Then given any s.o.p. $x_1, \dots, x_d$, the monomial conjecture holds for $x_1, x_2, x_3^t, \dots, x_d^t$ for $t \gg 0$.*

PROOF: Since A is normal height $J \geq 3$. Hence for $t \gg 0$, J cannot be contained in $x_1, x_2, x_3^t, \dots, x_d^t$.

Section 3.

(3.1) We have already mentioned in the introduction that from one point of view the study of C.E.C. reduces to the question of whether certain syzygies can have a free summand. By using (1.0) and extending it a bit further we resolve the whole situation down to understanding certain homologies of free complexes on $\bar{A} = A/pA$. By exploiting the Frobenius map it becomes clear in this case that one has to understand asymptotic behavior of the homologies of infinite free complexes with finite length homologies. This forces us to use the main theorem of [**D2**] (which we state in (3.2)) and by using this Theorem we see the problem reduces to understanding the asymptotic behavior of Tor_1 and Tor_0 of $(A/(p^s, x_2, \dots, x_d)$ against ${}^{f^n}\bar{A}$ and we are led to the following question:

Let A be a complete normal domain in mixed characteristic, let $p =$ characteristic of k and assume that k is perfect. Let $p, x_2, \dots, x_d$ be any s.o.p. of A. We write $\bar{A} = A/pA$; then characteristic of $\bar{A} = p > 0$ and $\dim A = d - 1$. Let $f^n : \bar{A} \to \bar{A}$ be given by $\bar{s}(x) = x^{p^n}$ and let us denote by ${}^{f^n}\bar{A}$ the $\bar{A}$ algebra via f^n. We ask for $s \gg 0$: Is

$$\text{(Q)} \quad \lim_{n \to \infty} \ell(\operatorname{Tor}_1^A(A/(p^s, x_2, \dots, x_d), {}^{f^n}\bar{A}))/p^{n(d-1)} = \lim_{n \to \infty} \ell(\operatorname{Tor}_0^A(A/(p^s, x_2, \dots, x_d), {}^{f^n}\bar{A}))/p^{n(d-1)} \ ?$$

An affirmative answer to the above question will establish the C.E.C. in mixed characteristic (3.3).

(3.2) We quote a theorem (Theorem 1.5) from [**D2**] without proof, which we will be using in this section to prove our next theorem. Let R be a complete local ring without any embedded components, characteristic $R = p > 0$. The Frobenius map $f : R \to R$ given by $f(x) = x^p$ for all $x \in R$ is a ring homomorphism. We denote by ${}^{f^n}R$ the bi-algebra R having the structure of an R-algebra from the left by f^n and from the right by the identity map, i.e. $\alpha \in R$, $x \in {}^{f^n}R$, $\alpha \cdot x = \alpha_x^{p^n}$ and $x \cdot \alpha = x\alpha$. For any module M on R we write $F^n(M)$ for $M \otimes_R {}^{f^n}R$ and for any complex $C\bullet$ we write $F^n(C\bullet)$ for $C\bullet \otimes_R {}^{f^n}R$. Let $\dim R = d$.

STATEMENT OF THE THEOREM [**D2**]. *Let $F\bullet$ be a free complex $\to F_2 \to F_1 \to F_0 \to 0$, with homologies of finite length and let N be a finitely generated module. Let $W_{j,n}$ denote the j-th homology of* $\operatorname{Hom}(F^n(F\bullet), N)$ *(or* $\operatorname{Hom}(F\bullet, {}^{f^n}(N))$*). We have the following:*

(i) *if* $\dim N < d$, $\lim_{n\to\infty} \ell(W_{j,n})/p^{nd} = 0$
(ii) *if* $\dim N = d$ *and:*
(a) $j < d$, $\lim_{n\to\infty} \ell(W_{j,n})/p^{nd} = 0$;
(b) $j = d$, $\lim \ell(W_{j,n})/p^{nd} = \lim_{n\to\infty} \ell(F^n(H_0(F\cdot) \otimes \tilde{N})/p^{nd}$, *which is positive;*
(c) $j > d$ *and* $\bar{S}^1(\tilde{N})$ *is* $\bar{S}^1R$ *free of finite rank,* $\lim_{n\to\infty} \ell(W_{j,n})/p^{nd} = \lim \ell(H_{j-d}(F^n(F\cdot)) \otimes \tilde{N})/p^{nd}$. *Here* $\tilde{N} = \operatorname{Hom}(H_m^d(N), E(k))$.

(3.3) Now A is a complete normal domain, $\dim A = d(> 2)$, $\operatorname{depth} A = d - r > 0$, $0 < p =$ characteristic of k, and k is perfect. Let $p, x_2, \ldots, x_d$ be any s.o.p. of A. Notations are as described in (3.1). Write $\underline{x} = (x_2, \ldots, x_d)$; note that p is a parameter on $A/\underline{x}$ and $H_m^0(A/\underline{x}) = \operatorname{Hom}(A/p^s, A/\underline{x})$ for $s \gg 0$ — denote it by H. Let I be the ideal $H + \underline{x}$. Then p is a non-zero-divisor on A/I. We denote the Koszul complex $K(x_2, \ldots, x_d; A)$ by $K\bullet(\underline{x})$.

THEOREM 3.3. *Let A be a complete normal domain, with $p, x_2, \ldots, x_d$ described as above. Suppose that the answer to (Q) in (3.1) is in the affirmative. Then $p, x_2, \ldots, x_d$ satisfies the C.E.C.*

PROOF: Since the proof is rather long we break it up into 4 steps.

Step 1. Let $F\bullet$ be an induced complex for H corresponding to $K\bullet(\underline{x})$

and $A/\underline{x}$ i.e. we have the following commutative diagram

$$\begin{array}{lccccccccccccccc}
F\bullet : \longrightarrow & A^{r_d} & \xrightarrow{\beta_d} & A^{r_{d-1}} & \xrightarrow{\beta_{d-1}} & A^{r_{d-2}} & \longrightarrow & \ldots & \longrightarrow & A^{r_1} & \xrightarrow{\beta_1} & A^{r_0} & \longrightarrow & H \\
(1) & \downarrow & & \downarrow{\scriptstyle \alpha_{d-1}} & & \downarrow{\scriptstyle \alpha_{d-1}} & & & & \downarrow{\scriptstyle \alpha_1} & & \downarrow{\scriptstyle \alpha_0} & & \downarrow \\
K\bullet(\underline{x}) : & 0 & \longrightarrow & A & \longrightarrow & A^{d-1} & \longrightarrow & \ldots & \longrightarrow & A^{d-1} & \longrightarrow & A & \longrightarrow & A/x
\end{array}$$

CLAIM. *If the C.E.C. does not hold then*

$$\operatorname{Im} \alpha_{d-1} = A.$$

PROOF OF THE CLAIM: Notice that, by definition of an induced complex, the mapping cone of (1), say $M\bullet$, is a free resolution of $\operatorname{coker}(H \hookrightarrow A/\underline{x}) = A/I$. Let $\theta\bullet : K\bullet(x) \to M\bullet$ be the natural injection.

Let $F'\bullet$ be an induced complex for $H_1(p^s, x_2, \ldots, x_d; A)$ corresponding to $K\overset{t}{\bullet}(p^s, x_2, \ldots, x_d; A)$ and $G_1(p^s, x_2, \ldots, x_d; A)$. We have the following commutative diagram ($F_i' = A^{s_i}$)

$$\begin{array}{ccccccccccccc}
\longrightarrow A^{s_d} & \longrightarrow & A^{s_{d-1}} & \longrightarrow & \ldots & \longrightarrow & A^{s-1} & \longrightarrow & A^{s_0} & \longrightarrow & H_1 & \longrightarrow & 0 \\
\downarrow & & \downarrow{\scriptstyle \psi'_{d-1}} & & & & \downarrow{\scriptstyle \psi'} & & \downarrow{\scriptstyle \psi'_0} & & \downarrow & & \\
0 & \longrightarrow & A & \longrightarrow & \ldots & \longrightarrow & A^{\binom{d}{2}} & \longrightarrow & A^d & \longrightarrow & G_1 & \longrightarrow & 0
\end{array}$$

The mapping cone of the above diagram gives a free resolution of $(p^s, x_2, \ldots, x_d)A$. We pointed out in [**D1**] that the claim in our theorem is true if and only if $\operatorname{Im} \psi'_{d-1} \neq A$.

Suppose on the contrary $\operatorname{Im} \psi'_{d-1} = A$.

We recall that the Koszul complex $K\bullet(p^s, x_s, \ldots, x_d; A)$ is the mapping cone of $K\bullet(x_2, \ldots, x_d; A) \xrightarrow{p^s} K\bullet(x_2, \ldots, x_d; A)$. Hence we get a projection from $K\overset{t}{\bullet}(p^s, x_2, \ldots, x_d; A) \xrightarrow{\eta\bullet} K\bullet(x_2, \ldots, x_d; A)$ (we abbreviate as $K\bullet(\underline{x})$) which maps G_1 onto $A/\underline{x}$ where $\underline{x} = (x_2, \ldots, x_d)$. Thus we get a map $\eta\bullet \cdot \psi'\bullet : F'\bullet \to K\bullet(\underline{x})$ such that $\eta_{d-1} \cdot {\psi'_{d-1}}^{(A^{s_{d-1}})} = A$.

Notice that $\eta \cdot \psi'(H_1) = (0 : p^s)A/\underline{x}$, and hence $\subset H$. Since $\theta(\eta(\psi'))(H_1) = 0$ in A/I, $(\theta\bullet)(\eta\bullet)(\psi'\bullet)$ is homotopically trivial, but since $\eta_{d-1}\psi'_{d-1}(A^{s_{d-1}}) = A$ and $M\bullet$ is the mapping cone of (1), we must have $\operatorname{Im} \alpha_{d-1} = A$.

Step 2. Now $\operatorname{Im} \alpha_{d-1} = A$ implies that we can choose a basis $e_1, \ldots, e_{r_{d-2}}$ of $A^{r_{d-2}}$ such that $\alpha_{d-1}(e_1) = 1$, $\alpha_{d-1}(e_i) = 0$ for $i > 1$. Let $\beta_{d-1}(e_1) = (a_1, \ldots, a_{r_{d-2}}) \in A^{r_{d-2}}$. Then from (1) $\alpha_{d-2}(a_1, \ldots, a_{r_{d-2}}) = (x_2, \ldots, \pm x_d)$.

We notice that since the homologies of the two complexes in (1) are isomorphic, $H_i(F\bullet) = 0$ for $i \geq r$ where depth $A = d - r$. Write $S_{d-1} = \operatorname{Im} \beta_{d-1}$. From (1) we get $\operatorname{Im} \beta_d \subset$ free submodule generated by $\{e_2, \ldots, e_{r_{d-1}}\} = \langle e_2, \ldots, e_{r_{d-1}} \rangle \subset A^{r_{d-1}}$. Hence

$$S_{d-1} = (a_1, \ldots, a_{r_{d-2}}) \oplus \beta_{d-1}\left(\langle e_2, \ldots, e_{r_{d-1}} \rangle\right) = A \oplus S'_{d-1}, \tag{2}$$

where $S'_{d-1} = \operatorname{Im}\langle e_2, \ldots, e_{r_{d-1}} \rangle$ in $A^{r_{d-2}}$. (These steps have been explained in (1.0)).

We consider the following exact sequences:

$$\begin{aligned} &0 \to A \to A^{r_{d-2}} \to T \to 0 \text{ and } 0 \to S'_{d-1} \to A^{r_{d-2}} \to W \to 0 \\ &1 \to (a_1, \ldots, a_{r_{d-2}}). \end{aligned} \tag{3}$$

Because of (2) we get an exact sequence

$$0 \to A^{r_{d-2}} \to T \oplus W \to S_{d-2} \to 0 \tag{4}$$

where $S_{d-2} = \operatorname{Im} \beta_{d-2}$.

Since in (1), $\alpha_{d-1}(e_1) = 1$ and $\beta_{d-1}(e_1) \in mA^{r_{d-2}}$, when we consider $M\bullet$, $W = \operatorname{syz}^{d-1}(A/I)$ (In $M\bullet$, $A^{r_{d-1}} \to A \oplus A^{r_{d-2}}$, e_1 goes to $(-1, a_1, \ldots, a_{r_{d-2}})$) and hence $S'_{d-1} = \operatorname{syz}^d(A/I)$ by (3) (here these syzygies are not necessarily minimal).

Applying $\operatorname{Hom}(-, {}^{f^n}\bar{A})$ to (4) we get a surjection

$$\begin{aligned} \operatorname{Ext}^1_A(S_{d-2}, {}^{f^n}\bar{A}) &\underset{\text{map}}{\overset{\text{onto}}{\longrightarrow}} \operatorname{Ext}^1_A(T, {}^{f^n}\bar{A}) \oplus \operatorname{Ext}^1_A(W, {}^{f^n}\bar{A}) \\ &= (A/\underline{a} \otimes {}^{f^n}\bar{A}) \oplus \operatorname{Ext}^d_A(A/I, {}^{f^n}\bar{A}). \end{aligned} \tag{5}$$

We write $\bar{C}\bullet$ to denote $C\bullet \otimes_A \bar{A}$ for any complex $C\bullet$ and $\bar{N}$ to denote $N \otimes_A \bar{A}$ for any module N. Since p is a non-zero-divisor on A/I, $\operatorname{Ext}^d_A(A/I, {}^{f^n}\bar{A}) = \operatorname{Ext}^d_{\bar{A}}(\bar{A}/\bar{I}, {}^{f^n}\bar{A})$.

Step 3. Now consider the following short exact sequences:

$$0 \to H \to A/\underline{x} \to A/I \to 0 \text{ and}$$

$$0 \to A/I \to A/\underline{x} \to A/(p^s, x_2, \ldots, x_d) \to 0$$

Tensoring the second one with ${}^{f^n}\bar{A}$ one gets an exact sequence:

$$\begin{aligned} \operatorname{Tor}_1^A(A/(x_2, \ldots, x_d), {}^{f^n}\bar{A}) \to \operatorname{Tor}_1^A(A/(p^s, x_2, \ldots, x_d), {}^{f^n}\bar{A}) \\ \to F^n(\bar{A}/\bar{I}) \to 0. \end{aligned}$$

Since $\lim_{n\to\infty} \ell(\mathrm{Tor}_1^A(A/\underline{x}, {}^{f^n}\bar{A}))/p^{n(d-1)} = 0$ (see [**D2**]), we get

$$(6)\quad \lim_{n\to\infty} \ell(\mathrm{Tor}_1^A(A/p^s, x_2, \ldots, x_d, {}^{f^n}\bar{A}))/p^{(d-1)n} = \lim_{n\to\infty} \ell(F^n(\bar{A}/\bar{I}))/p^{n(d-1)}.$$

Moreover tensoring the first exact sequence by $\bar{A}$ we get an exact sequence

$$(7)\qquad 0 \to \bar{H} \to \bar{A}/(\bar{x}_2, \ldots, \bar{x}_d) \to \bar{A}/\bar{I} \to 0.$$

Now $\mathrm{Ext}^1(S_{d-2}, {}^{f^n}\bar{A}) = H_{d-1}(\mathrm{Hom}(F\bullet, {}^{f^n}\bar{A})) = H_{d-1}(\mathrm{Hom}(\bar{F}\bullet, {}^{f^n}\bar{A})) = W_{d-1,n}$ (as in the statement of the Theorem in (3.2)). Hence (5) gives us a surjection

$$W_{d-1,n} \to \bar{A}/(\bar{a}_1^{p^n}, \ldots, \bar{a}_{r_{d-2}}^{p^n}) \oplus \mathrm{Ext}_{\bar{A}}^d(\bar{A}/\bar{I}, {}^{f^n}\bar{A})$$

since $\underline{a} \supset \underline{x}$, $\ell(\bar{A}/(\bar{a}_1, \ldots, \bar{a}_{r_{d-2}}))$ is finite and hence

$$\begin{aligned} \lim_{n\to\infty} \ell(W_{d-1}, n)/p^{n(d-1)} &\geq \lim_{n\to\infty} \ell(\bar{A}/(\bar{a}^{p^n})/p^{n(d-1)}) \\ &+ \lim_{n\to\infty} \ell(\mathrm{Ext}_{\bar{A}}^d(\bar{A}/\bar{I}, {}^{f^n}\bar{A})/p^{n(d-1)}) \\ &> \lim_{n\to\infty} \ell(\mathrm{Ext}_{\bar{A}}^d(\bar{A}/\bar{I}, {}^{f^n}\bar{A})/p^{n(d-1)}) \end{aligned}$$

By the stated theorem (3.2) we derive the following inequality from above:

$$(8)\quad \lim_{n\to\infty} \ell(F^n(\bar{H}) \otimes \tilde{\bar{A}})/p^{n(d-1)} > \lim \ell(\mathrm{Tor}_1^{\bar{A}}(\bar{A}/\bar{I}, {}^{f^n}\bar{A}) \otimes \tilde{\bar{A}})/p^{n(d-1)}$$

[by using (ii)(c) of (3.1)]. Here $\tilde{\bar{A}} = \mathrm{Hom}(H_m^{d-1}(\bar{A}), E(k))$. Now we show that (8) cannot occur.

Step 4. Applying $\otimes^{f^n}\bar{A}$ to (7) we get an exact sequence:

$$(9)\quad \mathrm{Tor}_1^{\bar{A}}(\bar{A}/(\bar{x}_i), {}^{f^n}\bar{A}) \to \mathrm{Tor}_1^{\bar{A}}(\bar{A}/\bar{I}, {}^{f^n}\bar{A}) \to F^n(\bar{H}) \to F^n(\bar{A}/(\bar{x}_i)) \to F^n(\bar{A}/\bar{I}) \to 0.$$

Here $(\bar{x}_i)$ stands for $(\bar{x}_2, \ldots, \bar{x}_d)$. From the hypothesis and (6) it follows that

$$\lim \ell(F^n(\bar{A}/(\bar{x}_i)))/p^{n(d-1)} = \lim \ell(F^n(\bar{A}/\bar{I}))/p^{n(d-1)}.$$

Thus from (9) we get an exact sequence

$$(10)\qquad \mathrm{Tor}_1^{\bar{A}}(\bar{A}/\bar{I}, {}^{f^n}\bar{A}) \to F^n(\bar{H}) \to D_n \to 0$$

where $\lim \ell(D_n)/p^{n(d-1)} = 0$ (follows from the above equality and (9)).

Applying $\otimes \bar{\bar{A}}$ to (10) and taking the lengths we get

$$\ell(F^n(\bar{H}) \otimes \bar{\bar{A}}) \leq \ell(D_n \otimes \bar{\bar{A}}) + \ell(\mathrm{Tor}_1^{\bar{A}}(\bar{A}/\bar{I}, {}^{f^n}\bar{A}) \otimes \bar{\bar{A}}).$$

Since $\lim \ell(D_n \otimes \bar{\bar{A}})/p^{n(d-1)} = 0$ ($\bar{\bar{A}}$ is a finitely generated $\bar{A}$-module), it follows that

$$\lim \ell(F^n(\bar{H})^{f^n}\bar{A})/p^{n(d-1)} \leq \lim \ell(\mathrm{Tor}_1^{\bar{A}}(\bar{A}/\bar{I}, {}^{f^n}\bar{A}) \otimes \bar{\bar{A}})/p^{n(d-1)}$$

which contradicts (8). Thus $\mathrm{Im}\, \psi'_{d-1} \neq A$ and hence the theorem.

References

[**D1**] S. P. Dutta, *On the canonical element conjecture*, Transactions of the American Mathematical Society **299, No. 2** (1987), 803–811.

[**D2**] S. P. Dutta, Ext *and Frobenius*, preprint.

[**EG**] E. G. Evans and P. Griffith, *The Syzygy Problem*, Annals of Math **114** (1981), 323–353.

[**H1**] M. Hochster, *Contracted ideals from integral extensions of regular rings*, Nagoya Math. J. **51** (1973), 25–43.

[**H2**] M. Hochster, *Canonical elements in local cohomology modules and the direct summand conjecture*, Journal of Algebra **84 No. 2** (1983), 503–553.

[**PS**] C. Peskine and L. Szpiro, *Syzygies and multiplicities*, C. R. Acad. Sci. Paris, Sér. A-B **278** (1974), 1421–1424.

[**R1**] P. Roberts, *Two applications of dualizing complexes over local rings,*, Ann. Sci. Ec. Norm. Sup., 4e Sér., t. 9 (1976), 103–106.

[**R2**] P. Roberts, *Le théorème d'intersection*, C. R. Acad. Sci. Paris, t. 304 Sér. I, n. 7 (1987).

[**SZ**] R. Y. Sharp and H. Zakeri, *Generalized fractions and the monomial conjecture*, Journal of Algebra **92** (1985), 380–388.

Department of Mathematics, University of Illinois, 1409 West Green Street,
Urbana IL 61801

This research was partially supported by a National Science Foundation Grant and the Mathematical Sciences Research Institute, Berkeley CA.

Remarks on Points in a Projective Space

DAVID EISENBUD AND JEE-HEUB KOH

Introduction.

In this paper we will survey some results and conjectures on the free resolutions of ideals of sets of points in projective r-space. For general sets of points, there are conjectures of Lorenzini. A weaker statement, relevant to sets containing between $r+1$ and $2r$ points has been proved by Green and Lazarsfeld [**GL2**], and they conjecture a necessary and sufficient condition on the set of points for the weaker statement to hold. As Green has noted, a part of their conjecture follows from a conjecture of ours (with Mike Stillman) on linear syzygies [**EKS**], and we explain this connection and the consequences of the known cases of the linear syzygy conjecture; in particular the conjecture holds for $r \leq 4$. We also extend the result of Green-Lazarsfeld to deal with some larger sets of points.

One special case of the Green-Lazarsfeld conjecture says that if X is a set of $2r$ points in $\mathbf{P}^r$ such that no $2k+1$ of them lie in a k-plane, then the homogeneous ideal of X is generated by quadrics. We give a proof of this part of the conjecture, independent of the linear syzygy conjecture, for $r \leq 4$. Using a result in Matroid Theory due to J. Edmonds we prove a corresponding result, but only "scheme-theoretically", for sets of dr points, and forms of degree d, for any d and r.

We work over an algebraically closed field K. Let $X \subset \mathbf{P}^r_K = \mathbf{P}^r$ be a set of (distinct and reduced) points, not contained in any hyperplane. We denote by $S = \oplus_{d \geq 0} S_d = K[x_0, \ldots, x_r]$ the homogeneous coordinate ring of $\mathbf{P}^r$ and by $|X|$ the cardinality of X. We say that X imposes independent conditions on forms of degree $d \geq 1$ if the following holds:

(i) If $|X| \leq \dim_K S_d$, then $I(X)_d = \dim_K S_d - |X|$.

(ii) If $|X| \geq \dim_K S_d$, then no $(\dim_K S_d)$-points of X lie on a hypersurface of degree d.

Let $I = I(X)$ be the homogeneous ideal of X and let $E_\bullet$ be a minimal graded free resolution of I over S:

$$0 \to E_r \to E_{r-1} \to \cdots \to E_2 \to E_1 \to S \to S/I \to 0,$$

where $E_i = \oplus_{j\geq 1}\{S(-i-j)\otimes \mathrm{Tor}_i^S(S/I,K)_{i+j}\}$. We deal with the question of when the first few terms of $E_\bullet$ are as simple as possible and we extend the property (N_p) of Green and Lazarsfeld [**GL1**] as follows: If $C(r+d-1,r) \leq |X| < C(r+d,r)$, where $C(a,b)$ denote the binomial coefficient $\binom{a}{b}$, we will say that X satisfies $(N_{d,p})$ (for $0 \leq p \leq r$) if X imposes independent conditions on forms of degree d and, if $p > 0$, $\mathrm{Tor}_i^R(S/I(X),K)_j = 0$ for $j \geq d+i$ and $i \leq p$. Thus X satisfies $(N_{d,1})$ if the ideal of X is generated by forms of degree d, X satisfies $(N_{d,2})$ if in addition all the relations among these generators are linear, etc. We note that $(N_{2,p})$ is the property (N_p) of Green and Lazarsfeld [**GL1**].

This paper concerns the following result and conjecture of Green and Lazarsfeld [**GL2**]:

THEOREM (Green-Lazarsfeld). *Let $X \subset \mathbb{P}^r$ be a set of $(r+1)+(r-p)$ $(1 \leq p \leq r)$ points in linear general position, i.e. no $r+1$ lying on a hyperplane. Then X satisfies $(N_{2,p})$.*

CONJECTURE 1 (Green-Lazarsfeld). *Let $X \subset \mathbb{P}^r$ be a set of $(r+1)+(r-p)$ $(1 \leq p \leq r)$ points. If X fails to satisfy $(N_{2,p})$, then there is an integer $k \leq r$, and a subset $Y \subset X$ consisting of at least $2k+2-p$ points, such that*

(a) *Y is contained in a linear subspace $\mathbb{P}^k \subset \mathbb{P}^r$, and*
(b) *$(N_{2,p})$ fails for Y in $\mathbb{P}^k$.*

M. Green has proved that Conjecture 1(a) is a consequence of the following conjecture on linear syzygies made by us in collaboration with M. Stillman: A graded S-module $M = \oplus_{d\geq t} M_d$, $M_t \neq 0$, is said to have a k-th linear syzygy if $\mathrm{Tor}_k^S(M,K)_{k+t} \neq 0$.

CONJECTURE 2 (Linear Syzygy Conjecture). *Let $M = \oplus_{d\geq t} M_d$, $M_t \neq 0$, be a graded S-module. Let $\mathcal{R}$ denote the kernel of the map $S_1 \otimes M_t \to M_{t+1}$ and let $\mathcal{R}_r$ denote the rank r locus of $\mathcal{R}$ (here an element of $\mathcal{R}$ is viewed as a linear transformation from M_t^* to S_1). If M has a linear k-th syzygy, then $\mathcal{R}$ satisfies:*

(i) *if $\dim_{\mathrm{K}} M_t \leq k$ then $\dim \mathcal{R}_1 \geq k$, and*
(ii) *if $\dim_{\mathrm{K}} M_t \geq k$ then $\dim \mathcal{R}_{m-k+1} \geq k$,*

where dim *denotes the dimension as an affine variety and $m = \dim_k M_t$.*

Some cases of Linear Syzygy Conjecture are known [**EKS**] and Conjecture 1(a) for $r \leq 4$ or $p \geq r-2$ follows from these cases. In Section 1 we give a modification of Green's argument to prove that Conjecture 2 implies Conjecture 1(a).

In Section 2 we give a direct argument for Conjecture 1 when $r \leq 4$ and $p = 1$ (part (b) of Conjecture 1 follows from part (a) in this case).

In Section 3 we use a result of J. Edmonds [**Ed**] to prove the following two theorems:

A form of degree d is called multilinear if it is a product of linear forms.

THEOREM 1. *Let X be a set of points in $\mathbb{P}^r$, and let $d \geq 2$ be an integer. If, for all $k \geq 1$, no $dk+1$ of the points of X lie in a projective k-plane, then X is scheme-theoretically the intersection of multilinear forms of degree d.*

THEOREM 2. *Let X be a set of points in $\mathbb{P}^r$, and let $d \geq 2$ be an integer. If, for all $k \geq 1$, no $dk+2$ of the points of X lie in a projective k-plane, then X impose independent conditions on forms of degree d; in fact there is a multilinear form of degree d containing any subset consisting of all but one of the points, but missing the last.*

Theorem 2 in case $d = 2$ is Conjecture 1(a) for the case $p = 0$. This case was also proved by Green and Lazarsfeld.

Since Edmonds's paper is somewhat obscurely published, and since his argument is very elegant, we will reproduce it in Section 3 (with minor modifications to clarify one point).

In Section 4 we extend the argument of Green and Lazarsfeld to generalize their theorem above to:

THEOREM 3. *Let $X \subset \mathbb{P}^r$ be a set of $\binom{r+d}{d} + (r-p)$ points $(0 \leq p \leq r)$ imposing independent condition on forms of degree d. Then*

(i) *X imposes independent conditions on forms of degree $(d+1)$, and*
(ii) $\mathrm{Tor}_i{}^S(S/I, K)_j = 0$, *for all $1 \leq i \leq p$ and $j \neq i + d$.*

The Green-Lazarsfeld theorem and conjectures should be contrasted with the best plausible conjectures for general points in $\mathbb{P}^r$, which have been worked out by Lorenzini [**L**]. Let M be a finitely generated graded module over S; there is a natural approximation β_{ij} to the graded Betti numbers $b_{ij} = \dim_K \mathrm{Tor}^S{}_i(M, K)_j$ which can be computed in terms of the Hilbert function $H(M,t) = \dim_K M_t$ of M. These are given as follows: there is a unique function $\phi : \mathbb{Z} \to \mathbb{N}$ such that $\phi(j) = 0$ for $j \ll 0$ and integers $\beta_{ij} \geq 0$ such that

(1) $H(M,t) = \sum_k (-1)^{\phi(k)} \beta_{\phi(k)k} C(r+t+1-k, r)$

(2) $\phi(j) \leq \phi(j+1) \leq \phi(j)+1$
(3) $\beta_{ij} = 0$ unless $i = \phi(j)$
(4) if $\phi(j) \neq \phi(j-1)$, then $\beta_{\phi(j),j} > 0$.

ϕ and the β_{ij} may be constructed inductively as follows: Supposing that $H(M,t) = 0$ for $t \leq t_0$, we set $\phi(t) = 0$ and all $\beta_{it} = 0$ for $t \leq t_0$. Having defined $\phi(t)$ and for all $t \leq$ some t_1, we define

$$\phi(t_1+1) = \begin{cases} \phi(t_1) & \text{if } (-1)^{\phi(t_1)}\{H(M,t_1+1) \\ & \qquad -\sum_k (-1)^{\phi(k)}\beta_{\phi(k)k}C(r+t_1+1-k,r)\} \geq 0 \\ \phi(t_1)+1 & \text{if not.} \end{cases}$$

and $\beta_{\phi(t_1+1),t_1+1} = \left|H(M,t_1+1) - \sum_k (-1)^{\phi(k)}\beta_{\phi(k)k}C(r+t_1+1-k,r)\right|$ and of course $\beta_{i,t_1+1} = 0$ for $i \neq \phi(t_1+1)$.

The reader may check that this is unique. Note however that it is easy to produce Hilbert functions for which the β_{ij} defined above cannot be equal to the b_{ij}. For example, let M be two copies of K in degrees 0 and 2, i.e. $M = M_0 \oplus M_2$ and $M_0 = M_2 = K$. One checks that $\beta_{02} = 0$ but $b_{02} = 1$ and M is the unique module whose Hilbert function is $H(t)$, where $H(t) = 1$ if $t = 0$ or 2 and 0 otherwise. However, in the case where M is the homogeneous coordinate ring of a general set of points in $\mathbb{P}^r$, and in many other geometric situations, such problems do not arise, and it is natural to conjecture that $b_{ij} = \beta_{ij}$. It would be very interesting to have general conditions under which this conjecture is plausible.

Lorenzini has worked out the numbers β_{ij} explicitly for the case of the homogeneous coordinate ring of a set of points in general position. Geramita and Maroscia [**GM**] have shown that the $b_{ij} = \beta_{ij}$ for general sets of points in $\mathbb{P}^3$. An easily stated consequence of her computation is:

General Points Conjecture. *If X is a general set of points in $\mathbb{P}^r$ with $C(r+d-1,r) \leq |X| < C(r+d,r)$, then X satisfies $N_{d,p}$ $(0 \leq p \leq r)$ iff*

$$(d/d+p+1)\,C(r+d,r) < |X| \leq (d/d+p)\,C(r+d,r).$$

The conclusion of this conjecture is much stronger than that given in the Theorem of Green-Lazarsfeld in the case $d = 2$ or by our Theorem 3 in case $d \geq 3$. For example, the the conjecture suggests that 10 general points in $\mathbb{P}^4$ satisfy $N_{2,1} = N_1$ (which is true since the hyperplane section of a general canonical curve in $\mathbb{P}^5$ clearly satisfies N_1), whereas the Green-Lazarsfeld theorem says only that 8 or fewer general points satisfy N_1. On the other hand, the hypothesis of the Green-Lazarsfeld theorem, or our Theorem 3, that the points impose independent conditions on forms of degree d, is also much weaker than the hypothesis of generality, and with this weaker hypotheses, the theorems are (at least sometimes) sharp.

For example, the Green Lazarsfeld theorem says that 7 or fewer points in linearly general position in $\mathbb{P}^3$ impose independent conditions on quadrics, while actually any number of general points have this property. However, if we choose 8 points on a twisted cubic curve, then they will be in linearly general position (no 4 on a plane) but will not impose independent conditions on quadrics, as every quadric containing 7 of the points contains the twisted cubic.

We thank M. Stillman and J. Harris for helpful discussions. We are very grateful to Neil White for telling us of Edmonds's theorem. Many of the results here were discovered, confirmed, or both with the help of the computer algebra program Macaulay written by D. Bayer and M. Stillman.

1. The Linear Syzygy Conjecture implies part (a) of the Green-Lazarsfeld Conjecture.

The result of this section was first proved by Mark Green (unpublished). We give a simplification of his proof.

Let X be as in Conjecture 1 and let $I = I(X)$. Let $R = S/I$ and let m denote the irrelevant maximal ideal R_d. Let $\omega = \omega_R$ denote the dualizing module $\mathrm{Ext}^r_S(R, S(-r-1))$. Let x be a linear form which does not vanish at any point of X. For each point P of X, choose a homogeneous coordinates so that $x(P) = 1$. We write $\mathrm{Hom}_K(R, K)$ for $\otimes_d \mathrm{Hom}_K(R_d, K)$, the graded dual of R.

LEMMA 1. *There is an exact sequence* $0 \to \omega \to \oplus_{P\in X} K[x, x^{-1}]P \xrightarrow{\phi} \mathrm{Hom}_K(R, K) \to 0$, *where* ϕ *is defined by* $\phi(x^d P)(r) = r(P)$ *if degree* $r = d$ *and 0 otherwise.*

REMARK: The above sequence can presumably be derived from the local cohomology exact sequence associated to the inclusion of the punctured cone U over X into the cone CX over X: $(H^0_m(\omega) = 0 \to H_0(CX, \omega) \to H^0(U, \omega) \to H^1_m(\omega) \to H^1(CX, \omega) = 0$. Here one can identify $H^0(U, \omega)$ with $\oplus_{P\in X} K[x, x^{-1}]P$, since U is a disjoint union of punctured lines, one for each point P of X, and one can identify $H^1_m(\omega)$ with

$$\mathrm{Hom}_R(\mathrm{Hom}_R(\omega, \omega), E_R(K)) \simeq \mathrm{Hom}_R(R, E_R(K)) \simeq \mathrm{Hom}_K(R, K),$$

$E_R(K)$ being the injective envelope of K as an R-module. However, there are so many identifications in this interpretation that we found it simpler to give a direct proof.

PROOF: We first prove that ϕ is onto. Since $\mathrm{Hom}_K(R, K)$ is generated by its elements of large negative degree, it suffices to show that ϕ is onto in degree $-n$ for large n. Since X imposes independent conditions on forms of degree n (n large), ϕ is one-to-one in degree $-n$. Since $\dim_K\{\oplus_{P\in X} K[x, x^{-1}]P\}_{-n} = |X| = \dim_K \mathrm{Hom}_K(R, K)_{-n}$, ϕ is onto in degree $-n$ for all large n.

We now complete our proof by proving: for any exact sequence

$$(*) \qquad 0 \to M \to \oplus_{P\in X} K[x, x^{-1}]P \to \mathrm{Hom}_K(R, K) \to 0,$$

$M \simeq \omega$.

To prove this we use the fact that $M \simeq \omega$ iff M is torsion-free and, for some non-zero divisor x in R, $M/xM(1) \simeq \omega_{R/xR}$, the dualizing module of R/xR. (One may check ($\Leftarrow$) as follows: Since M is torsion-free, $\mathrm{Ext}^1_R(M, \omega) \simeq \mathrm{Hom}_K(H^0_m(M), K) = 0$ and the map $\mathrm{Hom}(M, \omega) \to \mathrm{Hom}(M, \omega/x\omega)$ is onto. Hence we can lift the map $M \to M/xM \simeq \omega_{R/xR}(-1) \simeq \omega/x\omega$ to ω and Nakayama's lemma, with the torsion-freeness of M, shows that this lifting is an isomorphism.)

Let x be a non-zero divisor. M is clearly torsion-free and we only need to check that $M/xM \simeq \omega_{R/xR}(-1)$. Since the multiplication by x gives an isomorphism of $\oplus_{P\in X} K[x, x^{-1}]P$, $\mathrm{Tor}_i^R(\oplus_{P\in X} K[x, x^{-1}]P, R/xR) = 0$ for all $i \geq 0$. Thus from the long exact sequence of Tor modules associated with the exact sequence (*), we obtain $\mathrm{Tor}_1^R(\mathrm{Hom}_K(R, K), R/xR) \simeq M/xM$. But $\mathrm{Tor}_1^R(\mathrm{Hom}_K(R, K), R/xR) \simeq \mathrm{Hom}_R(R/xR, \mathrm{Hom}_K(R, K)(-1)) \simeq \mathrm{Hom}_R(R/xR, E_R(K))(-1) \simeq E_{R/xR}(K)(-1) \simeq \omega_{R/xR}(-1)$ and we are done. (Here $E_{R/xR}$ denotes the injective envelope of R/xR). □

We say that a homogeneous ring S/I is n-regular if $\mathrm{Tor}_i^S(S/I, K)_{i+j} = 0$ for all i and all $j \geq n$.

LEMMA 2. *Let $X \subset \mathbb{P}^r$ be a set of points with $|X| \leq \dim_K S_d$. If X imposes independent conditions on forms of degree d, then*

(i) *$S/I(X)$ is $(d+1)$-regular, and*
(ii) *$\mathrm{Tor}_i^S(S/I(X), K)_{i+d} \neq 0$ if and only if $\omega_{S/I(X)}$ has a $(r-i)$-th linear syzygy.*

PROOF: (i) is well known (and easy to prove: just note that modulo a general linear form $I(X)$ contains the $(d+1)$-rst power of the maximal ideal.) For (ii) let $E_\bullet$ be a minimal graded free resolution of $S/I(X)$ over S. Since $S/I(X)$ is Cohen-Macaulay, $\mathrm{Hom}(E_\bullet, S(-r-1))$ is a minimal graded resolution of $\omega_{S/I(X)}$. Since $S/I(X)$ is $(d+1)$-regular, the conclusion follows. □

Suppose now that X fails to satisfy (N_p), so that $\mathrm{Tor}_p^S(S/I, K)_{p+2} \neq 0$ and, by Lemma 2, ω has a $(r-p)$-th linear syzygy. Let $\mathcal{R}$ denote the kernel of the map $S_1 \otimes \omega_{-1} \to \omega_0$ and $\mathcal{R}_1$ its rank 1 locus. By Linear Syzygy Conjecture,

$$(**) \qquad \dim \mathcal{R}_1 \geq r - p.$$

We can describe $\mathcal{R}_1$ explicitly from the exact sequence of Lemma 1. For a subset Y of X, let

$$\begin{aligned} B(Y) &= \left\{ \sum c_P\, x^{-1} P \in \omega_{-1} \mid c_P = 0, \text{ for all } P \text{ not in } Y \right\}, \\ L(Y) &= \{ y \in S_1 \mid y(P) = 0, \text{ for all } P \in Y \}, \text{ and} \\ s(Y) &= \text{(projective) dimension of the linear space in } \mathbb{P}^r \text{ spanned by } Y. \end{aligned}$$

Let $y \otimes a \in \mathcal{R}_1$, where $y \in S_1$ and $a = \sum c_P\, x^{-1} P \in \omega_{-1}$. Then $ya = 0$ in ω and from the exact sequence in Lemma 1, $x(P) = 0$ whenever $c_P \neq 0$. Let $Y = \{P \in X \mid c_P \neq 0\}$. Then $y \otimes a \in L(Y) \otimes B(Y)$ and hence

$$\mathcal{R}_1 = \cup_{Y \subset X} \{L(Y) \otimes B(Y)\}_1,$$

where $\{L(Y) \otimes B(Y)\}_1$ denotes the rank 1 locus of $L(Y) \otimes B(Y)$. Since $\dim_K L(Y) = r - s(Y)$ and $\dim_K B(Y) = |Y| - s(Y) - 1$,

$$\dim\{L(Y) \otimes B(Y)\}_1 = r + |Y| - 2s(Y) - 2.$$

Hence, for some $Y \subset X$, $\dim \mathcal{R}_1 = \dim\{L(Y) \otimes B(Y)\}_1 = r + |Y| - 2s(Y) - 2$. Thus $|Y| \geq 2s(Y) + 2 - p$ by (**) and this is what we wanted to prove for Conjecture 1(a).

2. $2r$ Points in $\mathbb{P}^r$.

In this section we show that Conjecture 1 holds if $p = 1$ and $r \leq 4$. Because an ideal of $2r + 1$ points in $\mathbb{P}^r$ with $r \leq 3$ is never generated by quadrics, part (b) of Conjecture 1 follows from part (a) in the case $r \leq 4$. Let $X \subset \mathbb{P}^r_K = \mathbb{P}^r$ be a set of $2r$ points such that, for all $k \geq 1$, no $2k+1$ points of X lie in a projective k-plane. For part (a), we must show that if $r \leq 4$, X satisfies $(N_{2,1})$, i.e. $I = I(X)$ is generated by quadrics.

Suppose that I is not generated by quadrics and let J denote the ideal generated by the quadrics in I. We will show that $V(J)$ has a positive dimension and this will contradict Theorem 1 with $d = 1$. Since $V(J)$ has a positive dimension if and only if a general hyper plane meets $V(J)$, it will be enough to show that the height of the ideal $J + x_0 S$ doesn't exceed r for all general linear form x_0 of S.

We will use the following notation:

$$\begin{aligned} R &= S/I \\ T &= S/x_0 S \\ \bar{R} &= R/x_0 R \\ \omega = \omega_R &= \operatorname{Ext}^r_S(R, S(-r-1)) \\ \bar{\omega} = \omega_{\bar{R}} &= \operatorname{Ext}^{r+1}_S(\bar{R}, S(-r-1)) = \operatorname{Ext}^r_T(\bar{R}, T(-r)). \end{aligned}$$

Since X imposes independent conditions on quadrics by Theorem 2, $\dim_K I_2 = (1/2)r(r-1)$ and $\bar{R} = \bar{R}_0 \oplus \bar{R}_1 \oplus \bar{R}_2$ with $\dim_K \bar{R}_2 = r - 1$. By duality, $\bar{\omega} = \operatorname{Hom}_K(\bar{R}, K) = \bar{\omega}_{-2} \oplus \bar{\omega}_{-1} \oplus \bar{\omega}_0$. Since I is not generated by quadrics, $\operatorname{Tor}^S_1(R, K)_3 \neq 0$ and ω has a $(r-1)$-st linear syzygy by Lemma 2 of Section 1. Since $\bar{\omega} = (\omega/x\omega)(1)$, $\bar{\omega}$ also has a $(r-1)$-th linear syzygy over T. Let $\{x_1, \ldots, x_r\}$ be elements of S which form a basis in T. Using the Koszul resolution of K, we obtain

$$\operatorname{Tor}^T_{r-1}(\bar{\omega}, K)_{r-3} \simeq \operatorname{Ker}(\wedge^{r-1} T_1 \otimes \bar{\omega}_{-2} \xrightarrow{\partial} \wedge^{r-2} T_1 \otimes \bar{\omega}_{-1})$$

and a nonzero element of $\operatorname{Tor}^T_{r-1}(\bar{\omega}, K)_{r-3}$ can be expressed as

$$a = \sum e_i \otimes a_i, \text{ where } e_i = x_1 \wedge \cdots \wedge x_{i-1} \wedge x_{i+1} \wedge \cdots \wedge x_r.$$

Since $\partial(a) = \sum \pm e_{ij}(x_i\, a_j - x_j\, a_i)$, where $e_{ij} = x_1 \wedge \cdots \wedge x_{i-1} \wedge x_{i+1} \wedge \cdots \wedge x_{j-1} \wedge x_{j+1} \wedge \cdots \wedge x_r$, the 2×2 minors of the matrix

$$A = \begin{pmatrix} x_1 & \ldots & x_r \\ a_1 & \ldots & a_r \end{pmatrix}$$

are zero in $\bar{\omega}$. Since $\dim_K \bar{\omega}_{-2} = r-1$, we may change variables and assume $a_1 = 0$.

PROPOSITION. *Let W denote the subspace of $\bar{\omega}_{-2}$ spanned by $\{a_2, \ldots, a_{r-1}\}$.*

(a) $\dim_K W < r-1$.

(b) *If $\dim_K W = 1$, then the ideal $(x_0, x_1, \ldots, x_{r-1})$ contains J.*

(c) *If $\dim_K W = r-2$, then the ideal (x_0, x_1, x_2) contains $(2r-2)$-dimensional subspace of I_2, where I_2 is the vector space of quadrics in I.*

PROOF: (a) Because $a_1 = 0$ we get $x_1 a_i = 0$ for all $2 \le i \le r$, or equivalently, $a_i(x_1 \bar{R}_1) = 0$. Suppose that $\dim_K W = r-1$. Since $a_2, \ldots, a_r$ span the dual of $\bar{R}_2$, $x_1 \bar{R}_1 = 0$ and we can choose L_i in S_1 such that $x_1 x_i - x_0 L_i \in I$, for $1 \le i \le r$. Let

$$B = \begin{pmatrix} x_0 \, x_1 & \ldots & x_r \\ x_1 \, L_1 & \ldots & L_r \end{pmatrix}$$

Since any $2 \times (r+1)$ matrix of linear forms in $r+1$ variables can be transformed by row and column operations to make at least one entry 0 (see for example [**Ei**]), we can put B into the form

$$\begin{pmatrix} x_0 & \ldots & x_s & x_{s+1} & \ldots & x_r \\ L'_0 & \ldots & L'_s & 0 & \ldots & 0 \end{pmatrix},$$

for suitable s with $0 \le s < r$. Let $M = V(L'_0, \ldots, L'_s)$ and $N = V(x_{s+1}, \ldots, x_r)$. Since the 2×2 minors of B are contained in I, we get $X \subset M \cup N$. Since $\dim M + \dim N = r-1$, one of M and N, say M, must contain at least $2(\dim_K M) + 1$ points of X which contradicts our assumption.

Let $\mu : R_1 \otimes R_1 \to R_2$ be the multiplication map. To prove (b), we may assume that $a_1 = \cdots = a_{r-1} = 0$. Then $(x_1, \ldots, x_{r-1})a_r = 0$ in $\bar{\omega}$ and $\dim_K \mu((x_1, \ldots, x_{r-1}) \otimes R_1) \le r-2$. Hence $(x_0, x_1, \ldots, x_{r-1})$ contains $\{((1/2)r(r+1)-1)-(r-2)\}$-dimensional subspace of quadrics of I_2. But $((1/2)r(r+1)-1)-(r-2) = \dim_K I_2$ and we are done. To prove (c), we may assume that $a_1 = a_2 = 0$. Then $(x_1, x_2)(a_3, \ldots, a_r) = 0$ in $\bar{\omega}$ and $\dim_K \mu((x_1, x_2) \otimes R_1) \le 1$. Hence (x_0, x_1, x_2) contains $((2r-1)-1) = (2r-2)$-dimensional subspace of quadrics of I_2. □

We recall that we are trying to prove that the height of $(J + x_0 S) \le r$, $(r \le 4)$. (a) and (b) of the Proposition above prove the case when $r = 3$ and the case when $r = 4$ and $\dim_K W = 1$. It remains to check the case when $r = 4$ and $\dim_K W = 2$. Since (x_0, x_1, x_2) contains 6-dimensional subspace of quadrics of I_2, by (c) of the Proposition, and $\dim_K I_2 = 7$, (x_0, x_1, x_2)+(remaining quadric) is an ideal of height ≤ 4 which contains $J + x_0 S$.

3. Points Cut Out by Multilinear Forms of Degree d.

Theorem 1 and Theorem 2 follow easily from a result of Jack Edmonds [**Ed**]:

THEOREM E (Edmonds). *Let B be a set of points in projective space and let $d \geq 2$ be an integer. B may be divided into d disjoint sets of linearly independent points if and only if, for all $k \geq 0$, no $dk + d + 1$ of the points of B lie in a projective k-plane.*

In Theorem E we do not assume that the points of B are distinct–indeed they will not be in our applications. However, if either condition of the Theorem is satisfied, it is evident that no more than d of the points can be coincident at any one point.

The proof of Theorem E gives a little more information than we have stated: if we are given d distinct independent sets $A_i \subset A$ such that A_i has n_i elements, then there exists a decomposition of A into d disjoint subsets of independent vectors such that the i-th set has at least n_i elements. It would be nice to know under what circumstances one could guarantee a decomposition into independent sets corresponding to a given numerical decomposition of the number of points into d parts.

PROOF OF THEOREM E: For any subset C of B we write span C for the set of elements of B which are linearly dependent on the elements of C and rank C for the affine dimension of the linear space spanned by the points in C. It is easy to check that the first condition given in Theorem E is equivalent to the statement that for every subset C of B we have $(1/d)|C| \leq \operatorname{rank} C$.

Suppose we are given d (possibly empty) disjoint subsets B_i of B, each consisting of independent vectors. Let $B' = B - B_i$. If for some i the span of B_i does not contain some element x of B', then we can add x to B_i, preserving independence and completing the proof. Thus we may assume that for every i span of B_i contains B'. We will give a procedure for exchanging elements of various B_i for elements in B' in such a way as to change this situation. This will prove the Theorem.

Let $S_1 = \operatorname{span} B_1$. By our hypothesis we have $S_1 \neq \cup_{1\leq i\leq d} (S_1 \cap B_i)$. We will inductively define a strictly decreasing sequence of sets S_i, and a sequence of indices $m(i)$; in general, having defined S_{i-1}, we will define $m(i)$ and S_i iff S_{i-1} meets B', so that

$$S_{i-1} \neq \bigcup_1^d (S_{i-1} \cap B_i).$$

If this inequality is satisfied then for some index j we have $|S_{i-1} \cap B_j| < (1/d)|S_{i-1}|$, and choosing such a j we set $m(i) = j$ and $S_i = \operatorname{span}(S_{i-1} \cap B_{m(i)})$.

It is obvious from the definition that the sequence of S_i is weakly decreasing, but in fact the hypothesis of Theorem E gives the last of the string of inequalities $\operatorname{rank} S_i = |S_{i-1} \cap B_{m(i)}| < (1/d)|S_{i-1}| \leq \operatorname{rank} S_{i-1}$, so in fact the sequence is strictly decreasing.

Let h be the smallest number such that S_h does not contain B', and let $x \in B'$ be an element outside S_h. By hypothesis, $\{x\} \cup B_{m(h)}$ is a dependent set, and we let C be a minimal dependent subset, necessarily containing x. Let $k \leq h$ be the smallest index such that S_k does not contain C, and choose an element $y \in C$, $y \notin S_h$.

We will replace x by y in $B_{m(h)}$,obtaining a new collection of disjoint subsets

$$B'_i = \begin{cases} B_i & \text{if } i \neq m(h); \\ B_{m(h)} \cup \{y\} - \{x\} & \text{if } i = m(h). \end{cases}$$

We claim that if we now proceed as before, constructing a sequence of sets $S'_i = \operatorname{span}(S'_{i-1} \cap B'_{m(i)})$, then for $i \leq k$ we will have $S_i = S'_i$ so that in particular the defining inequality $|S'_{i-1} \cap B'_{m(i)}| < (1/d)|S'_{i-1}|$ will hold in this range. We prove this inductively: the case $i = 1$ being a consequence of the fact that the spans of $B_{m(h)}$ and $B'_{m(h)}$ both contain x and y, and thus coincide. We may thus assume that $i > 1$ and that $S'_{i-1} = S_{i-1}$. If $m(i) \neq m(h)$ then the desired inequality is immediate. If on the other hand $m(i) = m(h)$, then since $S_{i-1} \supset C$, we see that $S_{i-1} \cap B_{m(h)}$ and $S_{i-1} \cap B'_{m(h)}$ differ only in that the first does not contain x while the second does not contain y; since both contain the rest of C, they have equal spans $S_i = S'_i$ as required.

Finally, we claim that $k < h$; by induction, this will complete the proof. If on the contrary $k = h$, then by the equalities just established, $C \subset S_{h-i}$. But then $C - \{x\} \subset B_{m(h)}$, and $x \in \operatorname{span} C - \{x\}$, so $x \in S_h$, contradicting the definition of h. $\square$

PROOF OF THEOREM 1 FROM THEOREM E: We must show that the points of X are separated by multilinear forms of degree d from points not in X and from infinitely near points of X. To do this, it suffices, by adding some points in general position if necessary, to prove the Theorem in the case that X spans a projective r-space and contains exactly $d(r+1)+1$ points.

Note that if B is any set obtained from X by adding d points, then the hypothesis of Theorem 1 implies that the set B will satisfy the conditions of Theorem E, except in the case where all d points coincide with one of the points of X (the exception is essentially caused by the fact that in Theorem E we allow all $k \geq 0$, whereas in Theorem 1 k is constrained to be ≥ 1).

First, to prove that X is set-theoretically cut out by multilinear forms of degree d, let P be a point not in X, and let B be X with the point P adjoined d times. By Theorem E, B can be divided into d independent sets, and of course each of these will have $r+1$ elements. Clearly, each must contain one copy of P. Dropping these d copies of P, each of the resulting sets will span a hyperplane of $\mathbb{P}^r$, and these hyperplanes cannot contain P. Thus their union is a multilinear forms of degree d containing X but not containing P.

Finally, to show that X can be separated from an infinitely near point at $P \in X$, we let Q be a point distinct from P, but lying on the line through P and the infinitely near point. Let B be the result of adjoining Q and $d-1$ copies of P to X. Again by Theorem E, B is the union of d independent sets. Since again B contains a total of d copies of P, each of these sets must contain P, and in addition one — say B_1 — contains Q. Dropping Q from B_1 and dropping P from each of the other sets, we obtain d hyperplanes; exactly one of these hyperplanes, corresponding to B_1, contains P, and that hyperplane does not contain Q, so the union of the hyperplanes contains X but not the given infinitely near point at P. □

PROOF OF THEOREM 2 FROM THEOREM E: Adding some generally situated points if necessary, it suffices to prove Theorem 2 in the case where the number of points in X is $dr+1$, the maximum possible. For $P \in X$, we wish to construct a multilinear forms of degree d containing all the points of X except P. To this end add $d-1$ copies of P to X, obtaining a set B to which Theorem E may be applied. If we divide the $d(r+1)$ points of B into d independent sets, then each will contain a copy of P. The remaining points in each set span a hyperplane of $\mathbb{P}^r$, and the union of these hyperplanes is the desired form of degree d. □

We give here a very simple proof, due to Joe Harris, of a weakening of Theorem 2 in the case $d = 2$.

PROPOSITION. *Let X be a set of points in projective space. If, for all $k \geq 1$, no $2k+2$ of the points of X lie in a projective k-plane, then the points of X impose independent conditions on quadrics.*

PROOF: By induction, every proper subset of X impose independent condition on quadrics, so if X did not, then every quadric containing all but at most one element of X would contain X. It thus suffices to find a quadric containing all but exactly one element of X.

Suppose that the span of X is r-dimensional, so in particular $|X| \leq 2r+1$, and let $Y \subset X$ be a set of $r+1$ independent elements. The residual set $X - Y$ contains at most r elements, and is thus contained in a hyperplane H_1. If H_2 is the hyperplane spanned by the elements of Y besides P, then $H_1 \cap H_2$ is the desired quadric. □

4. $\binom{r+d}{d} + (r-p)$ points in $\mathbb{P}^r$.

In this section we prove Theorem 3 using descending induction on p. The proof uses the ideas of the proof in [**GL2**].

Let $p = r$. Then $|X| = \dim_K S_d$ and S/I is $(d+1)$-regular. Since $I_d = 0$, $\mathrm{Tor}_i^S(S/I, K)_j = 0$, for all i and all $j < i + d$ and (ii) follows. To prove (i), let P be a point of X. We want to find a form G of degree $d+1$ such that $G = 0$ on $X - \{P\}$ and $G(P) \neq 0$. We can do this by first finding such a form of degree d and then multiplying it by a general linear form.

Now let $p < r$ and let $X' = X - \{\text{point}\}$. Then by our induction hypothesis, X' imposes independent conditions on $(d+1)$-forms and $I(X')$ is generated by forms of degree $d+1$. Hence $I(X)_{d+1} \neq I(X')_{d+1}$ and this proves (i).

To prove (ii) it will suffice to show, by Lemma 2 in Section 1, that $\mathrm{Tor}_p^S(S/I, K)_{p+d+1} = 0$. Let $Y = \{P_0, \ldots, P_r\}$ be a subset of X in linear general position and let $\{x_0, \ldots, x_r\}$ be a basis of S_1 such that $x_i(P_i) = \delta_{ij}$. Let Q be the ideal of Y and J the ideal of $X - Y$. Then $I = I(X) = Q \cap J$ and we have an exact sequence $0 \to (S/I) \to (S/J) \oplus (S/Q) \to (S/J + Q) \to 0$. Since X imposes independent conditions on $(d+1)$-forms, the map $(S/I)_{d'} \to (S/J)_{d'} \oplus (S/Q)_{d'}$ is an isomorphism for all $d' \geq d+1$.

We use the Koszul resolution of K to compute Tor. We have the following diagram with obvious maps (see Figure 1). Since $|X-Y| \leq \dim_K S_d$, S/J is $(d+1)$-regular and $\operatorname{Tor}_p^S(S/J,K)_{p+d+1} = 0$. Since $\operatorname{Tor}_p^S(S/J,K)_{p+d+1}$ is the homology of

$$\left(\wedge^{p+1} S_1 \otimes (S/J)_d \to \wedge^p S_1 \otimes (S/J)_{d+1} \to \wedge^{p-1} S_1 \otimes (S/J)_{d+2}\right),$$

$\pi(\operatorname{Im}\delta) \supset \operatorname{Ker}\partial_2$. Thus it suffice to prove that $\operatorname{Ker}\partial_1 \subset \operatorname{Im}\delta$. Since Q is generated by $\{x_i x_j \mid 0 \leq i \neq j \leq r\}$, $\operatorname{Ker}\partial_1$ is generated by

$$\left\{x_{i_1} \wedge \cdots \wedge x_{i_p} \otimes x_j^{d+1} \mid 0 \leq i_1 < \cdots < i_p \leq r, j \notin \{i_1, \ldots, i_p\}\right\}.$$

Fix $x_{i_1} \wedge \cdots \wedge x_{i_p} \otimes x_j^{d+1}$ and let $F = \sum c_i x_i \pmod{Q}$ be a form of degree d such that $F = 0$ on $(X-Y) \cup \{P_{i_1}, \ldots, P_{i_p}\}$. Since X imposes independent conditions on forms of degree d and $\left|(X-Y) \cup \{P_{i_1}, \ldots, P_{i_p}\}\right| = \dim_K S_{d-1}$, $F(P_j) \neq 0$ (and hence $c_j \neq 0$) for all $j \notin \{i_1, \ldots, i_p\}$. Thus

$$\delta(x_{i_1} \wedge \cdots \wedge x_{i_p} \wedge x_j \otimes F) = e_{i_1} \wedge \cdots \wedge e_{i_p} \otimes x_j F = x_{i_p} \wedge \cdots \wedge x_{i_p} \otimes c_j x_j^{d+1}$$

and we are done. □

REMARK: It follows from Theorem 3 that if a set of $(r + \dim_K S_d)$ points imposes independent conditions on forms of degree d, then it imposes independent conditions on forms of degree $d+1$ also.

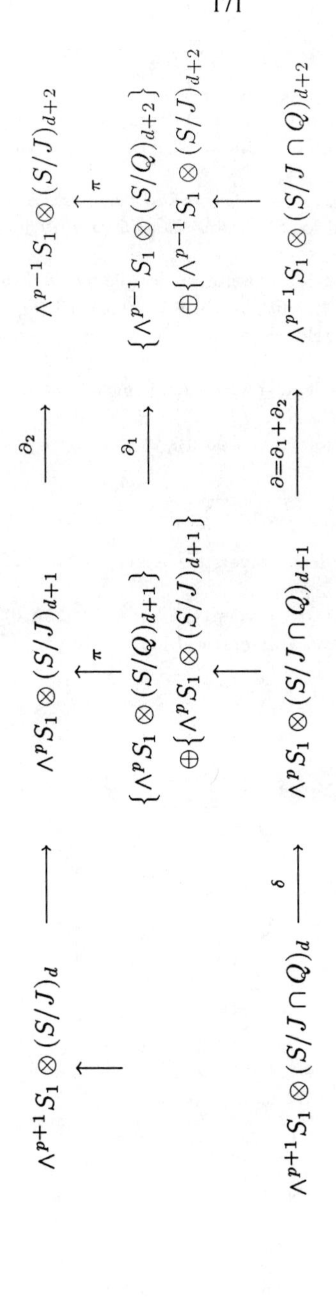

$\Lambda^{p+1}S_1 \otimes (S/J)_d \longrightarrow \Lambda^p S_1 \otimes (S/J)_{d+1} \xrightarrow{\partial_2} \Lambda^{p-1} S_1 \otimes (S/J)_{d+2}$
π
π
$\{\Lambda^p S_1 \otimes (S/Q)_{d+1}\} \oplus \{\Lambda^p S_1 \otimes (S/J)_{d+1}\} \xrightarrow{\partial_1} \{\Lambda^{p-1} S_1 \otimes (S/Q)_{d+2}\} \oplus \{\Lambda^{p-1} S_1 \otimes (S/J)_{d+2}\}$
$\Lambda^{p+1}S_1 \otimes (S/J \cap Q)_d \xrightarrow{\delta} \Lambda^p S_1 \otimes (S/J \cap Q)_{d+1} \xrightarrow{\partial=\partial_1+\partial_2} \Lambda^{p-1} S_1 \otimes (S/J \cap Q)_{d+2}$

Figure 1

References

[**Ed**] J. Edmonds, *Minimum partition of a matroid into independent subsets*, J. of Research of the National Bureau of Standards — B. Mathematics and Mathematical Physics **69B** (1965), 67–72.

[**Ei**] D. Eisenbud, *Linear sections of determinantal varieties*, Amer. J. Math (1988).

[**EKS**] D. Eisenbud, J. Koh, and M. Stillman, *The linear syzygy conjecture*, in preparation.

[**GL1**] M. Green and R. Lazarsfeld, *On the projective normality of complete linear series on an algebraic curve*, Invent. Math. **83** (1986), 73–90.

[**GL2**] M. Green and R. Lazarsfeld, *Some results on the syzygies of finite sets and algebraic curves*, preprint.

[**GM**] A. Geramita and P. Maroscia, *The ideal of forms vanishing at a finite set of points in* $\mathbb{P}^n$, J. of Algebra **90, No. 2** (1984).

[**L**] A. Lorenzini, *On the betti numbers of points in projective space*, Thesis, Queen's Univ. (1987).

Department of Mathematics, Brandeis University, Waltham MA 02254

Department of Mathematics, Harvard University, Cambridge MA 02138, and
Department of Mathematics, Indiana University, Bloomington IN 47405

Rank Varieties of Matrices

DAVID EISENBUD AND DAVID SALTMAN

Abstract. In this paper we extend work of Gerstenhaber [5], Kostant [9], Kraft-Procesi [10], Tanisaki [15], and others on orbit closures in the nilpotent cone of matrices by studying varieties of square matrices defined by conditions on the ranks of powers of the matrices, or more generally on the ranks of polynomial functions of them. We show that the irreducible components of such varieties are always Gorenstein with rational singularities (in particular they are normal). We compute their tangent spaces, and also their limits under deformations of the defining polynomial functions. We also study generators for the ideals of such varieties, and we compute the singular loci of the hypersurfaces in the space of $n \times n$ matrices given by the vanishing of a single coefficient of the characteristic polynomial.

Introduction.

Let V be an n-dimensional vectorspace over the complex numbers (or, with suitable scheme-theoretic interpretations of what is to come, over any field of characteristic 0), and let $X = \operatorname{End} V$ be the space of linear transformations of V into itself.

By analogy with the algebraic subsets of the affine line, it is natural to consider sets of transformations $A \in X$ defined by the vanishing of a collection of polynomial functions $p_i(A)$. Since the rank of a matrix in case $n > 1$ can take on values other than 0 and n, it is natural to extend consideration to sets of matrices of the form

$$\{A \in X \mid \operatorname{rank} p_i(A) \leq r_i, \ i = 1, \dots\}$$

for various polynomials $p_i(t)$ of one variable, and numbers r_i. In this paper we are concerned with sets defined in this way; we will call them *rank sets.*

A rank set is clearly invariant under the action of $\mathrm{PGL}(V)$ on X by conjugation. On the other hand, among the rank sets are the closures of the orbits of this action. Indeed, as shown by Gerstenhaber [5], the closure of the orbit of a transformation A is the set of those transformations B such that for every $i = 1, 2, \dots$ and for every $\lambda \in \mathbb{C}$ (or just every λ which is an eigenvalue of A),

$$\operatorname{rank}(B - \lambda)^i \leq \operatorname{rank}(A - \lambda)^i.$$

This case, and in particular the case of a nilpotent orbit closure to which it quickly reduces, has been studied by many authors from both topological and algebraic points of view; in particular, the results of this paper were already known for orbit closures.

At the opposite extreme of the rank sets are the determinantal varieties, each given by a single condition of the form rank $A \leq r$. Here there are many orbits, of many different dimensions, and infinitely many of these have maximal dimension.

Both these examples of rank sets are irreducible, but in the general case, irreducibility cannot be expected: a set of the form

$$\{A \mid \text{rank}[(A-\lambda)^i (A-\mu)^j]\} \leq r,$$

with $\lambda \neq \mu$, will consist in the union, over pairs of positive integers r_1, r_2 whose sum is r, of the set of A with rank $(A-\lambda)^i \leq r_1$ and rank$(A-\mu)^j \leq r_2$. We are thus led to consider first the rank varieties defined in terms of polynomials $p_i(t)$, each of which is a power of an irreducible polynomial, and we accordingly make the following definition:

For any sequence of complex numbers $\lambda = (\lambda_1, \ldots, \lambda_s)$, and for any doubly indexed set of integers $r(i,j)$, $1 \leq i \leq s$, $1 \leq j$ let $X_{r,\lambda}$ be the subset of $X = \text{End}(V)$ given by

$$X_{r,\lambda} = \{A \in X \mid \text{rank}(A-\lambda_i)^j \leq r(i,j) \text{ for all } 1 \leq i \leq s, 1 \leq j\}.$$

Extending the example above we easily derive:

Proposition. *Every rank set is a union of sets of the form $X_{r,\lambda}$.*

Proof Sketch: Since the intersection of two sets of the form $X_{r,\lambda}$ is again of that form, it is enough to treat the case of a rank set Y defined by a single condition rank $p(A) \leq u$. We may factor $p(t)$ into powers of irreducibles $p(t) = (t-\lambda_1)^{i_1}(t-\lambda_2)^{i_2}\ldots(t-\lambda_s)^{i_s}$. Given a tranformation $A \in X$, we may split V into eigenspaces of A with distinct eigenvalues, and from this splitting we see that the corank of $p(A)$ is simply the sum of the coranks of the transformations $(A-\lambda_j)^{i_j}$. Thus Y is the union, over all sequences $(u_1, \ldots, u_s)$ summing to u, of the sets $X_{r,\lambda}$ with $r(j,k) = n$, the dimension of V, for $k < i_j$, while $r(j,k) = u_j$ for $k \geq i_j$. □

This being so, we focus our attention on the $X_{r,\lambda}$. Of course the simplest of these are the ones given by conditions relative to a single eigenvalue, that is, those for which s is 1, and, after translating by a scalar matrix, we may even assume that the eigenvalue in question is 0. For convenience, we define, for any integer valued function $r(j)$,

$$X_r = \{A \in X \mid \operatorname{rank} A^j \leq r(j) \text{ for all } 1 \leq j\}.$$

Of course for general r there may not be any matrix A with rank $A^j = r(j)$ for all j. But it is not hard to show that such an A exists iff r is a decreasing concave non-negative integral function with $r(0) = n$; we will call such a function a rank function. Given an arbitrary non-negative function, r with $r(0) \geq n$, it is easy to see that there is a unique maximal rank function r' with $r'(j) \leq r(j)$ for every j, and $X_r = X_{r'}$.

The main results of Sections 1 and 2 of this paper may now be summarized as follows:

THEOREM 1. *The varieties $X_{r,\lambda}$ are Gorenstein, and are normal with rational singularities. Their limits under deformations of the λ_i are always varieties of the same form. Each $X_{r,\lambda}$ is a generically transverse intersection of translates by scalar matrices of varieties of the form $X_{r'}$, for suitable $r'(i)$.*

In some respects the hardest part of this is the normality statement, in whose proof we closely follow and augment ideas of DeConcini and Procesi [**3**]. The fundamental notions in their arguments are that of the variety of pairs of transformations whose composition is nilpotent, and that of a rather natural complete intersection of which it is the quotient, whose points correspond roughly to the transformations which are the possible products of the transformations in the original pair. In Section 1 we make the necessary changes to extend these ideas to the non-nilpotent situation, proving the normality of the varieties X_r. As in the argument of DeConcini and Procesi [**3**], the hypothesis of characteristic 0 enters to prove that the X_r are actually the desired quotients.

We return to the geometry of the $X_{r,\lambda}$ in Section 2. We construct a canonical resolution of singularities of the varieties $X_{r,\lambda}$. We then use the Grauert-Riemenschneider vanishing theorem to derive the normality of these varieties from that of the X_r and to establish the remaining properties given in the Theorem; of course this requires the characteristic 0 hypothesis again. Nevertheless, it seems reasonable to hope that the theorem remains true over an arbitrary algebraically closed field. We also derive an interesting representation for the tangent spaces to X_r at some smooth points.

In the third section of the paper we study the equations of the varieties $X_{r,\lambda}$. By the Theorem, it is enough to treat the X_r. Choosing a basis for V, we may identify $\operatorname{End} V$ with the set of $n \times n$ matrices. We give generators up to radical for the ideals of polynomials in the entries of these matrices which vanish on the varieties X_r. In a preliminary version of this paper we conjectured, extending previous conjectures of DeConcini-Procesi [**3**] and Tanisaki [**15**] which dealt with the nilpotent case, that these generators suffice to generate the ideals of the varieties themselves. After having read our preliminary version, J. Weyman [**16**] proved our conjectures in this direction.

The basic building blocks for the ideals defining the equations of the X_r are ideals identified by Tanisaki that we call $I(\lambda_d^t)$, for various t and d. $I(\lambda_d^t)$ is the ideal generated by the coefficients of x^{t-d} in the t-th exterior power of the matrix

$$x - A_{\text{gen}},$$

where A_{gen} is the generic matrix, whose entries are the coordinate functions on X, and x is an auxilliary variable; thus the generators of $I(\lambda_d^t)$ are polynomials of degree d in the entries of a matrix. By way of familiar examples, $I(\lambda_d^n)$ is generated by the single polynomial which is the d-th coeffient of the characteristic polynomial of the generic matrix, while $I(\lambda_d^d)$ is the ideal generated by all $d \times d$ minors of the generic matrix.

In the final section, solving the problem from which this paper originally arose, we identify for all k the singular locus S_k of the hypersurface in X defined by the vanishing of the coefficient of x^{n-k} in the characteristic polynomial

$$\det(x - A_{\text{gen}}):$$

THEOREM 4.10. *The matrix A is a singular point of the k-th coefficient of the characteristic polynomial iff writing $d :=$ rank A^n, we have $d \leq k-2$ and* rank $A^{k-d+1} \leq d$.

We also give an irredundant decomposition of S_k into irreducible components: these turn out to be rank varieties (Corollary 4.14).

These singular loci are related in an interesting way to the reduced varieties defined by the ideals $I(\lambda_k^t)$ introduced above. The variety defined by the vanishing of the k-th coefficient of the characteristic polynomial is $V(\lambda_k^n)$, and we have:

THEOREM 4.15. *The singular locus of* $V(\lambda_k^n)$ *is* $V(\lambda_{k-1}^{n-1})$.

This result is particularly suggestive in view of the well-known identification of the singular locus of a determinantal variety; in that case one has $\operatorname{Sing} V(\lambda_k^k) = V(\lambda_{k-1}^{k-1})$. One might at first conjecture that $\operatorname{Sing} V(\lambda_k^t) = V(\lambda_{k-1}^{t-1})$ in general, but examples show that this is false; however, it remains plausible that this conjecture becomes true if one replaces the two varieties by their normalizations, or even by the disjoint union of their irreducible components.

1. Normality.

1A) Linear Algebra. Let $r : Z^+ \to Z^+$ be a rank function. Our goal is to study $X_r = \{A \in M_n(F) \mid \operatorname{rank}(A^i) \leq r(i)\}$. To begin with, we consider the open subset $Y_r = \{A \in M_n(F) \mid \operatorname{rank}(A^i) = r(i)\}$. The stable value of r is zero if and only if Y_r consists of nilpotent matrices. The stable rank of an $A : U \to U$ is the stable value of its rank function.

Suppose $A : U \to U$ is arbitrary. If U has dimension n define the stable space, V, of A to be $A^n(U)$, a subspace of U. Observe that V is the direct sum of the eigenspaces of A associated to nonzero eigenvalues. It follows that $A(V) = V$ and A restricted to V is an isomorphism. Of course, the dimension of V is the stable rank of A. For any A, we define A' to be the induced linear transformation $U/V \to U/V$. Let s be the stable rank of A. Note that A' is nilpotent with rank function $r - s$, where r is the rank function of A. In general, if r is a rank function, define r' to be $r - s$, where s is the stable value of r. For any Y_r, choose U' to be a space with dimension $\dim(U) - s$, s being the stable value of r. Consider $Y_{r'}$ as a subset of $\operatorname{End}(U')$.

Let us state two elementary properties of the Y_r's.

LEMMA 1.1. Y_r *is irreducible and* $\operatorname{codim}_{\operatorname{End}}(U)(Y_r) = \operatorname{codim}_{\operatorname{End}}(U')(Y_{r'})$.

PROOF (outline): Let s be the stable value of r and consider the Grassmann variety G_s of s dimensional subspaces of U. Define $W \subseteq \operatorname{End}(U) \times G_s$ to be the locally closed subvariety of pairs (A, V) such that $A(V) = V$. Let $p : W \to \operatorname{End}(U)$ and $q : W \to G_s$ be the restrictions of the projection maps. Define $W_r \subseteq W$ to be the inverse image under p of Y_r. The map p restricted to W_r is an isomorphism. The inverse map sends A to (A, V), where V is the stable space of A.

Next consider the induced map $q_r : W_r \to G_s$. This is a surjection. If $V \in G_s$, $q_r^{-1}(V)$ consists of $A : U \to U$ with stable space V such that $A' : U/V \to U/V$ has rank function $r' = r - s$. We write $q_r^{-1}(V)$ as $Y_r(V)$. There is a regular map $t : Y_r(V) \to \mathrm{End}(U/V)$ defined by $t(A) = A'$. The image of t is precisely $Y_{r'}$. For any $C \in Y_{r'}$, $t^{-1}(C)$ consist of maps, A, whose matrix (with respect to a suitable basis) in block form is:

$$\begin{pmatrix} C & 0 \\ D & E \end{pmatrix}$$

where E is the matrix of A restricted to V, and so is nonsingular. It follows that all the $t^{-1}(C)$ are irreducible of dimension ns. By e.g., Shafarevich [**13**, p. 61], $Y_r(V)$ is irreducible of dimension $\dim(Y_{r'}) + ns$ and again W_r is irreducible of dimension $\dim(Y_{r'}) + ns + s(n-s)$. Thus $\mathrm{codim}_{\mathrm{End}}(U)(Y_r) = n^2 - \dim(Y_r) = n^2 - \dim(W_r) = n^2 - 2ns + s^2 - \dim(Y_{r'}) = \mathrm{codim}_{\mathrm{End}}(U')(Y_{r'})$. □

Our next subject is a theory for pairs of linear maps that parallels the one above for single ones. Formally speaking, a *pair* will be $\alpha = (A : U \to V; B : V \to U)$ where U, V are vector spaces and A and B are linear maps. For fixed U, V, then, a pair α is an element of $\mathrm{Hom}(U, V) \times \mathrm{Hom}(V, U)$ and we will write $\alpha = (A, B)$ for convenience. $\alpha = (A : U \to V; B : V \to U)$ and $\beta = (C : S \to T; D : T \to S)$ are isomorphic if there are linear isomorphisms $\varphi : U \to S$ and $\tau : V \to T$ such that $C = \tau A \varphi^{-1}$ and $D = \varphi B \tau^{-1}$. If we fix U and V, then these same equations define an action of $GL(U) \times GL(V)$ on $\mathrm{Hom}(U, V) \times \mathrm{Hom}(V, U)$, that is, on all pairs defined using U, V. If α and β are as above, we define the direct sum $\alpha \oplus \beta$ to be the pair $(A \oplus C : (U \oplus S) \to (V \oplus T);\ B \oplus D : (V \oplus T) \to (U \oplus S))$. We say α is a *nilpotent pair* if AB (and therefore BA) is nilpotent.

Suppose $\alpha = (A : U \to V; B : V \to U)$ is a pair, and let V'' be the stable space of $AB : V \to V$. Set $U'' = B(V'')$. If n is greater than the dimension of U and V, then $U'' = B(AB)^n(V) = (BA)^n B(V) \subseteq (BA)^n(U)$. On the other hand, $U'' \supseteq (BA)^n BA(U) = (BA)^{n+1}(U)$. But $n \geq \dim(U)$ so $(BA)^n(U) = (BA)^{n+1}(U)$ and U'' is the stable space of BA. Now B is injective on V'', and so B induces an isomorphism from V'' to U''. Dually, A restricts to an isomorphism from U'' to V''. If $U' = U/U''$, and

$V' = V/V''$, then A and B induce $A' : U' \to V'$ and $B' : V' \to U'$. If we define $\alpha' = (A' : U' \to V'; B' : U' \to V')$ then α' is a nilpotent pair. We call α' the reduced pair of α.

We next must recall the classification theory of nilpotent pairs, for which Kraft-Procesi [**10**] is a good source. Fix U and V, and let α be a nilpotent pair. Then α can be written uniquely (up to order and isomorphism) as a direct sum of indecomposable nilpotent pairs. There is a one to one correspondence between isomorphism classes of these indecomposables and strings of alternating symbols "a" and "b". For example, consider the string $\sigma = abab\dots ba$ of length $2n+1$. To this string we associate the indecomposable pair described as follows. U has basis $a_1, \dots, a_{n+1}$ and V has basis $b_1, \dots, b_n$. A is defined by $A(a_i) = b_i$ for $i \le n$, $A(a_{n+1}) = 0$, and $B(b_i) = a_{i+1}$. Such an A and B look like:

$$A: \quad \begin{array}{ccc} a_1 & \longrightarrow & b_1 \\ \vdots & & \vdots \\ a_n & \longrightarrow & b_n \\ a_{n+1} & \longrightarrow & 0 \end{array} \qquad\qquad B: \quad \begin{array}{ccc} b_1 & & a_1 \\ & \searrow & \\ \vdots & \vdots & \vdots \\ b_n & & a_n \\ & \searrow & \\ & & a_{n+1} \end{array}$$

We denote the pair we just defined by $\alpha(\sigma)$. If σ' is the string $bab\dots ab$ of length $2n+1$, we define the pair $\alpha(\sigma')$ associated to σ' just as above but with U, V; A, B; and the a's and b's switched.

We define the indecomposable nilpotent pairs associated with even length strings as follows. Let σ be the string $abab \cdots ab$ of length $2n$. Let U have basis $a_1, \dots, a_n$ and V have basis $b_1, \dots, b_n$. Define $A(a_i) = b_{i+1}$ for $i < n$, $A(a_n) = 0$, and $B(b_i) = a_i$. Such A and B look like:

$$A: \quad \begin{array}{ccc} a_1 & & b_1 \\ & \searrow & \\ \vdots & \vdots & \vdots \\ a_n & & b_n \\ & \searrow & \\ & & 0 \end{array} \qquad\qquad B: \quad \begin{array}{ccc} b_1 & \longrightarrow & a_1 \\ \vdots & & \vdots \\ b_n & \longrightarrow & a_n \end{array}$$

This defines a nilpotent pair we again denote by $\alpha(\sigma)$. By reversing a's, b's; A, B and U, V we define $\alpha(\sigma')$, for the string $\sigma' = baba\dots ba$. Note that our definition here is reversed from the one in Kraft-Procesi [**10**]. All indecomposable nilpotent pairs are isomorphic to $\alpha(\sigma)$ for some string σ, and different strings give nonisomorphic pairs.

Let Σ be the set of strings as above and let Λ denote the set of formal sums of elements of Σ with nonnegative integer coefficients. To any nilpotent pair α, the decomposition of α into indecomposables defines a unique $\eta(\alpha) \in \Lambda$ where for each string σ, the coefficient of σ in $\eta(\alpha)$ is the number of times $\alpha(\sigma)$ appears in α. Obviously, we have a one to one correspondence between isomorphism classes of nilpotent pairs and Λ. Also, $\eta(\alpha \oplus \alpha') = \eta(\alpha) + \eta(\alpha')$. If α is an arbitrary pair, let α' be the reduced nilpotent pair, and define $\eta(\alpha) = \eta(\alpha')$.

For any string $\sigma \in \Sigma$, we can define $\sigma(A, B)$ to be the linear map gotten by substituting A for a and B for b in the string and then computing the composition. For example, if $\sigma = ababa$, $\sigma(A, B)$ is $ABABA$. If $\alpha = (A, B)$ is a pair, we write $\sigma(A, B)$ as $\sigma(\alpha)$.

Suppose α is a pair, not necessarily nilpotent. Then α defines a rank function $r_\alpha : \Sigma \to Z^+$ as follows. If $\sigma \in \Sigma$, define $r_\alpha(\sigma) = \text{rank}(\sigma(\alpha))$. If σ, σ' are strings, and σ is strictly longer than σ', then $r_\alpha(\sigma) \leq r_\alpha(\sigma')$. There is an integer N such that if σ is longer than N, $r_\alpha(\sigma)$ has constant value, and this stable value of r_α. is the stable rank of α. A function $r : \Sigma \to Z^+$ is called a rank function if $r = r_\alpha$ for some α.

THEOREM 1.2.

a) *Let α, β be two pairs. Then $\eta(\alpha) = \eta(\beta)$ if and only if $r_\alpha = r_\beta$.*
b) *Let α be a pair and set $\eta(\alpha) = \eta$. Let $V_\eta \subseteq \text{Hom}(U, V) \times \text{Hom}(V, U)$ be all pairs β with $\eta(\beta) = \eta$. Then V_η is a locally closed subvariety of $\text{Hom}(U, V) \times \text{Hom}(V, U)$.*

PROOF: Part a) implies that V_η is defined by rank equations, so b) follows from a). As for a), suppose $\eta(\alpha) = \eta(\beta)$. If α' and β' are the respective reduced pairs, then α' and β' are isomorphic. In particular, $r_{\alpha'} = r_{\beta'}$, and α' and β' are defined on spaces of the same dimension. Thus α, β have equal stable rank, s, and $r_\alpha = r_{\alpha'} + s = r_{\beta'}, +s = r_\beta$.

Conversely, suppose $r_\alpha = r_\beta$. We must show that $\eta(\alpha) = \eta(\beta)$. As r_α and r_β have equal stable values, α and β have equal stable rank s. Letting α', β' be the reduced pairs again, we have $r_{\alpha'} = r_{\beta'}$. Thus we may assume α, β are both nilpotent pairs. Let σ be a string appearing in $\eta(\alpha)$ of maximal length m. It suffices by induction to show that σ appears in β.

To finish our argument, we consider the behavior of the rank function of an indecomposable nilpotent pair. Recall that $\alpha(\sigma)$ is the indecomposable associated with σ, and $\tau(\alpha(\sigma))$ is the linear map derived by plugging the pair of maps of $\alpha(\sigma)$ into the pattern τ and composing. One can compute that $\sigma(\alpha(\sigma))$ is the zero map, and if σ' is the string derived from σ by removing the rightmost symbol, $\sigma'(\alpha(\sigma)) = 0$ also. If m is the length of σ, then $\tau(\alpha(\sigma)) = 0$ for any string τ of length greater than or equal to m. However, if σ'' is the string with the leftmost symbol removed, $\sigma''(\alpha(\sigma))$ is not zero.

The fact that m is the length of a maximal string in $\eta(\alpha)$ is thus equivalent to the fact that $r_\alpha(\sigma) = 0$ for both strings of length m but $r_\alpha(\sigma') \neq 0$ for some σ' of length $m-1$. Thus m is also the maximal length of a string in β. Let σ have this maximal length m and let σ'' be the string with the leftmost symbol removed. The fact that σ appears in $\eta(\alpha)$ is equivalent to $r_\alpha(\sigma'') \neq 0$, and so if σ appears in $\eta(\alpha)$ it appears in $\eta(\beta)$. The theorem is proved.

Since we have shown V_η is defined by rank equations, it makes sense to change our notation and to define, for a rank function $r = r_\alpha : \Sigma \to Z^+$, V_r to be the subvariety of $\mathrm{Hom}(U,V) \times \mathrm{Hom}(V,U)$ consisting of pairs β with $r_\beta = r$. We have seen that if $\eta = \eta(\alpha)$, then $V_r = V_\eta$. We study the variety V_r by reducing to the case of nilpotent pairs, as follows.

If $r = r_\alpha$ is a rank function, let s be the stable value of r and set $r' = r - s$. We know that r' is the rank function of the reduced (nilpotent) pair α'. Fix u and v as the dimensions of U and V respectively. Choose spaces U', V' of dimension $u-s$ and $v-s$ respectively, and consider $V_{r'} \subseteq \mathrm{Hom}(U',V') \times \mathrm{Hom}(V',U')$. Since α' is a nilpotent pair, $V_{r'}$ is the orbit of α' under $GL(U') \times GL(V')$ and so $V_{r'}$ is irreducible. In order to state the next result, set $P = \mathrm{Hom}(U,V) \times \mathrm{Hom}(V,U)$ and $P' = \mathrm{Hom}(U',V') \times \mathrm{Hom}(V',U')$.

PROPOSITION 1.3. *For any rank function $r = r_\alpha$, V_r is irreducible and* $\mathrm{codim}_{P'}(V_r) = \mathrm{codim}_{P'}(V_{r'})$.

PROOF: Let s be the stable rank of α and let $G_s(U)$ and $G_s(V)$ be the Grassman varieties of subspaces of U respectively V of dimension s. Let $W \subseteq \mathrm{Hom}(U,V) \times \mathrm{Hom}(V,U) \times G_s(U) \times G_s(V)$ be the locally closed subvariety of points (A,B,U'',V'') such that $A(U'') = V''$ and $B(V'') = U''$. There are regular maps $p : W \to G_s(U) \times G_s(V)$ and $q : W \to \mathrm{Hom}(U,V) \times \mathrm{Hom}(V,U)$ induced by the respective projections. Define W_r to be $q^{-1}(V_r)$. Let q_r be the restriction of q to W_r. If $\beta = (A,B) \in V_r$, let U'' and V'' be the stable spaces of β in U, V. U'' and V'' necessarily have dimension s and we may define $f : V_r \to W_r$ by setting $f(\beta) = (A,B,U'',V'')$. The regular map f is the inverse to q_r and so $W_r \cong V_r$.

The restriction, $p_{r'}$, of p to W_r is a surjection. If $(U'', V'') \in G_s(U) \times G_s(V)$, then $p_r(U'', V'')$ consists of all pairs $\beta = (A, B)$ such that $A(U'') = V''$, $B(V'') = U''$, and the reduced pair $(A' : U/U'' \to V/V''; B' : V/V'' \to U/U'')$ lies in $V_{r'}$. Thus there is a surjective regular map $f : p_r^{-1}(U'', V'') \to V_{r'} \subseteq \mathrm{Hom}(U/U'', V/V'') \times \mathrm{Hom}(V/V'', U/U'')$. If $\beta' \in V_r$, then $f^{-1}(\beta')$ consists of all pairs of linear maps whose matrices (with respect to a suitable basis) look like:

$$\begin{pmatrix} A' & 0 \\ B & D \end{pmatrix}, \quad \begin{pmatrix} B' & 0 \\ E & G \end{pmatrix}$$

where D and G are nonsingular. Thus all $f^{-1}(\beta')$ are irreducible and isomorphic, of dimension $us + vs$. Hence $p^{-1}(U'', V'')$ is irreducible of dimension $\dim(V_{r'}) + us + vs$ and W_r is irreducible of dimension $\dim(V_{r'}) + us + vs + s(u-s) + s(v-s)$. Thus $\mathrm{codim}(V_r) = 2uv - us - vs - s(u-s) - s(v-s) - \dim(V_{r'}) = 2(u-s)(v-s) - \dim(V_{r'}) = \mathrm{codim}(V_{r'})$.

If $\alpha = (A : U \to V; B : V \to U)$ is a pair define $\pi(\alpha) = AB \in \mathrm{End}(V)$ and $\rho(\alpha) = BA \in \mathrm{End}(U)$. If α is a nilpotent pair, we can describe the Jordan normal form of $\pi(\alpha)$ and $\rho(\alpha)$ as follows. Let Π be the set of positive integers, and Ω the set of formal sums of elements of Π with nonnegative coefficients. We think of Ω as the set of all possible Jordan normal forms of a nilpotent matrix. Define a map $\mu_a : \Sigma \to \Pi \cup \{0\}$ by counting the number of a's in any string in Σ. μ_a induces a map, we also call μ_a, from Λ to Ω, where μ_a maps the string "b" to 0. We make an analogous definition of μ_b. It is observed in Kraft-Procesi [**10**] that $\mu_a(\eta(\alpha))$ is the Jordan block decomposition of $\pi(\alpha)$, and $\mu_b(\eta(\alpha))$ is the block decomposition of $\rho(\alpha)$.

Keep the spaces U, V fixed. Let r be a rank function $r : \Sigma \to Z^+$ with stable value s. To r is associated an element $\eta \in \Lambda$ such that $V_r = V_\eta$. Let $\theta = \mu_a(\eta)$. Let t be the rank function $t : Z^+ \to Z^+$ associated to θ. If $t = t' + s$, then t is the rank function of any $C : V \to V$ such that the reduction C' has form θ. Using reduction it is easy to see that the restriction of π to V_r maps onto $Y_t \subseteq \mathrm{End}(V)$.

LEMMA 1.4. *$\pi : V_r \to Y_t$ is smooth.*

PROOF: If the stable value $s = 0$, then this was shown in Kraft-Procesi [**10**]. If $U' \subseteq U$ and $V' \subseteq V$, then recall that $Y_t(V')$ is the subvariety with stable space V' and define $V_r(U', V')$ to be the subvariety with stable spaces U', V'. An easy argument using reduction shows that the map $\pi : V_r(U', V') \to Y_r(V')$ is smooth.

Let $P \subseteq GL(U)$ and $Q \subseteq GL(V)$ be parabolic subgroups such that $G_s(U) \cong GL(U)/P$ and $G_s(V) \cong GL(V)/Q$. The maps $GL(U) \to G_s(U)$ and $GL(V) \to G_s(V)$ have sections defined locally, and from this it follows that the maps $f : V_r \to G_s(U) \times G_s(V)$ and $g : Y_t \to G_s(V)$ are locally trivial. That is, in the second case say, $G_s(V)$ has an open cover $G_s(V) = \cup W_i$ such that $f^{-1}(W_i) \cong Y_t(V') \times W_i$ for some $V' \in W_i$. The lemma follows.

1B) Invariant Theory. Let U be a vector space, r a rank function with $r(0) = \dim(U)$, and $X_r \subseteq \mathrm{End}(V)$ the variety $\{A \mid \mathrm{rank}(A^i) \leq r(i)$ for all $i\}$. It is the purpose of this section to show that X_r is a normal variety. Thus for this section we fix r and consider $X = X_r$.

If r has stable value 0 then X_r is the closure of $Y_r = \{A \mid \mathrm{rank}(A^i) = r(i)\}$ and Y_r is the orbit of a single nilpotent linear map. In this case Kraft and Procesi have shown that X_r is normal, and a generalization of their argument proves our theorem. In this section we will outline the proof in our more general setting, only emphasizing those parts that differ significantly from Kraft and Procesi's paper.

Notice that we do not know the radical ideal defined by X; that is, the equations we have given only define X set theoretically. Thus it is difficult to imagine how to prove normality directly. One of the beautiful ideas of Kraft-Procesi is that normality can be proved by showing that X is the quotient of a normal variety under the action of a reductive algebraic group. We follow this tack closely.

Denote by s the stable value of r. Let n be the least integer such that $r(n) = s$. Note that X_r can equally well be defined by the finite set of inequalities $\mathrm{rank}(A^i) \leq r(i)$ for $i = 1, \ldots, n$. Let $U_0, U_1, \ldots, U_n$ be vector spaces such that $U_n = U$, and U_i has dimension $r(n-i)$. Note that we have arranged it so that the dimensions of the U_i form a strictly decreasing sequence. Let M be the affine space consisting of tuples $(A_0, B_0, A_1, \ldots, A_{n-1}, B_{n-1})$ where $A_i : U_i \to U_{i+1}$ and $B_i : U_{i+1} \to U_i$ are linear maps. M is the set of all tuples of maps that fit into the diagram:

$$U_0 \underset{B_0}{\overset{A_0}{\leftrightarrows}} U_1 \underset{B_1}{\overset{A_1}{\leftrightarrows}} \cdots \underset{B_{n-1}}{\overset{A_{n-1}}{\leftrightarrows}} U_n$$

Let $Z \subseteq M$ be the closed subscheme of M defined by the equations $A_{i-1}B_{i-1} = B_i A_i$ for $i = 1, \ldots, n-1$. Let G be the group $GL(U_0) \times \ldots \times$

$GL(U_n)$ and H the subgroup $GL(U_0) \times \ldots GL(U_{n-1})$. We define an action of G on M by setting $(g_0, \ldots, g_n)(A_0, \ldots, B_{n-1}) = (g_1 A_0 g_0^{-1}, g_0 B_0 g_1^{-1}, \ldots, g_n A_{n-1} g_{n-1}^{-1}, g_{n-1} B_{n-1} g_n^{-1})$. This action induces an action of G on Z.

The definition of M and Z can be trivially generalized to any sequence of spaces $U_0', \cdots, U_t'$. We will write these varieties as $M(U_0', \ldots, U_t') \supseteq Z(U_0', \ldots, U_t')$.

The definition of Z can be viewed in the following manner. Let N be the variety $\operatorname{End}(U_1) \times \ldots \operatorname{End}(U_{n-1})$ and define $\varphi : M \to N$ by setting:

$$\varphi(A_0, \ldots, B_{n-1}) = (B_1 A_1 - A_0 B_0, \ldots, B_{n-1} A_{n-1} - A_{n-2} B_{n-2}) .$$

Z is then the variety $\varphi^{-1}(0)$, where $0 \in N$ is the 0 tuple.

Next define a map $\psi : M \to \operatorname{End}(U)$ by setting $\psi(A_0, \ldots, B_{n-1}) = A_{n-1} B_{n-1}$. Note that if $A = A_{n-1} B_{n-1}$ is in the image of ψ, then $A^2 = A_{n-1} B_{n-1} A_{n-1} B_{n-1} = A_{n-1} A_{n-2} B_{n-2} B_{n-1}$. By induction, $A^i = A_{n-1} \cdots A_{n-i} B_{n-i} \cdots B_{n-1}$. In particular, A^i factors through U_{n-i} and so $\operatorname{rank}(A^i) \leq r(i)$. Thus $A \in X$. Conversely, it is elementary to see that if $A \in X$, then there is an $\alpha = (A_0, \ldots, B_{n-1}) \in Z$ such that all the A's are injections and $\psi(\alpha) = A$.

Let Z_{red} be the reduced scheme defined by Z. The action of G on Z induces an action on Z_{red}, and by restriction we get an action of H on Z_{red}. Since X is reduced by definition, there is an induced map $\psi : Z_{\text{red}} \to X$. The proof of the next fact follows the proof in Kraft and Procesi with only minor changes and so we omit it.

LEMMA 1.5. *ψ induces an isomorphism $Z_{\text{red}}/H \cong X$.*

We next turn to finding some smooth points on Z. Let $M^0 \subseteq M$ be the open subset defined by requiring for each $i = 1, \ldots, n-1$ that either A_i or B_i has maximal rank. Arguing as in Kraft and Procesi, we observe that the tangent space map $d\varphi : T(M) \to T(N)$ is surjective over each element of M^0. Note that when one follows the proof in Kraft and Procesi, one sees that no restrictions on A_0 or B_0 are required. Set $Z^0 = Z \cap M^0$ which is open in Z, and observe that we showed above that Z_0 maps onto X. Z^0 is certainly nonempty. We have shown:

LEMMA 1.6. *Z^0 is nonsingular of dimension* $\dim(M) - \dim(N)$.

We next want to show that $Z - Z^0$ has small dimension. More specifically, we show:

PROPOSITION 1.7. *$Z - Z^0$ has codimension ≥ 2 in Z.*

We begin the proof of the above proposition by noticing that elements of Z have stable spaces that generalize the stable spaces of section 1A. Let $(A_0, \dots, B_{n-1}) = \alpha$ be an element of Z. Define α_i to be the pair (A_i, B_i). Let Σ and Λ be as in section 1A, and define $\eta_i(\alpha) = \eta(\alpha_i)$. Set $U_i'' \subseteq U_i$ to be the stable space of $B_i A_i$. Since $B_i A_i = A_{i-1} B_{i-1}$, U_i'' is also the stable space of $A_{i-1} B_{i-1}$. Thus all U_i'' have equal dimension which we call the stable rank of α. In addition, $A_i(U_i'') = U_{i+1}''$ and $B_i(U_{i+1}'') = U_i''$.

For this fixed α, define $U_i' = U_i / U_i''$. Let $M' = M(U_0', \dots, U_n')$ and $Z' = Z(U_0', \dots, U_n')$. In the obvious way, α induces a point $\alpha' = (A_0', \dots B_{n-1}')$ such that all $A_i' B_i'$ are nilpotent and α' has stable rank 0. We call α' the reduction of α.

For each $\alpha \in Z$, define $\Gamma(\alpha) \in (\Lambda \times \dots \times \Lambda)$ ($n-1$ times) to be the sequence $\eta_i(\alpha)$. The relations defining Z force $\pi_a(\eta_i(\alpha)) = \pi_b(\eta_{i-1}(\alpha))$. Conversely, any sequence satisfying these relations can easily be seen to be $\Gamma(\alpha)$ for some α. Note that, by definition, $\Gamma(\alpha) = \Gamma(\alpha')$. Now fix $\Gamma = \Gamma(\alpha)$, let s be the stable rank of α, set $\eta = \eta_{n-1}(\alpha)$, and set $\pi_a(\eta) = \theta$. Let $t : Z^+ \to Z^+$ be the rank function of stable value s corresponding to θ. Choose U_i' to be a vector space of dimension $\dim(U_i) - s$ and define M' and Z' as above. Set Z_Γ to be the inverse image of Γ in Z, while Z'_Γ is the same for Z'. Define $N' = \mathrm{End}(U_1') \times \dots \times \mathrm{End}(U_{n-1}')$. We claim:

PROPOSITION 1.8. *Z_Γ is smooth and irreducible of dimension* $\dim(Z_\Gamma) = \dim(M) - \dim(N) - \dim(M') + \dim(N') + \dim(Z'_\Gamma)$.

PROOF: Let $Z* = Z(U_0, \dots, U_{n-1})$ and $Z*_\Gamma \subseteq Z*$ be the inverse image of Γ in $Z*$ (yes, the last component of Γ is ignored). There is a pullback diagram:

$$\begin{array}{ccc} Z_\Gamma & \longrightarrow & V_\eta \\ \downarrow & & \downarrow \\ Z*_\Gamma & \longrightarrow & Y_r \end{array}$$

The right column is smooth by 1.4, so induction shows that Z_Γ is smooth and irreducible. A similar pullback diagram holds for Z'_Γ, so the dimension equation also follows by induction.

Note that U_0' is not (0), so Z'_Γ is not quite one of the Z_λ studied by Kraft and Procesi. But like the variety defined by them, Z'_Γ consists of nilpotent pairs. In fact, by adding spaces with negative subscripts, one can see that

Z'_Γ is the "upper tail" of one of the varieties Z_λ defined in their paper. The key result can now be stated (compare Kraft-Procesi [**10**, p. 236]).

PROPOSITION 1.9. *Either* $Z_\Gamma \subseteq Z_0$ *or* $\dim(Z^0) - \dim(Z_\Gamma) \geq 2$.

PROOF: Let α be such that $\Gamma = \Gamma(\alpha)$. Note first of all that $Z_\Gamma \subseteq Z_0$ if and only if each $\eta_i(\alpha)$ has the property that either all strings in $\eta_i(\alpha)$ start with a "b" or all strings in $\eta_i(\alpha)$ end with an "a". Γ which violate the above we call *defective*. By 1.6 $\dim(Z^0) = \dim(M) - \dim(N)$, so by 1.8 to prove the Proposition we must show that:

(*) If Γ is defective then $\dim(M') - \dim(N') - \dim(Z'_\Gamma) \geq 2$.

It is necessary to give a formula for the dimension of Z'_Γ. Let t be the rank function of stable value 0 associated with $\mu_a(\eta_{n-1}(\alpha))$ and b the rank function associated with $\mu_b(\eta_0(\alpha))$. The point is that if $\alpha = (A'_0, \dots, B'_{n-1}) \in Z'_\Gamma$, then t is the rank function of $A'_{n-1}B'_{n-1}) \in \mathrm{End}(U'_n)$ and b is the rank function of $B'_0A'_0 \in \mathrm{End}(U'_0)$. Using the same argument as in [**10**, Corollary p. 241] we have that:

LEMMA 1.10. *Let* u_i *be the dimension of* U'_i. *Then:*

$$\dim(Z'_\Gamma) = \left(\sum u_i u_{i+1}\right) + 1/2(\dim(Y_b) + \dim(Y_t)).$$

Now $\dim(Y_b) \leq u_0(u_0 - 1)$ since b is the rank function of a nilpotent matrix. In order to study $\dim(Y_t)$, define $t' : Z^+ \to Z^+$ as follows. For $i \leq n$, set $t'(i) = u_{n-i}$. If $n < i < u_0 + n$, set $t'(i) = u_0 - (n - i)$. Finally, set $t'(i) = 0$ if $i \geq u_0 + n$. Since $u_1 - u_0 \geq 1$, t' is a rank function. Furthermore, $t(i) \leq t'(i)$ for all i. By [**10**, part a) of the Proposition p. 229], Y_t is in the closure of $Y_{t'}$, and so $\dim(Y_t) \leq \dim(Y_{t'})$. By part b) of that same proposition, $(1/2)\dim(Y_{t'}) =$

$$\sum t'(i)(t'(i-1) - t'(i)) = \left(\sum u_i(u_{i+1} - u_i)\right) + u_0(u_0 - 1)/2$$

It follows that $(1/2)(\dim(Y_b) + \dim(Y_t))$ is less than or equal to:

$$\left(\sum u_i(u_{i+1} - u_i)\right) + u_0(u_0 - 1)$$

Of course $\dim(M') - \dim(N') =$

$$2\sum u_i u_{i+1} - \sum u_i^2 = \sum u_i u_{i+1} + \sum u_i(u_{i+1} - u_i) + u_0 u_1$$

Combining these facts we have $\dim(M')-\dim(N')-\dim(Z'_\Gamma) \geq u_0$. Thus it suffices to prove (*) in the cases $u_0 = 1$ or 0. If $u_0 = 0$, then Z'_Γ *is* a Z_λ as in Kraft-Procesi and the result of their paper applies here. If $u_0 = 1$, then we can set $U_{-1} = (0)$ and Z'_Γ, M' and N' are isomorphic to the corresponding varieties in Kraft-Procesi, so the results there apply again. Thus (*) is proved.

Given (*), we have the Proposition 1.9. From 1.9, 1.7 immediately follows. Thus $\dim(Z) = \dim(Z_0) = \dim(M) - \dim(N)$. Since Z is defined by $\dim(N)$ equations, Z is a complete intersection. Thus Z is Cohen-Macaulay. As $Z_0 \subseteq Z$ is nonsingular and of codimension 2, Z is normal. As Z is a cone, it is also connected. By Serre's criterion (e.g., Matsumura [**11**, p. 125]), Z is reduced. Thus by 1.5 X_r is isomorphic to Z/H. It follows that X_r is normal.

2. Rational resolutions and the proof of the main theorem.

NOTATION: Throughout this section and the next we will adhere to the following notation: V will be a vectorspace of dimension n over $\mathbb{C}$, r will be a decreasing non-negative concave integral function (a "rank function") with $r(0) = n$. We will write $r(\infty)$ for the stable value of r (which is equal to $r(n)$). We write a for the partition $a(r) = (a_1(r), \ldots, 0, 0, \ldots)$ with $a_i(r) = r(i-1) - r(i)$. We write $b(r) = (b_1(r), \ldots, 0, 0, \ldots)$ for the dual partition to a, that is, $b_i(r)$ is the number of indices j with $a_j(r) \geq i$. It is easily verified that if $A \in \mathrm{End}(V)$ has rank function r, that is rank $A^i = r(i)$, then

$$a_i(r) = \dim((\ker A) \cap (\operatorname{im} A^{i-1}))$$

while $b_i(r)$ is the size of the i-th (in descending order) block with eigenvalue 0 in the Jordan normal form of A.

In the last section we saw that if r is a rank function, then

$$X_r := \{A \in \mathrm{End}(V) | \operatorname{rank} A^k \leq r(k)\}_{\mathrm{red}}$$

is a normal variety. In this section we will prove that it is in fact Gorenstein with rational singularities. We will also show that it fits into a flat family over $\mathbb{A}^m$ of normal varieties, whose fiber over a point $(\lambda_1, \ldots, \lambda_m)$ such that the λ_i are all distinct is

$$X_{r;\lambda_1,\ldots,\lambda_m} = \{A \in \mathrm{End}(V) | \operatorname{corank}(A - \lambda_i) \geq r(i-1) - r(i), i = 1, \ldots, m\}.$$

In fact we will construct a ("very weak") simultaneous resolution of singularities for this family of varieties, and we will construct the family from the resolution. We will use the resolution of X_r to give a description of the tangent space to X_r at a point corresponding to an endomorphism A such that rank $A^i = r_i$ for all i. As an application of the tangent space computation we show that the general fiber of this family can be written as the scheme-theoretic intersection of varieties like X_r.

To construct the resolution, suppose that m is the largest number such that $r(m) \neq r(m+1)$. Let $x_1, \ldots, x_m$ be coordinates on $\mathbb{A}^m$, and let $W \subset \mathbb{A}^m$ be a linear subvariety (that is, the translate of a subvectorspace). Let $F = \text{Flag}(V; r_1, \ldots, r_m)$ be the variety of flags

$$(V = V_0 \supset V_1 \supset \cdots \supset V_m \supset 0) \quad \dim V_i = r(i).$$

Let

$$\mathcal{X}_{r,W} \subset \text{End}(V) \times W \times F$$

be the subvariety of triples $(A, \lambda, \{V_i\})$ such that, with $\lambda_i := x_i(\lambda)$,

$$(A - \lambda_i)V_{i-1} \subset V_i \text{ for } i = 1, \ldots, m.$$

Write $\pi_1 = \pi_{1,W}$ for the projection of $\mathcal{X}_{r,W}$ to $\text{End}(V) \times W$. Let $X_{r,W}$ be the reduced image of π_1 and let $X'_{r,W}$ be $\operatorname{spec} \pi_{1*}\mathcal{O}_{\mathcal{X}_{r,W}}$. Since the flag manifold is complete, π_1 is proper, so $X_{r,W}$ is a closed affine subvariety of $\text{End}(V) \times W$, and $X'_{r,W}$ is also affine, and finite over $X_{r,W}$. If W is a single point $\lambda \in \mathbb{A}^m$, and the numbers λ_i are distinct, then $X_{r,W}$ is the same as the variety $X_{r;\lambda_1,\ldots,\lambda_m}$ defined previously.

THEOREM 2.1.

i) *$\mathcal{X}_{r,W}$ is smooth over W and irreducible, of dimension $n^2 - \sum_i a_i(r)^2 + \dim W$, with trivial canonical bundle, while π_1 is proper and birational, with*

$$R^i \pi_{1*}\mathcal{O}_{\mathcal{X}_{r,W}} = 0 \text{ for } i > 0.$$

ii) *$X_{r,W}$ is normal, so that $X_{r,W} = X'_{r,W}$, $\mathcal{X}_{r,W}$ is a rational resolution of singularities of $X_{r,W}$, and $X_{r,W}$ is Gorenstein.*

iii) *$X_{r,W}$ is the restriction of $X_{r,\mathbb{A}^m}$ to W, and it is flat over W.*

Taken together, statement ii) of the Theorem and the vanishing part of statement i) may be rephrased as saying that $\mathcal{X}_{r,W}$ is a rational resolution of the singularities of $X_{r,W}$; see Kempf et al [8].

We will postpone the proofs of this and our other results until after all the statements have been given.

Next we study the fibers of the family $X_{r,\mathbb{A}^m}$ over points $\lambda \in \mathbb{A}^m$ for which some of the λ_i may coincide. We partition the coordinates λ_i of λ by equality. Let p be a permutation of $\{1,\ldots,m\}$ such that

$$\begin{aligned}
\lambda_{p(1)} &= \cdots = \lambda_{p(m_1)}, \\
\lambda_{p(m_1+1)} &= \cdots = \lambda_{p(m_2)}, \\
&\cdots \\
\lambda_{p(m_s+1)} &= \cdots = \lambda_{p(m_{s+1})}.
\end{aligned}$$

and such that p preserves the order within each interval $m_i+1,\ldots,m_{i+1}$. Thus with $a_i = r(i-1)-r(i)$, we have $a_{p(m_i+1)} \geq \cdots \geq a_{p(m_{i+1})}$. We write

$$r(i,j) := \dim V - a_{p(m_i+1)} - \cdots - a_{p(m_i+j)} \qquad \text{for } i=1,\ldots,s \text{ and } j=1,\ldots,m_{i+1}-m_i$$

We have:

Corollary 2.2. *The scheme-theoretic fiber of $X_{r,W}$ over $\lambda \in W$ is*

$$X_{r,\lambda} = \{A \in \operatorname{End}(V) \mid \operatorname{rank}(A-\lambda_i)^j \leq r(i,j) \text{ for all } i,j \text{ as above}\}_{\mathrm{red}}.$$

In particular, this variety is Gorenstein with rational singularities for every $\lambda \in \mathbb{A}^m$, and $\mathcal{X}_{r,\lambda}$ is a rational resolution of its singularities.

Using this description of the desingularization we can give a description of some tangent spaces to X_r:

Proposition 2.3. *If $\operatorname{rank} A^i = r(i)$ for all i then A is a smooth point of $X_r = X_{r,0}$, and the tangent space to X_r at A is naturally equal to*

$$\{\alpha \in \operatorname{End}(V) \mid \sum_{i+j=k-1} A^i \alpha A^j \text{ maps } \ker A^k \text{ into } \operatorname{im} A^k \text{ for all } k\}.$$

A consequence is the following transversality result, which will play a role in the description to be given in the next section of the equations defining some of the varieties X_r:

COROLLARY 2.4. *For all $\lambda \in \mathbb{A}^m$, the variety $X_{r,\lambda}$ is scheme-theoretically the intersection*

$$\cap_i \left(\{A \in \text{End}(V) | \, \text{rank}(A - \lambda_i)^j \leq r(i,j) \text{ for all } j \text{ as above}\}_{\text{red}}\right).$$

PROOF OF THEOREM 2.1: We first show that $\mathcal{X}_{r,W}$ is smooth and irreducible. Write $\pi_2 : \mathcal{X}_{r,\mathbb{A}^m} \to \mathbb{A}^m \times F$ for the projection. Choose a cover of F by sets F' on which the bundles V_i are trivialized. The scheme-theoretic preimage by π_2 of $\mathbb{A}^m \times F'$ may be identified with the product of $\mathbb{A}^m \times F'$ with the set of block-upper-triangular-matrices of a suitable shape, and is thus smooth over $\mathbb{A}^m \times F'$ and irreducible. As $\mathbb{A}^m \times F$ is smooth over $\mathbb{A}^m$, we see that $\mathcal{X}_{r,W}$ is smooth for every W. Since $W \times F$ is irreducible, so is $\mathcal{X}_{r,W}$.

Next we use the adjunction formula to prove that the canonical bundle of $\mathcal{X}_{r,W}$ is trivial. Since $\mathcal{X}_{r,W}$ is a complete intersection with trivial normal bundle in $\mathcal{X} := \mathcal{X}_{r,\mathbb{A}^m}$, it suffices to treat the case $W = \mathbb{A}^m$.

Let V_F be the trivial bundle $V \times F$ on F, and let $W = (V_F = \mathcal{V}_0 \supset \cdots \supset \mathcal{V}_m \supset 0)$ be the tautological flag on F. Let $\mathcal{D} \subset \text{End}(V_F)$ be the vector bundle of "upper-triangular" endomorphisms preserving $\mathcal{V}$ — that is, those A with $A\mathcal{V}_i \subset \mathcal{V}_i$ for all i — and let D be the total space of $\mathcal{D}$. We may regard $\mathcal{X}_{r,\mathbb{A}^m}$ as a subvariety of $D \times A^m$. Inside $\mathcal{D}$ is the bundle of "strictly upper-triangular" endomorphisms, that is, those A with $A\mathcal{V}_{i-1} \subset \mathcal{V}_i$ for all i; as is well-known and easy to verify by direct computation, it may be identified with the cotangent bundle $\mathcal{T}^*(F)$ to F.

On the other hand, pulling $\mathcal{D}$ back to D we get a tautological endomorphism $\mathcal{A} \in \mathcal{D}_D \subset \text{End}(V_D)$ which preserves the pullback $\mathcal{V}_D = (V_D = \mathcal{V}_0 \supset \cdots \supset \mathcal{V}_m \supset 0)$ to D of the tautological flag on F. Writing $\mathcal{A}_i$ for the image of $\mathcal{A}$ in $\text{End}(\mathcal{V}_{i-1}/\mathcal{V}_i)$ on D we see that $\mathcal{X}$ is the zero locus of the section

$$(\mathcal{A}_1 - \lambda_1, \ldots, \mathcal{A}_m - \lambda_m) \in H^0\left(D, \oplus_1^m(\text{End}(\mathcal{V}_{i-1}/\mathcal{V}_i))_D\right).$$

Computing dimensions, we see that this section vanishes in codimension $= \text{rank}(\oplus_1^m(\text{End}(\mathcal{V}_{i-1}/\mathcal{V}_i))_D)$, so $\mathcal{X}$ is locally a complete intersection in D with normal bundle $\oplus_1^m(\text{End}(\mathcal{V}_{i-1}/\mathcal{V}_i))_{\mathcal{X}}$.

By the adjunction formula (Hartshorne [**7**, Ch. 2]), the canonical bundle of $\mathcal{X}$ is the canonical bundle ω_D of D tensored with the highest exterior power of the normal bundle to $\mathcal{X}$ in D. But the highest exterior power of the endomorphism bundle of any bundle is trivial, so this is simply the restriction of ω_D to $\mathcal{X}$. It thus suffices to show that $\omega_D \cong \mathcal{O}_D$.

Now the cotangent bundle $\mathcal{T}^*(D)$ of D is easily seen to be the pullback from F to D of $\mathcal{D}^* \oplus \mathcal{T}^*(F)$. From the description of $\mathcal{T}^*(F)$ in terms of upper triangular matrices we get an exact sequence

$$0 \to \mathcal{T}^*(F) \to \mathcal{D} \to \oplus_1^{m+1} \operatorname{End}(\mathcal{V}_{i-1}/\mathcal{V}_i) \to 0.$$

From the triviality of the top exterior powers of the $\operatorname{End}(\mathcal{V}_{i-1}/\mathcal{V}_i)$, we thus see that the highest exterior power of $\mathcal{D}^* \oplus \mathcal{T}^*(F)$ is the same as that of $\mathcal{D}^* \oplus \mathcal{D}$. Since this is trivial we get the desired isomorphism $\omega_{\mathcal{X}} \cong \mathcal{O}_{\mathcal{X}}$.

From the vanishing theorem of Grauert-Riemenschneider [6] we now obtain

$$R^i\pi_{1*}\mathcal{O}_{\mathcal{X}} = R^i\pi_{1*}\omega_{\mathcal{X}} = 0 \text{ for } i > 0,$$

completing the proof of part i) of the Theorem. Using duality (see for example Elkik [4]) we also get $\omega_{X'_{rW}} = \mathcal{O}_{X'_{rW}}$, where we have written X' for $X'_{r,\mathbb{A}^m}$. Thus this last variety is Gorenstein.

ii) Since $X_{r,W}$ is the image of $\mathcal{X}_{r,W}$, it is irreducible. On the other hand π_1 is one to one over the image of the open set of those $(A, \lambda, \{V_i\})$ such that $(A - \lambda_i)V_{i-1} = V_i$ for $i = 1, \ldots, m$, so π_1 is birational. It is proper because F is a projective variety. In particular, $X'_{r,W}$ is the normalization of $X_{r,W}$.

Consider, for any linear varieties $U \subset W \subset \mathbb{A}^m$, the commutative diagram (of rings, since all the varieties in question are affine):

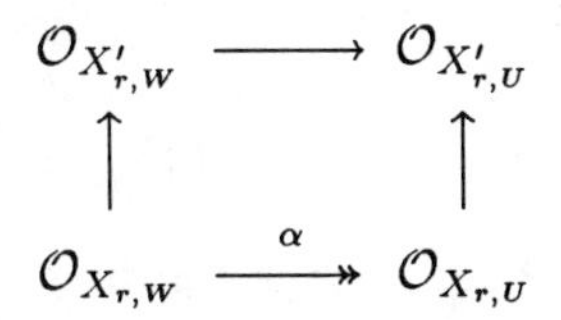

$$\begin{array}{ccc} \mathcal{O}_{X'_{r,W}} & \longrightarrow & \mathcal{O}_{X'_{r,U}} \\ \uparrow & & \uparrow \\ \mathcal{O}_{X_{r,W}} & \stackrel{\alpha}{\twoheadrightarrow} & \mathcal{O}_{X_{r,U}} \end{array}$$

The horizontal map on the bottom is an epimorphism since $X_{r,U}$, being a closed subset of $\operatorname{End}(V) \times W$ by the properness of $\pi_{1,W}$, is a closed subvariety of $X_{r,W}$.

Consider first the case $U = 0$. Since $\mathcal{O}_{X_{r,0}}$ is normal by the result of Section 1, the right hand vertical map is an epimorphism. We wish to prove that the left hand vertical map is an epimorphism. This will prove part ii) for spaces W containing 0.

Since the rings on the bottom of the diagram are the reduced rings of finitely generated graded rings, they are graded, and their normalizations on the top row of the diagram are graded too. Thus we may use Nakayama's lemma to conclude that the left hand vertical map is onto if we establish that $\mathcal{O}_{X'_{r,W}}$ modulo the kernel of α, or, even better, modulo $x_1,\ldots,x_m$, is $\mathcal{O}_{X'_{r,0}}$. This is the special case $W=0$ of the first statement of part iii) of the Theorem, applied to the X' in place of the X. We will prove it now in this form for arbitrary W:

Inductively, it suffices to show that if $W \subset W'$ are linear subvarieties of $\mathbb{A}^m$ such that W is cut out in W' by a single equation $y \in \mathcal{O}_{W'}$, then the sequence

$$0 \to \pi_1 * \mathcal{O}_{\mathcal{X}_{r,W'}} \xrightarrow{y} \pi_1 * \mathcal{O}_{\mathcal{X}_{r,W'}} \longrightarrow \pi_1 * \mathcal{O}_{\mathcal{X}_{r,W}}$$

is right exact, and this follows at once from the last statement of part i), proved above. This completes the proof of ii) for spaces W containing 0.

To prove ii) for arbitrary linear varieties $U \subset \mathbb{A}^m$, consider the diagram above with $W = \mathbb{A}^m$, so that by what we have just proved, the left-hand vertical map is onto. By the version of the first part of iii) established above, the top horizontal map is onto. Of course this implies that the right-hand vertical map is onto, proving ii) in general.

iii) We now see that $X'_{r,W} = X_{r,W}$ for each W, so that the first part of iii), the restriction statement, is already proved. As for flatness, it suffices to treat the case $W = \mathbb{A}^m$. Since $\mathcal{O}_{X_{r,\mathbb{A}^m}}$ is graded, it suffices to prove flatness locally at $0 \in W$; that is, by the "local criterion" (see for example Matsumura [**11**]), it suffices to show that $x_1,\ldots,x_m$ is a regular sequence on $\mathcal{O}_{X_{r,\mathbb{A}^m}}$. But since $\mathcal{O}_{X_{r,W}}$ is a domain for every W, this follows from the restriction statement. This concludes the proof of Theorem 2.1. □

PROOF OF COROLLARY 2.2: All but the first statement comes directly from the special case $W = \{\lambda\}$ of the Theorem. By the Theorem, the fiber $X_{r,\lambda}$ is reduced, so to prove the first statement it is enough to check that the given variety is the image of $\mathcal{X}_{r,\lambda}$ set-theoretically. This follows easily by decomposing V into eigenspaces. □

PROOF OF PROPOSITION 2.3: Smoothness at A follows from Zariski's Main Theorem since we know that X_r is normal and the fiber of the desingularization $\mathcal{X}_{r,0}$ over A consists of the single point corresponding to the flag $\{A^iV\}$.

We next show that the given set of endomorphisms is contained in the tangent space by showing that it is in the image of the tangent space of $\mathcal{X}_{r,0}$. For this it is enough to show that if $\varepsilon^2 = 0$ then the set $\{(A+\varepsilon\alpha)^i V_{\mathbb{C}[\varepsilon]}\}$

of subspaces of $V_{\mathbf{C}[\varepsilon]}$ is a $\mathbf{C}[\varepsilon]$-valued point of F over $\{A^i V\}$; it will follow that the pair

$$(A+\varepsilon\alpha, \{(A+\varepsilon\alpha)^i V_{\mathbf{C}[\varepsilon]}\})$$

is a $\mathbf{C}[\varepsilon]$-valued point of $\mathcal{X}_{r,0}$.

The $\mathbf{C}[\varepsilon]$-valued points of F over $\{A^k V\}$ are the sets $\{\mathcal{V}_k\}$ of subspaces of $\mathcal{V} := V_{\mathbf{C}[\varepsilon]}$ such that for each k, $\mathcal{V}_k$ reduces mod ε to V_k, $\mathcal{V}_k \supset \mathcal{V}_{k+1}$, and $\mathcal{V}_k$ is a direct summand of $\mathcal{V}$. The subspaces $(A+\varepsilon\alpha)^k\mathcal{V}$ obviously satisfy the first two conditions, and we must show that they satisfy the third. Since rank $A^k = r(k)$ the $r(k)\times r(k)$ minors of $(A+\varepsilon\alpha)^k$ generate the unit ideal of $\mathbf{C}[\varepsilon]$, so we need only show that the $(r(k)+1)-(r(k)+1)$ minors are 0, or equivalently that every exterior product

$$(A+\varepsilon\alpha)^k v_1 \wedge \cdots \wedge (A+\varepsilon\alpha)^k v_{r(k)+1}$$

vanishes. Since $\operatorname{codim}_V \ker A^k = r(k)$, we may assume that one of the v_i, say $v_{r(k)+1}$, is in $\ker A^k$. Writing $(A+\varepsilon\alpha)^k = A^k + \varepsilon B$, with $B = \sum_{i+j=k-1} A^i \alpha A^j$ we see that such an exterior product consists of a sum of terms of the form

$$(*) \qquad \varepsilon \cdot A^k v_1 \wedge \cdots \wedge B v_j \wedge \cdots \wedge A^k v_{r(k)+1},$$

each containing one factor with B in place of A^k. Of course all but the term with $j = r(k)+1$ vanishes. By our condition on α, $Bv_{r(k)+1} \in A^k V$, so the expression in (*) vanishes for $j = r(k)+1$ as well. This proves that α is in the tangent space to X_r at A.

To show that the given set of α is exactly the tangent space to X_r, it now suffices to show that it has at least the correct dimension; that is, that its codimension in $\operatorname{End}(V)$ is at most $\sum_i a_i(r)^2$. Inductively, suppose that the set

$$E_h = \{\alpha \in \operatorname{End}(V) | \sum_{i+j=k-1} A^i \alpha A^j \text{ maps } \ker(A^k) \text{ into } \operatorname{im}(A^k) \text{ for } k < h\}$$

has codimension $\leq \sum_{i<h} a_i(r)^2$ in $\operatorname{End}(V)$. The space E_{h+1} is defined in E_h by the condition that $\sum_{i+j=h} A^i\alpha A^j$ takes $\ker A^{h+1}$ into $\operatorname{im} A^{h+1}$. But for $\alpha \in E_h$ the transformation

$$\sum_{i+j=h} A^i \alpha A^j = A^h \alpha + \left(\sum_{i+j=h-1} A^i \alpha A^j \right) A$$

already takes $\ker A^h$ into $\operatorname{im} A^h$, and both the codimension of $\ker A^h$ in $\ker A^{h+1}$ and the codimension of $\operatorname{im} A^{h+1}$ in $\operatorname{im} A^h$ are equal to $a_{h+1}(r)$. Thus $\operatorname{codim}_{E_h} E_{h+1} \leq a_{h+1}(r)^2$, and we are done. □

PROOF OF COROLLARY 2.4: The given intersection is certainly equal to $X_{r,\lambda}$ set-theoretically, so it is enough to prove that it is reduced. By the dimension formula of Theorem 2.1, the intersection has codimension equal to the sum of the codimensions of the varieties being intersected. Since each of the varieties being intersected is Cohen-Macaulay, the scheme-theoretic intersection is too, so if it were not reduced, it would be everywhere nonreduced. Thus it is enough to show that the intersection is transverse at some one point.

It is easy to write down (for example by using the Jordan normal form) an endomorphism $A \in X_{r,\lambda}$ which satisfies $\operatorname{rank} A^i = r(i,j)$ for every appropriate i,j. The condition on an element $\alpha \in \operatorname{End}(V)$ for it to belong to the the tangent space at A to one of the varieties

$$\{A \in \operatorname{End}(V) | \operatorname{rank}(A - \lambda_i)^j \leq r(i,j) \text{ for all } j \text{ as above}\}_{\text{red}}$$

being intersected is, by Corollary 2.3, a condition only on the restriction of α to the kernel of a large power of $A - \lambda_i$. Since the direct sum of these kernels, for the different i, is embedded in V, such conditions are independant. Consequently the tangent spaces at A of the varieties being intersected meet properly, and the scheme-theoretic intersection is smooth at A as required. □

3. Equations.

Throughout this section we will make use of the notations V, n, r, a, b introduced in the beginning of Section 2. We will also write A_{gen} for the generic $n \times n$ matrix, defined over the polynomial ring $\mathbb{C}\big[\{x_{ij}\}_{1 \leq i,j \leq n}\big]$, which we identify with the coordinate ring of $\operatorname{End}(V)$.

The action of $\operatorname{GL}(V)$ on $\operatorname{End}(V)$ by conjugation induces an action on the coordinate ring $\mathbb{C}[x_{ij}]$ of $\operatorname{End}(V)$. Since the variety X_r is invariant, it is natural to wish for equations for X_r that are invariant under $\operatorname{GL}(V)$. However, as is well-known, the ring of $\operatorname{GL}(V)$-invariants is generated by the polynomials in x_{ij} that are the coefficients of the characteristic polynomial of the generic matrix, so if one requires that the individual equations be invariant, one is doomed to failure (except in the case of the "largest" nilpotent orbit closure X_r, that for which $r(i) \equiv n-i$, where the equations are generated by all the coefficients of the characteristic polynomial!) There seems little reason to expect in advance anything more than that the whole ideal $I(X_r)$ of functions vanishing on X_r will be invariant.

But in fact a stronger invariance does hold. One can construct matrices of functions which are conjugation invariant as matrices, in a suitable sense, and the entries of such a matrix generate a conjugation-invariant ideal; the ideals of functions vanishing on the X_r turn out to be sums of ideals constructed in this way.

Powers, exterior powers, and powers of exterior powers of the generic matrix A_{gen} are all examples of matrices F of functions which are conjugation invariant in a sufficiently strong sense: namely for some t and all endomorphisms A of V, $F(A)$ acts naturally on $\bigwedge^t V$; and for any invertible matrix B,

$$F(B^{-1}AB) = (\textstyle\bigwedge^t B)^{-1} F(A) (\bigwedge^t B).$$

De Concini and Procesi [**3**] noticed that there is a very general method of making such constructions, and this method seems to provide enough invariant ideals to define all the nilpotent X_r.

Tanisaki [**15**], following the work of De Concini and Procesi, found a more compact and convenient set of invariant matrices of functions to use: Let $A \in \operatorname{End}(V)$. Regarding $\bigwedge^t(x - A)$ as a polynomial in x whose coefficients are endomorphisms of $\bigwedge^t V$, we define $\lambda_d^t(A) \in \operatorname{End}(\bigwedge^t V)$ to be the coefficient of x^{t-d}. (Note that if $d = 0$ this is the identity map, while if $d > t$ it is the 0 map.) The index d in the notation $\lambda_d^t(A)$ is chosen in this way because if we choose a basis for V, so that we may regard A and $\lambda_d^t(A)$ as matrices, the entries of $\lambda_d^t(A)$ are polynomials of degree d in the entries of A. In particular,

$$\lambda_d^t := \lambda_d^t(A_{\text{gen}})$$

is a matrix of polynomial functions of degree d. We write $I(\lambda_d^t)$ for the ideal generated by the entries of the matrix λ_d^t, and $V(\lambda_d^t)$ for the (reduced) variety in $\mathbb{A}^{n^2}$ that they define.

For example, if $d = t$ then we get

$$\lambda_t^t = \textstyle\bigwedge^t A_{\text{gen}},$$

so the $I(\lambda_t^t)$ are the usual determinantal ideals. On the other hand, λ_d^n is the 1×1 matrix whose entry is (up to sign) the d-th coefficient of the characteristic polynomial.

In the nilpotent case, where $r(\infty) = 0$, X_r is the closure of a single orbit, so a collection of functions which arise as the entries of an invariant matrix has a nice property: to check its vanishing on the whole orbit closure, it is enough to check its vanishing at a single element of the orbit. In this case, Tanisaki showed that the vanishing of certain of the λ_d^t defines the variety X_r set-theoretically (and a little more); De Concini and Procesi had previously proved the corresponding results for their matrices.

In the non-nilpotent case the X_r are no longer closures of single conjugation orbits, and it is no longer true that if an invariant matrix of functions — even an invariant function like a coefficient of the characteristic polynomial — vanishes on an element of $X_r - \bigcup_{r'<r} X_{r'}$, then it vanishes on all of X_r. For example, taking $r(i) \equiv n$, the constant fuction, so that $X_r = \mathrm{End}(V)$, the trace function vanishes on lots of invertible matrices without vanishing everywhere. But one can hope for ideals of functions with a still stronger invariance property. To describe what is desirable, we will write A', for the restriction of A to its own 0-eigenspace, so that A' is the "nilpotent part" of A. Since the membership of an endomorphism A in one of the X_r can be determined from A', what one hopes for are conjugation invariant matrices of functions in the entries of A, or collections of such matrices, whose vanishing depends only on A'. Tanisaki's construction realizes this hope in a rather strong form.

It follows from an easy argument of Tanisaki that for any t and d the ideal generated by the entries of the collection of matrices $\lambda_d^t, \lambda_{d+1}^t, \ldots, \lambda_t^t$ has the strong invariance property just described. But in fact — this is perhaps the main new point in the discussion below — if $t < n$ then the vanishing of an individual λ_d^t on $A \in \mathrm{End}(V)$ depends only on A' (Theorem 3.1). Thus the λ_d^t are a natural source of equations for defining the varieties X_r for arbitrary r. We prove here that they suffice set-theoretically (Corollary 3.3), and as mentioned in the introduction, Weyman [**16**] has proved that suitable $I(\lambda_d^t)$ actually suffice to generate the ideal of X_r. (Weyman's result was proved for some special rank functions r by Strickland [**14**], and a preliminary version of this paper contained a proof in some further special cases.)

The vanishing loci of the λ_d^t are interesting varieties in their own right.

It turns out that they are unions of the X_r in all but the familiar case of λ_d^n (Corollary 3.4). In Section 4 we shall see that $V(\lambda_{d-1}^{n-1})$ arises as the singular locus of the hypersurface $V(\lambda_d^n)$.

Here is our main technical result, which give the invariance result in an explicit form:

THEOREM 3.1. *Suppose* $\operatorname{rank} A^n = \rho$ *and let* A' *be the nilpotent part of* A. *If* $\rho < d$, *then* $\lambda_d^t(A) = 0$ *iff* $\lambda_{d-\rho}^{t-\rho}(A') = 0$. *If* $t < n$ *then in addition* $\lambda_d^t(A) = 0$ *implies* $\rho < d$.

The example of $\lambda_1^n(A)$, the trace of A, shows, as remarked above, how badly this fails for $t = n$.

We postpone the proofs until after all the results have been stated.

Theorem 3.1 allows us to analyze the varieties X_r and $V(\lambda_d^t)$ in terms of each other, at least set-theoretically. First we give a formula telling exactly which $V(\lambda_d^t)$ contain which X_r. Some of these containments follow from others, as the second statement of the result shows. Here and later in this section we will use the notations $a_i(r)$ and $b_i(r)$ introduced at the beginning of Section 2:

THEOREM 3.2. $V(\lambda_d^t) \supseteqq X_r$ *iff* $d > t - \sum_{i>n-t} b_i(r)$. *If* $e \geq d$ *and* $s \geq t$ *then* $V(\lambda_e^s) \supseteqq V(\lambda_d^t)$ *unless* $s = t = n$.

The inclusions in Theorem 3.2 are inclusions of sets — that is, reduced schemes, and the X_r are indeed defined as such. However, we could also regard $V(\lambda_d^t)$ as a scheme, defined by the ideal $I(\lambda_d^t)$ generated by the entries of the matrix λ_d^t. Thus we may ask whether the assertions of Theorem 3.2 hold scheme-theoretically — that, whether the corresponding inclusions hold among the ideals $I(\lambda_d^t)$.

It is immediate from the definitions that Theorem 3.2 holds scheme-theoretically for $d = 1$ and $d = t$, and Weyman's result implies that the first statement holds scheme-theoretically, as already mentioned. Also, we have been able to prove (using the description of the λ_d^t given in part A) that the second statement is true scheme-theoretically for $d = e$. We have checked that these statements are true in all cases with $n \leq 6$ using the computer program Macaulay of Bayer and Stillman [**1**], and we thus conjecture that they hold scheme-theoretically in all cases.

COROLLARY 3.3. $X_r = \bigcap V(\lambda_d^t)$, *the intersection being taken over all those*

t and d such that $b_{n-t+1}(r) > 0$ and

$$t < n, \quad d = t + 1 - \sum_{i>n-t} b_i(r)$$

or

$$t = n, \quad r(\infty) + b_1(r) > d > t - \sum_{i>n-t} b_i(r).$$

Tanisaki proved the weaker result that the above formula holds if the intersection is extended over the (larger) set of indices with

$$d > t - \sum_{i>n-t} b_i(r).$$

PROOF (Tanisaki): The formula is obvious if one first diagonalizes $x - A$; that is, it suffices to consider the elementary divisors of $x - A$. □

Our result by contrast cannot be treated simply by diagonalizing the matrix $x - A$ over the ring $\mathbb{C}[x]$, because this process mixes together the ideal $I(\lambda_d^t)$ with $I(\lambda_e^t)$ for $e > d$. But by Theorem 3.2 the radical of $I(\lambda_d^t)$ contains $I(\lambda_e^t)$, and modulo this fact the results are equivalent.

Each λ_d^n is a 1×1 matrix whose entry is an irreducible polynomial, not a rank variety except for $d = n$. However for $t < n$ the irreducible components of the $V(\lambda_d^t)$ are rank varieties:

COROLLARY 3.4. *If $t < n$ then $V(\lambda_d^t)$ is the union, for $1 \leq k \leq 1 + [(d-1)/(n-t)]$, of the rank varieties*

$$\{A \mid \operatorname{rank} A^k \leq \min(n - k(n-t+1), d - 1 - (k-1)(n-t))\}.$$

We derive:

COROLLARY 3.5. *The following conditions are equivalent if $t < n$:*

i) *$V(\lambda_d^t)$ is irreducible.*
ii) *$V(\lambda_d^t) = V(\lambda_d^d) = \{A \mid \operatorname{rank} A \leq d - 1\}$.*
iii) *$d = t$ or $t + d \leq n$.*

REMARK: The ideal $I(\lambda_d^t)$ is given with $({}_nC_t)^2$ generating polynomials (where as usual ${}_nC_t$ denotes the binomial coefficient $n!/(n-t)!t!$). It is easy to see that they are each expressible as a linear combination of the $d \times d$ minors of the generic matrix, and thus at most $({}_nC_d)^2$ of them are linearly independent. Now the ${}_nC_d$ minors of the generic matrix are linearly independent (as one sees at once from the fact that one can construct a matrix of numbers with any given minor nonzero, and all other minors zero) so if ${}_nC_t < {}_nC_d$ then $I(\lambda_d^t) \subsetneqq I(\lambda_d^d)$. In this situation, Corollary 3.5 says that the two ideals even have different radicals.

On the other hand, $t+d \leq n$ iff ${}_nC_t \geq {}_nC_d$,and in this case the Corollary says that the two ideals have the same radical; but in fact, in a number of cases that we have checked by computer (with $n \leq 6$) the two ideals are actually equal when $t+d \leq n$. It would be interesting to know whether the span of the entries of λ_d^t in the space of all forms of degree d has dimension $= \min\big(({}_nC_t)^2, ({}_nC_d)^2\big)$, the "largest possible" value, in every case.

Our next goal is the proof of Theorems 3.1 and 3.2. We first show how to compute the minors of $x - A$ if A is in Jordan normal form. (Since we are interested in the coeffiecients of the different powers of x that appear, we cannot simply diagonalize $x-A$ over $\mathbb{C}[x]$; this would of course preserve the ideal in $\mathbb{C}[x]$ generated by the minors, but would not preserve the $\mathbb{C}$-vectorspace spanned by the minors.)

Assume, then, that A is in Jordan form; that is, decompose V as a direct sum $V = \oplus V_\alpha$ in such a way that $AV_\alpha \subset V_\alpha$ and $(A-\lambda_\alpha)\big|_{V_\alpha}$ is nilpotent of index equal to the dimension of V_α for each α. Choose bases $\mathcal{B}_\alpha$ for each V_α so that $A - \lambda_\alpha$ restricted to V_α becomes upper triangular with ones on the superdiagonal and zeros elsewhere, and let $\mathcal{B} = \cup\mathcal{B}_\alpha$. For subsets I and J of $\mathcal{B}$, we let $d(I,J) = \det_{I,J}(x - A)$ be the corresponding minor of $x - A$ (or 0 if I and J do not have the same cardinality). Writing $I_\alpha = I \cap \mathcal{B}_\alpha$ and $J_\alpha = J \cap \mathcal{B}_\alpha$, we obviously have $d(I,J) = \Pi_\alpha d(I_\alpha, J_\alpha)$ if the cardinality of I_α is the same as that of J_α for each α.

If I and J have the same cardinality, we let $\varphi : I \to J$ be the unique order preserving map from I to J, and set

$$e_\alpha = \operatorname{card}\{i \in I_\alpha \mid \varphi(i) = i\}.$$

We have

LEMMA 3.6. *With notation as above, suppose that I and J are t element subsets of $\mathcal{B}$. The minor $d(I,J)$ is zero iff $\varphi I_\alpha \neq J_\alpha$ for some α or $\varphi(i) \neq i, i+1$ for some i. Else*

$$d(I,J) = \Pi(x - \lambda_\alpha)^{e_\alpha}.$$

Thus if A consists of a single Jordan block, with eigenvalue λ, then the set of possible values of $d(I, J)$ is given by

$$\{d(I, J)\} = \{(x - \lambda)^e\}$$

where $e = t$ if $t = n$, and else e takes on all values from 0 to t. If A is nilpotent, with Jordan blocks of size $b_i(r)$, then

$$\lambda_d^t(A) = 0 \text{ iff } t - d < \sum_{i>n-t} b_i(r).$$

PROOF OF LEMMA 3.6: We first derive the formula for $d(I, J)$. Since $d(I, J) = \Pi_\alpha d(I_\alpha, J_\alpha)$ we see that $d(I, J)$ can only be nonzero if I_α and J_α have the same number of elements for each α, which is equivalent to the statement that $\varphi(I_\alpha) = \varphi(J_\alpha)$ for every α. Thus we may assume that this condition is satisfied. Let i be the smallest index occurring in a block I_α, and let $I' = I - \{i\}$, $J' = J - \{\varphi(i)\}$. If $\varphi(i) \neq i$, $i + 1$, then the row corresponding to i in the determinant $\det_{I,J}(x - A)$ is zero, so $d(I, J) = 0$. If $\varphi(i) = i + 1$, then this row has a 1 in the i-th column and zeros elsewhere, so $d(I, J) = d(I', J')$. If $\varphi(i) = i$ then the i-th column in the determinant $\det_{I,J}(x - A)$ has $x - \lambda_a$ in the i-th row and zeros elsewhere, so $d(I, J) = (x - \lambda_\alpha)\, d(I', J')$. In either case, since $\varphi \big|_{I'}$ is again the unique order preserving function from I' to J', we are done by induction.

It is clear in the one block case that $d(I, J)$ cannot take on any other values than those indicated. But if $0 \leq e \leq t < n$ then the choice

$$I = \{1, 2, \ldots, t\},$$
$$J = \{1, \ldots, e, e + 2, \ldots, t + 1\},$$

yields $d(I, J) = (x - \lambda)^e$.

Finally, in the nilpotent case, a choice of t rows I, that is, t elements of $\mathcal{B}$, excludes $n - t$ rows, and thus can exclude the elements of at most $n - t$ distinct $\mathcal{B}_\alpha$. But each $\mathcal{B}_\alpha$ which is entirely included contributes a factor of x^{b_α} to the determinants $d(I, -)$. Thus each of these determinants is divisible by the $\sum_{i>n-t} b_i(r)$ power of x, and from the one-block case it is clear that any power of x larger than this one, up to the t-th power, can be so obtained. Since by definition $\lambda_d^t(A) = 0$ iff the $(t - d)$-th power of x does not occur, we are done. □

PROOF OF THEOREM 3.2: If $t = n$ then $t - \sum_{i>n-t} b_i(r) = r(\infty)$. In this case the first statement of the theorem is elementary, while the second is vacuous.

We may thus assume $t < n$. We will first prove that the stated inclusions hold under the additional assumptions $r(\infty) = 0$, and only among the intersections of the given varieties with the set of nilpotent transformations. We will use this nilpotent case, below, to prove Theorem 3.1. But Theorem 3.1 shows that the general case of Theorem 3.2 follows from the nilpotent case of Theorem 3.2, so the proof of Theorem 3.2 will be completed when we complete the proof of Theorem 3.1.

Assume, then that $r(\infty) = 0$, and let A be a generic point of X_r, so that A is nilpotent. Of course, since the X_r are just the orbit closures in the nilpotent case, A can have any nilpotent Jordan form. The first statement of the Theorem in this case follows at once from the last statement of Lemma 3.6.

To prove the second statement of the Theorem in the nilpotent case — that is, with $V(\lambda_e^s)$ and $V(\lambda_d^t)$ replaced by their intersections with the set of nilpotent matrices — we use the fact that the X_r are orbit closures. Using the first statement of the Theorem, it is enough to show that $t \geq d > t - \sum_{i>n-t} b_i(r)$ implies $e > s - \sum_{i>n-s} b_i(r)$ in relevant cases. By induction we may further assume that e and s differ from d and t by at most 1. The cases $s = t$, $e = d + 1$ and $s = t + 1$, $e = d + 1$ are innocuous. If $s = t + 1$, $e = d$ then problems can only arise if $b_{n-t}(r) = 0$, in which case also $\sum_{i>n-t} b_i(r) = 0$, so $d > t$, a contradiction. This concludes the proof in the nilpotent case. □

PROOF OF THEOREM 3.1: First suppose that $\rho < d$. In this case we must show that $\lambda_d^t(A) \neq 0$ iff $\lambda_{d-\rho}^{t-\rho}(A') \neq 0$. Since A' is nilpotent, we may apply the last statement of Lemma 3.6 and obtain

$$\lambda_{d-\rho}^{t-\rho}(A') \neq 0$$

iff

$$d + \rho \leq (t + \rho) - \sum_{i>(n-\rho)-(t-\rho)} b_i(r),$$

that is, iff

$$d \leq t - \sum_{i>n-t} b_i(r).$$

Let A'' be the part of A with eigenvalues $\neq 0$, so that $A = A' \oplus A''$ and A'' is a $\rho \times \rho$ matrix of rank ρ. We may assume that A' is in Jordan normal form. The coefficient of x^0 in $\det(x - A'')$ is $\pm \det A''$, which is nonzero, and since by the nilpotent case of Theorem 3.2 and the above the $(t-\rho) \times (t-\rho)$ minors of $x - A'$ are the powers of x from $t - \sum_{i>n-t} b_i(r)$ to t, we see that the $t \times t$ minors of A containing all the rows and columns of A'' already have the powers of x in this range occurring with nonzero coefficient. Thus $\lambda_{d-\rho}^{t-\rho}(A') \neq 0$ implies $\lambda_d^t(A) \neq 0$. On the other hand, any $t \times t$ minor $d(I, J)$ of $x - A$ involves at least $t - \rho$ rows and columns of A', and thus has as a factor a minor of A' of size $s \geq t - \rho$. Applying the last statement of Lemma 3.6 to this minor, we see that it is divisible by x to the power

$$\sum_{i>n-\rho-s} b_i(r) \geq \sum_{i>n-t} b_i(r),$$

finishing the proof in the case $\rho < d$.

To finish the proof we must show that if $t < n$ and $\lambda_d^t(A) = 0$ then $\rho < d$. From Lemma 3.6, whose notation we adopt, we see that because $t < n$, we can find among the entries of $\bigwedge^t(x - A)$ polynomials of degree t of the form

$$p_\beta(x) = q(x)/(x - \lambda_\beta),$$

where

$$q(x) = \Pi(x - \lambda_\alpha)^{e_\alpha}$$

the λ_α running over all the eigenvalues (or any $t+1$ of them, if there are that many) and λ_β running over all the λ_α. To say that $\lambda_d^t(A) = 0$ thus implies that the d-th elementary symmetric function in t variables vanishes when applied to any t of the roots of q. In this situation the following Lemma, applied with $n = t + 1$, shows that at least $n - d + 1$ of the eigenvalues of A (counted with multiplicity) are 0; that is, $n - \rho \geq n - d + 1$, yielding $\rho < d$ as desired. □

Write $\sigma_d(x_1, \ldots, x_t)$ for the d-th elementary symmetric function in t variables. Since $\sigma_d(x_1, \ldots, x_t)$ is the sum of all multilinear monomials of degree d, it vanishes if at least $t - d + 1$ of the variables are 0. The converse of this is of course false except for $t = d$; however, if more than t variables are involved, a sort of converse is true:

Lemma 3.8. *If $1 \leq d \leq t < n$ are integers, and $\lambda_1, \ldots, \lambda_n$ are complex numbers such that the elementary symmetric function $\sigma_d(x_1, \ldots, x_t)$ vanishes on every set of t of the λ_i, then at least $n - d + 1$ of the λ_i are 0.*

PROOF OF LEMMA 3.8: If $d = t$ the result is obvious. If some $\lambda_i = 0$ then since

$$\sigma_d(x_1, \ldots, x_{t-1}, 0) = \sigma_d(x_1, \ldots, x_{t-1}),$$

the result follows by induction on n and t. Thus we may suppose that $\sigma_d(x_1, \ldots, x_t)$ vanishes on every set of t of the λ_i, but all the λ_i are nonzero, and we will derive a contradiction.

Note first that the sum $\sum_j \sigma_d(\lambda_{i_1}, \ldots, \hat{\lambda}_{i_j}, \ldots, \lambda_{i_{t+1}})$, leaving out one index at a time, is a positive integral multiple of $\sigma_d(\lambda_{i_1}, \ldots, \lambda_{i_{t+1}})$, so $\sigma_d(x_1, \ldots, x_{t+1})$ vanishes on every $t+1$ element subset of the λ_i. On the other hand,

$$\sigma_d(x_1, \ldots, x_{t+1}) = x_{t+1}\sigma_{d-1}(x_1, \ldots, x_t) + \sigma_d(x_1, \ldots, x_t),$$

(This formula also works for $d = 1$ if we interpret $\sigma_0(x_1, \ldots, x_t)$ as the constant function equal to 1.) Since by our hypothesis all the λ_i are nonzero, we see that $\sigma_{d-1}(x_1, \ldots, x_t)$ vanishes on every t-element subset of the λ_i. Continuing, we see that $\sigma_0(x_1, \ldots, x_t)$ vanishes on every t-element subset of the λ_i, which is the desired contradiction. □

As already remarked, Corollary 3.3 follows easily from the other results.

PROOF OF COROLLARY 3.4: The rank varieties X_r with $r(\infty) = 0$ are the minimal closed conjugation-invariant subvarieties of the set of nilpotent matrices. Since taking the nilpotent part commutes with conjugation, and the $V(\lambda_d^t)$ are invariant, it follows from Theorem 3.1 that each $V(\lambda_d^t)$ is the union of those X_r that are contained in it. It thus suffices to show that the maximal X_r contained in a given $V(\lambda_d^t)$ are among those on the list given in the Corollary.

From Theorem 3.2 we know that $X_r \subset V(\lambda_d^t)$ iff

$$d > t - \sum_{i > n-t} b_i(r) = t - \sum_{i \geq 1} (a_i(r) - (n-t))_+$$

where we have written $(s)_+$ for $\max(0, s)$. But we can make X_r larger without altering the validity of this inequality by replacing r with a rank function for which the $a_i(r)$ that are $\leq n-t$ are replaced by 0. Suppose that $a_k(r) > n-t$, but $a_{k+1}(r) = 0$, so that in particular

$$r(k) \leq n - k(n-t+1).$$

The condition given above then becomes

$$\begin{aligned} d &> t - \sum_1^k (a_i(r) - (n-t)) \\ &= t + k(n-t) - \sum_1^k a_i(r) \\ &= t + k(n-t) - (n - r(k)) \\ &= (k-1)(n-t) + r(k), \end{aligned}$$

that is,

$$r(k) < d - (k-1)(n-t).$$

But any X_r satisfying the above inequalities is contained in

$$\{A \mid \operatorname{rank} A^k \leq \min(n - k(n-t+1), d - 1 - (k-1)(n-t))\},$$

so this is the largest rank variety given our choice of k.

If the inequality on k in the Corollary is violated, then

$$d - 1 - (k-1)(n-t) < 0,$$

so the above rank variety is empty and can be dropped from the union.

□

In order to prove Corollary 3.5, we will need to know the rank of a generic member of the variety $\{A \mid \operatorname{rank} A^k \leq s\}$.

LEMMA 3.9. *A generic menber of the variety* $\{A \mid \operatorname{rank} A^k \leq s\}$ *has* rank $= n - a_1$ *where* a_1 *is the smallest integer* $\geq (n-s)/k$.

PROOF OF LEMMA 3.9: If A is such a generic member, then since

$$\begin{aligned} \operatorname{rank} A^{i-1} - \operatorname{rank} A^i &= \dim(\ker A) \cap (\operatorname{im} A^{i-1}) \\ &\leq \dim \ker A, \end{aligned}$$

we have $s = \operatorname{rank} A^k \geq n - k(\dim \ker A)$, so $\dim \ker A \geq a_1$. On the other hand it is easy to construct elements of the variety with $\dim \ker A = a_1$.

□

PROOF OF COROLLARY 3.5: Let

$$V_k = \{A \mid \operatorname{rank} A^k \leq \min(n - k(n-t+1), d - 1 - (k-1)(n-t))\}.$$

The generic element of $V_1 = V(\lambda_d^d)$ is an endomorphism of stable rank $d-1$, which is therefore not an element of any of the other V_k. This proves the equivalence of i) and ii). We also see that $V(\lambda_t^d)$ is irreducible iff V_1 contains all the other V_k with $k < (d-1)/(n-t)$, and this will be true iff the generic element of V_k has rank at most $d-1$. This rank of a generic element is computed for us by Lemma 3.9, and straightforward arithmetic leads to the desired result. □

REMARK: In Section 2 we saw that the variety $X_r = X_{r,0}$ could be realized as the flat limit of a family of varieties $X_{r,\lambda}$. Further, Corollary 2.4 shows that the equations of $X_{r,\lambda}$ are known if one knows the equations of $X_{r'}$ for certain $r' > r$. This gives an effective method of calculating the ideal of X_r (for example by computer) in any particular case.

4. Varieties defined by the coefficients of the characteristic polynomial.

A classical example of the construction λ_k^t of Section 3 occurs when $t = n$ as λ_k^n is a 1×1 matrix (i.e. a polynomial) such that $\lambda_k^n(A)$ is the coefficient of x^{n-k} in the characteristic polynomial of A. In this section we will consider the singular locus (Theorem 4.10) and its components (Corollary 4.14) of the variety of matrices defined by $\lambda_k^n = 0$. For simplicity, we will often write λ_k^n as λ_k. We let x_{ij} be the variable corresponding to the i, j entry of our square matrices, and consider λ_k as a polynomial in the x_{ij}'s. Occasionally, we will work basis free and consider $A \in \mathrm{End}_F(W)$. As λ_k is PGL_n invariant, we can unambiguously write $\lambda_k(A)$. Finally, λ_1 is just the linear, trace polynomial and so its zero set is trivial. We will henceforth assume that $k > 1$.

To begin with, we mention an alternative description of λ_k. If S, $T \leq \{l, \ldots, n\}$ are subsets of equal order, and $A \in M_n(F)$, denote by $A(S;T)$ the submatrix obtained by "crossing out" the rows in S and the columns in T. A *principal* submatrix is a submatrix of the form $A(S;S)$. The following is well known.

LEMMA 4.1. *$\lambda_k(A)$ is the trace of $\Lambda^k(A)$, which is the sum of the determinants of the principal $k \times k$ submatrices of A.*

To prove 4.1, one reduces to diagonal matrices using the Zariski density of diagonalizable matrices, and then computes. It follows from 4.1 or direct computation that λ_k is a multilinear polynomial.

LEMMA 4.2. *λ_k is an irreducible polynomial.*

PROOF: Write $\lambda_k = x_{11}q + r$ where x_{11} does not appear in q or r. Of course, λ_k is irreducible if and only if q and r have no common factors. If $X = (x_{ij})$ is the generic matrix, let $X' = X(1;1)$ be the submatrix. By inspection, q is the coefficient of x^{n-k} in the characteristic polynomial of X'. By induction (and the trivial trace polynomial case), we may assume that q is irreducible. Thus it suffices to show that q does not divide r, that is, that q does not divide λ_k. It is easy to construct a diagonal A with $q(A) = 0$ but $\lambda_k(A) \neq 0$, so the lemma is proved. □

Let $W_k \subseteq M_n(F)$ be the zero set of λ_k. As λ_k is irreducible, W_k is irreducible and is scheme theoretically defined by $\lambda_k = 0$. Let $S_k = \mathrm{Sing}\,(W_k)$ be the reduced subvariety consisting of the singular points of W_k. As F has characteristic 0, S_k is the zero set of the n^2 partial derivatives of λ_k. In order to express these partials, we introduce the following notation. If $S \subseteq \{1, 2, \ldots, n\}$ is a subset, consider the complementary set $\{1, \ldots, n\} - S$ and order this complementary set as usual. For $i \in S$, define $i' = 0_S(i)$ if i is in the i' position in $\{1, \ldots, n\} - S$ (lowest first). Note that if $X' = X(S;T)$ is a submatrix of the generic matrix, $i' = 0_S(i)$, and $j' = 0_T(j)$ then x_{ij} is the i', j' entry of X'. If $S = T$, *i.e.* if X' is principal, then $0_S(i) + 0_S(j)$ is congruent to $i + j$ modulo 2.

The proof of the next result is a calculus exercise.

LEMMA 4.3.

$$\frac{\partial \lambda_k}{\partial x_{ij}} = \sum_{S \in \Gamma} (-1)^{i+j} \det\bigl(X(S \cup \{i\}; S \cup \{j\})\bigr)$$

where Γ is the set of subsets of $\{1, 2, \ldots, n\}$ with cardinality $n - k$ and not containing i and j.

We begin our investigation of S_k by examining the restriction of the polynomials $\partial \lambda_k / \partial x_{ij}$ to upper triangular matrices. Recall from 3.8 that $\sigma_k(y_1, \ldots, y_s)$ is the k degree elementary symmetric function in $y_1, \ldots, y_s$. An immediate consequence of 4.3 is:

LEMMA 4.4. *Suppose $A \in M_n(F)$ is upper triangular with diagonal entries $a_{11}, a_{22}, \ldots, a_{nn}$. Then $\partial \lambda_k / \partial x_{jj}(A) = \sigma_{k-1}(a_{11}, a_{22}, \ldots, \hat{a}_{jj}, \ldots, a_{nn})$.*

COROLLARY 4.5. *If $A \in S_k$ and d is the stable rank of A then $d \leq k - 2$.*

PROOF: Since S_k and the stable rank are invariant under conjugation, we may assume that A is upper triangular. By 3.8 and 4.4, A has at least $n-k+2$ zeroes on its diagonal. The result follows. □

To derive more information about the elements of S_k we must make use of the equations $\partial\lambda_k/\partial x_{ij} = 0$ for $i \neq j$. If A is upper triangular, half of these equations automaticaly hold.

LEMMA 4.6. *Let A be upper triangular. Then $\partial\lambda_k/\partial x_{ij}(A) = 0$ if $j > i$.*

PROOF: If $A = (a_{ij})$ then by assumption $a_{ij} = 0$ for $i > j$. Let $B = A(S;S)$ be any principal $k \times k$ submatrix with $i, j \notin S$. It suffices to show that $\det(B(S \cup \{i\};\ S \cup \{j\} = 0$. If $B' = B(S \cup \{i\};\ S \cup \{j\}) = (b_{ij})$ then a little work shows that B' is upper triangular with $b_{mm} = 0$ for $0_S(i) \leq m \leq 0_S(j)$. □

Once again, S_k is invariant under conjugation and so it suffices to describe the $A \in S_k$ which are in Jordan canonical form. So for the next computation assume that A has the form:

$$\begin{pmatrix} B & 0 \\ 0 & C \end{pmatrix} \tag{1}$$

where B,C are in Jordan canonical form, B is nilpotent, and C is nonsingular.

LEMMA 4.7. *Let A have the form (1) and suppose $i > j$. Then $\det(A(\{i\};\{j\})) = 0$ unless B is a single Jordan block and the i, j position in A is the lower left corner of B. Under these circumstances, $\det(A(\{i\};\{j\})) = \det(C) \neq 0$.*

PROOF: B has a zero row containing the bottom row of each Jordan block, and a zero column containing the leftmost column of each Jordan block. Thus if B has at least 2 Jordan blocks, $A(\{i\};\{j\})$ has a zero row and zero determinant. If B is a single Jordan block, and $\det(A(\{i\};\{j\})) = 0$, $A(\{i\};\{j\})$ must contain neither the first column nor the last row of B. Hence i, j is the position claimed. For such a B and i, j, $A(\{i\};\{j\})$ has the form:

$$\begin{pmatrix} I & 0 \\ 0 & C \end{pmatrix}$$

and the result is clear. □

Given that A has form (1), the next lemma describes the principal submatrices of A. Its proof is both straightforward and elementary, and so is left to the reader.

LEMMA 4.8. *Let A be as in (1) and let D be a principal $k \times k$ submatrix of A.*

a) *D has the form:*

$$\begin{pmatrix} B' & 0 \\ 0 & C' \end{pmatrix}$$

where B', C' are in Jordan normal form, B' is nilpotent, and C' is nonsingular.

b) *A row or column of B' comes from one in B. Similarly for C and C'.*

c) *The maximum size of a Jordan block in B' is less than or equal to the maximum size of a Jordan block in B.*

We are ready to state and prove the key result, which describes the matrices in form (1) which lie in S_k.

PROPOSITION 4.9. *Suppose $A \in M_n(F)$ has the form (1), and that within B, the size of the Jordan blocks is nondecreasing as you go from top left to bottom right. Let d be the stable rank of A. The following are equivalent:*

a) *$A \in S_k$.*

b) *$d \leq k-2$ and all Jordan blocks in B have size less than or equal to $k-d-1$.*

c) *All principal $k \times k$ submatrices of A have rank less than or equal to $k-2$.*

PROOF: a) $\Rightarrow$ b): Assume a). By 4.5, $d \leq k-2$. Assume to the contrary that B has a Jordan block of size $r > k-d$. We can choose this block to be the one in the lower right corner of B. Let $i = n-d$, so B is an $i \times i$ matrix. Set $j = i-(k-d-1) = n-k+1$. Let $D = A(S;S)$ be a principal $k \times k$ submatrix of A with i, $j \in S$, and set $i' = 0_S(i)$, $j' = 0_S(j)$. Set $D' = D(\{i\};\{j\})$. Assume $\det(D') \neq 0$. By 4.7 and 4.8, i', j', must be the position in D in the lower left hand corner of E where $D =$

$$\begin{pmatrix} E & 0 \\ 0 & G \end{pmatrix} \tag{2}$$

and where E, G are in Jordan canonical form, E is nilpotent, and G is nonsingular. Since $j' = 1$; $1, 2, \ldots, j-1 \in S$. Since $j-1 = n-k$, $S = \{1, \ldots, j-1\}$, $G = C$, and $\det(D') = \det(C)$. Conversely if $S = \{1, \ldots, j-1\}$, E is a single Jordan block, and $\det(D(\{i'\};\{j'\}) = \det(C) \neq$

0. In other words, in the sum of 4.3 for $\partial\lambda_k/\partial x_{ij}(A)$, all terms but one are zero and that term is $\pm\det(C) \neq 0$. This contradicts our assumption that $\partial\lambda_k/\lambda x_{ij}(A) = 0$, and b) is proved.

b) $\Rightarrow$ c): Assume b). If D is any principal $k \times k$ submatrix of A, write D as in (2). Since G has size less than or equal to d, E has size greater than or equal to $k - d$. By 4.8, E contains at least 2 Jordan blocks, and so D has rank less than or equal to $k - 2$.

c) $\Rightarrow$ a): Let D be a principal $k \times k$ submatrix of A and D' a $k-1 \times k-1$ submatrix of D. Part c) says precisely that all such D' have determinant zero. In the sum of 4.3 for $\partial\lambda_k/\partial x_{ij}(A)$, all terms have the form $\pm\det(D')$ for such D'. Thus c) implies that $\partial\lambda_k/\partial x_{ij}(A) = 0$ for all i, j. That is, $A \in S_k$. $\square$

We are ready to give our description of S_k.

THEOREM 4.10. *Suppose $A \in M_n(F)$ has stable rank d. Then $A \in S_k$ if and only if $d \leq k - 2$ and rank $(A^{k-d-1}) = d$.*

PROOF: As both sides of the equivalence are conjugation invariant, we may assume that A is in the form (1). Thus rank $(A^r) = \text{rank}(B^r) + \text{rank}(C^r)\,\text{rank}(B^r) + d$. Now part b) of 4.9 is equivalent to $d \leq k - 2$ and $k - d - 1 = \text{rank}(B^{k-d-1}) = \text{rank}(A^{k-d-1}) - d$. $\square$

Let V_d^k be the variety $\{A | \text{rank}(A^{k-d-1}) \leq d\}$. By the observation in the introduction, $V_d^k = \{A \mid \text{rank}(A^i) \leq r(i)\}$ for some rank function r. We next explicitly determine the rank function r.

Since $V_d^k = X_r$ is irreducible, r is the rank function of a generic element in V_d^k. Intuitively, such a generic element, A, should have, in the appropriate sense, 0-eigenvalue block sizes as large as possible subject to the conditions rank $(A^n) = d$ and $A \in V_d^k$. To this end write $n - d = (k - d - 1)q + f$ where q, f are integers and $0 \leq f < k - d - 1$. Let A be a matrix in Jordan normal form with stable rank d, q many 0-eigenvalue blocks of size $k - d - 1$ and one 0-eigenvalue block of size f. Let r be the rank function of A.

LEMMA 4.11. *$V_d^k = X_r$. That is, r is the rank function of a generic point of V_d^k.*

PROOF: That $X_r \subseteq V_d^k$ is obvious. Suppose $B \in V_d^k$ has rank function s. Form the rank function $t(i) = \max(r(i), s(i))$. Note that $r(i) \leq t(i)$ for all i and that $t(k - d - 1) = d = r(k - d - 1)$. It suffices to show that $t(i) = r(i)$ for all i. We can suppose, by way of contradiction, that $t(i+1) > r(i+1)$ but $t(j) = r(j)$ for all $j \leq i$. Since $t(j) = r(j) = d$ for $j > k-d-1$, we have $i+1 < k-d-1$. Set $h = t(i) - t(i+1)$. Then $d = t(k-d-1) \geq t(i) - h(k-d-1-i) = r(i) - h(k-d-1-i)$. If $i \geq f$, this equals $n - iq - f - h(k-d-1-i)$, so $n-d-f \leq iq + h(k-d-1-i)$ implying that $h \geq q$ or $t(i+1) = t(i) - h = r(i) - h < r(i) - q = r(i+1)$, a contradiction. If $i < f$, $r(i) - h(k-d-1-i) = n - (q+1)i - h(k-d-1-i)$ so we have $n-d \leq (q+1)i + h(k-d-1-i) = (q+1)(k-d-1) - (q+1-h)(k-d-1-i)$. Thus $(q+1-h)(k-d-1-i) \leq (k-d-1-f)$. Since $i < f$, $q+1-h < 1$ or $h \geq q+1$. But $t(i+1) = t(i) - h = r(i) - h \leq r(i+1)$, a contradiction is again reached, and the lemma is proved. $\square$

We will use the V_d^k's to give the decomposition of S_k into irreducible components. To begin, we use 4.10 to express S_k in terms of the V_d^k's.

LEMMA 4.12. $S_k = \bigcup_{d=0}^{k-2} V_d^k$.

PROOF: Clearly, from 4.10, S_k is contained in this union. But from 4.11 a generic point of V_d^k is in S_k and the result is proved. $\square$

In order to purge the union in 4.12 of redundancies, we next investigate when $V_d^k \subseteq V_{d'}^k$. Write $V_d^k = X_r$ as in 4.11, and assume $V_d^k \subseteq V_{d'}^k$ with $d \neq d'$. Then $r(k-d'-1) \leq d'$. To express r, let $n - d = (k-d-1)q + f$ where q, f are integers and $0 \leq f < k-d-1$. First of all, r has stable value d so $d < d'$. If $k - d' - 1 < f$ then $r(k-d'-1) = n - (k - d' - 1)(q+1) = (d'-d)(q-1) + 2d' + (1-k+f)$ after substituting $n = (k-d-1)q + f + d$. Still assuming $k - d' - 1 < f$, then $1 - k + f > -d'$ so $(d'-d)(q-1) + 2d' + (1-k+f) > (d'-d)(q-1) + d' \geq d'$ a contradiction. Thus $k - d' - 1 \geq f$ and $r(k-d'-1) = n - (k-d'-1)q - f = d + (d'-d)q$ after substituting the same expression for n. Since $r(k-d'-1) \leq d'$, we have $q = 1$. Altogether, if $d \neq d'$ and $V_d^k \subseteq V_{d'}^k$, then $d < d'$ and $q = 1$. This is part of the proof of:

LEMMA 4.13. *Let d, $d \leq k-2$. The following are equivalent:*

a) $V_d^k \subseteq V_{d+1}^k$
b) $V_d^k \subseteq V_{d'}^k$ *for some* $d \neq d'$
c) $q = 1$

PROOF: That a) implies b) is obvious, and that b) implies c) was argued above. Assume c). As $f < k-d-1$ we have that $d+1 \leq k-f-1$ or

$k-(d+1)-1 > f$. Hence $r(k-(d+1)-1) = d+(d+1-d)q = d+1$ and so a) holds. □

With 4.13 in hand, we can give a more precise version of 4.12.

COROLLARY 4.14. *Let m be the maximum of 0 and $2k-2-n$. Then*

$$S_k = \bigcup_{d=m}^{k-2} V_d^k$$

and S_k is not the union of any subset of these V_d^k's.

PROOF: Since $q > 0$, we have $q = 1$ if and only if $(n-d)/(k-d-1) < 2$. This last inequality is just $d < 2k-2-n$. Thus it is precisely the V_d^k with $0 \le d < k-2-n$ that are redundant in 4.12, and 4.14 is proved. □

We close this section with a result relating the singular set S_k and the matrices λ_k^t studied in section 3. Recall that $\lambda_k = \lambda_k^n$ and $V(\lambda_k^t)$ is the zero set of the entries of λ_k^t. In particular, $V_k = V(\lambda_k^n)$.

THEOREM 4.15. $S_k = \mathrm{Sing}(V(\lambda_k^n)) = V(\lambda_{k-1}^{n-1})$.

PROOF: According to 3.4, $V(\lambda_{k-1}^{n-1})$ is the union, for $0 < s \le k-1$, of the rank varieties $\{A \mid \mathrm{rank}(A^s) \le \min(n-2s, k-s-1)\}$. Substituting $k-d-1$ for s, we have that $V(\lambda_{k-1}^{n-1})$ is the union, for $0 \le d \le k-2$, of the varieties $\{A \mid \mathrm{rank}(A^{k-d-1}) \le \min(d, n-2k+2d+2)\}$. As this minimum is certainly less than or equal to d, we have by 4.12 that $S_k \supseteq V(\lambda_{k-1}^{n-1})$. Thus by 4.14 it suffices to show that if d satisfies $0 \le d \le k-2$ and $d \ge 2k-1-n$, then $V_d^k \subseteq V(\lambda_{k-1}^{n-1})$. But $d \ge 2k-2-n$ implies that $d \le n-2k+2d+2$, so V_d^k appears in the union for $V(\lambda_{k-1}^{n-1})$ and the result is proved. □

ACKNOWLEDGEMENTS: Both authors are grateful to the NSF for partial support during the preparation of this work and to the Mathematical Sciences Research Institute, where the work was done, for providing a congenial atmosphere in a beautiful setting.

References

1. D. Bayer and M. Stillman, *Macaulay, a computer algebra program*, Available free from the authors, for many machines including the Macintosh, IBM-PC, Sun, Vax, and others (1986).
2. W. Borho and H.-P. Kraft, *Über Bahnen und deren Deformationen bei linearen Aktionen reduktiver Gruppen*, Comment. Math. Helv. **54** (1979), 61–104.
3. C. De Concini and C. Procesi, *Symmetric functions, conjugacy classes, and the flag variety*, Invent. Math. **64** (1981), 203–219.
4. R. Elkik, *Singularités rationelles et déformations*, Invent. Math. **47** (1978), 139–147.
5. M. Gerstenhaber, *On dominance and varieties of commuting matrices*, Ann. of Math. **(2) 73** (1961), 324–348.
6. H. Grauert and O. Riemenschneider, *Verschwindungssätze für analytische Kohomologiegruppen auf komplexen Räumen*, Invent. Math. **11** (1970), 263–292.
7. R. Hartshorne, "Algebraic Geometry," Springer-Verlag, New York, 1977.
8. G. Kempf, F. Knudson, D. Mumford and B. Saint-Donat, "Toroidal embeddings, I," Springer Lecture Notes in Math., vol. 338, 1973.
9. B. Kostant, *Lie group representations on polynomial rings*, Amer. J. Math. **85** (1963), 327–404.
10. H.-P. Kraft and C. Procesi, *Closures of conjugacy classes of matrices are normal*, Invent. Math. **53** (1979), 227–247.
11. T. Matsumura, T., "Commutative Algebra," Benjamin/Cummings Publishing Co., 1980.
12. C. Procesi, C., *A formal inverse to the Cayley-Hamilton Theorem*, preprint (1986).
13. I. R. Shafarevich, "Basic Algebraic Geometry," Springer-Verlag, Berlin/Heidelberg/New York, 1977.
14. E. Strickland, *On the variety of projectors*, J. Alg. **106** (1987), 135–147.
15. T. Tanisaki, *Defining ideals of the closures of the conjugacy classes and representations of the Weyl groups*, Tohoku Math. J. **34** (1982), 575–585.
16. J. Weyman, *Equations of conjugacy classes of nilpotent matrices*, preprint (1987).

Department of Mathematics, Brandeis University, Waltham MA 02254

Department of Mathematics, University of Texas, Austin TX 78713

Order Ideals

E. GRAHAM EVANS, JR., AND PHILLIP A. GRIFFITH

The purpose of this report is to discuss the notion of order ideals as it has been used in the past and to present a new result (Theorem 2.4) on order ideals of k-th syzygies which extends results of Auslander-Buchsbaum [**2**] as well as those of the authors [**8**].

Let R be a commutative ring, let M be an R-module and let m be an element of M. Then the order ideal of m in M is defined to be $\{f(m) \mid f \in \text{Hom}(M, R)\}$. We denote the order ideal by $O_M(m)$. This notion goes back at least to Bass's [**3**] fundamental paper on K-theory and stable algebra.

The organization for this paper is as follows. In Section 1 we discuss the case in which the module in question is a finitely generated projective module as well as the failure of these results when the module is not projective. In Section 2 we discuss how the theorems need to be modified in order to be valid for arbitrary finitely generated modules. We also discuss in some detail the special situation of k-th syzygies of finite projective dimension. Finally, in Section 3 we consider bizarre consequences (as related to order ideals) should our syzygy theorem [**8**] fail to be true in mixed characteristic.

Throughout R will be a commutative noetherian ring. In Section 1, which is directed towards projective modules over global rings, we will be assuming that R is a domain so that rank is well defined. In subsequent sections, when our deliberations are largely restricted to local rings and modules of finite projective dimension, the domain assumption will usually be dropped. For unexplained terminology or notation one should consult our monograph (Evans-Griffith [**10**]).

1. Order ideals of projective modules versus those of nonprojective modules.

Bass [**3**] used the concept of order ideal as a means of proving Serre's theorem [**17**] that a projective module of rank larger than the dimension of R must have a free summand. It was in this context that the formal notion of order ideal really manifested itself as a useful theoretical tool.

Bass actually achieved his proof of Serre's theorem by producing an element whose order ideal was R itself. There were three crucial results in this connection which we record here. In the first of these the module in question need not be projective.

THEOREM 1.1. *Let R be a noetherian domain and let P be a finitely generated R-module. If $m \in P$ and if Q is a prime ideal of R, then Q contains $O_P(m)$ if and only if m is not a free generator of P_Q.*

PROOF: The proof is obvious once we remark that forming $O_P(m)$ commutes with localization and that the order ideal of m in P_Q is R_Q precisely when m generates a free summand of P_Q.

THEOREM 1.2. *Let R be a commutative domain and let P be a finitely generated projective R-module. If $m \in P$ is such that $O_P(m)$ is a proper ideal, then the height of $O_P(m)$ is at most the rank of P.*

PROOF: Let d be the rank of P as an R-module. Then P_Q is free of rank d as an R_Q-module for any prime ideal Q of R. Therefore, if $Q \supseteq O_P(m)$, then $(O_P(m))_Q$ is a proper ideal which can be generated by d elements. Thus by Krull's Altitude Theorem the height of $(O_P(m))_Q$ is at most d.

THEOREM 1.3. *Let R be a commutative domain and let P be a finitely generated projective R-module. Then there exists an element m in P such that $O_P(m)$ has height equal to the rank of P (where we understand that, if the rank of P exceeds the dimension of R, then $O_P(m)$ is R itself).*

The proof of this result is rather long but is given in detail in many places (c.f. Eisenbud-Evans [**6**]). The argument presented in [**6**] allows one to choose a prime ideal Q of height equal to the rank of P and to pick the element m inside QP to ensure that the height is not accidently bigger than the rank of P. The mildly weaker version that the height of $O_P(m)$ is at least the rank of P is implicit in Serre's [**17**] original treatment as well as in Bass' [**3**] treatment.

In short there is a nice connection between the height of $O_P(m)$ and the rank of P, for the case of P being a projective module. If we drop the assumption that P is projective and only ask that it be finitely generated then most of the above statements are no longer true.

By way of example let I be an ideal of grade at least two and let M be a direct sum of n copies of I for $n > 2$. Then I contains $O_M(m)$ for every $m \in M$ and thus no element can be such that its order ideal has height equal to the rank of M. This is contrary to what Theorem 1.3 implies in the projective case.

Let R be a regular local ring of dimension at least four. Bruns [4] shows that there exists a three generated ideal $I = (a, b, c)$ such that the depth of R/I is zero. Hence there is an element $d \in R - I$ such that $\mathfrak{m}d \subseteq I$ where $\mathfrak{m}$ is the maximal ideal of R. Let $M = R^4/\langle a, b, c, d\rangle R$ and let $m \in M$ be the coset represented by $\langle 0, 0, 0, 1\rangle$. Then an elementary computation shows that $O_M(m) = \mathfrak{m}$ and therefore has height equal to the dimension of R while rank $M = 3$. These properties of m and M are contrary to those predicted by Theorem 1.2 for projective modules.

The statement of Theorem 1.1 says that the element m in the module P has the property $O_P(m) \subseteq Q$, Q a prime ideal of R, if and only if $m \in QP_Q$. If P is a projective module we can strengthen this statement to say: $O_P(m) \subseteq Q$ if and only if $m \in QP$. In order to observe the degree of failure of the latter statement for non-projective modules we establish the following general statements about graded torsion free modules over polynomial rings. For additional information and detail concerning the forthcoming hypotheses and conclusions one should consult our monograph [**10**, pp. 41–43].

PROPOSITION 1.4. *Let k be an infinite field and let $R = k[X_1, \ldots, X_n]$ with $n \geq 4$. Let M be a graded submodule of a finitely generated graded free R-module F such that $N = F/M$ is a torsion free R-module, i.e., M is a finitely generated graded R-module which is a second syzygy. If $M \cap QF = QM$ for each graded prime ideal Q of height two then M is a free R-module.*

PROOF: The hypothesis on M translates into the fact $\mathrm{Tor}_1^R(R/Q, N) = 0$ for each graded prime ideal Q of height two. Since N is torsion free it is a first syzygy for a module, say W. Then homological dimension shifting gives $\mathrm{Tor}_2^R(R/Q, W) = \mathrm{Tor}_1^R(R/Q, N) = 0$ for all graded prime ideals of height two. However, shifting dimensions in the first argument of Tor_1 gives $\mathrm{Tor}_1^R(Q, W) = 0$ for all graded prime ideals of Q of height two. Let K be the second syzygy of the residue field $k = R/(X_1, \ldots, X_n)$. We now apply Theorem 2.17 [**10**, p. 41] and obtain a graded short exact sequence $0 \to G \to K \to Q \to 0$ in which G is free and Q is a prime ideal of height two. It follows from above that $0 = \mathrm{Tor}_1^R(K, W) = \mathrm{Tor}_3^R(k, W) = \mathrm{Tor}_1^R(k, M)$. The vanishing of $\mathrm{Tor}_1^R(k, M)$ for $k = R/(X_1, \ldots, X_n)$ and M finitely generated and graded is equivalent to the statement that M is a graded free R-module.

The above result gives rise to a dichotomy for the class of finitely generated graded and reflexive R-modules, that is, the class of finitely generated, graded second syzygies over R, where $R = k[X_1, \ldots, X_n]$ as above. As noted previously, if P is projective then, for $m \in P$, we have that a prime Q contains $O_P(m)$ if and only if $m \in QP$. The situation for nonfree graded modules is described in the following statement.

PROPOSITION 1.5. *Let $R = k[X_1, \ldots, X_n]$ where k is an infinite field and $n \geq 4$. If M is a nonfree, graded and reflexive R-module, then there is $m \in M$ and a prime ideal Q of height two such that $m \notin QM$ but $O_M(m) \subseteq Q$.*

PROOF: Since M is a second syzygy over R, then there is a (graded) universal embedding $M \to F$, where F is a graded free R-module and where the natural map $\operatorname{Hom}_R(F, R) \to \operatorname{Hom}_R(M, R)$ is surjective (see [**10**, Exercise 3, p. 62]). It follows that, after identifying M with its image in F, if $m \in M$ then $O_M(m) = O_F(m)$. From Proposition 1.4 there is a graded prime ideal Q of height two such that $M \cap QF \neq QM$, i.e., there is $m \in M \cap QF$ with $m \notin QM$. It follows that such an element m has the property that $O_M(m) \subseteq Q$ while $m \notin QM$.

One way of obtaining a remedy for the above discrepancies betweeen free and nonfree modules is to employ the notion of basic elements (see Eisenbud-Evans [**6**]). In this theory one forgets about $O_M(m)$ and asks about the set of prime ideals Q such that the element $m \in M$ is part of a minimal generating set of M_Q. This approach in effect treats Theorem 1.1 as a definition and is able to conclude the analogue of Theorem 1.3. This approach has general interesting consequences and is the underlying theme in the remainder of this article.

2. The general case of finitely generated modules and their order ideals.

As suggested at the end of Section 1, one can modify the approach to order ideals of projective modules in order to obtain rather general results. The first such result is found in Eisenbud-Evans [**6**] and is the natural analogue of Theorem 1.2. One should consult Bruns [**5**] for the case in which the ring in question does not contain a field. We state the result.

THEOREM 2.1. *Let R be a local ring with maximal ideal $\mathfrak{m}$, let M be a finitely generated R-module and let $m \in \mathfrak{m}M$. Then height $O_M(m) \leq$* rank M.

The analogue of Theorem 1.3 is somewhat more difficult. One cannot hope to find an element so that $O_M(m)$ has height exactly the rank of M since in the cases sighted in Section 1 the height of all order ideals can be kept less than or equal to two even though the rank of M can be made arbitrarily large. The known cases in which the module has some order ideals $O_M(m)$ with height $O_M(m) > \operatorname{rank} M$ also have some satisfying height $O_M(m) \leq \operatorname{rank} M$. This is the essence of the next theorem which can be found in Evans-Griffith [**9**]. Unfortunately, the method of proof (at the time of this writing) needs the assumption that the residue field is algebraically closed.

THEOREM 2.2. *Let R be a local ring with algebraically closed residue field. Let M be a finitely generated nonfree R-module. Then there is an element $m \notin \mathfrak{m}M$ such that* height $O_M(m) \leq \operatorname{rank} M$.

The proof of Theorem 2.2 makes use of the notion of the "generic" order ideal. This is constructed as follows. Let M be generated by $m_1, \ldots, m_n$ and consider the module $S \otimes_R M$ where S is the polynomial ring $S = R[X_1, \ldots, X_n]$. In $S \otimes_R M$ we single out the element $v = X_1 m_1 + \cdots + X_n m_n$. The order ideal I of v with respect to $S \otimes_R M$ over S is then referred to as a "generic order ideal" of M. From the result of Eisenbud-Evans [**6**] we have that

$$\operatorname{height} I \leq \operatorname{rank}_S S \otimes_R M = \operatorname{rank}_R M.$$

Thus I is contained in a prime ideal Q of S such that $\operatorname{height}_S Q \leq \operatorname{rank}_R M$. If $\mathfrak{m}$ is the maximal ideal of R then one next considers the S-ideal $J = \mathfrak{m}S + Q$ and argues that S/J has positive dimension. This fact allows one to obtain ideals of the form $(X_1 - a_1, \ldots, X_n - a_n) + \mathfrak{m}S$ which contain J and for which some $a_i \notin \mathfrak{m}$. An elementary argument then shows that by setting $x_i - a_i = 0$, for each i, one obtains an element $m = a_1 m_1 + \cdots + a_n m_n$ in M whose order ideal has height (in R) not exceeding rank M. For more details one should consult Evans-Griffith [**9**].

The requirement that the residue field should be algebraically closed seems unnecessary in that the prime ideal Q should be essentially linear in the X's since the generators of $O_M(V)$ are linear in the X's. So we shouldn't need the residue field to be algebraically closed.

Next we want to discuss the case of M being a finitely generated k-th syzygy of finite projective dimension. If Q is a prime ideal of height less than or equal to k, then M is free. Thus, such an M acts as though it is projective for a large part of Spec(R). In this case one could hope that theorems more like those for the projective case would hold. The connection that we wish to explore is the one between the rank of M and the size of $O_M(m)$. The earliest such result is implicit in Gröbner's proof of Hilbert's Syzygy Theorem (Gröbner [**13,14**]). It can be stated as follows:

THEOREM 2.3. *Let R be a local ring, let M be a finitely generated k-th syzygy of finite projective dimension and let m be a minimal generator of M which does not generate a free summand of M. Then $O_M(m)$ is not contained in any ideal I such that the projective dimension of R/I is less than k.*

PROOF: We assume the contrary and let

$$0 \to M \to R^{n_{k-1}} \to \cdots \to R^{n_1} \to R^{n_0} \to V \to 0$$

be an exact sequence which exhibits M as a k-th syzygy in a minimal free resolution. Using a result of Auslander-Bridger [**1**] (see also Evans-Griffith [**10**, p. 49]) we may assume the above exact sequence remains exact after applying the functor $\operatorname{Hom}_R(-, R)$, that is, we may assume the dual sequence is also exact. As in the proof of Proposition 1.5, the natural surjection $\operatorname{Hom}_R(R^{n_{k-1}}, R) \to \operatorname{Hom}_R(M, R)$ implies that $O_M(m) = O_{R^{n_{k-1}}}(m)$ after viewing M as a submodule of the free module $R^{n_{k-1}}$. Furthermore, in extending the above complex to a minimal free resolution of V we may assume this is arranged so that the first basis element $\langle 1, 0, \ldots, 0 \rangle \in R^{n_k}$ is mapped onto m. We next compute $\operatorname{Tor}_k^R(V, R/I)$. On the one hand, if we compute $\operatorname{Tor}_k^R(V, R/I)$ from the resolution of R/I, we see that it is zero. However, if we compute it from the resolution of V, we see that the element $\langle 1, 0, \ldots, 0 \rangle$ goes to zero after tensoring with R/I. This follows because, as discussed above, the order ideal of m in M is identical with that of m viewed as an element in the free module $R^{n_{k-1}}$; hence $m \in IR^{n_{k-1}}$. At the same time we also see that $\langle 1, 0, \ldots, 0 \rangle$ cannot be an image from $R^{n_{k+1}}$ since all relations on the given generators of M lie in $\mathfrak{m}R^{n_k}$. This computation shows that $\operatorname{Tor}_k^R(V, R/I)$ is nonzero. This contradiction establishes the proof.

Needless to say this is not the way that Gröbner argued. Indeed homological algebra had not yet been invented when he gave his first proof! He gave a completely elementary proof for the case that R is a polynomial ring over a field and I is generated by a subset of the indeterminants. Later he extended his proof to cover the case that I is a perfect ideal. The Tor interpretation of the proof was shown to us by David Eisenbud.

One can push this a bit further since, if we had a finitely generated module P so that the projective dimension of P is less than k and P is annihilated by $O_M(m)$ we would get the same contradiction. The next step in this series was to show that, if R contains a field (so there are "big" Cohen-Macaulay modules over all the factor rings of R), then the height of $O_M(m)$ is at least k. This can be accomplished by computing $\mathrm{Tor}_k^R(V, C)$, where C is a big Cohen-Macaulay module over $R/O_M(m)$. This argument and result are discussed in detail in our monograph [**10**]. If one combines this result with the existence of minimal generators having height $O_M(m) \leq \operatorname{rank} M$, then one obtains that the rank of a nonfree k-th syzygy of finite projective dimension is at least k. This series of ideas can be pushed even a bit further to show the following:

THEOREM 2.4. *Let R be a local ring which contains a field, let M be a finitely generated k-th syzygy of finite projective dimension and let m be a minimal generator of M. Then the grade of $O_M(m)$ is at least k (that is, $O_M(m)$ contains an R-sequence of length at least k).*

Before we prove this result we would like to mention one special case of independent interest. If we take the case of M being an ideal of finite projective dimension (and hence $k = 1$), we get that every minimal generator of M is a nonzero divisor. This generalizes the classical result of Auslander-Buchsbaum [**2**] that every ideal of finite projective dimension contains a nonzero divisor. This fact is elementary in the case that I itself is generated by an R-sequence and thus is another example of how ideals generated by an R-sequence are typical of ideals of finite projective dimension.

We will give two proofs of this theorem. The first in case R contains a field of characteristic $p > 0$ was suggested to us by Craig Huneke. It uses the ideas that he and Melvin Hochster have developed recently under the heading of "tight closure" of ideals. A more technical version would allow the reduction from "containing a field of characteristic 0" to "characteristic p". The second proof is a reduction to showing that a minimal generator of an ideal of finite projective dimension is a non-zero divisor followed by an argument generalizing our proof that $O_M(m)$ has height at least the syzyginess of M.

PROOF (Huneke): Assume that grade $O_M(m) < k$. Let $r_1, \ldots, r_g$ be a maximal R-sequence in $O_M(m)$ with $g < k$. Then there is an $r \in R$ which is not in $(r_1, \ldots, r_g)$ so that $rO_M(m) \subseteq (r_1, \ldots, r_g)$. Let

$$0 \to M \to R^{n_{k-1}} \to \cdots \to R^{n-1} \to R^{n_0} \to V \to 0$$

be an exact sequence giving M as a k-th syzygy. As in the previous argument we may assume the above complex comes from the minimal free resolution of V and that the map $R^{n_k} \to R^{n_{k-1}}$ sends the first basis element $e = \langle 1, 0, \ldots, 0 \rangle \in R^{n_k}$ to $m \in M \subseteq R^{n_{k-1}}$. As before we have that $\mathrm{Tor}_k^R(V, R/(r_1, \ldots, r_g)) = 0$ since $g < k$. Therefore, after tensoring the resolution of V with $R/(r_1, \ldots, r_g)$, we must have that re_1 is sent to 0. Thus re_1 must be the image of some element from $R^{n_k} \otimes R/(r_1, \ldots, r_g)$. This means that $re_1 = \langle r, 0, \ldots, 0 \rangle$ is in the image of the map $R^{n_k} \xrightarrow{\phi_k} R^{n_{k-1}}$ modulo the ideal $(r_1, \ldots, r_g)$. The map ϕ_k is given by a matrix, say $\phi_k = (a_{ij})$. We next apply the Frobenius to the finite free resolution of V. This gives a new exact sequence where the modules are the same but the maps are changed by replacing their entries by their p-th powers. The new resolution resolves a new module which we denote by $V^{(p)}$. The homology obtained from this resolution after tensoring with $R/(r_1, \ldots, r_g)$ is $\mathrm{Tor}_k^R(N^{(p)}, R/(r_1, \ldots, r_g))$ and is still zero. Thus $\langle r, 0, \ldots, 0 \rangle$ is in the image of the new map modulo the ideal $(r_1, \ldots, r_g)$ where the new map is $\phi_k^{(p)} = (a_{ij}^p)$, that is, $\langle r, \ldots, 0 \rangle$ is in $\mathfrak{m}^p R^{n_{k-1}}$ modulo the ideal $(r_1, \ldots, r_g)$, where $\mathfrak{m}$ denotes the maximal ideal of R. After applying the Frobenius n times we conclude from the above argument that $\langle r, 0, \ldots, 0 \rangle$ is in $\mathfrak{m}^{p^n} R^{n_{k-1}}$ modulo $(r_1, \ldots, r_g)$. Thus, by the Krull Intersection Theorem, $\langle r, 0, \ldots, 0 \rangle$ must be zero modulo $(r_1, \ldots, r_g)R^{n_{k-1}}$, that is, $r \in (r_1, \ldots, r_g)$, which is contrary to the original hypothesis on r.

The second proof is a reduction to the case of an ideal of finite projective dimension. It needs three lemmas. The first allows one to reduce the syzyginess by one to get down to the case of first syzygies. The second allows one to cut the rank down by one to get to the case of a rank one first syzygy. The third lemma handles the case of an ideal of finite projective dimension in the case that R contains a field.

LEMMA 2.5. *Let R be a local ring such that Theorem 2.4 is true for the case of $(k-1)$-th syzygies, $k \geq 2$, for R and all of its homomorphic images. Then Theorem 2.4 is true for R in the case of k-th syzygies.*

PROOF: Let M be a finitely generated k-th syzygy of finite projective dimension and let m be a minimal generator of M. Assume that $O_M(m)$ has grade less than k. Then the grade of $O_M(m)$ must be exactly $k-1$ since M is also a $(k-1)$-th syzygy. Thus $O_M(m)$ contains a nonzero divisor on R, say x. It follows that x is necessarily a nonzero divisor on M since M is at least a first syzygy. Furthermore, $\bar{M} = M/xM$ is a $(k-1)$-th syzygy of finite projective dimension over $R = R/(x)$. Thus $J = O_{\bar{M}}(m)$ contains an R-sequence of length at least $k-1$, where m is the image of m in M. Letting K denote the inverse image of J in R we see that K is an ideal of grade at least k. We want to compare K with $O_M(m)$. We consider the exact sequence $0 \to R \xrightarrow{x} R \to R/xR \to 0$ and apply $\operatorname{Hom}_R(M, -)$ to it. This gives a long exact sequence which starts

$$0 \to \operatorname{Hom}_R(M,R) \to \operatorname{Hom}_R(M,R) \to \operatorname{Hom}_R(M,R/xR) \to \operatorname{Ext}^1_R(M,R).$$

If Q is a prime ideal in R of grade less than or equal to k, then M_Q is free since it is a k-th syzygy of finite projective dimension in a ring of depth less than or equal to k. Thus $\operatorname{Ext}^1(M,R)$ is zero when localized at Q. It follows that $O_M(m)$ equals K after localizing at Q. However this would yield a contradiction if Q were a prime ideal of grade $k-1$ which contained $O_M(m)$.

LEMMA 2.6. *Let R be a local ring such that the conclusion of Theorem 2.4 holds for all modules of rank less than d with $d \geq k+1$. Then the conclusion of Theorem 2.4 holds for all R-modules of* rank d.

PROOF: Let M be an R-module of rank d. Then applying the technique of Bruns [4] we can find an element w in M such that M/Rw is still a k-th syzygy of finite projective dimension, has rank $d-1$ and such that m remains a minimal generator in M/Rw. Letting $N = M/Rw$ one sees that $O_N(m') \subseteq O_M(m)$ where m' is the image of m in N. Thus the inequality $\operatorname{grade} O_M(m) \geq \operatorname{grade} O_N(m') \geq k$ completes our argument.

Our final lemma here, before presenting a second proof of Theorem 2.4, makes use of maximal Cohen-Macaulay modules. A fundamental resuslt of Hochster [**15**] established the existence of these modules (not necessarily finitely generated) over local rings which contain a field. Such modules C over R have the property that $\operatorname{depth} C = \dim R$ (see Evans-Griffith [**10**, pp. 12-18] for a discussion on computing depths of these modules). Using Hochster's discovery Griffith [**12**] constructed a more specialized version of maximal Cohen-Macaulay modules which Sharp [**18**] referred to as "balanced" maximal Cohen-Macaulay modules. The balanced maximal Cohen-Macaulay modules have the important property that their localizations are maximal Cohen-Macaulay.

LEMMA 2.7. *Let R be a local ring, let M be a finitely generated module of finite projective dimension and let $P \in \operatorname{Ass}(R)$. If C is a balanced maximal Cohen-Macaulay module over R/P, then $\operatorname{Tor}_i^R(C, M) = 0$ for $i > 0$.*

PROOF: We proceed by induction on the dimension of R/P. If $\dim R/P = 0$, then M must be free since $\operatorname{depth} R = 0$ and M has finite projective dimension. Then of course $\operatorname{Tor}_i^R(C, M) = 0$ for $i > 0$.

We next consider the case of $\dim R/P > 0$. Let

$$0 \to F_d \to F_{d-1} \to \cdots \to F_1 \to F_0 \to M \to 0$$

be a minimal finite free resolution of M. We apply the functor $C \otimes_R -$ to this resolution. If the support of $\operatorname{Tor}_i^R(C, M)$ contains primes other than the maximal ideal for $i > 0$, then we can localize so that its support is exactly the maximal ideal. (This is, of course, the reason for picking C to be balanced). The resulting complex

$$0 \to C \otimes F_d \to C \otimes F_{d-1} \to \cdots \to C \otimes F_1 \to C \otimes F_0 \to 0$$

is a complex over R/P having length equal to d (the projective dimension of M over R). Using the acyclicity lemma of Peskine-Szpiro [**16**] as refined by Foxby [**11**] (see Evans-Griffith [**10**] for a detailed discussion) we get that $\operatorname{Tor}_i^R(C, M) = 0$ for $i > 0$ as desired.

SECOND PROOF OF THEOREM 2.4: Using Lemmas 2.5 and 2.6 we reduce to the case of an ideal. That is, we reduce the problem to one of showing: if I is an ideal of finite projective dimension and if x is a minimal generator of I then x is not a zero divisor of R. To this end suppose that $P \in \operatorname{Ass} R$ and that $x \in P$. We set $M = R/I$ and choose C to be a balanced Cohen-Macaulay module over R/P. A typical Gröbner calculation, as in the proof of Theorem 2.3, shows that $\operatorname{Tor}_1^R(C, M) \neq 0$ while an application of Lemma 2.7 gives that $\operatorname{Tor}_1^R(C, M) = 0$. This contradiction completes the argument.

3. Should the syzygy theorem fail in mixed characteristic.

In concluding our article we expand on the observation following Theorem 2.4. In particular the following statement is the essence of Lemmas 2.5, 2.6, 2.7 and the "second proof" of Theorem 2.4. Moreover the statement is independent of the characteristic of the local ring.

THEOREM 3.1. *Let R be a local ring. Then the syzygy theorem holds for R and all of its homomorphic images if and only if the ideals of finite projective dimension for these rings satisfy the criterion: a minimal generator is necessarily a nonzero divisor.*

Since the syzygy theorem holds for local rings containing a field then their ideals of finite projective dimension satisfy the zero divisor criterion stated in Theorem 3.1. In the remainder of this section we wish to explore the possibilities of pathological behavior that will arise in mixed characteristic should the syzygy theorem fail to hold in this context. In particular the next theorem demonstrates, that should the zero divisor criterion fail for ideals of finite projective dimension, then it fails in a rather excessive manner.

THEOREM 3.2. *Let (A, m) be a local ring of mixed characteristic $p > 0$ and assume that p is a nonzero divisor on A. Furthermore, assume that A has a finitely generated nonfree k-th syzygy module M of finite projective dimension for which its rank is less than k. Then for $n > 0$ there is a homomorphic image R of A (depending on M and n) and an ideal I of R having finite projective dimension such that I has at least n elements in $I - \mathfrak{m}I$ which are zero divisors on R. Moreover, the n elements in the above statement can be arranged to include any finite number of generating sets for I.*

PROOF: Since the syzygy theorem is true for M/pM over A/pA it follows from results in Evans-Griffith [8] that M can be taken to have rank $k - 1$ and be free locally on the punctured spectrum of A. Further, it follows from Theorem 2.4 that each $e \in M - \mathfrak{m}M$ has order ideal $O_M(e)$ of grade $k - 1$. Let $e_1, \ldots, e_n$ be any finite set of minimal generators of M. It follows that the intersection $J = \bigcap_{i=1}^{n} O_M(e_i)$ has grade $k - 1$ as well. Let $b_1, \ldots, b_{k-1}$ be a maximal A-sequence in J. Since M necessarily satisfies the Serre condition S_k (see Auslander-Bridger [1] or Evans-Griffith [10, Theorem 3.8]) and since M has finite projective dimension, we have that $b_1, \ldots, b_{k-1}$ is also an M-regular sequence.

We have arrived at the situation $\bar{M} = M/(b_1, \ldots, b_{k-1})M$ is an S_1-module of finite projective dimension over the factor ring $R = A/(b_1, \ldots, b_{k-1})$. We let $\bar{e}_1$ denote the images of the e_i in M, respectively. The observation that each order ideal $O_{\bar{M}}(\bar{e}_i)$ consists of zero divisors follows from repeated application of the following lemma whose proof parallels that of Lemma 2.5.

LEMMA 3.3. *Let A be a local ring and let M be an finitely generated A-module. If $e \in M$ and if $b \in O_M(e)$ is a nonzero divisor on A and M, then:*

(a) $\text{grade}_{\bar{A}}\, O_{\bar{M}}(\bar{e}) \geq \text{grade}_A\, O_M(e) - 1.$
(b) $\text{grade}_A\, O_M(e) \geq \min(\text{grade}_A \text{Ext}^1_A(M, A), \text{grade}_{\bar{A}}\, O_{\bar{M}}(\bar{e}) + 1).$

(Here "($^-$)" denotes reduction modulo b).

With regard to our application of Lemma 3.3 we note that grade_A $\text{Ext}^1_A(M, A) = \text{grade}\, A \geq k$, since M is a nonfree k-th syzygy of finite projective dimension which is free locally on the punctured spectrum of A. Since each e_i has $\text{grade}_A\, O_M(e_i) = k - 1$, it then follows from (a) and (b) in Lemma 3.3 that $\text{grade}_{\bar{A}}\, O_{\bar{M}}(e_i) = \text{grade}_A\, O_M(e_i) - 1$, for $i = 1, \ldots, n$.

Returning to our construction we let N denote the R-module $M/(b_1, \ldots, b_{k-1})M$ with specified minimal generators $\bar{e}_1, \ldots, \bar{e}_n$. At this point we apply Bruns's theorem [4] on basic elements in order to obtain a free submodule F of N such that $I = N/F$ satisfies:

i) I is a first syzygy
ii) I has rank one.

Thus I is isomorphic to an ideal in R and has finite projective dimension over R since N does. We note that the images of the $\bar{e}_i$ under any embedding of I into R must be contained in $O_N(\bar{e}_i)$ and thus must be zero divisors. As a final remark we note that none of the $\bar{e}_i$ could be lost via the construction of the quotient $I = N/F$ since none of the $\bar{e}_i$ are able to satisfy the basicity requirement in the Bruns [4] construction.

The authors would like to thank the referee for pointing out several technical deficiencies in the original version of the manuscript.

References

1. M. Auslander and M. Bridger, Mem. Amer. Math. Soc., no. 94 (1969), "Stable Module Theory," American Math. Soc., Providence R.I,.
2. M. Auslander and D. Buschbaum, *Codimension and multiplicity*, Ann. of Math. **68** (1958), 625-657.
3. H. Bass, *K-Theory and stable algebra*, Publications Mathématiques, No. 22, IHES, Paris (1964).
4. W. Bruns, *"Jede" endliche freie Auflösung is freie Aufl sung eines von drei Elementen erzeugten Ideals*, J. Algebra **39** (1976), 429-439.
5. W. Bruns, *The Eisenbud-Evans Principal Ideal Theorem and Determinantal Ideals*, Proc. Amer. Math. Soc. **83** (1981), 19-24.
6. D. Eisenbud and E. G. Evans, *Generating modules efficiently: theorems from algebraic K-theory*, J. Algebra **27** (1973), 278-305.
7. D. Eisenbud and E. G. Evans, *A generalized principal ideal theorem*, Nagoya Math. J. **62** (1976), 41-53.
8. E. G. Evans and P. Griffith, *The syzygy problem*, Annals of Math. **114** (1981), 323-353.
9. E. G. Evans and P. Griffith, *Order ideals of minimal generators*, Proc. Amer. Math. Soc. **86** (1982), 375-378.
10. E. G. Evans and P. Griffith, "Syzygies," London Math. Soc. Lecture Notes Series, no. 106, Cambridge University Press, 1985.
11. H.-B. Foxby, *On the μ^i in a minimal injective resolution II*, Math. Scand. **41** (1977), 19-44.
12. P. Griffith, *A representation theorem for complete local rings*, J. Pure and Applied Alg. **7** (1976), 303-315.
13. W. Gröbner, "Moderne Algebraische Geometrie: die ideal theoretischen Grundlagen," Sprnger Verlag, Berlin, 1949.
14. W. Gröbner, *Über die Syzgyien-Theorie der Polynomideale*, Monatsh. Math. **53** (1949), 1-6.
15. M. Hochster, "Topics in the Homological Theory of Modules over Commutative Rings," C.B.M.S. Regional Conference Ser. Math., no. 14, American Math. Soc., Providence, R.I, 1976.
16. C. Peskine and L. Szpiro, *Dimension projective finie et cohomologie locale*, Publ. Math. I.H.E.S. **42** (1973), 47-119.
17. J.-P. Serre, *Modules projectifs et espace fibres à fibre vectorielle*, Seminaire P. Dubreil, Éxpose 23, Paris (1957).
18. R. Y. Sharp, *Cohen-Macaulay properties for balanced big Cohen-Macaulay modules*, Proc. London Comb. Phil. Soc. **90** (1981), 229-238.

Department of Mathematics, University of Illinois, 1409 West Green Street,
Urbana IL 61801

Department of Mathematics, University of Illinois, 1409 West Green Street,
Urbana IL 61801

Both authors were partially supported by the National Science Foundation during the preparation of this paper.

A Characterization of F-Regularity in Terms of F-Purity

RICHARD FEDDER AND KEI-ICHI WATANABE

0. Introduction.

In recent years, some very interesting theorems have been proven independently using complex analytic techniques or, alternatively, using reduction to characteristic p techniques (relying on special properties of the Frobenius homomorphism). In particular, Hochster and Roberts [**12**] proved that the ring R^G of invariants of a group G acting on a regular ring R is necessarily Cohen-Macaulay by an argument which exploits the fact that R^G is a direct summand of R in characteristic 0 and that, therefore, after reduction to characteristic p, the Frobenius homomorphism is especially well-behaved for "almost all p". Not long after, using the Grauert-Riemenschneider vanishing theorem, Boutôt [**1**] proved an even stronger result— in the affine and analytic cases, a direct summand (in characteristic 0) of a ring with rational singularity necessarily has a rational singularity.

A key point in the analogy between proofs which depend on complex variables and proofs which depend on reduction to characteristic p seems to be that F-purity gives an algebraic "analogue" to the geometric notion of a rational singularity. Since an F-pure ring need not even be Cohen-Macaulay, this analogy is somewhat loose. In fact, it has been shown [**20**] that a ring which has a rational singularity in characteristic 0 need not give rise to an F-pure ring in any characteristic. However, the "bad" examples so constructed are not Gorenstein, and the analogy between rational singularity and F-pure (type) in Gorenstein rings is still strongly conjectured and quite striking in dimension ≤ 2.

Hochster and Huneke have recently introduced the notion of F-regularity, a significant modification of the notion of F-purity. They proved under some mild conditions [**9**] that the direct summand of an F-regular ring must be F-regular and that an isolated F-regular singularity must be a rational singularity in characteristic 0.

Postponing the formal definitions, we will summarize what is known about these related notions of singularity in the following theorem.

THEOREM 0.1. *Consider the following propositions about the ring (R, m) which is assumed to be either local with maximal ideal m or positively graded with R_0 a field and R_+ the maximal ideal m:*

(1) *R is F-regular.*
(1′) *R is F-rational.*
(2) *R has a rational singularity.*
(3) *R is Cohen-Macaulay, graded, with $a(R) < 0$.*
(4) *R is F-pure and F-unstable.*
(4′) *R is F-injective and F-unstable.*

Then, (1) $\Rightarrow$ (1′), (1′)$\Rightarrow$(2) if R is finitely generated over a field of characteristic 0 and has an isolated singularity at m [**9**]. *(2)$\Rightarrow$(3) if R is graded* [**6** or **19**], *(3)$\Rightarrow$(4) if R is a complete intersection ring with an isolated singularity at m* [**4**], *(4)$\Rightarrow$(4′)* [**3**] *(4')$\Rightarrow$(4) if R is Gorenstein* [**3**], *and (3)$\Rightarrow$(4′) if R is 2-dimensional and R_Q is F-pure for all primes Q of height one in R* [**5**].

The work of Hochster and Huneke [**9,10,11**] focuses attention on the problem of finding effective criteria for determining F-regularity in broad classes of rings. Specifically, the motivation for this paper was the question of Hochster and Huneke - Which graded hypersurfaces are F-regular? We prove that in the statement of Theorem 0.1 above, (1′)$\Leftrightarrow$(4′) under the conditions that:

(a) R has characteristic p.
(b) R has an isolated singularity at m and $H^i_m(R)$ has finite length for every $i < n$.
(c) R is an equidimensional quotient of a Cohen-Macaulay ring.
(d) $R^{1/p}$ is a finite R-module. (It suffices, for example, for R to be essentially of finite type over a perfect field k.)

Of course, this result in characteristic p implies, via the usual reduction to characteristic p techniques, the corresponding characteristic 0 result.

It follows from (1′)$\Leftrightarrow$(4′) that all six notions of singularity in Theorem 0.1 are equivalent (in characteristic 0) for a graded complete intersection ring which has an isolated singularity at m or for a graded Gorenstein ring of dimension ≤ 2. Since, for a positively graded complete intersection ring $R = k[X_1, \ldots, X_n]/(f_1, \ldots, f_d)$, it is easily calculated that $a(R) = \sum_{i=1}^{d} \deg(f_i) - \sum_{i=1}^{n} \deg(X_i)$; it follows from our theorem here and previous work of Fedder [4] partially classifying F-pure complete intersections that if R has an isolated singularity and if R has characteristic 0, then R is F-regular if and only if $\sum \deg(f_i) < \sum \deg(X_i)$. It should be noted that this result is not true in characteristic p. For example, the ring $R = k[X,Y,Z]/(X^2+Y^3+Z^5)$ has $\deg(X^2+Y^3+Z^5) = 30$ and $\deg X + \deg Y + \deg Z = 31$. It therefore satisfies the criterion above for F-regularity in characteristic 0. However, in characteristic 2,3, or 5, R is not even F-pure (immediate from the criterion $f^{p-1} \in m^{[p]}$ derived in [**3**]); much less F-regular. However, it is true that R is F-regular in characteristic p for all sufficiently large primes.

We also illuminate a speculation of DeConcini, Eisenbud, and Procesi [**2**] by showing that if R is a square-free Hodge *domain* with an isolated singularity at H, the maximal ideal determined by the poset of indeterminates which generate R, and if $(R_0)_H$ is Cohen-Macaulay, then $(R)_H$ is F-rational. Here R_0 is the "discrete Hodge Algebra" associated to R. In particular, if k has characteristic 0, then R has a rational singularity as speculated in [**2**]. The proof is particularly interesting because, as pointed out in [**2**] for the case of a rational singularity, the most natural deformation argument from the well-understood algebra $(R_0)_H$ to $(R)_H$ cannot possibly work. $(R_0)_H$ is essentially never F-rational because it is essentially never a domain. The deformation argument only tells us that R is F-injective. We then use the fact that if R is a Hodge domain, the poset H has a unique minimal element X and R/XR is also Hodge. From this we can deduce that R is not only F-injective but actually F-rational.

In proving that F-rationality is equivalent to F-injectivity plus F-instability, we discover another fact which is interesting in its own right. Let (R,m) be an F-pure (respectively F-injective) ring with an isolated singularity and let I be an n-generated ideal (respectively an n-generated system of parameters ideal). Then, $m(\overline{I^n}) \subseteq I$. (Here $\overline{}$ denotes the integral closure.)

ACKNOWLEDGEMENT: We wish to thank C. Huneke for his careful reading of this manuscript and helpful discussions. The first author would also like to thank Purdue University for its fine hospitality during the period when he was preparing this manuscript.

I. Definitions and Notations.

Let R be a ring of characteristic p. Denote by $\mathbf{^eR}$, the ring R viewed as an R-module via the eth power of the Frobenius map. Then, by definition, $R \to {}^eR$ is an R-linear map. Furthermore, for any R-module M, we denote by $^e\mathbf{M}$ the module $M \otimes_R {}^eR$. Note that if R is reduced and $q = p^e$, there is a natural identification of maps $R \xrightarrow{F^e} {}^eR$ with $R \subseteq R^{1/q}$ and with $R^q \subset R$ where R^n denotes the ring $\{x^n | \, x \in R\}$ for $n = q$ or $1/q$. In particular, the second identification makes clear that applying the eth power of the Frobenius is really the same as adjoining the (unique!) $(p^e)^{th}$ roots of the elements of R. For $q = p^e$, we will denote by $I^{[q]}$ the ideal generated in R by the qth powers of all the elements of I.

By convention throughout this paper, we will use the notation (R, m) to signify *either* a local ring with maximal ideal m or a positively graded ring with $R_0 = k$ a field and $m = R_+$. Except for comments about applications (e.g. the discussion above about complete intersection rings), all our results will take place in characteristic p. Therefore, we need not discuss here the technique of reduction from characteristic 0 to characteristic p which is described in [**3**], [**4**], [**13**], and [**18**].

DEFINITION 1.1. *R is* **F-pure** *if for every R-module M (equivalently, every finitely generated R-module M [**13**]) $0 \to M \to M \otimes_R {}^1R$ is exact (equivalently, for some $e > 0$, $0 \to M \to M \otimes_R {}^eR$ is exact).*

REMARK 1.2: In the main cases where either R is complete or ^1R is a finite R-module, R is F-pure if and only if R is a direct summand of ^1R as an R-module ([**3**] and [**13**]).

DEFINITION 1.3. *Let R be a ring of characteristic p. Let R^0 denote the complement (in R) of the union of minimal primes of R. Then, the* **tight closure** *of the ideal $I \subset R$ is the set $\{f \in R |$ there exists $c \in R^0$ such that $cf^q \in I^{[q]}$ for all sufficiently large $q = p^e\}$. We denote the tight closure by I^*.*

DEFINITION 1.4. *R is* **weakly F-regular** *if every ideal in R is tightly closed. R is* **F-regular** *if R_p is weakly F-regular for every $p \in$ Spec(R).*

DEFINITION 1.5. *The map $R \xrightarrow{\phi} S$ is* **cyclically pure** *[**8**] if, for every ideal $I \subset R$, $\{x \in R | \, \phi(x) \in IS\} = I$. Note that the fact that ϕ must be injective follows from the case where $I = 0$. Let $S = {}^1R$ and ϕ be the Frobenius map. Then, since $I({}^1R) = {}^1(I^{[p]}R)$, it follows that $R \to {}^1R$ is cyclically pure if and only if $f^p \in I^{[p]}$ implies $f \in I$.*

REMARK 1.6: Weak F-regularity implies F-purity. To be precise, weak F-regularity always implies that the map $R \to {}^1R$ is cyclically pure because $f^p \in I^{[p]}$ implies $1 \cdot f^q \in I^{[q]}$ for every $q = p^e$; whence, $f \in I^* = I$. But, in [**8**], Hochster shows that if R is "approximately Gorenstein", then $R \to S$ is cyclically pure if and only if it is pure. In addition Hochster proves that any normal ring is approximately Gorenstein. Since weakly F-regular rings are normal, it follows that weak F-regularity implies F-purity.

DEFINITION 1.7. *(R, m) is* **F-injective** *if the Frobenius map induces an injective map of local cohomology modules $H^i_m(R) \to H^i_m({}^1R)$ for each $0 \le i \le \dim(R)$.*

REMARK 1.8: Hochster and Roberts point out [**13**] that F-purity implies F-injectivity. The results of Watanabe [**20**], characterizing 2-dimensional F-pure rings, and Fedder [**5**], characterizing 2-dimensional F-injective rings, (in the graded, isolated singularity case) give rise to an abundance of examples which are F-injective but not F-pure .

REMARK 1.9: If (R, m) is Cohen-Macaulay, then $H^i_m(R) = 0$ for $i < n$ and $H^n_m(R) = \varinjlim R/(y_1^d, \ldots, y_n^d)$ where $(y_1, \ldots, y_n)$ is any system of parameters. Since $H^n_m({}^1R)$ can be identified with $\varinjlim {}^1(R/(y_1^{dp}, \ldots, y_n^{dp}))$ in such a way that the Frobenius map sends $\bar{r} \in R/(y_1^d, \ldots, y_n^d)$ to $\bar{r}^p \in R/(y_1^{dp}, \ldots, y_n^{dp})$ and since the direct limit system here is injective, we can deduce a kind of "cyclic purity criterion" for F-injectivity. That is, a Cohen-Macaulay ring (R, m) is F-injective if and only if the contractedness condition, $f^p \in I^{[p]}$ implies $f \in I$, holds for every ideal I generated by a system of parameters. As in Remark 1.6, $I = I^*$ necessarily implies the contractedness condition holds for I.

We define a weakening of the notion of F-regularity which is sufficient to imply F-injectivity:

DEFINITION 1.10. *R is* **F-rational** *if every ideal generated by a system of parameters is tightly closed.*

DEFINITION 1.11. *Let k be a field of characteristic 0, let Y be a normal variety over k, and let $f : Y' \to Y$ be a resolution of the singularities of Y. A point y of Y is called a* **rational singularity** *if all the higher direct image sheaves vanish at y, that is, $R^q f_*(\mathcal{O}_{Y'})_y = 0$ for all $q > 0$. This definition is well known to be independent of the resolution f and to reflect certain properties of the local ring $\mathcal{O}_{Y,y}$. For example, if y is an isolated* *singularity, then* $R^q f_*(\mathcal{O}_{Y'})_y \simeq H_y^{q+1}(\mathcal{O}_Y)$*, the local cohomology group, for* $0 < q < dim(\mathcal{O}_{Y,y}) - 1$*. Thus, it follows that* $\mathcal{O}_{Y,y}$ *is Cohen-Macaulay.*

For convenient reference, we will state the major results that we will use from the theory of F-regularity in the following summary theorem:

THEOREM 1.12 (HOCHSTER-HUNEKE) [**9**].

(a) *Let* R *be a ring which can be represented as the quotient of a Cohen-Macaulay ring. If* R *is* F*-rational, then* R *is normal and Cohen-Macaulay.*

(b) *Let* $f \in R^0$ *be such that* R_f *is a regular (or just strongly* F*-regular) ring. Assume further that* ${}^1R$ *is a finite* R*-module. Then, there exists* $d \geq 0$ *such that for any ideal* $I \subseteq R$ *and any* $x \in I^*$, $f^d \in \bigcap_{q=p^e}(I^{[q]} : x^q)$.

(c) *Let* I *be an* n*-generated ideal of height at least one in* R*. Then,* $\overline{I^n} \subseteq I^*$*. (Here* $\bar{\ }$ *denotes integral closure.)*

(d) *Let* R *be a Cohen-Macaulay ring of dimension* n *with an isolated singularity. Assume further that* R *is essentially of finite type over a field of characteristic* 0*. If, for every system of parameters ideal* I, $\overline{I^n} \subseteq I$*, then* R *has a rational singularity. In particular, if* R *is* F*-rational* ($I^* = I$)*, then* R *has a rational singularity.*

DEFINITION 1.13. *Let* R *be a ring of characteristic* $p > 0$*. An element* $c \in R^0$ *is said to be a test element if for all* $x \in R$ *and ideals* I *of* R, $x \in I^*$ *iff* $cx^q \in I^{[q]}$ *for all* $q = p^e$, $e \geq 0$.

The element f^d in (b) of Theorem 1.12 is a test element.

In this paper, we show that F-regular Gorenstein rings differ from F-pure ones and, likewise, F-rational Cohen-Macaulay rings differ from F-injective ones by one additional condition. This condition was suggested by work of Hartshorne and Speiser [**7**].

DEFINITION 1.14. *Let* (R, m) *have dimension* n *and let* $\mathcal{S}_i$ *denote the socle of* $H_m^i(R)$*. We say that* $H_m^i(R)$ *is* **F-unstable** *if there exists* $N > 0$ *such that* $\mathcal{S}_i \cap F^e(\mathcal{S}_i) \neq 0$ *for every* $e \geq N$*. We say that* R **is F-unstable** *if* $H_m^i(R)$ *is* F*-unstable for each* $0 \leq i \leq n$*. The reader should note that, in discussing* $\mathcal{S}_i \cap F^e(\mathcal{S}_i)$*, we are making an identification, since* $F^e(\mathcal{S}_i)$ *is really (as an* R/m *vector space) contained in* $H_m^i({}^eR)$ *rather than* $H_m^i(R)$*. It is standard that* F *acts on* $H_m^i(R)$*,* F*-linearly (see* [**12**]*).*

Note also that if R is Gorenstein, $H^i_m(R) = 0$ for $i < n$ and $H^n_m(R)$ is an injective module. Therefore, the socle $\mathcal{S}$ is principly generated, and so the question of F-instability simplifies to just testing whether or not $F(\mathcal{S}) \cap \mathcal{S} = 0$.

DEFINITION 1.15. *Let (R, m) be a positively graded ring of dimension n. Then* $\mathbf{a_i(R)} = \max\{d \mid [H^i_m(R)]_d \neq 0\}$ *where $[M]_d$ denotes the dth graded piece of the graded module M. Denote* $\mathbf{a(R)} = a_n(R)$.

Note that $a_i(R) = a_i({}^1R)$ for every i since $R \simeq {}^1R$ as rings. The fact that $H^i_m(R)$ satisfies the descending chain condition guarantees that each $a_i(R)$ is finite.

REMARK 1.16: For an F-injective graded ring, $a_i(R) \leq 0$ because the Frobenius map induced on $H^i_m(R)$ multiplies degrees by p and is injective; whence, repeated application of the Frobenius necessarily embeds all positively graded pieces of $H^i_m(R)$ into 0.

REMARK 1.17: Assume that (R, m) is F-injective and graded. On the one hand, if $a_i(R) = 0$, there exists $0 \neq \eta \in [H^i_m(R)]_0$. Hence, $0 \neq F^e(\eta) \in [H^i_m({}^eR)]_0$. If $\deg(r) > 0$, then $\deg(r \cdot F^e(\eta)) > 0$, so we must have $r \cdot F^e(\eta) = 0$ for every $e > 0$ and every $r \in m$. That is, $\eta \in \operatorname{socle}(H^i_m(R))$ and $F^e(\eta) \in \operatorname{socle}(H^i_m({}^eR))$; so R is not F-unstable. On the other hand, if $a_i(R) < 0$, then we can pick a finite set of homogeneous elements $\{\eta_1, \ldots, \eta_t\}$ which generate the socle of $H^i_m(R)$ and which satisfy $\deg(\eta_i) = d_i < 0$. Choose N sufficiently large that $p^N > \max\{-d_1, \ldots, -d_t\}$. Then, for arbitrary homogeneous $\eta \in H^i_m(R)$, $\deg(F^e(\eta)) = p^e \deg(\eta) \leq -p^e$. It follows that $F^e(\eta) \notin \operatorname{socle}(H^i_m({}^eR))$ whenever $e \geq N$. (Here, we use the fact that since R is identical with ${}^eR$ as graded rings, ${}^e(H^i_m(R))$ is isomorphic with $H^i_m({}^eR)$ as graded modules; wherefore, the socle of $H^i_m({}^eR)$ is generated by elements of the same degree as the socle of $H^i_m(R)$.

In short, for a **graded F-injective ring, R is F-unstable if and only if $\mathbf{a_i(R)} < 0$ for every i.**

REMARK 1.18: If (R, m) is F-injective and $F^e(\eta)$ lies in the socle of $H^i_m(R)$, then $F^j(\eta)$ lies in the socle of $H^i_m(R)$ for every $j \leq e$. For, $F^{e-j}(m \cdot F^j(\eta)) = m^{p^{e-j}} F^e(\eta) = 0$. But, the F-injectivity says that $F^{e-j}(m \cdot F^j(\eta))$ cannot be 0 unless $m \cdot F^j(\eta) = 0$. Thus, $F^j(\eta) \in \operatorname{Socle}(H^i_m(R))$ as desired.

II. F-injective + F-unstable = F-rational for Cohen-Macaulay rings.

LEMMA 2.1. *Let $(y_1, \ldots, y_n)$ be a system of parameters for (R, m) and assume that the maps, $R/(y_1^d, \ldots, y_n^d) \xrightarrow{\prod y_i} R/(y_1^{d+1}, \ldots, y_n^{d+1})$, in the direct limit system for $H_m^n(R)$ are injective. Let $f \in R$, and denote by η the image of $\bar{f} \in R/(y_1^d, \ldots, y_n^d)$ in $H_m^n(R)$. Then, for fixed $N > 0$ and $Q = p^N$,*

$$\bigcap_{e>N} (0 :_R F^e(\eta)) = \bigcap_{q=p^e>Q} \left((y_1^d, \ldots, y_n^d)^{[q]} : f^q\right).$$

PROOF: Since $F^e(\eta)$ can be represented by $\bar{f}^q \in R/(y_1^d, \ldots, y_n^d)^{[q]}$ viewed inside $H_m^n({}^eR) = \varinjlim_d R/(y_1^d, \ldots, y_n^d)^{[q]} = H_m^n(R)$, the containment $\bigcap \left((y_1^d, \ldots, y_n^d)^{[q]} : f^q\right) \subseteq \bigcap (0 : F^e(\eta))$ is evident. Conversely, if $c \in \bigcap_{e>N} (0 : F^e(\eta))$, then $\overline{cf^q} \in R/(y_1^d, \ldots, y_n^d)^{[q]}$ is 0 in $H_m^n(R)$ for every $q = p^e > Q$. But, since the direct limit system is injective by hypothesis, we must have $\overline{cf^q} \equiv 0$ in $R/(y_1^d, \ldots, y_n^d)^{[q]}$ for every $q > Q$. That is, $c \in \bigcap_{q>Q} \left((y_1^d, \ldots, y_n^d)^{[q]} : f^q\right)$. □

Lemma 2.1 relates the test condition for tight closure of a given system of parameters to an annihilator condition inside the local cohomology module $H_m^n(R)$. Since the construction of $H_m^n(R)$ is independent of the particular choice of system of parameters, we conclude:

PROPOSITION 2.2. *Let (R, m) be a Cohen-Macaulay ring. Then, the following are equivalent:*

1. *For every $0 \neq \eta \in H_m^n(R)$, $\bigcap_{e>0}(0 :_R F^e(\eta)) \subset R \setminus R^0$.*
2. *Every system of parameters ideal is tightly closed.*
3. *There exists a system of parameters ideal which is tightly closed.*

PROOF: (1)⇒(2). Assume that $\bigcap_{e>0} (0 :_R F^e(\eta)) \subset R \setminus R^0$ for every nonzero η in $H_m^n(R)$. Let $(y_1, \ldots, y_n) = I$ be any system of parameters ideal in R. We must show that $I^* \subseteq I$. Let $f \in I^*$ be arbitrary. Then there exists $c \in R^0$ and $Q = p^N \geq 1$ such that $cf^q \in I^{[q]}$ for every $q = p^e \geq Q$. Let $\overline{f^Q} \in R/I^{[Q]}$ represent an element γ of $H_m^n(R)$. If $\gamma \neq 0$ then $c \in \bigcap_{q>Q}(I^{[q]} : f^q) = \bigcap_{e\geq 1} \left((I^{[Q]})^{[p^e]} : (f^Q)^{p^e}\right) = \bigcap_{e\geq 1} (0 : F^e(\gamma)) \subseteq R \setminus R^0$. This contradicts our assumption that $\gamma \neq 0$. If $\gamma = 0$, then $f^q \in I^{[Q]}$; and so $f^{p^e} \in I^{[p^e]}$ for all $p^e \geq Q$. In particular, $\bigcap_{q\geq Q}(I^{[q]} : f^q) = R$.

So $\bigcap_{q>1}(I^{[q]} : f^q) \supseteq I^{[Q]}$. Let δ denote the image of $\bar{f} \in R/I$ in $H^n_m(R)$. Then, again applying Lemma 2.1, $I^{[Q]} \subseteq \bigcap_{q>1}(I^{[q]} : f^q) = \bigcap_{e>0}(0 : F^e(\delta))$. Since dim $R \geq 1$ and $I^{[Q]}$ is m-primary, $I^{[Q]} \not\subseteq R \setminus R^0$. Our hypothesis, that $\bigcap(0 : F^e(\eta)) \subseteq R \setminus R^0$ whenever $\eta \neq 0$, leaves only the possibility that $\delta = 0$. But $\delta = 0$ if and only if $f \in I$ (since the direct limit system $\varinjlim R/(y_1^d, \ldots, y_n^d)$ is exact because R is Cohen-Macaulay). Thus $I^* \subseteq I$ as desired.

(3)$\Rightarrow$(1). Let $(y_1, \ldots, y_n)$ be a system of parameters ideal which is tightly closed and construct $H^n_m(R)$ as $\varinjlim R/(y_1^d, \ldots, y_n^d)$. Let $0 \neq \gamma \in H^n_m(R)$ be arbitrary. Choose $r \in R$ such that $r\gamma = \eta$ lies in the socle of $H^n_m(R)$. Because R is Cohen-Macaulay, the injective maps $R/(y_1^d, \ldots, y_n^d) \to R/(y_1^{d+1}, \ldots, y_n^{d+1})$ actually induce (R/m)-isomorphisms between the respective socles. (See [4], Proposition 1.4. The point is that the dimensions of the respective socles as R/m vector spaces is given by the Cohen-Macaulay types which, for $R/(y_1^d, \ldots, y_n^d)$ and $R/(y_1^{d+1}, \ldots, y_n^{d+1})$, are identical.) Thus, we can represent η by $\bar{f} \in R/(y_1, \ldots, y_n)$. Since $(y_1, \ldots, y_n)^* = (y_1, \ldots, y_n)$ and $f \notin (y_1, \ldots, y_n)$, it follows from the definition of tight closure that $\bigcap_{q=p^e}((y_1, \ldots, y_n)^{[q]} : f^q) \subseteq R \setminus R^0$. By Lemma 2.1, then, $\bigcap_e(0 :_R F^e(\eta)) \subseteq R \setminus R^0$. But, $\bigcap(0 :_R F^e(\eta)) = \bigcap(0 :_R F^e(r\gamma)) = \bigcap(0 : r^q F^e(\gamma))$. In particular, $\bigcap(0 :_R F^e(\gamma)) \subseteq \bigcap(0 :_R F^e(\eta))$. Hence, $\bigcap(0 :_R F^e(\gamma)) \subseteq R \setminus R^0$. □

LEMMA 2.3. *Let (R, m) be an F-injective ring of dimension n which is not F-unstable. Denote the socle of $H^i_m(R)$ by $\mathcal{S}_i$. Then, for each $\mathcal{S}_i$ which does not satisfy the F-unstable property (i.e. for which $\mathcal{S}_i \cap F^e(\mathcal{S}_i) \neq 0$ holds for infinitely many choices of $e > 0$), there exists $0 \neq \eta \in \mathcal{S}_i$ such that $F^e(\eta) \in \mathcal{S}_i$ for every $e \geq 0$.*

PROOF: Fix i such that $H^i_m(R)$ is not F-unstable. Denote $\mathcal{S}_i$ simply by $\mathcal{S}$. To be precise, let us denote by $\mathcal{S}_e$ the socle of $H^i_m({}^eR)$. (Of course, $\mathcal{S}_e$ can be identified with $\mathcal{S}_0 = \mathcal{S}$, but not via an R-linear map. Since linearity is essential in this argument, we postpone the identification of $\mathcal{S}_e$ with $\mathcal{S}$ until the end.) Let M_e denote $\mathcal{S}_e \cap F^e(\mathcal{S})$ which is a subset of $H^i_m({}^eR)$ for each $e \geq 0$. We first claim that $M_{e+1} \subseteq F(M_e)$. To see this, let $\mu = F^{e+1}(\gamma)$ be an arbitrary element of M_{e+1} ($\mu \in \mathcal{S}_{e+1}$ and $\gamma \in \mathcal{S}$). By Remark 1.18, $F^j(\gamma) \in M_j$ for every $0 \leq j \leq e+1$. In particular, $F^e(\gamma) \in M_e$ and so $\mu = F(F^e(\gamma)) \in F(M_e)$ as claimed.

Now denote R/m by K. For each fixed $e > 0$, both $\mathcal{S}_e$ and $F^e(\mathcal{S})$ have the structure of K-vector spaces under the scalar multiplication $k \cdot \gamma = k^{p^e}\gamma$ in $H^i_m({}^eR)$. Moreover, using this definition of scalar multiplication, the map F from M_e to M_{e+1} becomes a K-linear map for each $e \geq 0$. Since $\mathcal{S}$ is a finite dimensional K-vector space, M_e is finite dimensional for every e. Since F is K-linear, $F(M_e) \supseteq M_{e+1}$, and F is injective, we must have $\dim_K(M_{e+1}) \leq \dim_K F(M_e) = \dim_K(M_e)$ with equality holding if and only if F is a K-isomorphism between M_e and M_{e+1}. It follows from the descending finite dimensionality that there exists $N \geq 0$ such that $\dim_k(M_e) = \dim_k(M_N)$ for every $e \geq N$. Hence, F^{e-N} is a K-isomorphism between M_N and M_e for every $e \geq N$. If $M_N = 0$, then $M_e = 0$ for every $e \geq N$, contradicting our assumption that R was not F-unstable. If, on the other hand, there exists $0 \neq \gamma \in M_N$, then $0 \neq F^e(\gamma) \in M_{N+e}$ for every $e \geq 0$. Since $\gamma \in M_N = \mathcal{S}_N \cap F^N(\mathcal{S})$, there exists $0 \neq \eta \in \mathcal{S}$ such that $F^N(\eta) = \gamma \in \mathcal{S}_N$. It follows from Remark 1.18 that $F^j(\eta) \in \mathcal{S}_j$ for every $0 \leq j \leq N$. More importantly, for $j \geq N$, $F^j(\eta) = F^{j-N}(\gamma) \in M_j \subseteq \mathcal{S}_j$. That is, $0 \neq F^j(\eta) \in \mathcal{S}_j$ for every $j \geq 0$ as required. □

PROPOSITION 2.4. *If (R, m) is F-regular, (respectively, F-rational) and R is a quotient of a Cohen-Macaulay ring, then:*

(a) *$R \to {}^1R$ is cyclically pure (respectively, R is F-injective).*

(b) *R is F-unstable.*

PROOF: As pointed out in Theorem 1.12 (a), the hypotheses guarantee that R is Cohen-Macaulay.

(a) For the F-regular case, see Remark 1.6. For the F-rational case, use the same argument as in Remark 1.6, after noting the "weak cyclic purity" criterion for F-injectivity in Remark 1.9.

(b) If, by way of contradiction, R were not F-unstable, then there exists $\eta \in \mathcal{S}$ such that $F^e(\eta) \in \mathcal{S}$ for every e by Lemma 2.3. (Here $\mathcal{S}$ denotes the socle of $H^n_m(R)$, where $n = \dim(R)$. The fact that R is Cohen-Macaulay means $H^i_m(R) = 0$ for $i < n$, so there is nothing to check for $i < n$). Hence, $\bigcap(0 :_R F^e(\eta)) = m$. Let $(y_1, \ldots, y_n)$ be a system of parameters for m. Since η is in the socle of $H^n_m(R)$, then η can be represented as $\bar{f} \in R/(y_1^d, \ldots, y_n^d)$ for some d. The map $R/(y_1^d, \ldots, y_n^d) \stackrel{\Pi y_i}{\to} R/(y_1^{d+1}, \ldots, y_n^{d+1})$ must be injective because R is Cohen-Macaulay. By Lemma 2.1, $\bigcap((y_1^d, \ldots, y_n^d)^{[q]} : f^q) = m$; and so $f \in (y_1^d, \ldots, y_n^d)^* = (y_1^d,$

$\ldots, y_n^d)$. This leads to the contradiction that $\eta = \bar{f} = 0$. $\square$

PROPOSITION 2.5. *Let R be a reduced ring. Let $I \subset R$ be an ideal such that $I^{[q]}$ is contracted with respect to the Frobenius map for every $q = p^e$. Then, for any x in the tight closure of I, $J(x) = \bigcap_{q=p^e}(I^{[q]} : x^q)$ is a radical ideal.*

PROOF: We will show that $c^2 \in J(x)$ implies $c \in J(x)$. If $c^2 \in J(x)$, then $c^p \in J(x)$. Hence, $(cx^{q/p})^p = c^p x^q \in I^{[q]} = (I^{[q/p]})^{[p]}$ for every $q = p^e$. Since, $q/p = p^{e-1}$ it follows from the contractedness hypothesis that $cx^{q/p} \in I^{[q/p]}$. Since p is fixed while q varies, $c \in \cap(I^{[q]} : x^q)$ as desired. $\square$

COROLLARY 2.6. *Let (R, m) be an F-pure (respectively, Cohen-Macaulay F-injective) ring with an isolated singularity at m. Assume further that ${}^1R$ is module finite over R. Let $I = (y_1, \ldots, y_n)$ be an n-generated ideal (respectively, a system of parameters ideal). Then:*

(a) *For every x in the tight closure of I, $\cap(I^{[q]} : x^q) = m$.*
(b) *$m\overline{I^n} \subseteq I$ (Here $\bar{\ }$ denotes integral closure.)*

PROOF: Given Remarks 1.6 and 1.9, characterizing F-pure (respectively F-injective Cohen-Macaulay) rings in terms of contractedness of ideals (respectively, system of parameters ideals) with respect to the Frobenius map, (a) is an immediate consequence of Theorem 1.12(b) and Proposition 2.5. Then, (b) follows from the fact that $\overline{I^n} \subseteq I^*$. (See Theorem 1.12(c)). $\square$

REMARK 2.7: The conclusion $m\overline{I^n} \subseteq I$ resembles a result due to Huneke [**14**] that, in a Cohen-Macaulay local F-pure ring of dimension n, every system of parameters ideal I satisfies $\overline{(I^{n+1})} \subseteq I$. It shows just how close F-injective rings come to satisfying the Briançon-Skoda condition, $\overline{I^n} \subseteq I$. However, one cannot eliminate the m in general.

We are now ready to prove a partial converse to Proposition 2.4. We first use the local cohomology criterion in Proposition 2.2 to obtain the weaker result (that R is F-rational). Then, an application of local duality leads to the stronger result (that R is F-regular), provided that $H_m^n(R)$ is an injective module.

THEOREM 2.8. *Let (R, m) be a ring of dimension n with an isolated singularity. If R is an equidimensional quotient of a Cohen-Macaulay ring, if ${}^1R$*

is a finite R-module, and if $H^i_m(R)$ *has finite length (possibly* 0*) for every* $i < n$*, then:*

(a) *R is F-rational if and only if R is F-injective and F-unstable.*

(b) *If R is Gorenstein, R is F-regular if and only if R is F-pure and F-unstable.*

PROOF: (a) We proved the "only if" part in Proposition 2.4.

To prove the "if" direction, we first show that the F-injective and F-unstable hypothesis, together with the fact that $H^i_m(R)$ has finite length for all $0 \leq i < n$, imply that $H^i_m(R) = 0$ for $i < n$; whence, R is Cohen-Macaulay. If, by way of contradiction, $0 \neq H^i_m(R)$ has finite length, then there exists N such that $m^N H^i_m(R) \neq 0$ but $m^{N+1} H^i_m(R) = 0$. Note that if $N \geq 1$, then

$$F(m^N H^i_m(R)) \subseteq F(m^N) H^i_m(R) \subseteq m^{N+1} H^i_m(R) = 0.$$

Since $m^N H^i_m(R) \neq 0$, this violates our assumption that F acts injectively. On the other hand, if $N = 0$, then $H^i_m(R)$ is the same as its socle. Since F acts injectively, $0 \neq F^e(H^i_m(R)) \subseteq H^i_m(R)$ for every e. Hence $F^e(H^i_m(R)) \cap H^i_m(R) \neq 0$ for every e. This violates the F-unstable hypothesis. It follows that $H^i_m(R) = 0$. Hence, R is Cohen-Macaulay.

Now let $I = (y_1, \ldots, y_n)$ be any system of parameters ideal. Assume by way of contradiction, that there exists $x \in I^* \setminus I$. Since R is Cohen-Macaulay and F-injective, we can apply Corollary 2.6(a) to deduce that $\bigcap_{q=p^e}(I^{[q]} :_R x^q) = m$. If we identify $\overline{x} \in R/I$ with $\eta \in H^n_m(R)$, then by Lemma 2.1, $\bigcap_{e>0}(0 :_R F^e(\eta)) = m$. This means that $F^e(\eta)$ lies in the socle $\mathcal{S}$ of $H^n_m(R)$ for every $e > 0$. Thus, $0 \neq \gamma = F(\eta) \in \mathcal{S}$ and $0 \neq F^e(\gamma) = F^{e+1}(\eta) \in \mathcal{S}$ for every $e > 0$. It follows that R is not F-unstable, which is the desired contradiction.

(b) If R is Gorenstein, F-pure (equivalently, F-injective), and F-unstable, then every ideal I in R (not just every system of parameters ideal) is tightly closed. For, if $I^* \supset I$ properly, then $I^* \cdot (0 :_E I) \neq 0$ by local duality. (Here $H^n_m(R) = E$ which is the injective hull of the residue class field R/m [**17**].) We can therefore choose $y \in I^*$ and $x \in (0 :_E I)$ such that $\eta = yx$ is a generator for the socle of E. Since $y \in I^*$, there exists $c \in R^0$ such that $c \cdot y^q \in I^{[q]}$ for all $q = p^e > p^N$ for some fixed $N > 0$. Since $x \in (0 :_E I)$, $F^e(x) \in (0 :_E I^{[q]})$. Thus, $0 = (cy^q)F^e(x) = c \cdot F^e(yx) = c \cdot F^e(\eta)$ for every $e > N$. That is, $c \in \bigcap_{e>N}(0 : F^e(\eta))$. Since $\bigcap_{0<e\leq N}(0 : F^e(\eta)) \supset$

$(y_1, \ldots, y_n)^{[p^N]}$ and since $c \in m$ whereby there exists t such that $c^t \in (y_1, \ldots, y_n)^{[p^N]}$, we have $c^t \in \bigcap_{e>0}(0 : F^e(\eta))$ and $c^t \in R^0$. We know that R is F-rational by the result above (whether or not R is Gorenstein). But then, by Proposition 2.2, $\bigcap_{e>0}(0 : F^e(\eta)) \subseteq R \setminus R^0$. This contradiction proves that I must have been tightly closed. □

REMARK 2.9: The condition that $H^i_m(R)$ should have finite length for $i < dim(R)$ is equivalent to assuming that the m-adic completion of R is equidimensional and locally Cohen-Macaulay except possibly at the completion of the maximal ideal [**13**]. More generally it suffices for R to be an equidimensional quotient of a Gorenstein ring and locally Cohen-Macaulay at every $P \neq m$.

REMARK 2.10: It was originally proven by Hochster and Huneke that, for Gorenstein rings, R is F-regular if and only if there exists a system of parameters ideal which is tightly closed.

EXAMPLE 2.11: As mentioned in the introduction, if $R = k[X_1, \ldots, X_n]/(f_1, \ldots, f_d)$ is a graded complete intersection ring with an isolated singularity and if $\sum \deg(f_i) < \sum \deg(X_i)$, then R is F-regular for all sufficiently large prime characteristics p. (For a precise description of the bound on p, see [4].) That R is F-pure was proven in [4]. That R is F-unstable follows immediately because in this case $a(R) = \sum \deg(f_i) - \sum \deg(X_i) < 0$ (See Remark 1.17).

EXAMPLE 2.12: After looking explicitly at the equations which define the F-unstable condition on each $H^i_m(R)$, it seems tempting to try to deduce the F-instability of $H^i_m(R)$ (for $i < n$) inductively from the F-instability of $H^n_m(R)$. It would follow that if $H^i_m(R)$ has finite length for each i, if R is F-injective, and if $H^n_m(R)$ is F-unstable, then R is Cohen-Macaulay. The following example shows that this is not the case. Let $R = k[X, Y, Z]/(f)$ be an F-pure ring where f is a cubic form. (For example, if $p \equiv 1 \pmod 3$), take $f = X^3 + Y^3 + Z^3$). Let $S = k[u, v]$. Let T be the Segre product, $T = R\#S = \bigoplus_{d \geq 0} (R_d \otimes_k S_d)$. Then T is F-pure, being a direct summand of $R \otimes_k S$. Let m, n, and μ denote the obvious choice of maximal ideal for R, S, and T respectively. Since (T, μ) has an isolated singularity and $a(T) = -2$ (because $a(R) = 0$ and $a(S) = -2$), we can conclude that $H^3_\mu(T) \simeq H^2_m(R)\#H^2_n(S)$ is F-unstable. But certainly R is not Cohen-Macaulay since $H^2_\mu(T) \simeq H^2_m(R)\#S \neq 0$. (See [**15**]).

Our last result, while not explicitly depending on Theorem 2.8, again illustrates the principle of dividing the F-rational question into its F-injective and F-unstable components. We show that F-injective rings quite often deform into F-rational rings. While the proof is ridiculously easy, the consequences are striking. For example, in studying Hodge Algebras, it enables us to deform the F-injective property from the reduced square free case (which is essentially never a domain) until we get to a Hodge domain with an isolated singularity, at which point we can deduce the stronger F-rational conclusion. In particular, such Hodge domains must have a rational singularity if they are defined over a field of characteristic 0.

PROPOSITION 2.13. *Let (R, m) be a Cohen-Macaulay ring such that ${}^1R$ is a finite R-module. Let $f \in m$ be a non-zero divisor of R such that:*

(a) *$R/(f)$ is F-injective.*
(b) *R_f is regular.*

Then, R is F-rational.

PROOF: The fact that R_f is regular guarantees that some power of f, say f^d, is a test element (see Theorem 1.12(b)). Since f is a non-zero divisor of R, f can be extended to a system of parameters $(f, y_2, \ldots, y_n)$. Now if $x \in (f, y_2, \ldots, y_n)^*$, then we can solve $f^d x^q = a_{1q} f^q + a_{2q} y_2^q + \cdots + a_{nq} y_n^q$ for all $q = p^e \gg 0$. For $q > d$, $f^d(x^q - a_{1q} f^{q-d}) \in (y_2^q, \ldots, y_n^q)$. Since R is Cohen-Macaulay, f^d is a non-zero divisor on $R/(y_2^q, \ldots, y_n^q)$. Hence, $x^q \in (f^{q-d}, y_2^q, \ldots, y_n^q)$. Let $\bar{\ }$ denote reduction modulo (f). Then $\overline{x}^q \in (\overline{y}_2^q, \ldots, \overline{y}_n^q)$. Since $R/(f)$ is F-injective, it follows that $\overline{x} \in (\overline{y}_2, \ldots, \overline{y}_n)$. That is, $x \in (f, y_2, \ldots, y_n)$. Therefore, the ideal $(f, y_2, \ldots, y_n)$ is tightly closed. By Proposition 2.2, then, every system of parameters ideal is tightly closed; so R is F-rational. □

THEOREM/EXAMPLE 2.14. *Let H be a finite partially ordered set (of indeterminates) and let $R = k[H]/I$ be a square-free Hodge algebra. Let R_0 denote the discrete Hodge algebra associated to R, and let $(R_0)_H$ and R_H respectively denote the localization of R_0 and R at the ideal generated by the elements of H. Assume further that k is a perfect field. Then:*

(A) *If $(R_0)_H$ is Gorenstein, if the poset H has a unique minimal element X which is not a zero-divisor on R, and if R_H has an isolated singularity at H, then R_H is F-regular.*
(B) *If $(R_0)_H$ is Cohen-Macaulay, if the poset H has a unique minimal element X which is not a zero-divisor on R, and if R_H has an isolated*

singularity at H, then R_H is F-rational.

(C) *If k has characteristic* 0, *and R satisfies the conditions of (A) or (B), then R_H has a rational singularity.*

PROOF: First, we fix notation and review some basic results from the theory of Hodge algebras (See [**2**] for more details). By [2, Proposition 1.26], $T = R/XR$ is also a Hodge algebra whose generating poset is $I = H \setminus \{X\}$. Moreover, the "ideal" of non-standard monomials of I, denoted Σ_I, is just $\{M \in \Sigma_H | X \text{ does not divide } M\}$. (Here, of course, Σ_H denotes the "ideal" of non-standard monomials of H.) In fact, under our hypotheses, the same set of non-standard monomials that generates Σ_I will generate Σ_H. (If $M = XM_1$ is non-standard in H, then M_1 is nonstandard. For, in any Hodge Algebra a minimal element times any standard monomial is either standard or 0. But if $XM_1 = 0$, then X is a zero divisor in R, contradicting our hypothesis. Hence, M is not a generator.) We can deform any Hodge algebra in a finite number of steps to its associated discrete algebra. (A more detailed description will appear below.) The equations defining the discrete algebra are just square-free monomials, namely the generators of Σ. Thus, $T_0 = k[I]/\Sigma_I k[I]$ and $R_0 = k[H]/\Sigma_H k[H]$. As the generators for Σ_I are the same as the generators for Σ_H and do not involve X, it is clear that $T_0 = R_0/XR_0$.

Before beginning the proof, let us review the process by which we stepwise deform a Hodge algebra to its discrete algebra. Let S be some intermediate step in this deformation. S is a Hodge algebra generated by some partially ordered set H and governed by some ideal of non-standard monomials Σ. The indiscrete part of S, Ind(S) is the set

$$\{Y \in H | \text{there exists a generator } N \text{ of } \Sigma \text{ whose "straightening relation" } N = \sum_{i=1}^{t} k_i M_i, \text{ where each } M_i \text{ is a standard monomial, has the property that } Y|M_i \text{ for at least one } i\}.$$

Pick Y to be a minimal element in Ind(S). Form the Rees algebra

$$\mathcal{R} = S[t] \oplus (y/t)\, S \oplus (y^2/t^2)\, S \oplus \cdots \oplus (y^n/t^n)\, S \oplus, \cdots,$$

where y denotes the image of Y in S viewed as a subring (in degree 0) of $\mathcal{R}$ and t is an indeterminant. Then $S_1 = \mathcal{R}/t\mathcal{R}$ is the next step in the deformation of S. It is shown in [**2**], that both $\mathcal{R}$ and S_1 are Hodge algebras generated by the same H as S and governed by the same Σ as S. However, $\mathrm{Ind}(S_1) = \mathrm{Ind}(S) \setminus \{Y\}$ which is, in that sense, one step closer to the discrete algebra (i.e if the indiscrete part is empty, then every element in Σ must equal 0). Although each of these algebras is "generated" by H, one has to be careful to detach the set H from the way by which it is included in the Hodge algebra. The embeddings of H into S, $\mathcal{R}$, and S_1 are not the same. Specifically, let $\phi(H)$ be the given embedding of H into S. For any $X \in H$, let $x = \phi(X)$ denote its image in S. There is a natural inclusion of S onto the degree 0 piece of $\mathcal{R}$. We will also denote by x the image of $\phi(X)$ under this inclusion. However, the Hodge algebras structure of $\mathcal{R}$ does not devolve from this embedding of H. Instead, it comes from the embedding $\widetilde{\phi}$ defined by

$$\left\{ \begin{array}{ll} \widetilde{\phi}(X) = x & \text{if } X \neq Y \\ \widetilde{\Phi}(Y) = \frac{y}{t} & \end{array} \right\}.$$

Similarly, the embedding $\phi^{\#}$ of H into

$$S_1 = \mathcal{R}/t\mathcal{R} = S/(y_0) \oplus (y_0)/(y^2) \oplus \cdots \oplus (y^n)/(y^{n+1}) \oplus \cdots$$

is defined by

$$\left\{ \begin{array}{ll} \phi^{\#}(X) = \overline{x} \in S/(y) & \text{if } X \neq Y \\ \phi^{\#}(Y) = \overline{y} \in (y)/(y^2) & \end{array} \right\}.$$

In particular, note that the maximal ideal m of $\mathcal{R}$ generated by t and $\widetilde{\phi}(H)$ is just $\pi^{-1}(\phi^{\#}(H))$ where $(\phi^{\#}(H))$ is the maximal ideal generated by the image of H in S_1 and π is the natural surjection of $\mathcal{R}$ onto S_1. Also note that $m = (t, y/t, \phi(H))$ where $\phi(H)$ is the image of H in S, extended to $\mathcal{R}$.

The last piece of background information that we need is the fact that if the discrete algebra $(S_0)_H$ is Cohen-Macaulay (respectively, Gorenstein), then $(S)_H$ is necessarily Cohen-Macaulay (respectively, Gorenstein). Of course, localizing at H means localizing via the particular embedding of H which gives rise to the Hodge structure.

We now proceed with the proof. Since, for Gorenstein rings, F-regular is equivalent to F-rational (See [**9**] or Theorem 2.8), (A) is just a special case of (B). By [**9**] in the case of an isolated singularity essentially of finite type over a field k of characteristic 0, the F-rational property (after reducing to characteristic p) is sufficient to imply rational singularity. Therefore, we need only prove (B).

To prove (B), it is enough to show that $T = R/XR$ is F-injective at its maximal ideal generated by $I = H \setminus \{X\}$. The result is then immediate from Proposition 2.13.

To see that T is F-injective, note first that T_0 is F-injective. In fact, $T_0 = k[I]/\Sigma_I k[I]$ is well-known to be F-pure. (One quick way to see this is to let γ be the product of every distinct element of I which appears in one of the generators for Σ_I. Since Σ_I is generated by square-free monomials, $\gamma^{p-1} \in (\Sigma_I^{[p]} : \Sigma_I)$. By construction $\gamma^{p-1} \notin I^{[p]}$. Hence, T_0 is F-pure by the criterion in [**3**]). Essentially, we would like to deform the F-purity back to T. The deformation as described above, is made in two basic steps - first embedding S as a direct summand in $\mathcal{R}$, then killing a non-zero divisor in the maximal ideal of $\mathcal{R}$ to get S_1. On the one hand, F-purity behaves well with respect to passing to direct summand [**11**]. On the other hand, in Cohen-Macaulay rings, F-injectivity behaves well with respect to killing a non-zerodivisor [**3**]. Unfortunately, in the non-Gorenstein case, neither behaves well with respect to the other operation. In our case, however, we have more information than just that S is a direct summand of $\mathcal{R}$.

Let $\{u_1, \ldots, u_n\}$ be a system of parameters for S_H. Let m denote the maximal ideal of $\mathcal{R}$ generated by $\widetilde{\phi}(H)$ and t. Then, $m = (t, y/t, \phi(H))$. We claim that $\{u_1, \ldots, u_n, t - y/t\}$ is a system of parameters for $\mathcal{R}_m$. (Of course, $\dim(\mathcal{R}_m) = \dim(S_H) + 1$). Clearly $(\phi(H))^d \subset (u_1, \ldots, u_n, t - y/t)\mathcal{R}_m$ for some d since $\{u_1, \ldots, u_n\}$ is a system of parameters for $\phi(H)$ in S_H. It is therefore enough to show that t and y/t are both in the radical ideal generated by $(u_1, \ldots, u_n, t - y/t)$ in $\mathcal{R}_m$. But,

$$t^2 = t(t - y/t) + y$$

and, likewise,

$$(\frac{y}{t})^2 = \frac{y}{t}(t - \frac{y}{t}) - y.$$

Hence, $m^{2d} \subset (u_1, \ldots, u_n, t - y/t)$.

Now, by induction, we may assume that $(S_1)_H$ is F-injective. Since S_1 is Cohen-Macaulay and t is a non-zerodivisor in $\mathcal{R}_m$, $\mathcal{R}_m$ must be Cohen-Macaulay and F-injective. It follows that every system of parameters for m is contracted with respect to the Frobenius map. In particular, this applies to $(u_1, \ldots, u_n, (t - y/t)^d)$ for every $d > 0$.

Suppose that $f^p \in (u_1, \ldots, u_n)^{[p]} S_H$. Then, embedding this equation into $\mathcal{R}$, $f^p \in (u_1, \ldots, u_n)^{[p]} \mathcal{R}_m$. Hence $f^p \in (u_1, \ldots, u_n, (t-y/t)^d)^{[p]} \mathcal{R}_m$ for every $d > 0$. It follows from the discussion above that $f \in (u_1, \ldots, u_n, (t - y/t)^d)\mathcal{R}_m$ for every $d > 0$. Therefore, $f \in (u_1, \ldots, u_n)\mathcal{R}_m$. Since the map $S_H \hookrightarrow \mathcal{R}_m$ is split, $(u_1, \ldots, u_n)\mathcal{R}_m \cap S_H = (u_1, \ldots, u_n)S_H$ as desired.

$\square$

References

1. J.-F. Boutôt, *Singularités Rationelles et Quotient par les Groupes Reductifs*, Invent. Math. **88** (1987), 65–68.
2. C. DeConcini, D. Eisenbud, C. Procesi,, *Hodge Algebras*, Société Mathématique de France, Paris, 1982, Astérisque 91.
3. R. Fedder, *F-purity and Rational Singularity*, Trans. of the Amer. Math. Soc. **278** (1983), 461–480.
4. R. Fedder, *F-purity and Rational Singularity in Graded Complete Intersection Rings*, Trans. of the Amer. Math. Soc. **301** (1987), 47–62.
5. R. Fedder, *Rational Singularity Implies F-injective Type for Graded Cohen-Macaulay Rings of Dimension 2*, (in preparation).
6. H. Flenner, *Quasihomogene Rationale Singularitaten*, Arch. Math. **36** (1981), 35–44.
7. R. Hartshore and Speiser, *Local Cohomological Dimension in Characteristic p*, Ann. of Math **105** (1977), 45–79.
8. M. Hochster, *Cyclic purity versus purity in excellent Noetherian rings*, Trans. Amer. Math. Soc. **231** (1977), 463–488.
9. M. Hochster and C. Huneke, *Tightly Closed Ideals*, Bull. Amer. Math. Soc. **18** (1988), 45–48.
10. M. Hochster and C. Huneke, *Tight Closure and strong F-regularity*, to appear, Mémoire de la Societé Math. de France, v. dedicaté à P. Samuel.
11. M. Hochster and C. Huneke, *Tight Closure*, to appear, these proceeding.
12. M. Hochster and J. L. Roberts, *Rings of Invariants of Reductive Groups Acting on Regular Rings are Cohen-Macaulay*, Advances in Math. **13** (1974), 114–175.
13. M. Hochster and J. L. Roberts, *The Purity of the Frobenius and Local Cohomology*, Advances in Math. **21** (1976), 117–172.
14. C. Huneke, *Hilbert Functions and Symbolic Powers*, Michigan Math. J. **34** (1987), 293–317.
15. S. Goto and K.-i. Watanabe, *On Graded Rings I*, J. Math. Soc. Japan **30** (1978), 179–213.
16. J. Lipman and B. Teissier, *Pseudo-rational Local Rings and a Theorem of Briancon-Skoda about Integral Closure of Ideals*, Michigan Math. J. **28** (1981), 97–116.
17. E. Matlis, *Injective Modules over Noetherian rings*, Pacific J. Math **8** (1958), 511–528.
18. C. Peskine and L. Szpiro, *Dimension projective finie et cohomologie locale*, I.H.E.S. Publ. Math. **42** (1973), 323–395.
19. K.-i. Watanabe, *Rational singularities with $K*$-Action*, Lecture Notes in Pure and Applied Mathematics, Dekker **84** (1983), 339–351.
20. K.-i. Watanabe, *Study of F-purity in Dimension 2*, Algebraic Geometry and Commutative Algebra (in Honor of Masayoshi Nagata) (1987), 1000-1009.

100 Parkview Dr., Columbia MO 65201

Department of Mathematical Sciences, Tokai University, Hiratsuka, 259-12 JAPAN

Surface Singularities of Finite Buchsbaum-Representation Type

SHIRO GOTO

1. Introduction.

The purpose of this paper is to determine the structure of surface singularities of finite Buchsbaum-representation type and the main results are summarized into the following:

THEOREM (1.1). *Let R be a Noetherian complete local ring of $\dim R = 2$ and assume that the residue class field of R is algebraically closed. Let $e(R)$ and $v(R)$ denote, respectively, the multiplicity and the embedding dimension of R. Then the following three conditions are equivalent.*

(1) *R has finite Buchsbaum-representation type, that is R possesses only finitely many isomorphism classes of indecomposable maximal Buchsbaum R-modules.*
(2) *$e(R) = 1$ and $v(R) \leqq 3$.*
(3) *$R \cong P/XI$ where P is a three-dimensional complete regular local ring with maximal ideal $\mathfrak{n}$, $X \in \mathfrak{n} \backslash \mathfrak{n}^2$ and I is an ideal of P such that $\mathrm{ht}_P I \geqq 2$.*

Hence as an immediate consequence of Theorem (1.1) we get the following:

COROLLARY (1.2). *With the same hypothesis as in (1.1), R is a regular local ring if and only if R is unmixed and of finite Buchsbaum-representation type.*

Now let us recall some basic definitions. Throughout let R denote a Noetherian local ring with maximal ideal $\mathfrak{m}$ and let M be a finitely generated R-module. Then we say that M is a *Buchsbaum* R-module if the difference

$$I_R(M) := \ell_R(M/qM) - e_q(M)$$

is independent of the choice of a parameter ideal q for M (here $\ell_R(M/qM)$ and $e_q(M)$ denote, respectively, the length of M/qM and the multiplicity of M relative to q). Hence M is a Cohen-Macaulay R-module if and only if M is a Buchsbaum R-module of $I_R(M) = 0$. A Buchsbaum R-module M is called *maximal* when $\dim_R M = \dim R$. The ring R is said to be a Buchsbaum ring if R is a Buchsbaum module over itself[1].

We say that R has finite Buchsbaum- (resp. CM-) representation type if there are only finitely many isomorphism classes of indecomposable maximal Buchsbaum (resp. Cohen-Macaulay) R-modules. The Buchsbaum-representation theory was posed by J. Herzog to the author in 1985 and first introduced by [**9**]. (See [**1,2,3,4,6,10,11,13,16,18**] for the recent drastic progress of the research on the Cohen-Macaulay singularities of finite CM-representation type.) However the *fundamental theorem* is due to D. Eisenbud and the author [**5**] (cf. [**7**] too), which claims that any regular local ring R has finite Buchsbaum-representation type. (More precisely, the syzygy modules of the residue class field $R/\mathfrak{m}$ of R are the representatives of indecomposable maximal Buchsbaum R-modules and any maximal Buchsbaum R-module is a direct sum of them as well as a vector space over $R/\mathfrak{m}$.) The author and K. Nishida [**9**] just succeeded, by the help of recent results of [**4**] and [**13**] about maximal Cohen-Macaulay modules on hypersurface singularities, in proving that the converse of the *fundamental theorem* is also true, provided R is a Cohen-Macaulay complete local ring of $\dim R \geqq 2$ and R contains an algebraically closed coefficient field k of ch $k \neq 2$.

On the other hand, in the previous paper [**8**], the author has determined *all* the one-dimensional complete local rings R of finite Buchsbaum-representation type in the case where the residue class fields of R are infinite. Consequently the structure of Cohen-Macaulay complete local rings R of finite Buchsbaum-representation type is now completely known, if R contains an algebraically closed coefficient field k of ch $k \neq 2$.

Nevertheless, according to the author's experiences in [**8**] the hypothesis in [**9**] that R is *Cohen-Macaulay* seems too strong and might be removed (or should be replaced by a weaker one). Also it would be quite interesting if one can remove the restriction on the characteristic of the coefficient fields. The present research was motivated by such anticipations and, by Theorem (1.1) and Corollary (1.2) we are assured of them at least for the case of dimension two. The author suspects that the assertion (1.2) remains

[1] The readers may consult the monumental book [**17**] of J. Stückrad and W. Vogel for the general references on Buchsbaum rings and modules, where the research of the last ten years is gathered together.

true even in the case $\dim R \geqq 3$, which he would like to pose as a conjecture.

Let us explain how to organize this paper. The proof of the implication (3) $\Rightarrow$ (1) (resp. (1) $\Rightarrow$ (2) and (2) $\Rightarrow$ (3)) in Theorem (1.1) shall be given in Section 3 (resp. Section 5). Sections 2 and 4 are devoted to some preliminary steps. In Section 2 we will show that certain local integral domains of finite Buchsbaum-representation type are always regular (Theorem (2.1)). We will prove in Section 4 that a Cohen-Macaulay complete local ring R of $\dim R = 2$ is regular if the field $R/\mathfrak{m}$ is algebraically closed and if R has finite Buchsbaum-representation type (Theorem (4.1)). We need both the results (2.1) and (4.1) as key lemmas for the proof of the implication (1) $\Rightarrow$ (2) in Theorem (1.1).

Throughout this paper let R be a Noetherian local ring of $\dim R = d$ and $\mathfrak{m}$ the maximal ideal of R. We denote by $H^i_{\mathfrak{m}}(\cdot)$ the i-th local cohomology functor of R relative to $\mathfrak{m}$. For each finitely generated R-module M, let $\mu_R(M)$ denote the number of elements in a minimal system of generators for M.

2. Integral domains of finite Buchsbaum-representation type.

In this section we assume that R is a Noetherian local integral domain of $\dim R = d \geqq 2$. Let $\bar{R}$ denote the derived normal ring of R. We assume that $\bar{R}$ is a regular local ring and that $\bar{R}$ is module-finite over R. Let $\mathfrak{n}$ denote the maximal ideal of $\bar{R}$ and assume that $R/\mathfrak{m} = \bar{R}/\mathfrak{n}$.

The aim of this section is to prove the following:

THEOREM (2.1). *Assume that R is complete and the field $R/\mathfrak{m}$ is infinite. Then R is regular if R has finite Buchsbaum-representation type.*

We divide the proof of (2.1) into several steps. Let R be as above. Assume that R has finite Buchsbaum-representation type and the field $R/\mathfrak{m}$ is infinite. (Here we don't assume R to be complete.) We begin with the following:

LEMMA (2.2). $\ell_{\bar{R}}((\mathfrak{m}\bar{R} + \mathfrak{n}^2)/\mathfrak{n}^2) \geqq d - 1$.

PROOF: Assume that $\ell_{\bar{R}}(\mathfrak{n}/(\mathfrak{m}\bar{R} + \mathfrak{n}^2)) \geqq 2$ and let $\mathfrak{F}$ be the set of ideals I in $\bar{R}$ satisfying $\mathfrak{m}\bar{R} + \mathfrak{n}^2 \subset I \subset \mathfrak{n}$. Then $\mathfrak{F}$ is infinite as the field $\bar{R}/\mathfrak{n}$ is infinite. Let $I \in \mathfrak{F}$ and apply the functors $H^i_{\mathfrak{m}}(\cdot)$ to the exact sequence

$$0 \longrightarrow I \longrightarrow \bar{R} \longrightarrow \bar{R}/I \longrightarrow 0.$$

Then we have $H^i_{\mathfrak{m}}(I) = (0)$ for $i \neq 1$, d and $H^1_{\mathfrak{m}}(I) \cong \bar{R}/I$. Therefore, as $\mathfrak{m} \cdot H^1_{\mathfrak{m}}(I) = (0)$ and $\operatorname{rank}_R I = 1$, we get by [**17**, Ch. I, (2.12)] that I is an indecomposable maximal Buchsbaum R-module. Hence there is an R-isomorphism $\phi : I \longrightarrow J$ for some $I, J \in \mathfrak{F}$ such that $I \neq J$. As I and J are ideals in $\bar{R}$, the map ϕ is an $\bar{R}$-isomorphism too and so we get $I = J$. This is a contradiction.

We put $e = e(R)$, the multiplicity of R.

COROLLARY (2.3). *There exist elements $s_1, s_2, \ldots, s_{d-1}$ of $\mathfrak{m}$ and an element t of $\mathfrak{n}$ such that $\mathfrak{n} = (s_1, s_2, \ldots, s_{d-1}, t)\bar{R}$ and $\mathfrak{m}\bar{R} = (s_1, s_2, \ldots, s_{d-1}, t^e)\bar{R}$. Hence $\mu_R(\bar{R}) = e$ and $\bar{R} = \sum_{i=0}^{e-1} Rt^i$.*

PROOF: By (2.2) we may choose $s_1, s_2, \ldots, s_{d-1} \in \mathfrak{m}$ and $t \in \mathfrak{n}$ so that $\mathfrak{n} = (s_1, s_2, \ldots, s_{d-1}, t)\bar{R}$. Let $I = (s_1, s_2, \ldots, s_{d-1})R$. Then because $\bar{R}/I\bar{R}$ is a DVR, we may write $\mathfrak{m}\bar{R} = I\bar{R} + t^n\bar{R}$ with $n \geqq 1$ an integer. Hence $\mathfrak{m}\bar{R}$ is a parameter ideal of $\bar{R}$ and so we have $e(R) = \ell_{\bar{R}}(\bar{R}/\mathfrak{m}\bar{R})$. Thus $e = n$.

PROPOSITION (2.4). $e \leqq 3$.

PROOF cf. [**11**, Proof of Satz 1.6, b]: Assume that $e \geqq 4$. Let $s_1, s_2, \ldots, s_{d-1}, t$ be as in (2.3). For each $\lambda \in R/\mathfrak{m}$ we choose $c_\lambda \in R$ so that $\lambda = c_\lambda \bmod \mathfrak{m}$ and put $M_\lambda = R + R(t + t^2 + c_\lambda t^3) + \mathfrak{m}\bar{R}$. Then applying $H^i_{\mathfrak{m}}(\cdot)$ to the exact sequence

$$0 \longrightarrow M_\lambda \longrightarrow \bar{R} \longrightarrow \bar{R}/M_\lambda \longrightarrow 0,$$

we get

$$\begin{aligned} H^i_{\mathfrak{m}}(M_\lambda) &= \bar{R}/M_\lambda \quad (i = 1), \\ &= (0) \qquad\quad (i \neq 1, d). \end{aligned}$$

Since $\operatorname{rank}_R M_\lambda = 1$ and $\mathfrak{m} \cdot H^1_{\mathfrak{m}}(M_\lambda) = (0)$, we have M_λ to be an indecomposable maximal Buchsbaum R-module. Hence there is an R-isomorphism $\phi : M_\lambda \longrightarrow M_\mu$ with $\lambda \neq \mu$. Choose a unit θ of $\bar{R}$ so that $\phi(x) = \theta x$ for each $x \in M_\lambda$. Then because $\theta M_\lambda = M_\mu$, we see $\theta \in M_\mu$ and

$$(*) \qquad t + t^2 + c_\mu t^3 = \theta \left\{ \alpha + \beta(t + t^2 + c_\lambda t^3) + \gamma \right\}$$

for some $\alpha, \beta \in R$ and $\gamma \in \mathfrak{m}\bar{R}$. We write $\theta = a + b(t + t^2 + c_\mu t^3) + c$ with $a, b \in R$ and $c \in \mathfrak{m}\bar{R}$ (hence $a \notin \mathfrak{m}$). Let $J = (s_1, s_2, \ldots, s_{d-1})\bar{R} + t^4\bar{R}$.

Then $\mathfrak{m}\bar{R} \subset J$ by (2.3), since $e \geqq 4$. Let $\bar{\cdot}$ denote the reduction mod$J$. Then we have by (*) that

$$(**) \quad \bar{t} + \bar{t}^2 + \bar{c}_\mu \bar{t}^3$$
$$= \bar{a}\bar{\alpha} + \bar{a}\bar{\beta}(\bar{t} + \bar{t}^2 + \bar{c}_\lambda \bar{t}^3) + \bar{b}\bar{\alpha}(\bar{t} + \bar{t}^2 + \bar{c}_\mu \bar{t}^3) + \bar{b}\bar{\beta}(\bar{t}^2 + 2\bar{t}^3).$$

Notice that $\{\bar{t}^i\}_{0 \leqq i \leqq 3}$ is a basis of $\bar{R}/J$ over $R/\mathfrak{m}$. Then comparing the coefficients in the above identity (**), we find

$$\bar{a}\bar{\alpha} = 0, \quad \bar{a}\bar{\beta} + \bar{b}\bar{\alpha} = 1 \text{ and } \bar{a}\bar{\beta} + \bar{b}\bar{\alpha} + \bar{b}\bar{\beta} = 1.$$

Therefore as $a \notin \mathfrak{m}$, we get $\bar{\alpha} = 0$ whence $\bar{a}\bar{\beta} = 1$ and $\bar{b}\bar{\beta} = 0$. Thus

$$\bar{t} + \bar{t}^2 + \bar{c}_\mu \bar{t}^3 = \bar{t} + \bar{t}^2 + \bar{c}_\lambda \bar{t}^3$$

by (**) and so we have $\lambda = \mu$, which is a contradiction.

Let $s_1, s_2, \ldots, s_{d-1}$ and t be as in (2.3). We put $I = (s_1, s_2, \ldots, s_{d-1})R$, $P = I\bar{R}$ and $p = P \cap R$. Then $p \in \operatorname{Spec} R$, since $P \in \operatorname{Spec} \bar{R}$.

LEMMA (2.5). *The elements $s_1, s_2, \ldots, s_{d-1}$ of $\mathfrak{m}$ may be chosen so that $\bar{R}/P$ is the derived normal ring of R/p.*

PROOF: We may assume $R \neq \bar{R}$. Let $\mathfrak{c} = R : \bar{R}$ denote the conductor ideal and put $r = \dim R/\mathfrak{c}$. Then $r \leqq d - 1$. Hence by (2.2) we may choose the elements $s_1, s_2, \ldots, s_{d-1}$ of $\mathfrak{m}$ so that $s_1, s_2, \ldots, s_r$ form a system of parameters for $R/\mathfrak{c}$. If $p \supset \mathfrak{c}$ for this choice, then $p \supset (s_1, s_2, \ldots, s_r)R + \mathfrak{c}$ and we have $p = \mathfrak{m}$, which is impossible because $\dim R/p = \dim \bar{R}/P = 1$. Consequently $p \not\supset \mathfrak{c}$ and so the canonical map $R/p \longrightarrow \bar{R}/P$ is birational. Hence $\bar{R}/P$ is the derived normal ring of R/p.

Let

$$\text{(F)} \qquad 0 \longrightarrow M \longrightarrow R^e \xrightarrow{\varepsilon} \bar{R} \longrightarrow 0$$

denote the initial part of a minimal free resolution of the R-module $\bar{R}$. Then because $s_1, s_2, \ldots, s_{d-1}$ is an $\bar{R}$-regular sequence, we get from (F) an exact sequence

$$0 \longrightarrow M/IM \longrightarrow (R/I)^e \xrightarrow{\bar{\varepsilon}} \bar{R}/P \longrightarrow 0$$

of R/I-modules. Choose $\tau \in \operatorname{Hom}_{R/p}((R/p)^e, \bar{R}/P)$ so that the diagram

$$\begin{array}{ccc} (R/I)^e & \xrightarrow{\bar{\varepsilon}} & \bar{R}/P \\ \downarrow{\scriptstyle\sigma} & & \| \\ (R/p)^e & \xrightarrow{\tau} & \bar{R}/P \end{array}$$

is commutative, where $\sigma : (R/I)^e \longrightarrow (R/p)^e$ denotes the canonical epimorphism. We put $M' = \operatorname{Ker}\tau$. Then we clearly have

$$\mu_{R/p}(M') \leqq \mu_R(M) \text{ and } \mu_{R/p}(\bar{R}/P) = e. \tag{2.6}$$

LEMMA (2.7). $\mu_R(M) \leqq e$.

PROOF: We put $F = R^e$ and take $\rho \in \operatorname{End}_R F$ so that ρ lifts $t 1_{\bar{R}}$. Then $\rho^e(M) \subset \mathfrak{m}M$. In fact, since $t^e \in \mathfrak{m}\bar{R}$ and $\bar{R} = \sum_{i=0}^{e-1} Rt^i$ by (2.3), we may write $t^e = \sum_{i=0}^{e-1} a_i t^i$ with $a_i \in \mathfrak{m}$. Then both the endomorphisms ρ^e and $\sum_{i=0}^{e-1} a_i \rho^i$ of F lift $t^e 1_{\bar{R}}$ and so we have

$$\rho^e = \sum_{i=0}^{e-1} a_i \rho^i + j \circ \delta$$

for some $\delta \in \operatorname{Hom}_R(F, M)$, where $j : M \longrightarrow F$ denotes the inclusion map. As $M \subset \mathfrak{m}F$, we get $\delta(M) \subset \mathfrak{m}M$ whence $\rho^e(M) \subset \mathfrak{m}M$.

Now assume that $\mu_R(M) > e$. Let $k = R/\mathfrak{m}$ and $V = M/\mathfrak{m}M$. We denote by ξ the k-endomorphism of V induced from ρ. Then because $\dim_k V > e$ and $\xi^e = 0$, there exist elements f, g of M such that the classes $\bar{f}, \bar{g}$ of f, g in V are linearly independent over k and $\xi(\bar{f}) = \xi(\bar{g}) = 0$. For each $\lambda \in k$, take $c_\lambda \in R$ so that $\lambda = c_\lambda \bmod \mathfrak{m}$ and put

$$N_\lambda = \mathfrak{m}M + R(f + c_\lambda g).$$

Then F/N_λ is an indecomposable maximal Buchsbaum R-module (see [8, (2.3)(1)][2]). Hence $F/N_\lambda \cong F/N_\mu$ for some $\lambda, \mu \in k$ such that $\lambda \neq \mu$. Consequently by [8, (2.3)(2)] we have $\phi(M) = M$ and $\phi(N_\lambda) = N_\mu$ for some $\phi \in \operatorname{Aut}_R F$. Let $\bar{\phi}$ denote the R-automorphism of $\bar{R}$ induced from ϕ. Then $\bar{\phi} = u 1_{\bar{R}}$ (u, a unit of $\bar{R}$) and so we may write

$$\bar{\phi} = \left(\sum_{i=0}^{e-1} b_i t^i \right) 1_{\bar{R}}$$

[2] The proof of [8, (2.3)] still works even in the case $\dim R \geqq 2$.

with $b_i \in R$ $(0 \leqq i \leqq e-1$; hence $b_0 \notin \mathfrak{m})$. As ϕ and $\sum_{i=0}^{e-1} b_i \rho^i$ lift $\bar{\phi}$, we have

$$\phi = \sum_{i=0}^{e-1} b_i \rho^i + j \circ \delta$$

for some $\delta \in \operatorname{Hom}_R(F, M)$. Hence

$$\phi(f + c_\lambda g) \equiv b_0(f + c_\lambda g) \bmod \mathfrak{m}M,$$

because $\delta(M) \subset \mathfrak{m}M$ and $\rho(f + c_\lambda g) \in \mathfrak{m}M$ by choice of f, g. Therefore $f + c_\lambda g \in N_\mu$ (as $b_0 \notin \mathfrak{m}$) and so we have

$$\bar{f} + \lambda \bar{g} \in k(\bar{f} + \mu \bar{g}).$$

Thus $\lambda = \mu$, because $\bar{f}, \bar{g}$ are linearly independent over k. This is the required contradiction.

THEOREM (2.8). *R is an abstract hypersurface of $e(R) \leqq 3$.*

PROOF: By (2.5) we may assume that $\bar{R}/P$ is the derived normal ring of R/p. Hence $\operatorname{rank}_{R/p} M' = e - 1$ as $\operatorname{rank}_{R/p} \bar{R}/P = 1$ and so, by (2.6) and (2.7), we see

$$(*) \qquad e - 1 \leqq \mu_{R/p}(M') \leqq \mu_R(M) \leqq e.$$

Therefore if $\mu_{R/p}(M') < e$, then M' must be R/p-free. Consequently $\bar{R}/P$ has finite projective dimension over R/p and so we get $R/p = \bar{R}/P$, that is $e = 1$ by (2.6). As R is analytically irreducible, the assertion that $e = 1$ implies R is regular.

Let us assume $\mu_{R/p}(M') \geqq e$ and we see by the above inequalities (*) that

$$(**) \qquad \mu_{R/p}(M') = \mu_R(M) = e.$$

Recall the commutative diagram

$$\begin{array}{ccccccccc}
0 & \longrightarrow & M/IM & \longrightarrow & (R/I)^e & \xrightarrow{\bar{\varepsilon}} & \bar{R}/P & \longrightarrow & 0 \\
 & & & & \downarrow{\scriptstyle \sigma} & & \| & & \\
0 & \longrightarrow & M' & \longrightarrow & (R/p)^e & \xrightarrow{\tau} & \bar{R}/P & \longrightarrow & 0
\end{array}$$

with exact rows. Then clearly $M' = \sigma(M/IM)$ and $\operatorname{Ker} \sigma \subset M/IM$. Notice that $\mu_{R/p}(M') = \mu_{R/I}(M/IM)$ by (**) and we find $\operatorname{Ker} \sigma \subset$

$\mathfrak{m} \cdot (M/IM)$; hence $\operatorname{Ker} \sigma \subset \mathfrak{m}^2 \cdot (R/I)^e$. As $\operatorname{Ker} \sigma = p \cdot (R/I)^e$, we have $p \cdot (R/I)^e \subset \mathfrak{m}^2 \cdot (R/I)^e$ which implies $p \subset I + \mathfrak{m}^2$. Therefore $p + \mathfrak{m}^2 = I + \mathfrak{m}^2$ and we get

$$v(R/p) = v(R/I) = v(R) - (d-1),$$

because $s_1, s_2, \ldots, s_{d-1}$ are part of a minimal basis of $\mathfrak{m}$. Thus

$$v(R) = v(R/p) + d - 1.$$

We will show $v(R/p) \leqq 2$.

Assume that $v(R/p) \geqq 3$. Then $v(R/p) = e(R/p) = e = 3$ by (2.4), because $e(R/p) = e$ by (2.6) and $v(R/p) \leqq e(R/p)$ by [**15**]. Let $\bar{\mathfrak{m}}$ denote the maximal ideal of R/p. Then as the field $R/\mathfrak{m}$ is infinite, we may choose $x \in \bar{\mathfrak{m}}$ so that $\bar{\mathfrak{m}}^2 = x\bar{\mathfrak{m}}$ (cf. [**15**]). Consider the exact sequence

$$\text{(E)} \qquad 0 \longrightarrow M'/xM' \longrightarrow \left[(R/p)/x(R/p)\right]^3 \longrightarrow (\bar{R}/P)/x(\bar{R}/P) \longrightarrow 0$$

derived from the exact sequence $0 \longrightarrow M' \longrightarrow (R/p)^3 \xrightarrow{\tau} \bar{R}/P \longrightarrow 0$. Recall that

$$\ell_{R/p}((R/p)/x(R/p)) = \ell_{R/p}((\bar{R}/P)/x(\bar{R}/P)) = 3,$$

because $e(R/p) = 3$. Then by the above exact sequence (E) we get

$$\ell_{R/p}(M'/xM') = 6.$$

On the other hand, since $M' \subset \bar{\mathfrak{m}} \cdot (R/p)^3$ and $\bar{\mathfrak{m}}^2 = x\bar{\mathfrak{m}}$, we find by the exact sequence (E) that $\mathfrak{m} \cdot (M'/xM') = (0)$ and so

$$\ell_{R/p}(M'/xM') = \mu_{R/p}(M'/xM') = 3$$

by the equality (**). This is a contradiction. Hence we have $v(R/p) \leqq 2$ and therefore $v(R) \leqq d+1$. Because R is analytically irreducible, the condition that $v(R) \leqq d+1$ implies R is an abstract hypersurface. This completes the proof of (2.8), since $e = e(R) \leqq 3$ by (2.4).

We are now ready to prove Theorem (2.1). Assume furthermore that R is *complete*. Then by (2.8) R is a Cohen-Macaulay complete local ring. Therefore because R has finite CM-representation type, by virtue of a theorem due to M. Auslander [**2**] the ring R is an isolated singularity. Hence R is normal, that is $R = \bar{R}$ and so R is a regular local ring.

COROLLARY (2.9). *Let R be a Noetherian complete local integral domain of $\dim R \geqq 2$ and assume that R contains an algebraically closed coefficient field k of ch $k \neq 2$. Then R is regular if the derived normal ring $\bar{R}$ of R is Cohen-Macaulay and if R has finite Buchsbaum-representation type.*

PROOF: Let M be an indecomposable maximal Buchsbaum $\bar{R}$-module and assume that $\operatorname{depth}_{\bar{R}} M \geqq 1$. Let q be a parameter ideal for the R-module M. Then we have

$$\ell_R(M/qM) = \ell_{\bar{R}}(M/qM) \text{ and } e_q(M) = e_{q\bar{R}}(M)$$

because $R/\mathfrak{m} = \bar{R}/\mathfrak{n}$ (here $\mathfrak{n}$ denotes the maximal ideal of $\bar{R}$). Hence M is a maximal Buchsbaum R-module. As $\operatorname{depth}_{\bar{R}} M \geqq 1$, we see by [**17**, Ch. I, (1.10)] that M is $\bar{R}$-torsion-free and so we get $\operatorname{Hom}_{\bar{R}}(M, M) = \operatorname{Hom}_R(M, M)$ (cf. [**10**, Lemma 1]). Thus M is indecomposable as an R-module, too. Let M' be another indecomposable maximal Buchsbaum $\bar{R}$-module of $\operatorname{depth}_{\bar{R}} M' \geqq 1$. Then as both M and M' are $\bar{R}$-torsion-free, we have $\operatorname{Hom}_{\bar{R}}(M, M') = \operatorname{Hom}_R(M, M')$; hence $M \cong M'$ as $\bar{R}$-modules if $M \cong M'$ as R-modules. Accordingly $\bar{R}$ possesses only finitely many isomorphism classes of indecomposable maximal Buchsbaum $\bar{R}$-modules M of $\operatorname{depth}_{\bar{R}} M \geqq 1$. Therefore $\bar{R}$ is a regular local ring by the proof of [**9**, Theorem (1.1)], whence by (2.1) we find that R is regular.

COROLLARY (2.10). *Let k be an algebraically closed field of ch $k \neq 2$ and $P = k[[x_1, x_2, \ldots, x_n]]$ a formal power series ring over k. Let $f_1, f_2, \ldots, f_r$ be monomials in x_i's of positive degree and put $R = k[[f_1, f_2, \ldots, f_r]]$. Then R is regular if $\dim R \geqq 2$ and if R has finite Buchsbaum-representation type.*

PROOF: Since the derived normal ring $\bar{R}$ of R is Cohen-Macaulay (cf. [**12**]), the assertion follows from (2.9).

3. Proof of the implication (3) $\Rightarrow$ (1) in Theorem (1.1).

Let P denote a three-dimensional regular local ring with maximal ideal $\mathfrak{n}$. Let $X \in \mathfrak{n} \setminus \mathfrak{n}^2$ and let I be a *proper* ideal of P such that $\operatorname{ht}_P I \geqq 2$. We put $R = P/XI$, $\mathfrak{p} = XR$ and $\mathfrak{m} = \mathfrak{n}R$. The aim of this section is to prove the following:

THEOREM (3.1). *R has finite Buchsbaum-representation type and*

$$\mathfrak{m}/\mathfrak{p}, R/\mathfrak{m}\mathfrak{p} \text{ and } R/\mathfrak{p}$$

are the representatives of indecomposable maximal Buchsbaum R-modules.

PROOF: As $\mathfrak{m}/p$ is the maximal ideal in the regular local ring R/p, $\mathfrak{m}/p$ is an indecomposable maximal Buchsbaum R-module (cf. [**7**]). Since

$$\begin{aligned} H^i_{\mathfrak{m}}(R/\mathfrak{m}p) &= p/\mathfrak{m}p \quad (i=0), \\ &= (0) \qquad (i \neq 0,2), \end{aligned}$$

we have by [**17**, Ch. I, (2.12)] that $R/\mathfrak{m}p$ is a maximal Buchsbaum R-module. Because the depth is different, the R-modules $\mathfrak{m}/p$, $R/\mathfrak{m}p$ and R/p are not isomorphic to each other.

Conversely let M be an indecomposable maximal Buchsbaum R-module. We put $\bar{M} = M/H^0_{\mathfrak{m}}(M)$. Then $\bar{M}$ is again a maximal Buchsbaum R-module (cf. [**17**, Ch. I, (2.22)]).

CLAIM 1. $p\bar{M} = (0)$.

PROOF OF CLAIM 1: As p is a unique prime ideal of R with $\dim R/p = 2$, any element f of $R \setminus p$ is a non-zero-divisor on $\bar{M}$ (cf. [**17**, Ch. I, (1.10)]). Hence we have an embedding $\bar{M} \subset \bar{M}_p$ and so $p\bar{M} = (0)$, because $pR_p = (0)$.

By this claim we find $\bar{M} = M/H^0_{\mathfrak{m}}(M)$ is a maximal Buchsbaum R/p-module. Therefore because R/p is a two-dimensional regular local ring, we get by [**7**, (1.1)] a decomposition

$$\text{(D)} \qquad M/H^0_{\mathfrak{m}}(M) \cong (R/p)^{\alpha} \oplus (\mathfrak{m}/p)^{\beta}$$

with integers $\alpha, \beta \geqq 0$ (here $(\mathfrak{m}/p)^{\beta}$ stands for the direct sum of β copies of $\mathfrak{m}/p$).

We choose elements Y, Z of P so that $\mathfrak{n} = (X,Y,Z)P$ and Y, Z form a system of parameters for R. Let $x = X \bmod XI$, $y = Y \bmod XI$ and $z = Z \bmod XI$. Let $\bar{\cdot}$ denote the reduction mod p. Then $\bar{y}, \bar{z}$ is a regular system of parameters of R/p and so there is a presentation

$$R^3 \xrightarrow[\begin{pmatrix} x & 0 & y \\ 0 & x & z \end{pmatrix}]{} R^2 \xrightarrow{\varepsilon} \mathfrak{m}/p \longrightarrow 0$$

with $\varepsilon\left(\binom{a}{b}\right) = \overline{az - by}$. Let L be the R-submodule of R^2 generated by $\binom{x}{0}$, $\binom{0}{x}$ and $\binom{y}{z}$. Then because $H^0_{\mathfrak{m}}(M) \subset \mathfrak{m}M$ by [**9**, Claim in the proof of

Theorem (5.3)][3], we get by the above decomposition (D) a commutative diagram

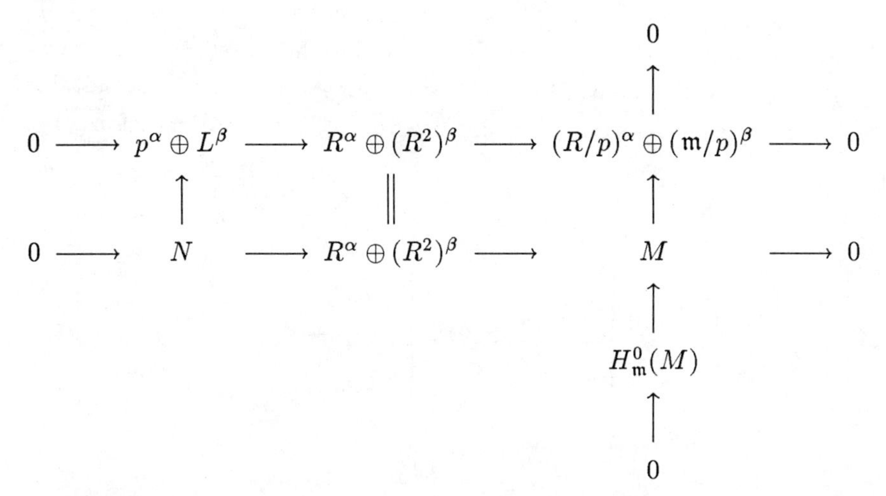

with exact rows and columns. (Here p^α denotes the direct sum of α copies of p.) Let $F = R^\alpha \oplus (R^2)^\beta$ and $W = p^\alpha \oplus L^\beta$. Then $W/N \cong H^0_{\mathfrak{m}}(M)$ and hence $\mathfrak{m}W \subset N$ by [**17**, Ch. I, (2.4)]. We note the next

CLAIM 2. *Let q be a parameter ideal of R. Then $W \cap qF \subset N$.*

PROOF OF CLAIM 2: Since $F/N(\cong M)$ is a maximal Buchsbaum R-module, q is a parameter ideal for F/N too. Because $H^0_{\mathfrak{m}}(F/N) = W/N$, we find by [**17**, Ch. I, (1.14)] that $(W/N) \cap q(F/N) = (0)$, that is $W \cap qF \subset N$.

Because y, z form, by the choice, a system of parameters of R, we see $\dim R/yR = \dim R/zR = 1$. Hence we may choose elements a, b of R such that $x + ay, z$ and $x + bz, y$ are systems of parameters of R. Let $q_1 = (y, z)R$, $q_2 = (x + ay, z)R$ and $q_3 = (x + bz, y)R$. Then clearly we have $\binom{y}{z} \in L \cap q_1 \cdot R^2$, $\binom{x}{0} + a\binom{y}{z} \in L \cap q_2 \cdot R^2$, and $\binom{0}{x} + b\binom{y}{z} \in L \cap q_3 \cdot R^2$. Hence it follows from Claim 2 that $N \supset (0) \oplus L^\beta$ and so we get $N = N_1 \oplus L^\beta$ with N_1 an R-submodule of p^α. Consequently

$$M \cong R^\alpha/N_1 \oplus (R^2/L)^\beta.$$

Therefore if $\beta \geqq 1$, we have $M \cong \mathfrak{m}/p$ because M is indecomposable.

Assume now that $\beta = 0$. Let $\tau : p^\alpha \longrightarrow (p/\mathfrak{m}p)^\alpha$ denote the canonical epimorphism. We put $V = \tau(N)$ and $\gamma = \dim_k V$ (here $k = R/\mathfrak{m}$). Recall

[3]The proof of [**9**, Claim in the proof of Theorem (5.3)] still works even in the case $\dim R \geqq 2$.

that $N \supset \mathfrak{m} \cdot p^\alpha$. Then if $\gamma = 0$, we have $N = \mathfrak{m} \cdot p^\alpha$ and so $M \cong (R/\mathfrak{m}p)^\alpha$. Thus $M \cong R/\mathfrak{m}p$ in this case. Consider the case $\gamma \geqq 1$. We regard each element of $(p/\mathfrak{m}p)^\alpha$ as a column vector with entries in $p/\mathfrak{m}p$. Choose a k-basis $\{v_i\}_{1 \leqq j \leqq \gamma}$ of V and consider the α by γ matrix $A = (v_1\, v_2 \ldots v_\gamma)$. Notice that $p/\mathfrak{m}p \cong k$ since $p = xR$. Then (via suitable elementary transformations with coefficients in k) we can transform the matrix A into the form

$$(*) \qquad \left(\begin{array}{cccc} x & & & \\ & x & & \\ & & \ddots & \\ & & & x \\ \hline & & 0 & \end{array} \right) \bmod \mathfrak{m}p.$$

Namely with some $\phi \in \operatorname{Aut}_R R^\alpha$, we may write $\phi(N) = U + \mathfrak{m} \cdot p^\alpha$, where U is the R-submodule of R^α generated by the columns of the above matrix (*). Therefore $M \cong R^\alpha/(U + \mathfrak{m} \cdot p^\alpha)$ and so we find that $M \cong R/p$ since M is indecomposable. This completes the proof of (3.1).

Because any regular local ring has finite Buchsbaum-representation type, the implication (3) $\Rightarrow$ (1) in Theorem (1.1) now readily follows from (3.1).

4. Two-dimensional Cohen-Macaulay rings of finite Buchsbaum-representation type.

The aim of this section is to prove the following:

THEOREM (4.1). *Let R be a Cohen-Macaulay complete local ring of* $\dim R = 2$ *and assume that the field $R/\mathfrak{m}$ is algebraically closed. Then R is regular if R possesses only finitely many isomorphism classes of indecomposable maximal Buchsbaum R-modules M of* $\operatorname{depth} M \geqq 1$.

We divide the proof into a few steps. Notice that the above theorem is already proved by the author and K. Nishida [9] when R contains an algebraically closed coefficient field k of ch $k \neq 2$. Our method of proof is almost the same as theirs and so the very detail shall be left to the reader.

Let R be as in (4.1) and assume that R has only finitely many isomorphism classes of indecomposable maximal Buchsbaum R-modules M of

$\operatorname{depth}_R M \geqq 1$. Then because R has finite CM-representation type too, we have by [**2**] that R is normal. Hence by [**9**, 2.5] R is a UFD. Therefore by J. Lipman [**14**, Theorem 25.1] we readily get the following:

LEMMA (4.2). *Let $p = \operatorname{ch} R/\mathfrak{m}$ and assume that R is not regular. Then $R \cong R/fP$ where P is a three-dimensional complete regular local ring with the regular system X, Y, Z of parameters and f is one of the following:*

$$X^3 + Y^5 + Z^2, \quad X^3 + Y^5 + Z^2 - XY^4 \qquad (p = 5),$$
$$X^3 + Y^5 + Z^2 - X^2Y^2, \quad X^3 + Y^5 + Z^2 - X^2Y^3 \qquad (p = 3),$$
$$X^3 + Y^5 + Z^2 - Y^3Z, \quad X^3 + Y^5 + Z^2 - XYZ,$$
$$X^3 + Y^5 + Z^2 - XY^2Z, \quad X^3 + Y^5 + Z^2 - XY^3Z \qquad (p = 2).$$

We now assume that R is *not* regular. We write $R = P/fP$ where P and f are as in (4.2). Let $\mathfrak{n} = (X, Y, Z)P$. Notice that $f = AX + BY^2 + Z^2$ with $A, B \in P$ such that $A \equiv X^2 \bmod YP$ and $B \in \mathfrak{n}$. Let the small letters x, y, z, a and b denote, respectively, the reduction of X, Y, Z, A and B mod fP.

We denote by L the R-submodule of R^2 generated by

$$f_1 = \begin{pmatrix} a \\ z \end{pmatrix}, \quad f_2 = \begin{pmatrix} y \\ 0 \end{pmatrix}, \quad f_3 = \begin{pmatrix} z \\ -x \end{pmatrix}, \quad f_4 = \begin{pmatrix} 0 \\ -y \end{pmatrix}.$$

Let L' be the second syzygy module of $R/\mathfrak{m}$:

$$0 \longrightarrow L' \longrightarrow R^3 \xrightarrow[(x \;\; y \;\; z)]{} R \longrightarrow R/\mathfrak{m} \longrightarrow 0.$$

Then L' is a maximal Cohen-Macaulay R-module of rank 2. Hence L' is indecomposable because R is a non-regular UFD. On the other hand, as is well-known, L' is generated by

$$\begin{pmatrix} a \\ by \\ z \end{pmatrix}, \begin{pmatrix} y \\ -x \\ 0 \end{pmatrix}, \begin{pmatrix} z \\ 0 \\ -x \end{pmatrix}, \begin{pmatrix} 0 \\ z \\ -y \end{pmatrix}.$$

Therefore because $\operatorname{rank}_R L = 2$ and L is a homomorphic image of L' (via the map $\phi : R^3 \longrightarrow R^2$ defined by $\phi\left(\begin{pmatrix} x_1 \\ x_2 \\ x_3 \end{pmatrix}\right) = \left(\begin{smallmatrix} x_1 \\ x_3 \end{smallmatrix}\right)$), we have $L \cong L'$; hence L is an indecomposable maximal Cohen-Macaulay R-module.

We put $\Lambda = \operatorname{End}_R L$ and denote by J the Jacobson radical of Λ. Let $k = R/\mathfrak{m}$. Then similarly as in [**9**, Section 4] we get $L = \Lambda f_2$ and so $\dim_k L/JL = 1$ (because $k = \Lambda/J$). We furthermore have

LEMMA (4.3). $\dim_k JL/(J^2L + \mathfrak{m}L) \geqq 2$.

PROOF: We will go along the same course as in the proof of [**9**, (4.1)(3)]. First check that L has a presentation (P) of the following form

$$\text{(P)}\qquad R^4 \xrightarrow{\begin{pmatrix} z & 0 & -x & -y \\ 0 & z & -by & a \\ -a & -y & -z & 0 \\ -by & x & 0 & -z \end{pmatrix}} R^4 \longrightarrow L \longrightarrow 0.$$

We put $T = R/yR$ and denote by $\bar{\cdot}$ the reduction mod yR. Let I be the maximal ideal $(\bar{x}, \bar{z})T$ of T. Then T/I has a resolution

$$T^2 \xrightarrow{\begin{pmatrix} \bar{z} & \bar{x} \\ -\bar{a} & \bar{z} \end{pmatrix}} T^2 \xrightarrow{(\bar{z} \quad -\bar{x})} T \longrightarrow T/I \longrightarrow 0.$$

Therefore because the matrix $\begin{pmatrix} z & 0 & -x & -y \\ 0 & z & -by & a \\ -a & -y & -z & 0 \\ -by & x & 0 & -z \end{pmatrix}$ mod yR is similar to the direct sum of two copies of $\begin{pmatrix} \bar{z} & \bar{x} \\ -\bar{a} & \bar{z} \end{pmatrix}$, we find by (P) that $L/yL \cong I \oplus I$.

Now recall that $(f, Y)P = (X^3 + Z^2, Y)P$. Then we have T to be an integral domain. Let $\bar{T}$ denote the derived normal ring of T and put $t = \bar{z}/\bar{x}$. Then t is a regular parameter of $\bar{T}$ and $\bar{x} = -t^2$, $\bar{z} = -t^3$. We put $\bar{L} = L/yL$. Because $I \cong \bar{T}$ as T-modules, we may identify $\bar{L} = \bar{T} \oplus \bar{T}$.

For each $f \in L$ let $\bar{f} = f \bmod yL$.

CLAIM 1. *$\bar{f}_2$ and $\bar{f}_3$ form a $\bar{T}$-free basis of $\bar{L}$.*

PROOF OF CLAIM 1: As $xf_1 = -((by)f_2 + zf_3)$ and $zf_4 = -yf_1 + af_2$, we get $\bar{f}_1 = -t\bar{f}_3$ and $\bar{f}_4 = -t\bar{f}_2$. Hence the $\bar{T}$-module $\bar{L}$ is generated by $\bar{f}_2$ and $\bar{f}_3$.

Because $\operatorname{End}_T \bar{L} = \operatorname{End}_{\bar{T}} \bar{L}$, we identify $\operatorname{End}_T \bar{L}$ with $\Gamma = M_2(\bar{T})$ (the matrix algebra) via the $\bar{T}$-free basis $\bar{f}_2, \bar{f}_3$. Let $\bar{\Lambda} = \Lambda/y\Lambda$. Then $\bar{\Lambda}$ is a subalgebra of $\operatorname{End}_T \bar{L}$ and so we have a homomorphism

$$\phi : \Lambda \longrightarrow \operatorname{End}_T \bar{L} = \Gamma$$

of R-algebras. Hence via ϕ we may write each element of Λ as a 2 by 2 matrix with entries in $\bar{T}$. For example, since

$$\begin{pmatrix} a & y & z & 0 \\ z & 0 & -x & y \end{pmatrix} \begin{pmatrix} 0 & 0 & 0 & 1 \\ 0 & 0 & -b & 0 \\ 0 & 1 & 0 & 0 \\ -b & 0 & 0 & 0 \end{pmatrix} \begin{pmatrix} z & 0 & -x & -y \\ 0 & z & -by & a \\ -a & -y & -z & 0 \\ -by & x & 0 & -z \end{pmatrix} = 0,$$

by the presentation (P) of L the matrix $\begin{pmatrix} 0 & 0 & 0 & 1 \\ 0 & 0 & -b & 0 \\ 0 & 1 & 0 & 0 \\ -b & 0 & 0 & 0 \end{pmatrix}$ induces an element ρ of J. As $\rho(f_2) = f_3$ and $\rho(f_3) = -bf_2$, we get

$$\phi(\rho) = \begin{pmatrix} 0 & -\bar{b} \\ 1 & 0 \end{pmatrix}.$$

This endomorphism ρ plays a key role in the proof of the next

CLAIM 2. *Let $\xi \in J$ and write $\phi(\xi) = \begin{pmatrix} \alpha & \gamma \\ \beta & \delta \end{pmatrix}$ with $\alpha, \beta, \gamma, \delta \in \bar{T}$. Then $\alpha, \delta \in t\bar{T}$ and $\gamma \in t^2 \bar{T}$.*

PROOF OF CLAIM 2: See [**9**, Proof of (4.6)].

We put $W = \mathfrak{a}^2 \bar{f}_2 + \mathfrak{a} \bar{f}_3$ where $\mathfrak{a} = t\bar{T}$. Let $\varepsilon : L \longrightarrow \bar{L} = L/yL$ denote the canonical epimorphism. Then by the same reason as in [**9**, Section 4] we have $\varepsilon(J^2L + \mathfrak{m}L) \subset W$. Hence there is an epimorphism

$$L/(J^2L + \mathfrak{m}L) \longrightarrow \bar{L}/W.$$

Consequently $\dim_k L/(J^2L + \mathfrak{m}L) \geqq \dim_k \bar{L}/W = 3$, by which we find that $\dim_k JL/(J^2L + \mathfrak{m}L) \geqq 2$ because $\dim_k L/JL = 1$ as was noted before.

By (4.3) as well as [**9**, (2.3)] we can construct, from L, a family $\{M_\lambda\}_{\lambda \in R/\mathfrak{m}}$ of indecomposable maximal Buchsbaum R-modules such that $\operatorname{depth}_R M_\lambda = 1$ for each $\lambda \in R/\mathfrak{m}$ and $M_\lambda \not\cong M_\mu$ if $\lambda \neq \mu$. This contradicts the hypothesis on R and hence we conclude that R is a regular local ring. This completes the proof of Theorem (4.1).

5. Proof of Theorem (1.1).

Let R be a Noetherian complete local ring of $\dim R = 2$ and assume that the field $R/\mathfrak{m}$ is algebraically closed. The aim of this section is to prove the implications (1) $\Rightarrow$ (2) and (2) $\Rightarrow$ (3) in Theorem (1.1).

We begin with

PROOF OF (1) $\Rightarrow$ (2): First we deal with the case where R is an integral domain. Let $\bar{R}$ denote the derived normal ring of R. Then because R is complete and $\dim R = 2$, $\bar{R}$ is a Cohen-Macaulay complete local ring. Let $\mathfrak{n}$ be the maximal ideal of $\bar{R}$. Then $R/\mathfrak{m} = \bar{R}/\mathfrak{n}$ since $R/\mathfrak{m}$ is algebraically closed. Therefore by the same reason as in the proof of (2.9) there are only finitely many isomorphism classes of indecomposable maximal Buchsbaum $\bar{R}$-modules M of $\operatorname{depth}_{\bar{R}} M \geqq 1$. Hence by (4.1) $\bar{R}$ is regular and so by (2.1) R is regular too.

Let us consider the general case. Choose $p \in \operatorname{Spec} R$ so that $\dim R/p = 2$. Then the integral domain R/p also has finite Buchsbaum-representation type. Hence R/p is regular as was proved above. Because R/p is Cohen-Macaulay, we get by [**8**, (2.4)][4] that $\mu_R(p) \leqq 1$. Thus $v(R) \leqq 3$. We take a three-dimensional complete regular local ring P and an ideal J of P so that $R \cong R/J$. Then as $\operatorname{ht}_P J = 1$, we may write $J = XI$ with $X \in P$ and an ideal I in P of $\operatorname{ht}_P I \geqq 2$. Notice that the Cohen-Macaulay local ring P/XP has finite Buchsbaum-representation type because it is a homomorphic image of R. Hence by (4.1) P/XP is a regular local ring. Since $\dim_R XR \leqq 1$, we have the equality $e(R) = e(R/XR)$ and therefore we get $e(R) = 1$ as $R/XR\,(\cong P/XP)$ is a regular local ring. This completes the proof of (1) $\Rightarrow$ (2).

Proof of (2) $\Rightarrow$ (3): Choose a three-dimensional complete regular local ring P and an ideal J of P so that $R \cong P/J$. Let $\mathfrak{n}$ denote the maximal ideal of P. Then $\operatorname{ht}_P J = 1$ so we may write $J = XI$ with $0 \neq X \in \mathfrak{n}$ and an ideal I in P of $\operatorname{ht}_P I \geqq 2$. Because $P/XP \cong R/XR$ and $\dim_R XR \leqq 1$, we have $e(P/XP) = e(R) = 1$ and hence $X \notin \mathfrak{n}^2$. This completes the proof of (2) $\Rightarrow$ (3).

---

[4] The proof of [**8**, (2.4)] works even in the case $\dim R \geqq 2$.

REFERENCES

1. M. Artin and J.-L. Verdier, *Reflexive modules over rational double points*, Math. Ann. **270** (1985), 79–82.
2. M. Auslander, *Isolated singularities and existence of almost split sequences*, Proc. ICRA IV, Springer Lecture Notes in Math. **1178** (1986), 194–241.
3. M. Auslander and I. Reiten, *The Cohen-Macaulay type of Cohen-Macaulay rings*, preprint (1987).
4. R.-O. Buchweitz, G.-M. Greuel and F.-O. Schreyer, *Cohen-Macaulay modules on hypersurface singularities II*, Invent. Math. **88** (1987), 165–182.
5. D. Eisenbud and S. Goto, *Linear free resolutions and minimal multiplicity*, J. Algebra **88** (1984), 89–133.
6. D. Eisenbud and J. Herzog, *The classification of homogeneous Cohen-Macaulay rings of finite representation type*, Math. Ann. **280** (1988), 347–352.
7. S. Goto, *Buchsbaum modules over regular local rings and a structure theorem for generalized Cohen-Macaulay modules*, Advanced Studies in Pure Mathematics **11** (1987), 39–64.
8. S. Goto, *Curve singularities of finite Buchsbaum-representation type*, preprint (1987).
9. S. Goto and K. Nishida, *Rings with only finitely many isomorphism classes of indecomposable maximal Buchsbaum modules*, J. Math. Soc. Japan **40** (1988), 501–518.
10. G.-M. Greuel and H. Knörrer, *Einfache Kurvensingularitäten und torsionfreie Moduln*, Math. Ann. **270** (1987), 417–425.
11. J. Herzog, *Ringe mit nur endlich vielen Isomorphieklassen von maximalen unzerlegbaren Cohen-Macaulay Moduln*, Math. Ann. **233** (1978), 21–34.
12. M. Hochster, *Rings of invariants of tori, Cohen-Macaulay rings generated by monomials, and polytopes*, Ann. of Math. **96** (1972), 318–337.
13. H. Knörrer, *Cohen-Macaulay modules on hypersurface singularities I*, Invent. Math. **88** (1987), 153–164.
14. J. Lipman, *Rational singularities with applications to algebraic surfaces and unique factorization*, IHES Publ. Math. **36** (1969), 195–279.
15. J. D. Sally, *On the associated graded rings of a local Cohen-Macaulay ring*, J. Math. Kyoto Univ. **17** (1977), 19–21.
16. Ø. Solberg, *Hypersurface singularities of finite Cohen-Macaulay type*, preprint (1986).
17. J. Stückrad and W. Vogel, *Buchsbaum rings and applications*, VEB Deutscher Verlag der Wissenschaften (1987).
18. Y. Yoshino, *Brauer-Thrall type theorem for maximal Cohen-Macaulay modules*, J. Math. Soc. Japan **39** (1987), 719–739.

Department of Mathematics, Nihon University, Sakura-josui, Setagaya-ku, Tokyo, 156 JAPAN

Partially supported by Grant-in-Aid for Cooperative Research.

Some Transcendence Degree Questions

ELOISE HAMANN

This paper addresses some open questions from [**1**]. The general concern is to study the behavior of transcendence degree over an arbitrary commutative ring R with identity with particular interest in those R algebras which are contained in a polynomial ring over R. To be precise, let $S = R[X_n]$ where $X_n = \{x_1, \dots, x_n\}$, a set of independent indeterminates and the objects of interest are R-algebras $B \subseteq S$ with B finitely generated over R. The paper gives two generalizations of theorems from [**1**] which give conditions which guarantee that $[S : B] + [B : R] = n$. ($[C : D]$ denotes the size of a maximal set contained in C which is algebraically independent over D.) S is a B-algebra about which little can be said except that S is finitely generated over R. If $[S : B] = d$, it does not follow that there exists a minimal generating set of S over B containing d algebraically independent elements. In fact, it is easy to construct examples where S can be minimally generated over B by algebraic elements, but $[S : B] > 0$. This paper will show that such pathological behavior cannot happen for the R-algebras B.

The paper will also show that given any k elements with $k < n$, which are contained in S, there exist $n - k$ elements such that the set of n elements is an algebraically independent set over R. An example is given of an R-algebra S which is not a polynomial algebra such that $[S : R] = 2$, for which there exists z in S with $\{z\}$ an algebraically independent set, but no w exists so $\{z, w\}$ is an algebraically independent set.

To save space let us use a.i. to denote algebraically independent and a.d. to denote algebraically dependent.

DEFINITION 1: Define $[C : D]_G = d$ where C is a D-algebra containing D if there exists a minimal generating set of C over D which has a subset of d elements a.i. over D and there does not exist a minimal generating set with an a.i. subset with more than d elements. $[C : D]_G$ will be called the generating transcendence degree of C over D.

DEFINITION 2: Define $[C : D] = d$ if there exists a set of d a.i. elements in C over D and no larger such set exists.

It might be noted that $[C : D]$ was called the complete transcendence degree and denoted $[C : D]_C$ in [**1**]. Since the author believes this notion of transcendence degree is the natural one, the adjective complete will not be used in this paper. In fact the reader should view Definition 1 as an artifice which allows the statement $[B : R]_G = [B : R]$ to replace the awkward statement "B has a minimal generating set which contains $[B : R]$ a.i. elements."

There are two natural goals in the study of transcendence degree over an arbitrary commutative ring R with identity. One is to recover as many of the properties of transcendence degree over an integral domain and the other is to reduce questions about particular rings to a domain case. Questions about $[B : R]$ can in many cases be reduced to questions about $[B/P : R/P]$ where P is a minimal prime P of R or an associated prime of zero of R. By an abuse of notation B/P denotes the image of B in S/PS. ($R \subseteq B \subseteq S$). One of the purposes of this paper is to distinguish the role of P as minimal prime or an embedded prime of zero.

DEFINITION 3: If R is an arbitrary commutative ring (with identity) call P an associated prime of zero if P is minimal over $\operatorname{Ann} r$ for some nonzero r in R. We will use $\operatorname{Ass}_R\{0\}$ to denote the set of all such primes P.

The author will not distinguish via notation between an element s of S an R-algebra and its image in S/PS. It will always be clear from context where s lives. Further, S/P will be used instead of S/PS. The phrase modulo P should make the context clear.

1. The General Case.

The following lemmas show that a set of elements of B or S is a.i. over R iff the set of images is a.i. over R/P for all P in $\operatorname{Ass}_R\{0\}$ under the assumption that S is flat as an R-module.

LEMMA 1. *If* $\{y_1, \ldots, y_m\} \subseteq S$, *an arbitrary* R*-algebra containing* R, *is* a.d. *over* R/P *modulo some* P *in* $\operatorname{Ass}_R\{0\}$, *then* $\{y_1, \ldots, y_m\}$ *is* a.d. *over* R.

PROOF: Suppose $\bar{F}(y_1, \ldots, y_m) = 0$ modulo P, minimal over $\operatorname{Ann} r$. Let $\bar{R} = R/P$, $R^* = R/\operatorname{Ann} r$, $P^* = P/\operatorname{Ann} r$. P^* is a minimal prime in R^*. $\bar{F}$ is a polynomial with coefficients in $\bar{R}$. Let F^*, F lift F so F^*, F have coefficients in $R^* - P^*$, $R - P$ respectively. Let $y_i{}^*$ be images of y_i in $S^* = S/(\operatorname{Ann} r)S$. $F(y_1, \ldots, y_m) \in PS$ and $= \sum a_i s_i$ where each s_i is a distinct element in S, and $a_i \in P$. Let $I = (a_1, \ldots, a_K)$ and $I^* = (a_1^*, \ldots, a_K^*)$ where I is an ideal of $R \subseteq P$ and I^* is an ideal of $R^* \subseteq P^*$. $I^*{}_{P^*}$ is nil in $R^*{}_{P^*}$ since it is finitely generated. Thus, $(I^*{}_{P^*})^N = 0$ for some N and there exists $s^* \notin P^*$ so $s^* I^{*N} = 0$ in R^*. $F(y_1, \ldots, y_m) \in IS = (a_1, \ldots, a_K)$ so $F^N(y_1, \ldots, y_m) \in I^N S$. Let s lift s^*, $s \notin P$ so the coefficients of sF^N are not all in P, but $sF^N(y_1, \ldots, y_m) \in sI^N \subseteq \operatorname{Ann} r$. Thus, $rsF^N(y_1, \ldots, y_m) = 0$ but the coefficients of rsF^N are not all 0 and the claim follows.

LEMMA 2. *Let S be an R-algebra containing R which is flat as an R-module. If $\{y_1, \ldots, y_m\} \subseteq S$ is a.d. over R, then there exists $P \in \operatorname{Ass}_R\{0\}$ such that $\{y_1, \ldots, y_m\}$ is a.d. over R/P.*

PROOF: Let $0 = \sum e_i m_i$ where m_i are monomials in $\{y_i\}$, and no $e_i = 0$. Since S is flat, let $m_i = \sum a_{ji} N_j$ where $\{N_j\}$ are elements of S and $\{a_{ji}\}$ are in R. We have $Ae = 0$ where $e = (e_1, \ldots, e_K)^t$ and $A = [a_{ji}]_{L \times K}$. If $L < K$, the K columns of A must be linearly dependent modulo any prime P. If $L \geq K$, then e_1 (or any e_i) kills all K by K minors of A. If P is a prime minimal over $\operatorname{Ann} e_1$, then modulo this P all K by K minors are 0 so the rank of A modulo P is $< K$ and the K columns are linearly dependent modulo P. Thus, in either case there exists P such that the K columns of A are linearly dependent modulo P so there exist b_i not zero in R/P so $\sum b_i m_i = 0$. Since the $\{m_i\}$ are distinct monomials in $\{y_j\}$, $\{y_1, \ldots, y_m\}$ is a.d. over R/P.

THEOREM 3. *Let S be an R-algebra which contains R and is a flat R-module, then $\{y_1, \ldots, y_m\} \subseteq S$ is a.i. over R iff $\{y_1, \ldots, y_m\}$ is a.i. over R/P for all P in $\operatorname{Ass}_R\{0\}$. If R is Noetherian or satisfies the ascending chain condition on prime ideals, $\{y_1, \ldots, y_m\}$ is a.i. iff it is a.i. over R/P with P a maximal prime of 0 in R, that is to say P is maximal among the primes P which are minimal over $\operatorname{Ann} r$ for some nonzero r in R.*

PROOF: The first statement is immediate from the lemmas. Lemma 1 implies the necessity of the second claim. Thus, it suffices to show that if $\{y_1, \ldots, y_m\}$ is a.d. modulo some P in $\operatorname{Ass}_R\{0\}$ that it is a.d. modulo some maximal such P. The proof of Lemma 2 shows that if $F(y_1, \ldots, y_m) = 0$ then $\{y_1, \ldots, y_m\}$ is a.d. modulo any P such that $\operatorname{Ann} c \subseteq P$ for some coefficient c of F. If R is Noetherian or of finite Krull dimension some P maximal with respect to the property of being minimal over $\operatorname{Ann} r$ for some nonzero r will exist.

The following theorem addresses the question of the relations between $[S:R], [S:R]$, and $[S/P:R/P]$.

THEOREM 4. *If R is Noetherian or if R satisfies the ascending chain condition on prime ideals with a finite number of P maximal in $\mathrm{Ass}_R\{0\}$ and if S is an R-algebra containing R which is flat as an R-module, then $[S:R] = \min_{P\in \mathrm{Ass}_R\{0\}} [S/P:R/P] = \min_{\max P\in \mathrm{Ass}_R\{0\}} [S/P:R/P]$ where S need not be finitely generated over R as an R-algebra. (S/P denotes S/PS) If we assume that S has a minimal generating set over R, then we also have $[S:R]_G = [S:R]$.*

PROOF: First assume that S has a minimal generating set over R. $[S:R]_G \leq [S:R]$ by definition. Similarly, $[S:R] \leq \min_{P\in \mathrm{Ass}_R\{0\}} [S/P:R/P]$ by Lemma 1. Thus, it suffices to show that if $[S/P:R/P] \geq d$ for all $\max P$ in $\mathrm{Ass}_R\{0\}$ where d is finite that there exists d algebraically independent elements part of a minimal generating set of S over R. For each such P there exists $P' \supseteq PS$ where P' is a prime ideal of S such that $[S/P':R/P] \geq d$. To see this, choose $A = \{y_1, \ldots, y_d\} \subseteq S/P$ (actually S/PS) such that A is a.i. over R/P. Then $R/P[y_1, \ldots, y_d]$ is a domain $\subseteq S/P$ so 0 is a minimal prime of $R/P[y_1, \ldots, y_d]$. Thus, there exists P' so that P'/P lies over 0. The images of $y_1, \ldots, y_d$ in S/P' form an a.i. set over R/P since $\bar{F}(y_1, \ldots, y_d) = 0$ in S/P' implies that if F lifts $\bar{F}$ to S/P that $F(y_1, \ldots, y_d) \in P'/P \cap R/P[y_1, \ldots, y_d] = 0$ which implies the coefficients of f are 0 in R/P since $\{y_1, \ldots, y_d\}$ is a.i. set of S/P. Now we show by induction that there exists $\{y_1, \ldots, y_d\}$ whose image in S/P' for each $\max P$ in $\mathrm{Ass}_R\{0\}$ is an a.i. set over R/P. If $d = 0$ there is nothing to prove as the empty set satisfies the condition. Now suppose j elements have been found with $j \leq d$ so the image of $\{y_1, \ldots, y_j\}$ is an a.i. set over R/P in each S/P'. Since $[S/P':R/P] \geq d$ and both S/P' and R/P are domains, there exists $m_P \in M$ such that the image of $\{y_1, \ldots, y_j, m_P\}$ is an a.i. set in S/P' over R/P. Choose m as an m_P which works for a maximal number of $\max P$ in $\mathrm{Ass}_R\{0\}$. If m does not work for all P, let Q be maximal in $\mathrm{Ass}_R\{0\}$ for which the image of $\{y_1, \ldots, y_j, m\}$ is an a.d. set in S/Q'. Choose $r \in R$ such that $r \in$ each P for which m works but not in Q and let $m' = m + rm_Q$. Clearly the image of $\{y_1, \ldots, y_j, m'\}$ is an a.i. set in S/P' for each original P and is an a.i. set in S/Q'. Since the number of maximal primes of 0 is finite, it is clear that we can find m such that the image of $\{y_1, \ldots, y_j, m\}$ is a.i. over R/P in S/P' for all such P. It is easy to see that the image of the set in S/P is also an a.i. set over R/P so by Theorem 3, $\{y_1,\ldots,y_j,m\}$ is a.i. over R. To see that $\{y_1,\ldots,y_j,m\}$ is part of a minimal generating set of S over R, note that if $M^* = \{y_1,\ldots,y_j\} \cup \{m_p\}$ where M^* is the subset of the minimal generating set, M, consisting of j independent elements of M and each m_p is as described above, that the substitutions of $m_p + rm_Q$ for m_P retain the generating power of M. Further $C = \{y_1,\ldots,y_j\} \cup \{m_P + rm_Q\}$ minimally generates $R[C]$ over R since a relation $c_i = F(c_1,\ldots,\hat{c}_i,\ldots,c_{j+1})$ with c_k in C would survive modulo any prime P contradicting the fact that C is a.i. modulo some primes P. If $m_Q \notin R[C]$, let $C' = C \cup \{m_Q\}$; otherwise let $C' = C$. In any case $R[C'] = R[\{y_1,\ldots,y_j,m_P,m_Q\}] = S'$ with C' minimally generating S' over R and $M - \{y_1,\ldots,y_j,\ldots,m_P,m_Q\} = M'$ minimally generating S over S'. Thus, $C' \cup M'$ minimally generates S over R.

If S does not have a minimal generating set over R, the above proof applied to any generating set of S over R shows $[S:R] = \min_{\max P \in \mathrm{Ass}_R\{0\}} [S/P : R/P]$ and the intermediate equalities also follow.

One of the natural properties that a.i. sets contained in an integral domain S over an integral domain R enjoy is the capability of being extended to a maximal a.i. set which is the size of the transcendence degree of S over R. This can be restated as "all maximally a.i. sets of S over R have the same cardinality." The following example shows that this result does not hold in the non-domain case without some assumptions on S and R.

EXAMPLE 1: Let $A = k[\{U_f\},\{V_g\}]$ where k is any field and f,g are in $k[Y_1,Y_2,X] - k[X]$. Let $R = A/J$ where $J = (\{U_fV_g, U_fU_g, V_fV_g\}\ f \neq g)$, $C = R[X,Y_1,Y_2]$ and $S = C/I$ where $I = (\{U_f f + V_f X\})$.

CLAIMS.

(1) *R injects into S so $S = R[y_1,y_2,x]$ with y_1,y_2,x images of Y_1,Y_2,X.*
(2) *$\{y_1,y_2\}$ is an a.i. set in S.*
(3) *$\{x\}$ is an a.i. set in S.*
(4) *$\{x\}$ cannot be extended to an a.i. set with 2 elements.*

PROOFS: (1) Clearly R injects into C and since $I \subseteq (Y_1,Y_2,X)$, $I \cap R = 0$ so R injects into S.

(2) If $p(y_1,y_2) = 0$ in B, coefficients of p in R, let $\hat{p}(Y_1,Y_2)$ lift p to C so $\hat{p}(Y_1,Y_2) \in I$. Say $\hat{p}(Y_1,Y_2) = \sum l_{f_i}(U_{f_i} + V_{f_i}X)$ where we can assume

$l_{f_i} \in C - (\{U_f\}, \{V_f\})$ where f varies over all $f \neq f_i$. View C as a free module over $k[X, Y_1, Y_2]$ with basis $\{U_f^k V_f^l\}$. Comparing coefficients of basis elements with V_{f_i} as a factor we obtain a multiple of $X =$ to a polynomial in Y_1, Y_2 with coefficients in k unless $l_{f_i} = 0$. But this yields $\hat{p}(Y_1, Y_2) = 0$ in C so $\hat{p} \equiv 0$ whence $p \equiv 0$ in S.

(3) If $p(x) = 0$ in S, let $\hat{p}(X)$ lift the situation to C so $\hat{p}(X) = \sum l_{f_i}(U_{f_i} f_i + V_{f_i} X)$ as above. This time compare coefficients of basis elements with U_{f_i} as a factor to get a pure polynomial in $X =$ to a polynomial involving at least one of Y_1, Y_2 unless $l_{f_i} = 0$ which gives $\hat{p}(X) = 0$, $\hat{p} \equiv 0$, and $p \equiv 0$.

(4) If $\{w, x\}$ is an a.i. set, we can assume $w \in (x, y_1, y_2)S$ so $w = p(x, y_1, y_2) + q(x, y_1, y_2)$ where p has coefficients in k and q has coefficients in $(\{U_f\}, \{V_g\})S$. Choose U_f with $f = p^i$ if $p \notin k[x]$, but in any event so no U_f^k or V_f^l occurs in a coefficient of q. Then $U_f w^i = U_f p^i$ which is a dependency relation if $p \in k[x]$. Otherwise $U_f w^i + V_f x = U_f p^i + V_f x = 0$ so w and x are a.d.

2. The Polynomial Ring Case.

Throughout this section $S = R[X_n]$ for $X = (x_1, \ldots, x_n)$ and $R \subseteq B \subseteq S$. We assume B is finitely generated over R. The following two theorems are slight generalizations of theorems which appear in [**1**]. The hypotheses have been relaxed to assumptions only on the minimal primes of R rather than on all P in $\mathrm{Ass}_R\{0\}$.

THEOREM 5. *If $R \subseteq B \subseteq S = R[X_n]$ with B finitely generated over R such that*

1. *$NS \cap B = NB$ where N is the ideal of nilpotents of R*
2. *$[B/P : R/P] = k$ for all minimal primes P of R (B/P denotes image of B in S/PS by an abuse of notation.)*
3. *S can be generated over B by $n-k$ elements, then $[S : B] = [S : B] = n - k$ and S is a polynomial ring over B in $n - k$ variables.*

THEOREM 6. *If $R \subseteq B \subseteq S = R[X_n]$ with B finitely generated over R such that (2) and (3) of Theorem 5 above hold, then $[S : B] = n - k$.*

PROOFS OF BOTH THEOREMS: It clearly suffices to show that $[B/P : R/P] = k$ for all P in $\mathrm{Ass}_R\{0\}$ under the hypotheses of the theorems by the corresponding theorems from [**1**] for P in $\mathrm{Ass}_R\{0\}$. If $\mathrm{Ass}_R\{0\}$ consists of only minimal primes we are done. So suppose $P \in \mathrm{Ass}_R\{0\}$ is not minimal. Let Q be minimal $\subseteq P$, and let t be a zero divisor $\notin Q$. Then $Q \supseteq \operatorname{Ann} t$. Suppose $\{y_1, \ldots, y_{k+1}\}$ is an a.i. set over R/P. This set must be a.d. over R/Q and therefore dependent over R. By the proof of Lemma 2, $\{y_1, \ldots, y_{k+1}\}$ is a.d. over R/P^* where $P^* \supseteq \operatorname{Ann} e$ for any e, a coefficient of the relation. By the proof of Lemma 1, e can be assumed to have the form st where $s \notin Q$. Let $r \in \operatorname{Ann} st$ so $rst = 0$ implies $r \in Q$ since neither s nor t are in Q. Thus $\operatorname{Ann} st \subseteq P$ and $[B/P : R/P] \leq k$. However, $[B/P : R/P]$ cannot be $< k$ since $[B/P : R/P] + [S/P : B/P] = n$ and $[S/P : B/P] \leq n - k$ since S can be generated over B by $n - k$ elements.

COROLLARY 7 (of the proof). *If $P, Q \in \operatorname{Ass}_R\{0\}$ with Q contained in P, then $[B/P : R/P] \leq [B/Q : R/Q]$ whenever $R \subseteq B$ where B is flat over R.*

The following lemma is used in showing that all maximally a.i. sets over R which are contained in a polynomial ring over R have the same cardinality.

LEMMA 8. *Given $\{y_1, \ldots, y_k\}$ a.i. contained in $S = R[X_n]$ with $k < n$, there exists a change of variables Z_n and for each P in $\operatorname{Ass}_R\{0\}$ a set $\{y_1^{(P)}, \ldots, y_k^{(P)}\}$ with the following properties:*

(1) *$\{y_1, \ldots, y_k\} \cup T$ is an a.i. set modulo P in $\operatorname{Ass}_R\{0\}$ iff $\{y_1^{(P)}, \ldots, y_k^{(P)}\} \cup T$ is an a.i. set modulo P for any set $T \subseteq S$.*
(2) *$\{\operatorname{Deg} y_i^{(P)}\}$ is bounded as P ranges over $\operatorname{Ass}_R\{0\}$*
(3) *If $g_i^{(P)}$ denotes the pure $\{z_1, \ldots, z_k\}$ part of $y_i^{(P)}$, $\{g_1^{(P)}, \ldots, g_k^{(P)}\}$ is an a.i. set modulo P.*

PROOF (Induction on k): If $k = 1$, choose $\{N_i\}$ such that if $z_1 = x_1$, $z_i = x_i + {x_1}^{N_i}$, $i \geq 2$, and $y_1 = \sum a_i m_i$, m_i monomials in x_i, terms of the form $a_i {z_i}^{p_i}$ for each a_i with distinct p_i occur when y_1 is expressed in terms of z_i. Let $y_1^{(P)} = y_1$ for all P. Since y_1 is not a zero divisor, $\{a_i\} \not\subseteq P$ for any P in $\operatorname{Ass}_R\{0\}$. Thus, $g_1^{(P)} = g_1$ for all P cannot be 0 modulo any P. It is now clear that (1), (2), and (3) hold.

Now assume the existence of Z_n and sets $\{y_1^{(P)}, \ldots, y_{k-1}^{(P)}\}$ so (1), (2) and (3) hold. Let $y_k^{(P)} = y_k$ for all P be an initial choice of $y_k^{(P)}$. Let $g_j^{(P)}$ denote the pure $\{z_1, \ldots, z_{k-1}\}$ part of y_j. Formally, write G_j as $\sum A_{ij} m_{ij}$ where if $j < k$, the m_{ij} are all the monomials in $\{z_1, \ldots, z_{k-1}\}$ which might occur and if $j = k$, the m_{ik} are all the monomials which actually occur. Recall that the degrees of $y_i^{(P)}$ are bounded. $\{G_1, \ldots, G_k\}$ is an a.d. set in $B[Z_n]$ over B where $B = Z[\{A_{ij}\}]$ with A_{ij} to be viewed as a.i. indeterminates over the integers. Let $F(G_1, \ldots, G_k) = 0$ and suppose that

$F(G_1, \ldots, G_k) = \sum b_j W_j$ where W_j are K monomials in the G_j. Thus the monomials $\{W_j\}$ are linearly dependent over B. If $W_j = \sum c_{jk}\, m_k$ where the m_k are monomials in z_i, then the matrix $C = [c_{jk}]$ with entries in B has rank at most $K-1$. Thus, either $N < K$ or every N by N minor is 0. If $\hat{B}$ is any homomorphic image of B and $\hat{C}$ is the corresponding homomorphic image of the matrix C, then $\hat{C}$ has rank at most $K-1$. For each P in $\mathrm{Ass}_R\{0\}$ there is a homomorphism from B into R/P which takes Z to the subring of R generated by $\{1\}$, and takes A_{ij} to a_{ij} in R/P where $a_{ij}\, m_{ij}$ occurs in y_i. If w_j denotes the monomial corresponding to W_j, we replace G_i in W_j with $g_i^{(P)}$, then $\{w_j\}$ must be a.d. modulo P. Thus, there exists $F^{(P)}$ so that $F^{(P)}(\{g_i{}^{(P)}\}) = 0$ modulo P and the degree of $F^{(P)}$ is at most the maximum of the degrees of W_j. By (1) $F^{(P)}(y_1^{(P)}, \ldots, y_k^{(P)})$ is not 0, so let this be the new $y_k^{(P)}$. Since $y_k^{(P)}$ must occur in $F^{(P)}$, (1) holds. Since $\{\mathrm{Deg}\, F^{(P)}\}$ is bounded, (2) holds.

Now, define $x_i = z_i$ for $i = 1, \ldots, k$, and $x_i = z_i + x_k^{N_i}$, $i \geq k+1$, where N_i are chosen so that even if $y_k^{(P)}$ should consist of every possible monomial m_i in z_j with small enough degree that if $\sum A_i\, m_i$ is an "abstraction" of $y_k^{(P)}$, $A_i\, x_k^{P_i}$ occur with distinct p_i in a rewriting in terms of the new variables x_i. Thus, the pure $\{x_1, \ldots, x_k\}$ part of any $y_k^{(P)}$ which we denote $h_k^{(P)}$ is not 0 and is divisible by x_k. The pure $\{x_1, \ldots, x_k\}$ parts of $y_i^{(P)}$ with $i < k$ are of the form $h_i^{(P)} = g_i^{(P)} + x_k f_i$ with the $\{g_i^{(P)}\}$ a.i. It is now straightforward to show that $\{h_i^{(P)}\}$ is a.i. by observing that $h_k^{(P)}$ is not a zero divisor so if there is a relation of dependence, there are terms not containing $h_k^{(P)}$. Thus, setting $x_k = 0$ contradicts $\{g_i^{(P)}\}$ a.i. Thus, (3) holds, and the lemma follows.

THEOREM 9. *If $\{y_1, \ldots, y_k\} \subseteq S = R[X_n]$ is a.i., there exists $y_{k+1}, \ldots, y_n$ so $\{y_1, \ldots, y_n\}$ is a.i. Further $\{y_{k+1}, \ldots, y_n\}$ can be chosen part of a set of variables of S over R.*

PROOF: If $k = n$ there is nothing to prove. If $k < n$, it suffices to argue that you can always find one more element a.i. of the rest as long as the number of elements is $< n$. Choose Z_n and sets $\{y_1^{(P)}, \ldots, y_k^{(P)}\}$ as in Lemma 8. Let $y_i = z_i$ for $i = k+1$. If $\{y_1, \ldots, y_{k+1}\}$ is a.d. it must be a.d. modulo some P in $\mathrm{Ass}_R\{0\}$ by Theorem 3. By property (1) $\{y_1^{(P)}, \ldots, y_k^{(P)}, y_{k+1}\}$ would be a.d. modulo P. However, that this set is a.i. can be argued exactly as in the final argument of Lemma 8.

The following theorem shows that the finiteness assumptions on R in

Theorem 4 can be dropped if the R-algebra is contained in a polynomial ring over R and is finitely generated over R.

THEOREM 10. *Let $R \subseteq B \subseteq S = R[X_n]$ with B finitely generated over R, then $[B:R]_G = [B:R] = \min_{P \in \mathrm{Ass}_R\{0\}} [B/P : R/P]$. As usual B/P denotes the image of B in S/PS.*

PROOF: $[B:R]_G \leq [B:R] \leq \min_{P \in \mathrm{Ass}_R\{0\}} [B/P : R/P]$. The first inequality follows from the fact that $[\ ]_G$ is a maximum over a smaller set than the set over which $[\ ]$ is computed. The second inequality follows from Lemma 1. Thus, it suffices to show that if $[B/P : R/P]$ is $\geq k$ for all $P \in \mathrm{Ass}_R\{0\}$ that there exist k elements which are a.i. contained in a minimal generating set of B over R. For this it suffices to show that given s elements which are a.i. and part of a minimal generating set of B over R with $s < k$ that the set can be extended by one element while preserving its properties. To spare the reader concern over vacuous cases, we argue the case $k = 1$. It is clear that there is nothing to prove if $k = 0$. If $k = 1$, let $\{y_1, \ldots, y_t\}$ be a minimal generating set of B over R. Without loss of generality we can assume the $\{y_i\}$ have zero constant terms as polynomials in $\{x_j\}$. Since $k = 1$, no element of R can kill all y_i. Choose N_i such that $z_1 = y_1 + y_2{}^{N_2} + \cdots + y_t{}^{N_t}$ where each $y_i{}^{N_i}$ consists of distinct monomials. z_1 cannot be a zero divisor lest each y_i be a zero divisor. (It is well known that an element of a polynomial ring which is a zero divisor will be killed by an element of the coefficient ring which necessarily kills each coefficient of the zero divisor polynomial.) Thus, replacing y_1 by z_1 gives us an independent element which is part of a m.g.s. Now suppose that we have found $y_1, \ldots, y_s$ part of a m.g.s. of B over R which are a.i. and $s < k$. For each $P \in \mathrm{Ass}_R\{0\}$ one of $y_{s+1}, \ldots, y_t$ is a.i. over $R/P[y_1, \ldots, y_s]$. Let us call these elements of which there are a finite number a cover of $\mathrm{Ass}_R\{0\}$. Now suppose $\{w_1, \ldots, w_m\}$ forms a minimal cover of $\mathrm{Ass}_R\{0\}$ where $\{y_1, \ldots, y_s, w_1, \ldots, w_m\}$ is part of a m.g.s. of B over R. If $m > 1$, let $u = w_1$ and $v = w_2$. We will derive a contradiction by finding N so that $u + v^N$ is a.i. over $R/P[y_1, \ldots, y_s]$ if one of u or v is a.i. over $R/P[y_1, \ldots, y_s]$. If $s < n-1$, apply Lemma 8 to $y_1, \ldots, y_s$. If $T = \{u\}$ (respectively $\{v\}$), and u is a.i. over $R/P[y_1, \ldots, y_s]$, exactly as in Lemma 8 there exists $u^{(P)}$ so that $y_1^{(P)}, \ldots, y_s^{(P)}, u^{(P)}$ are a.i. and a change of variables Z_n so that the pure $z_1, \ldots, z_{s+1}$ parts are also a.i. all modulo P. Modulo P $y_1^{(P)}, \ldots, y_s^{(P)}, u^{(P)}, z_{s+2}, \ldots, z_n$ are a.i. when $u^{(P)}$ is a.i. over $R/P[y_1, \ldots, y_s]$. A similar argument can be made for the existence of $v^{(P)}$

whenever v is a.i. over $R/P[y_1,\ldots,y_s]$. Since $y_1,\ldots,y_s,u$ or $y_1,\ldots,y_s,v$ are a.i. over $R/P[z_{s+2},\ldots,z_n]$ exactly when their upper P versions are a.i. we have the following situation. $R[y_1,\ldots,y_s,z_{s+2},\ldots,z_n,u,v] = B_1$ is such that $[B/P : R/P] = n$ if P is a prime which u or v covers. Clearly, it suffices to show that there exists N so $u + v^N$ is a.i. over $R/P[y_1,\ldots,y_s,z_{s+2},\ldots,z_n]$. Relabel these elements as $y_1,\ldots,y_{n-1}$. If $s = n-1$, we have the same setup. Let Y_i be an "abstraction" $\sum A_{ij} m_{ij}$ of y_i where m_{ij} are monomials in Z_n or X_n which occur in y_i. Similarly, abstract u and v to U and V. Y_i and U,V are elements which live in a polynomial ring of n variables over A where A is a polynomial ring over the integers. The $n+1$ elements $Y_1,\ldots,Y_{n-1},U,V$ must be a.d. over this ring so F exists such that $F(Y_1,\ldots,Y_{n-1},U,V) = 0$. Thus, a set of monomials $\{W_j\}$ in the Y_i and U,V are linearly dependent over A. Thus, the corresponding monomials are linearly dependent over any homomorphic image of A. If we map $A[z_1,\ldots,z_n]$ into $R/P[z_1,\ldots,z_n]$ we get a relation on $\{w_j\}$ so there exists $F^{(P)}$ such that $F^{(P)}(y_1,\ldots,y_{n-1},u,v) = 0$ and a bound exists on the degrees of $\{F^{(P)}\}$. Now let P be a prime such that both u and v are individually a.i. over $R/P[y_1,\ldots,y_{n-1}]$. Let N be such that the monomials occurring in u and v^N are distinct (modulo any P) and such that $N >$ the bound on the total degree of $F^{(P)}$. Modulo P the polynomials which u satisfies over $R/P[y_1,\ldots,y_{n-1},v]$ are all multiples of a single polynomial. We can assume the single polynomial is $F^{(P)}$ without altering the fact that the total degrees of $F^{(P)}$ are bounded and that $N >$ the total degree bound. If $G(u+v^N) = 0$ where G has coefficients in $R/P[y_1,\ldots,y_{n-1}] = C$, then we can assume $G(Y)$ is irreducible in $C[Y]$. Since v is a.i. over C we must have $G(Y)$ irreducible in $D[Y]$ where $D = C[v]$. Thus, $G(v^N+X)$ is irreducible in $D[X]$. However, $G(v^N+u) = 0$ so we must have $G(v^N+X) = (c(v)/d(v))F^{(P)}(X)$ where $c(v)$ and $d(v)$ have coefficients in C. Let $t = \deg F^{(P)}(X) = \deg G(X)$ and $b_t =$ the coefficient of X in $G(V^N + X)$. $b_t \in C$, so if $f_t(v) =$ the coefficient of X^t of $F^{(P)}(X)$ we have $c(v)/d(v) = b_t/f_t(v)$ and $G(v^N + X) = (b_t/f_t(v))F^{(P)}(X)$. Since the coefficients of $F^{(P)}$ are polynomials in v over C of degree $< N$ we have arrived at a contradiction. Thus, $u+v^N$ must be a.i. over $R/P[y_1,\ldots,y_{n-1}]$ whenever P is a prime such that each of u and v are individually a.i. over $R/P[y_1,\ldots,y_{n-1}]$. If only one of u or v is a.i. over $R/P[y_1,\ldots,y_{n-1}]$ it is clear that $u+v^N$ is also a.i. over $R/P[y_1,\ldots,y_{n-1}]$ in this case. Thus, we can reduce the size of the cover of $\mathrm{Ass}_R\{0\}$ by replacing $\{u,v\}$ with $\{u+v^N\}$

contradicting the minimality of the cover. Thus, the size of the cover is one and there exists an element which is part of a minimal generating set of B over R which is a.i. over $R[y_1, \ldots, y_{n-1}]$ whence over $R[y_1, \ldots, y_s]$ so the theorem holds.

The preceding theorem settles a question raised in [**1**]. It was shown that $[B:R]_G \leq [B:R] \leq$ the minimum of $[B/P : R/P]$ over the minimal primes of R under the assumption that $R \subseteq B \subseteq S = R[X_n]$ and that B is finitely generated over R. The question raised was whether the inequalities could be strict. We have $[B:R]_G \leq [B:R] \leq \min_{P \in \mathrm{Ass}_R\{0\}} [B/P : R/P] \leq \min_{P \in \mathrm{Min}_R\{0\}} [B/P : R/P]$, and the preceding theorem shows that the first two inequalities cannot be strict or that the question was raised over the wrong set of primes. That the last inequality can be strict follows from the last example of the next section.

3. $\mathrm{Ass}_R\{0\}$ and Examples.

The following example shows that some finiteness assumptions are necessary in order that $[B:R] = \min_{P \in \mathrm{Ass}_R\{0\}} [B/P : R/P]$.

EXAMPLE 2: Let $R = k[\{a_i\}, \{n_i\}]/I$ where i is a positive integer and $I = (\{n_i^2\}, \{a_i n_i\})$. Let $B = R[\{a_i x_1^i\}] \subseteq S = R[x_1]$.

CLAIM. *$[B/P : R/P] = 1$ for all $P \in \mathrm{Ass}_R\{0\}$ but $[B:R] = 0$.*

PROOF: It is clear that every element of B with 0 constant term as a polynomial in x_1 is a zero divisor so $[B:R] = 0$. $[B/P : R/P] = 0$ iff $P \supseteq A = (\{a_i\})$. The only prime $\supseteq A$ is $M = (\{a_i\}, \{n_i\})$. We see $M \notin \mathrm{Ass}_R\{0\}$ by observing that (1) every ideal generated by $\{n_i\}$ and a finite number of the a_i is prime and (2) for any r, $\mathrm{Ann}\, r$ is contained in such an ideal. Thus, M is not minimal over $\mathrm{Ann}\, r$ for any r and $M \notin \mathrm{Ass}_R\{0\}$.

The following example from [**1**] shows that in the Noetherian non-polynomial case that the assumption in Theorem 4 that S be a flat module over R cannot be relaxed.

EXAMPLE 3: Consider S over R where $A \subseteq R \subseteq S$ with $A = Z/4Z$, $R = A[2X]$, and $S = A[X]$. X is a.d. over R but X^2 is a.i. over R.

The astute reader may have noted that $[B/M : R/M] = 0 = [B : R] = [B:R]_G$ in Example 2 and raised a question about the definition of $\mathrm{Ass}_R\{0\}$. An alternative definition would be any prime P contained in the zero divisors of R and with this definition $[B : R] = [B : R]_G = \min_{P \in \mathrm{Ass}_R\{0\}} [B/P : R/P] = 0$ for this example. Of course the issue is not which definition is most appropriate, but which set of primes best reflect $[B : R]$. The following two observations are made to argue for the set of primes which are minimal over $\mathrm{Ann}\, r$ for some r. First, the proof of Lemma 1 depended on P minimal over $\mathrm{Ann}\, r$ for some r. Second, the following example shows that the minimum of $[B/P : R/P]$ over all $P \subseteq Z(R)$ can be too low.

EXAMPLE 4: Let $R = k[a, b, \{n_f\}]/I$ where $f \in k[a, b] - k$ and $I = (\{n_f f\}, \{n_f n_g\})$. Let $B = R[w] \subseteq R[X_2]$ where $w = ax_1 + bx_2$. Note that w is not a zero divisor since no element of R kills both a and b. Thus, $[B : R] = 1$. Every element of $P = (a, b, \{n_f\})R$ is a zero divisor by construction. Since P is prime, the minimum of $[B/P : R/P]$ over all $P \subseteq Z(R)$ is 0.

The final example shows that it is insufficient to consider only the minimal primes.

EXAMPLE 5: Let $R = Z[a, b, c, d, N]/I$ where $I = (N^2, Na - Nc, Nb - Nd)$. Let $y_1 = ax_1 + bx_2$, $y_2 = cx_1 + dx_2$ where $S = R[X_2]$ and $B = R[y_1, y_2]$.

CLAIM 1. *Modulo each minimal prime P of 0 in $R[B/P : R/P] = 2$.*

PROOF: $(\{N\})$ is the only minimal prime so the claim is clear.

CLAIM 2. *The minimum of $[B/P : R/P]$ over all the associated primes P of 0 is 1.*

PROOF: It is easy to check that $N(ad - bc) = 0$ so that $ad - bc \in \mathrm{Ann}\, N$ and $[B/P : R/P] = 1$ if $P \supseteq \mathrm{Ann}\, N$, and $P \in \mathrm{Ass}_R\{0\}$.

ACKNOWLEDGEMENTS: The author would like to thank Mel Hochster and Richard Swan for useful comments and conversations.

REFERENCES

1. E. Hamann, *Transcendence degree over an arbitrary commutative ring*, J. Algebra **101** (1986), 110–119.
2. M. Nagata, "Local Rings," Tracts in Pure and Applied Mathematics, no. 13, Interscience, New York, 1962.
3. O. Zariski and P. Samuel, "Commutative Algebra," vol. I, Van Nostrand, New York, 1958.

Department of Mathematics, San Jose State University, San Jose CA 95192

Exceptional Prime Divisors of Two-dimensional Local Domains

WILLIAM HEINZER AND DAVID LANTZ

1. Introduction.

Throughout this paper, (R, M, k) will denote a 2–dimensional excellent normal (i.e., integrally closed) local (Noetherian) domain and K its field of fractions. We consider the normal local domains between R and K that are spots over R. Any such spot lies on the blow–up of some ideal I of R and on the normalized blow–up $\text{Proj}(R[It]')$ (where t is an indeterminate and $'$ denotes integral closure) of an ideal generated by two elements. The purpose of the present article is to study containment relations among these spots and conditions under which they can appear on the normalized blow–up of a specific ideal I; in particular, we study conditions under which I can be chosen to be M–primary.

The following example illustrates the context in which we are interested: Suppose that (S, N) is a two–dimensional normal local domain between R and K such that $MS = P$ is a height one prime of S. If P is not the radical of a principal ideal of S, then standard properties of blowing–up imply that S is not on the blow–up of any M–primary ideal of R. A specific example of this kind is:

EXAMPLE: Suppose the variables x, y, and z over the complex field k are related by $zy^2 + z^2y - x^3 + z^2x = 0$, and $w = z(y + x)$ (so that $w^2 + y^2w - x^3(y + x) = 0$). Then the spot $S = k[x, y, z]_{(x,y,z)}$ over $R = k[x, y, w]_{(x,y,w)}$ is on the normalized blow–up of the height one prime ideal $(w, y + x)$. But since the height one prime ideal (x, y) of S is not the radical of a principal ideal [**T**] and is the center of the only exceptional prime divisor of the ring extension S/R, S is not on the normalized blow–up of any M–primary ideal of R.

2. Exceptional primes of a ring extension.

Recall that a "spot" over R is a localization of a finitely generated R–algebra. The normal local domains between R and K, not equal to K, that are spots over R fall into three classes. Two of these classes consist of discrete rank–one valuation rings, the "prime divisors" of R. The first class, the "essential prime divisors" of R, are the localizations of R (a Krull domain) at height one prime ideals. The second, which we will follow [**A2**] in calling the "hidden prime divisors" of R, are those whose maximal ideals meet R in M (i.e., these spots "dominate" R) and whose residue fields are not algebraic over the residue field k of R. The third class of normal local spots between R and K consists of those of dimension 2. We denote these three classes by $\text{epd}(R), \text{hpd}(R)$ and $\mathbf{s}_2(R)$, respectively.

Since R is universally catenary [**Ma**, pages 111 and 124, Theorems 33 and 38], the dimension formula [**Ma**, page 85, Theorem 23] holds for finitely generated R–algebras and hence for spots over R. It follows that (1) the union of these three classes and $\{K\}$ is the set of all normal local spots between R and K, (2) the transcendence degree of the residue field of any hidden prime divisor over k is 1, (3) each domain in $\mathbf{s}_2(R)$ dominates R, and (4) the residue field of any domain in $\mathbf{s}_2(R)$ is (finite) algebraic over k. Since R is an excellent and hence a Nagata ring [**Ma**, page 257, Theorem 78], it is not hard to verify that a valuation domain between R and K is a spot if and only if it is either a localization of R at a height one prime or a DVR that dominates R and whose residue field is transcendental over k. Corollary 2 to Theorem 1 of [**HHS**] shows that every 2–dimensional normal local domain dominating R and contained in K is a spot over R.

Now the elements of $\mathbf{s}_2(R)$ are themselves Krull domains, and their essential prime divisors are prime divisors, either essential or hidden, of R. For S in $\mathbf{s}_2(R)$, we call the hidden prime divisors of R that are essential prime divisors of S the "exceptional prime divisors" of the ring extension S/R, and we denote the set of all such by $\text{xpd}(S/R)$. (Since the elements of $\text{xpd}(S/R)$ are localizations of S at height one primes that contain the nonzero ideal $MS, \text{xpd}(S/R)$ is a finite set.)

Since an element S of $\mathbf{s}_2(R)$ satisfies the same hypotheses as does R [**Ma**, page 257], we can use the notations $\text{epd}(S), \text{hpd}(S), \mathbf{s}_2(S)$ and, for T in $\mathbf{s}_2(S), \text{xpd}(T/S)$.

The following proposition is a local version of [**Go**, Lemma 2.1 and Proposition 2.2].

PROPOSITION 1. *Let S_1, S_2, and T be elements of $\mathbf{s}_2(R)$.*

(a) *Suppose S_1 is contained in T. Then* $\operatorname{xpd}(T/S_1) = \emptyset$ *if and only if* $T = S_1$.

(b) *Suppose T is a localization of the integral closure of the compositum* $S_1[S_2]$. *Then*

$$\operatorname{xpd}(T/R) \subseteq \operatorname{xpd}(S_1/R) \cup \operatorname{xpd}(S_2/R),$$
$$\operatorname{xpd}(T/S_2) \subseteq \operatorname{xpd}(S_1/R), \text{ and } \operatorname{xpd}(T/S_1) \subseteq \operatorname{xpd}(S_2/R).$$

(c) *If* $\operatorname{xpd}(S_1/R) = \operatorname{xpd}(S_2/R)$, *then either* $S_1 = S_2$ *or no element of* $\mathbf{s}_2(R)$ *dominates both* S_1 *and* S_2.

PROOF: (a) The condition that $\operatorname{xpd}(T/S_1) = \emptyset$ is equivalent to the condition that the extension N_1T to T of the maximal ideal N_1 of S_1 is not contained in any height one prime of T, and hence that N_1T is primary for the maximal ideal N of T. By Zariski's Main Theorem [**N**, (37.4)], this is equivalent to the condition that $S_1 = T$. (Remark: Nagata's statement of ZMT seems to require a check that the residue field extension is finite, and, using the dimension formula, it is not difficult to show that this condition holds in our case. But it is worth noting for future reference that, by [**Co**, Theorem 3], the other hypotheses of ZMT as stated in [**N**, (37.4)] imply the residue field extension is finite.)

(b) Assume $\operatorname{xpd}(T/R)$ is not contained in $\operatorname{xpd}(S_1/R) \cup \operatorname{xpd}(S_2/R)$. Then there is a height one prime P of T that meets each S_i in its maximal ideal N_i. Let Q denote the intersection of P with $S_1[S_2]$. Then $S_1[S_2]/Q$ is generated by the images of S_1/N_1 and S_2/N_2 and so is a field algebraic over k; so Q is maximal. But P is not maximal in T, a contradiction to Going–Up. The other containments follow easily: An essential prime divisor of T that is hidden for (i.e., that dominates) S_2 must also be hidden for R, but it cannot be hidden for S_1, because $\operatorname{xpd}(T/R)$ is contained in the union.

(c) If an element of $\mathbf{s}_2(R)$ dominates both S_1 and S_2, then that element can be taken to be a localization T of $S_1[S_2]'$; but then the hypothesis and parts (a) and (b) show $S_1 = T = S_2$. □

As usual, if a ring A is contained in a quasilocal ring B, we call the intersection with A of the maximal ideal of B the "center" of B on A.

PROPOSITION 2. *Let S_1 and S_2 be elements of $\mathbf{s}_2(R)$. The following statements are equivalent:*

(a) *There is an element T of $\mathbf{s}_2(R)$ containing both S_1 and S_2.*
(b) *The dimension of the compositum $S_1[S_2]$ is at least 2.*
(c) *There is a prime of $S_1[S_2]$ meeting each of S_1 and S_2 in its maximal ideal.*

When these conditions hold, then the intersection S of S_1 and S_2 is again an element of $\mathbf{s}_2(R)$, and

$$\operatorname{xpd}(S/R) \subseteq (\operatorname{xpd}(S_1/R) \cup \operatorname{xpd}(S_2/R)) \setminus (\operatorname{xpd}(T/S_1) \cup \operatorname{xpd}(T/S_2)).$$

PROOF: (a) $\Rightarrow$ (b): Let Q be the center of T on $S_1[S_2]$. If $\operatorname{ht}(Q) = 1$, then the containment of $S_1[S_2]_Q$ in T contradicts the Krull–Akizuki theorem [**N**, (33.2)].

(b) $\Rightarrow$ (c): Let Q be a prime of height two in the compositum. If Q met, say, S_1 in a height one prime P, then the containment of $(S_1)_P$ in $S_1[S_2]_Q$ would again contradict Krull–Akizuki.

(c) $\Rightarrow$ (a): Let Q be a prime of $S_1[S_2]$ meeting each S_i in its maximal ideal. Then, as in the proof of Proposition 1, $S_1[S_2]/Q$ is algebraic over k, so the dimension formula shows that $\operatorname{ht}(Q) = 2$. Since each S_i is a localization of a finitely generated R–algebra, the same is true of $S_1[S_2]$ and also of its integral closure, because R is Nagata. Localizing at a height two prime of the integral closure lying over Q gives an element of $\mathbf{s}_2(R)$ as desired.

For the last sentence, we note that, if S_1 and S_2 are dominated by an element of $\mathbf{s}_2(R)$, then (a) their intersection S has only one maximal ideal (see the next paragraph). Also, (b) S cannot have dimension one by Krull–Akizuki, (c) S is clearly normal, and (d) S is Noetherian by [**He**, Theorem 9], so it is in $\mathbf{s}_2(R)$ by [**HHS**, Corollary 3 to Theorem 1]. The family $\operatorname{epd}(S_1) \cup \operatorname{epd}(S_2)$ of discrete valuation rings is of "finite character" (i.e., a nonzero element is a unit in all but finitely many), and their intersection is S, so $\operatorname{epd}(S) \subseteq \operatorname{epd}(S_1) \cup \operatorname{epd}(S_2)$. Thus, $\operatorname{xpd}(S/R) \subseteq \operatorname{xpd}(S_1/R) \cup \operatorname{xpd}(S_2/R)$. And if a prime divisor of R dominates either S_1 or S_2, then it also dominates S, so $\operatorname{xpd}(S/R)$ contains no element of $\operatorname{xpd}(T/S_1)$ or $\operatorname{xpd}(T/S_2)$. □

If both S_1 and S_2 are dominated by a quasilocal ring T (not necessarily a spot), then their intersection has only one maximal ideal. (For, a nonunit in their intersection is a nonunit in one of them, hence in T, hence in both;

so the set of nonunits is the intersection of the maximal ideals of S_1 and S_2, an ideal.) So if the intersection of the elements S_1 and S_2 of $\mathbf{s}_2(R)$ has two maximal ideals – for example, if they are distinct localizations of the same finitely generated R–algebra – then the conditions in Proposition 2 do not hold; i.e., $\dim(S_1[S_2]) \leq 1$.

The last sentence of Proposition 2 cannot be improved to "xpd$(S/R) \subseteq$ xpd$(S_1/R) \cap$ xpd(S_2/R)", as the following example shows: Let $R = k[x,y]_{(x,y)}$, V, V', and W be the prime divisors associated with the ideals (x,y), (x,y^2) and (x^2,xy^2,y^3) respectively [**ZS**, page 391, (E)]. Let S_1, S_2, and T be the localizations of $R[x/y, y^2/x]$, $R[x^2/y^3, x/y]$, and $R[y^2/x, x^2/y^3]$ at the ideals generated in these rings by M and the adjoined elements, and $S = S_1 \cap S_2$. Then xpd$(S_1/R) = \{V, V'\}$ and xpd$(S_2/R) = \{W\}$, but xpd$(S/R) = \{V\}$.

It is useful to know that we can "globalize" the set of exceptional prime divisors without enlarging it, in the following sense:

PROPOSITION 3. *If S is in $\mathbf{s}_2(R)$, there is a finitely generated normal R–subalgebra A of S for which the height one primes of A that contain M are precisely the centers on A of the exceptional prime divisors of S over R.*

PROOF: Write xpd$(S/R) = \{V_1, \cdots, V_n\}$. Since the residue field of each V_i is not algebraic over k and is the field of fractions of the image of S, there are elements $b_1, \cdots, b_n$ of S such that the image of b_i in the residue field of V_i is not algebraic over k. Set $B = R[b_1, \cdots, b_n]'$, a finitely generated R–algebra because R is Nagata. Let P_i be the center of V_i on B; then since P_i lies over M in R and B/P_i is a finitely generated k–algebra that is not algebraic over k, B/P_i is not a field, so P_i is a height one prime. Let I be the intersection of the height one primes of B that lie over M in R but are distinct from $P_1, \cdots, P_n$, and let A be the I–transform of B, i.e., the intersection of the localizations of B at the height one primes that either *are* one of the P_i or lie over a height one prime in R. Then clearly A is a normal R–algebra, and A is contained in all the essential prime divisors of S, so it is contained in S.

By [**He**, Theorem 9] A is Noetherian, and by [**Ma**, Theorem 79, page 258] R satisfies Onoda's "condition (C)" [**O**]. Since $\dim A \leq \dim R = 2$ by [**Gi**, Theorem (30.9) and Corollary (30.10), pages 360–361], A is finitely generated over R by [**O**, Corollary 4.4]. □

3. Exceptional primes of an ideal.

For an ideal I of R, the "normalized blow–up" X_I of I is the set of localizations at prime ideals of the integral closures of the rings $R[I/b]$ as b varies over the nonzero elements of I (or just over a finite set of generators of I); i.e., X_I is the union of the affine models $\mathrm{Spec}(R[I/b]')$ [**ZS**, page 116; **A1**, page 25]. The result is precisely the set of domains minimal with respect to *domination* among the normal local domains between R and K in which the extension of I is a principal ideal. (Thus, $X_I = \mathrm{Spec}(R)$ if and only if I is principal.) In particular, every valuation domain between R and K dominates exactly one element of X_I. Because R is a Nagata ring, the elements of X_I are spots over R. Conversely, any spot over R is a localization of, say, $R\left[a_1/b, \cdots, a_n/b\right]$; setting $I = (a_1, \cdots, a_n, b)$, we see that if the spot is normal it is an element of X_I for this ideal I.

Recall that the "completion" or "integral closure" I' of an ideal I of R is the set of all elements c of R that satisfy an equation of the form

$$c^n + b_1c^{n-1} + \cdots + b_{n-1} + b_n = 0,$$

where each b_j is an element of the j–th power of I; or, equivalently, I' is the intersection of R and all the extensions of I to valuation domains between R and K [**ZS**, page 350, Theorem 1]. The normalized blow–up X_I can also be described as the set $\mathrm{Proj}(R[It]')$ of homogeneous localizations of the integral closure $R[It]'$ of the (small) Rees algebra $R[It]$ of I (t an indeterminate) at homogeneous prime ideals that do not contain It. Since this integral closure has the form $R + I't + \left(I^2\right)' t^2 + \cdots$ we see that X_I does not change if I is replaced by any ideal having the same integral closure as one of the powers of I. Now since R is Nagata, $R[It]'$ is finitely generated over $R[It]$. If we let n be the largest degree of an element in a set of homogeneous generators, then $(I^n)' \left(I^k\right)' = \left(I^{n+k}\right)'$ for every k, so in particular $(I^{ns})' = \left((I^n)'\right)^s$ for every s. It is often convenient to replace I with $(I^n)'$, a normal ideal (i.e., all its powers are integrally closed), for then the Rees algebra of I is itself normal.

Alternatively, it may be convenient, when $(I^n)' = (a,b)'$, to replace I with the two–generated ideal (a,b); for then the elements of X_I are localizations of the integral closures of the simple extensions $R[a/b]$ and $R[b/a]$ of R. To see that, for any ideal I, we can find a positive integer n and a two–generated ideal (a,b) for which $(I^n)' = (a,b)'$, (i.e., (a,b) is a "reduction" of I^n), consider the graded ring $A = R/M + I/MI + I^2/MI^2 + \cdots$: There are elements a^* and b^* in the same homogeneous piece I^n/MI^n of A such that (a^*, b^*) is not in any height one prime of A. If a and b are preimages in I^n of a^* and b^*, then I^n is integral over (a, b), so (a, b) is a reduction of I^n.

The "exceptional prime divisors" of an ideal I of R are the hidden prime divisors of R that lie on X_I (colloquially, that "come out on X_I"); the set of all such is denoted by $T(I)$. For any 2–dimensional S in X_I, an element of xpd(S/R) is a localization of S, so it is contained in $T(I)$. Indeed, $T(I)$ is the union of all xpd(S/R) as S varies over the two–dimensional elements of X_I. To see this, we use:

PROPOSITION 4. *For a two–generated ideal (a, b), a hidden prime divisor V of R is exceptional for (a, b) if and only if the quotient a/b is "residually transcendental for V", i.e., it is an element of V and its image in the residue field of V is transcendental over k.*

PROOF: Suppose V is exceptional for (a, b), i.e., V is a localization of the integral closure of either $R[a/b]$ or $R[b/a]$, say the former. Then the residue field of V is algebraic over the field of fractions of the domain $k[c]$, where c denotes the residue of a/b; but the residue field of V is transcendental over k, so c must be transcendental over k. Also, since c is a nonzero element of the residue field of V, a/b is a unit in V, and hence so is b/a, and it also has transcendental residue. (Thus, V is also a localization of $R[b/a]'$.)

Conversely, suppose a/b is in V and has transcendental residue c. Then $k[c]$ is not a field, so the center of V on $R[a/b]$ is not maximal and therefore is the height one prime ideal $MR[a/b]$. Thus the center of V on the integral closure $R[a/b]'$ is a height one prime; since the localization of $R[a/b]'$ at that prime is a DVR contained in V, it is V. □

To prove the claim made before the proof, we let V be an exceptional prime divisor of an ideal I; we assume $I = (a, b)$. Then since a/b is residually transcendental for V, there is a height two prime of $R\,[a/b]'$ containing the center of V on $R\,[a/b]'$. Letting S denote the localization at that prime, we find that V is in xpd(S/R).

Since the exceptional prime divisors of an ideal are of importance to us, let us note some other ways to describe them. (1) In order to find valuation domains between $R[It]$ and its field of fractions $K(t)$ so that the integral closure of $R[It]$ is the intersection of these valuation domains, we need to add to the ones that intersect in $R[t]$ the ones defined as follows: For a valuation v on K, extend to $v\sharp$ on $K(t^{-1},t)$ by setting $v\sharp(t^{-1}) = v(I)$ $(= \min(\{v(b)|b \in I\}))$ and assigning to a polynomial in t and t^{-1} the minimum value of its terms. The *hidden* prime divisors of R for which the corresponding $v\sharp$ are essential for $R[It]'$ are the exceptional prime divisors of I. (2) In the natural map $X_I \to \mathrm{Spec}(R)$ given by domination, we consider the fiber Y over the maximal ideal M of R; i.e., Y consists of the localizations of the rings $R[I/b]'$ at primes that contain M. Each irreducible component of Y is the set of domains in X_I contained in a given prime divisor of R. These (necessarily hidden) prime divisors are the exceptional prime divisors of I. (3) The completion I' of I can be written as the intersection of R and the extensions of I to finitely many prime divisors of R [**ZS**, page 354, proof of Theorem 3]; the same is true of the powers of I. If we select the set of prime divisors for which this intersection is irredundant for all powers of I, the hidden prime divisors in the set are the exceptional prime divisors. (4) Define $v_I^* : K \to \mathbb{R} \cup \{\infty\}$ (where $\mathbb{R}$ denotes the reals) by $v_I^*(x) = \lim(v_I(x^n)/n)$, limit taken as n increases without bound, where $v_I(y) = j$ if y is in I^j but not in I^{j+1}. (See [**Sm**].) Then v_I^* is a homogeneous pseudovaluation, the infimum of a finite number of valuations $v_1, \cdots, v_n$ on K. The exceptional prime divisors of I are the hidden prime divisors of R whose corresponding valuations are among $v_1, \cdots, v_n$.

In three of the descriptions above, it was necessary to specify the *hidden* prime divisors, since otherwise the sets include the essential prime divisors of R associated with height one primes that contain I. But if I is M–primary, the sets in question do not include essential ones; so the exceptional prime divisors of an M–primary ideal give a substantial amount of information about the ideal and its normalized blow–up. For instance, if $S_1, \cdots, S_n$ are 2–dimensional elements of X_I, where I is M–primary, and $\mathrm{xpd}(S_i/R) = T(I)$ for each i, then, replacing I by a power if necessary, we can find a single element b of I (one that attains the minimum V–value of elements of I for every exceptional prime divisor V) that generates all the extensions IS_i, so all the S_i are "on the same affine piece $\mathrm{Spec}(R[I/b]')$ of X_I"; i.e., all the S_i are localizations of the same finitely generated R–algebra.

Of particular interest is the case of an M–primary ideal I having only one exceptional prime divisor; Muhly and Sakuma [**MS**] called such an ideal "asymptotically irreducible", while Sally [**Sy1, Sy2**] calls it "one–fibered". In view of our descriptions of $T(I)$ above, if the M–primary ideal I has only one exceptional prime divisor, then (1) there is only one minimal prime ideal of $MR[It]'$; (2) the fiber Y is irreducible; (3) The normal ideal $(I^n)'$ and its powers are valuation ideals for this prime divisor; and (4) v_I^* is a valuation, with associated valuation domain the localization of $R[It]'$ at the unique minimal prime of $MR[It]'$.

By Proposition 4, the two–generated ideal (a, b) has unique exceptional prime divisor V if a/b is residually transcendental for V but not for any other hidden prime divisor of R. In a partial converse, Huneke and Sally [**HS**, Remark 3.5] note that, if R is regular and the residue field of the hidden prime divisor V of R is a simple transcendental extension $k(c)$ of k, then there is a two–generated M–primary ideal (a, b) of which V is the only exceptional prime divisor and the image of a/b in $k(c)$ is c. It would be interesting to know how close this statement remains to the truth if the hypothesis of regularity is removed.

Our next results approach the situation from the point of view of the divisor rather than the ideal. Given a hidden prime divisor V of R, to find an element of $\mathbf{s}_2(R)$ for which V is essential, it is reasonable to invert an element of R (yielding a Dedekind domain not contained in any hidden prime divisor of R), to intersect with V to yield a domain B (and making V a localization of B at its center P, a height one prime), and to localize B at a height two prime containing P. (Such a height two prime exists because B/P is a finitely generated k–algebra with field of fractions the residue field of V, so it is not algebraic over k and hence not a field.) The next results give necessary and sufficient conditions under which some of these spots are on the normalized blow–up of an M–primary ideal of which V is the only exceptional prime divisor.

LEMMA 5. *Let V be a hidden prime divisor of R. Assume that, for every hidden prime divisor W of R other than V, there is a nonzero element b of M for which $V \cap R[1/b]$ is contained in W. Then for every valuation domain U between R and K, there is a nonzero element b of M such that $V \cap R[1/b] \subseteq U$.*

PROOF: Let U be a valuation domain between R and K. If the height of the center of U on R is 1, then for any b in R not in this center, U contains $R[1/b]$; so we may assume that U dominates R. Take any nonzero b in M; if $B = V \cap R[1/b]$ is contained in U, we are finished, so assume not. As with

A in the proof of Proposition 3, B is a finitely generated R–algebra by [**O**, Corollary 4.4]. Thus, there is a positive integer n for which $B = R[J/b^n]$, where $J = b^n V \cap R$. There is an element b_1 of J of smallest U–value, and U contains the localization S_1 of the integral closure $R[J/b_1]'$ at the center of U. If $\dim(S_1) = 1$, then S_1 is a DVR, so $S_1 = U$ is a hidden prime divisor of R and we are finished by hypothesis. If $\dim(S_1) = 2$, then there is a hidden prime divisor W of S_1 (and hence also of R). By hypothesis, there is a nonzero element c of M such that $R[1/c] \cap V = C$ is contained in W. Let S_2 be the localization of C at the center of W. If that center had height 1, then W would be the localization there, in which case W would be either V or essential for R, neither of which is possible; so $\dim(S_2) = 2$ also. Let T be the localization of $S_1[S_2]'$ at the center of W. Then since the images of S_1 and S_2 in the residue field of W are algebraic over $k, \dim(T) = 2$. Now V is the only exceptional prime divisor for the extensions S_1/R and S_2/R, so $\operatorname{xpd}(T/S_1) = \emptyset$. Thus $S_1 = T = S_2$, and hence $C \subseteq S_2 = S_1 \subseteq U$. □

THEOREM 6. *For a hidden prime divisor V of R, the following statements are equivalent:*

(a) *There is an M–primary ideal of which V is the only exceptional prime divisor (i.e., in Muhly's terminology* [**Mu**], *V is "Noetherian").*

(b) *There is a two–generated M–primary ideal (a, b) of R for which the intersection of V and $R[1/b]$ is the integral closure of $R[a/b]$ and the intersection of V and $R[1/a]$ is the integral closure of $R[b/a]$.*

(c) *There is a two–generated M–primary ideal (a, b) of R for which the intersection of V and $R[1/b]$ is the integral closure of $R[a/b]$.*

(d) *For every hidden prime divisor W of R other than V, there is a nonzero element b of M for which the intersection of V and $R[1/b]$ is contained in W.*

PROOF: (a) $\Rightarrow$ (b): Let (a, b) be a two–generated ideal such that $T((a, b)) = \{V\}$. Then clearly the intersection of V and $R[1/b]$ is integrally closed and contains $R[a/b]$, so it remains to show that every essential prime divisor of $R[a/b]'$ also contains the intersection. Now (a, b) is also M–primary, so a is not contained in any of the height one primes of R that contain b. Thus, the essential prime divisors of R that contain $R[a/b]$ also contain $R[1/b]$. If a hidden prime divisor of R is essential in $R[a/b]'$, then it is in $T((a, b))$, so it is V. The proof of the other assertion is symmetric.

(b) $\Rightarrow$ (d): Clear, since each W contains either a/b or b/a.

(d) $\Rightarrow$ (a): By the lemma, every valuation ring between R and K contains the intersection B of V and $R[1/b]$ for some nonzero b in M, and again by [**O**, Corollary 4.4] any such intersection is a finitely generated R–algebra. Thus, in the topology on the set ("Riemann surface") of valuation rings between R and K [**ZS**, page 110], the set of valuation rings containing a given B is open, and the collection of all such sets is an open cover. Since the topology is quasicompact [**ZS**, page 113, Theorem 40], there are finitely many elements $b_1, \cdots, b_m$ for which every valuation ring between R and K contains at least one of the corresponding $B_1, \cdots, B_m$. Now for each j, there is an integer $n(j)$ for which $B_j = R\left[I_j/b_j^{n(j)}\right]$, where I_j is the intersection with R of $\left(b_j^{n(j)}\right)V$. And there are integers $e(1), \cdots, e(m)$ with $e(j) \geq n(j)$ for which $\left(b_1^{e(1)}\right)V = \cdots = \left(b_m^{e(m)}\right)V$; let I denote the intersection of this common V–ideal with R. Then for each j, $I_j b_j^{e(j)-n(j)}/b_j^{e(j)}$ is contained in $I/b_j^{e(j)}$, so B_j is contained in $R\left[I/b_j^{e(j)}\right]$, which in turn is contained in both $R[1/b_j]$ and V, and hence in B_j. Thus the union of the sets of localizations of the B_j at prime ideals is X_I, and V is the unique exceptional prime divisor of I.

(b) $\Leftrightarrow$ (c): Clear, in view of Proposition 4. □

The import of (d) $\Rightarrow$ (a) in this theorem is that, under these hypotheses, finiteness and patching in the construction of a normalized blow–up "take care of themselves". When the divisor class group of R is torsion, this fact becomes even more striking:

COROLLARY 7. *Assume that R has torsion divisor class group, and let V be a hidden prime divisor of R. If, for every element W of* hpd(R) *not equal to V, there is an element S of* $\mathrm{s}_2(R)$ *for which W dominates S and* xpd$(S/R) = \{V\}$, *then there is an M–primary ideal I of R for which $T(I) = \{V\}$.*

PROOF: By the theorem, we only need to find, for each W different from V, a nonzero element b of R for which the intersection of $R[1/b]$ and V is contained in W. Since the S dominated by W in the hypothesis above is not equal to R, there is a height one prime P of R for which S is not contained in R_P. The fact that R has torsion divisor class group means that P is the radical of a principal ideal, say bR. Since every essential prime divisor of S contains the intersection of $R[1/b]$ and V, this intersection is contained in S and hence also in W. □

Remark 8: When the equivalent conditions of Theorem 6 hold, then for *every* nonzero b in $M, V \cap R[1/b]$ is the integral closure of the extension of R by a single element of K.

Proof: By hypothesis there exists a normal M–primary ideal I such that V is the unique exceptional prime divisor of I. It is enough to show that a power of b is one of two generators of a reduction of a power of I. Replacing b and I by powers, we may assume that $v(b) = v(I)$. Defining $v\sharp$ as above, we see that $R[It]$ is the intersection of $R[t]$ and the corresponding valuation ring $V\sharp$, and that the center of $V\sharp$ on $R[It]$ is the radical of the ideal generated by M. Since $v(b) = v(I), bt$ is not in the unique minimal prime of $R[It]/MR[It]$. By Noether normalization, for some s (for $s = 1$ if k is infinite) there is an element a of I^s for which at^s and $b^s t^s$ are not both contained in any height one prime of $R[It]/MR[It]$. It follows that (a, b^s) is a reduction of I^s, and hence that $V \cap R[1/b]$ is the integral closure of $R\left[a/b^s\right]$. □

Remark 9: If V and W are hidden prime divisors of R and b is a nonzero element of M, then $V \cap R[1/b]$ is contained in W if and only if $w(b)/v(b) \leq w(x)/v(x)$ for every nonzero x in M.

Proof: Note first that $V \cap R[1/b]$ is contained in W if and only if, for every a in $R, v\left(a/b^n\right) \geq 0$ implies $w\left(a/b^n\right) \geq 0$; and since these inequalities are trivial for a zero or a unit, this is true if and only if, for every nonzero a in $M, v(a) \geq nv(b)$ implies $w(a) \geq nw(b)$, i.e., $v(a)/v(b) \geq n$ implies $w(a)/w(b) \geq n$.

Now if $w(a)/v(a) \geq w(b)/v(b)$, then $w(a)/w(b) \geq v(a)/v(b)$, and so $v(a)/v(b) \geq n$ implies $w(a)/w(b) \geq n$. Thus the implication $(\Leftarrow)$ is clear.

Conversely, suppose there is a nonzero x in M for which $w(x)/v(x) < w(b)/v(b)$. Then $w(x)/w(b) < v(x)/v(b)$, so for some positive integer $m, m\left[(v(x)/v(b)) - (w(x)/w(b))\right] > 1$. Thus there is a positive integer n for which $mw(x)/w(b) < n \leq mv(x)/v(b)$. It follows that $v\left(x^m\right)/v(b) \geq n$, but $w\left(x^m\right)/w(b) < n$. i.e., x^m/b is in both $R[1/b]$ and V, but not in W. □

It is a consequence of Izumi's theorem [**R**] that, for hidden prime divisors V and W of R, the set of quotients $w(x)/v(x)$, as x varies over the nonzero elements of M, is bounded below by a positive real number. Remark 9 shows that $V \cap R[1/b]$ is contained in W if and only if $w(b)/v(b)$ attains the greatest lower bound of this set of rational numbers. Thus, the hidden prime divisor V of R is the unique exceptional prime divisor of an M–primary ideal if and only if, for every other hidden prime divisor W, the set $\{w(x)/v(x) | x \text{ in } M, x \neq 0\}$ includes its greatest lower bound.

4. Conditions (N) and (E).

The ring R (with the standing hypotheses) is said to satisfy "condition (N)" if, for every hidden prime divisor V of R, there is an M–primary ideal of which V is the unique exceptional prime divisor. It is immediate from Theorem 6 and Remark 9 that:

REMARK 10: The ring R satisfies condition (N) if and only if, for all hidden prime divisors V and W of R, the set $\{w(x)/v(x) | x \text{ in } M, x \neq 0\}$ includes its greatest lower bound. □

A (two–dimensional excellent) regular local ring satisfies condition (N). Muhly [**Mu**] showed that condition (N) implies a factorization theory for M–primary ideals that extends some aspects of and gives another proof for Zariski's factorization theory for complete ideals in two–dimensional regular local rings. Göhner [**Go**] related condition (N) to the property that the divisor class group is torsion. An initial connection is that, if the divisor class group is torsion, then the normalized blow–up of any ideal is the normalized blow–up of an M–primary ideal. (For, in this case, any ideal I has a power that is the product of an M–primary ideal J and a principal ideal, so that $X_I = X_J$.) The following is an alternative proof for Göhner's Theorem 4.6 (page 426), in which he showed that more is true when R is complete:

THEOREM 11. *(Göhner) If R is complete and has torsion divisor class group, then R has condition (N).*

REMARK: The excellence of R is one of our standing hypotheses, but it also follows from the assumption of completeness in this case.

PROOF: Let V be a hidden prime divisor of R. Take a nonzero b in M, and let $A = V \cap R[1/b]$; as in the proof of Proposition 3, A is a finitely generated R–algebra. Let P denote the center of V on A. Since A/P is a finitely generated k–algebra with field of fractions the residue field of V, it is not algebraic over k, so it is not a field, so P is not maximal in A. Let Q

be a maximal ideal of A containing P, and set $S = A_Q$. Then S is in $\mathbf{s}_2(R)$ and $\operatorname{xpd}(S/R) = \{V\}$. Denote the maximal ideal of S by N.

We claim that it will suffice to find a in R so that (replacing b by one of its powers) a/b is an element of N residually transcendental for V and the principal ideal aR has prime radical that does not contain b. Let us first show that it is possible to find such an a: Take an element a/b^n in Q but not in P. Since R has torsion divisor class group, there are elements $a_1, \cdots, a_s$ of R and a positive integer m such that $a^m = a_1 \cdots a_s$ and each principal ideal $a_i R$ has prime radical. Let $t, u(1), \cdots, u(s)$ be the V–values of $b, a_1, \cdots, a_s$ respectively; then since a/b^n has V–value $0, nmt = u(1) + \cdots + u(s)$, so we get

$$(a/b^n)^{mt} = \left(a_1^t/b^{u(1)}\right) \cdots \left(a_s^t/b^{u(s)}\right).$$

Each $a_i^t/b^{u(i)}$ is in A, none is in P, and at least one is in Q. That one is a nonzero nonunit in S/PS and hence residually transcendental for V. Finally, the localization W of R at the radical of $a_i R$ is the only essential prime divisor of R at which $a_i^t/b^{u(i)}$ can have positive value. If b has positive W–value, then W is not an essential prime divisor of A; so that $a_i^t/b^{u(i)}$ is a unit of A, contradicting the fact that $a_i^t/b^{u(i)}$ is in Q. Thus, b is not in the radical of $a_i R$. Write a for a_i^t and b for $b^{u(i)}$.

Our next objective is to show that the ideal of $R[a/b]$ generated by M and a/b is contained in only one maximal ideal M^* of the integral closure $R[a/b]'$. By [**EGA**, Lemma 23.2.7.1, page 219], it is enough to show that the completion of $R[a/b]_{(M,a/b)}$ has quasilocal integral closure; and for this we only need to show that this completion is a domain [**N**, (30.3) and (43.12)]. (See also [**K**].) Now since R is complete and $R[a/b] = R[x]/(bx - a)$, this completion is $R[[x]]/(bx - a)$; so we want to show that $(bx - a)R[[x]]$ is a prime ideal. For this, we consider the height one primes p of R: If p is not the radical of aR, then $bx - a$ is a unit in $R_p[[x]]$; and if p is that radical, then b is a unit in R_p, so $bx - a$ is a regular parameter in the two–dimensional regular local ring $R_p[[x]]$. Thus, the intersection $(bx - a)R[[x]]$ of the sets $(bx - a)R_p[[x]]$ (in $K[[x]]$, where K is the field of fractions of R) is prime.

Now the center on $R[a/b]'$ of any exceptional prime divisor of (a, b) meets $R[a/b]$ in the extension of M, so this center is contained in M^*. Similarly, $R[a/b]'$ is contained in S, and the intersection of N with $R[a/b]'$ is a maximal ideal lying over $(M, a/b)R[a/b]$, so this intersection is M^*; and by ZMT, $R[a/b]'_{M^*} = S$. Thus, every exceptional prime divisor of (a, b) is one for the extension S/R, and V is the only such. □

The ideas used in the proof above also yield a result on an individual two–generated ideal:

PROPOSITION 12. *Let R be complete and equicharacteristic, and (a,b) be M–primary. If R contains a field F such that, for infinitely many elements f of F, $(a+fb)R$ has prime radical in R, then (a,b) has a unique exceptional prime divisor.*

PROOF: It suffices to prove that there is only one prime of the integral closure $R[a/b]'$ lying over $MR[a/b]$, so assume not. Then the sum J of the primes of $R[a/b]'$ lying over $MR[a/b]$ is contained in only finitely many maximal ideals. But any maximal ideal of $R[a/b]'$ that contains $MR[a/b]$ and is unique in lying over its intersection with $R[a/b]$ contains J, by Lying Over; so all but finitely many of the maximal ideals of $R[a/b]$ that contain $MR[a/b]$ split in $R[a/b]'$. On the other hand, the proof of the theorem shows that, if the radical of $(a+fb)R$ is prime, then $(M,(a+fb)/b)R[a/b]$ does not split in $R[a/b]'$, a contradiction. $\square$

Göhner conjectured that, if R has condition (N), then it has torsion divisor class group. (Cutkosky [**Cu**] has recently shown that this conjecture is true if the residue field k of R is algebraically closed.) Göhner showed that, when R is not complete, it may have torsion, or even trivial, divisor class group and not satisfy condition (N) (as in Example 16 below); but R satisfies condition (N) if and only if its completion does. In this connection we would like to ask: If R has torsion divisor class group, must the same be true of an element S of $\mathbf{s}_2(R)$? It *is* true if $\mathrm{xpd}(S/R)$ has only one element. (For, there is a height one prime P of R for which S is not contained in R_P, and so $PS \cap R$ is M–primary. Now since R has torsion divisor class group, P is the radical of a principal ideal aR. The radical Q of aS is a height one prime in S, the center of the exceptional prime divisor of S/R. The divisor class group of S is the extension of the divisor class group of $S[1/a]$ by the class of Q [**F**, Corollary 7.2, page 36], and both of the latter are torsion (since $S[1/a]$ is a localization of $R[1/a]$); so the divisor class group of S is torsion.) Thus, if R satisfies condition (N) and has torsion divisor class group, then any S in $\mathbf{s}_2(R)$ also has torsion divisor class group. For, if $\mathrm{xpd}(S/R) = \{V_1, \cdots, V_n\}$, then using condition (N) we can find elements $S_2, \cdots, S_n$ of $\mathbf{s}_2(R)$ for which $R = S_1 \subset S_2 \subseteq \cdots \subseteq S_n \subseteq S_{n+1} = S$ and $\mathrm{xpd}\,(S_{i+1}/S_i) = \{V_i\}$. Then we can apply repeatedly the fact for a single exceptional prime divisor.

Condition (N) guarantees the existence of M*-primary* ideals with given exceptional prime divisors; we would like to consider the apparently weaker condition:

(E) for every hidden prime divisor V of R, there is an ideal of R of which V is the unique exceptional prime divisor.

In view of Proposition 4 and the fact that we need only look at two–generated ideals, condition (E) is equivalent to the condition that, for every hidden prime divisor V of R, there are elements a and b of R for which a/b is residually transcendental for V but not for any other hidden prime divisor of R. Also, if R satisfies condition (E) and S is in $\mathbf{s}_2(R)$, then S also satisfies condition (E). If R has torsion divisor class group and satisfies condition (E), then R satisfies condition (N), because, as mentioned above, in this case every nonprincipal ideal shares its normalized blow–up with an M–primary ideal. Hence if R satisfies condition (E), then any S in $\mathbf{s}_2(R)$ with torsion divisor class group satisfies condition (N). Cutkosky has recently shown that if R is complete, then R satisfies condition (E).

Let us diagram some of the implications among these concepts, those we have just mentioned and one that is discussed below. We denote by (A) the property that every domain in $s_2(R)$ except R is in the normalized blow–up of some M–primary ideal. Then:

$$\begin{array}{ccccc} \text{regular} & \longrightarrow & (N) & \longrightarrow & (A), \\ & & \downarrow & & \\ \text{complete} & \longrightarrow & (E) & & \end{array}$$

and if the divisor class group is torsion, then

$$\text{complete} \rightarrow (N) \leftrightarrow (E).$$

Condition (E) has some of the useful properties of condition (N). For example:

(1) Suppose R satisfies condition (E) and V and W are hidden prime divisors of R. Then

(a) choosing an ideal I for which $T(I) = \{V\}$, the element S of X_I dominated by W is such that $\operatorname{xpd}(S/R) = \{V\}$; or, on the other hand,

(b) by Zariski's Connectedness Theorem [**Ha**, Corollary 11.4, page 280 and Exercise 11.4, page 281], there is an element S of $\mathbf{s}_2(R)$ for which $\operatorname{xpd}(S/R) = \{V, W\}$.

(We remark that the S in (a) is unique, but the S in (b) need not be. Also, it is not always possible, given *three* hidden prime divisors, to find an element of $\mathbf{s}_2(R)$ for which all three are exceptional.)

(2) Suppose R satisfies condition (E), and consider any finite set of hidden prime divisors of R. Taking for each element of the set an ideal of which it is the unique exceptional prime divisor, we find that the product of these ideals has as its set of exceptional prime divisors precisely the given set of hidden prime divisors.

(3) Suppose R satisfies condition (E) and S_2 is an element of $\mathbf{s}_2(R)$ for which $\operatorname{xpd}(S_2/R) = \{V\}$ and $W \in \operatorname{hpd}(S_2)$. Let U be a rank two valuation ring contained in V that dominates S_2, choose an ideal I for which $T(I) = \{W\}$ and let S_1 denote the element of X_I dominated by U. Then by the last sentence of Proposition 2, $S_1 \cap S_2 = S$ is an element of $\mathbf{s}_2(R)$ for which $\operatorname{xpd}(S/R) = \emptyset$, so $S = R$.

(4) Suppose R satisfies condition (E). Then we can add to the last sentence of Proposition 2: If S_1 and S_2 in $\mathbf{s}_2(R)$ satisfy the conditions of that proposition and $S = S_1 \cap S_2$, then $\operatorname{xpd}(S_1/R) \cap \operatorname{xpd}(S_2/R) \subseteq \operatorname{xpd}(S/R)$. To see this, pick an ideal I for which $T(I) = \operatorname{xpd}(S_1/R) \cap \operatorname{xpd}(S_2/R)$. Since every exceptional prime divisor of S_1 and S_2 is an exceptional prime divisor of I, S_1 and S_2 dominate elements, say T_1 and T_2, of X_I [**Go**, Proposition 2.2, page 412]. Take hidden prime divisors W_1 and W_2 of T_1 and T_2 respectively, and choose b in I^n so that $bW_i = I^n W_i$ for $i = 1, 2$ and $bV = I^n V$ for each V in $T(I)$. Then T_1 and T_2 are both localizations of $R[I^n/b]'$. But since S is local, so is $T_1 \cap T_2$, and this is only possible if $T_1 = T_2$. Thus every exceptional prime divisor of I is also exceptional for S/R.

If, for all hidden prime divisors V and W of R, there is an S in $\mathbf{s}_2(R)$ for which V is in $\operatorname{xpd}(S/R)$ and W dominates S, must R satisfy condition (E)? If R has torsion divisor class group, then R satisfies condition (E) if and only if it satisfies condition (N); so, by Corollary 7, this question has a positive answer if R has torsion divisor class group.

Further investigation of condition (E) seems to require more information about an ideal that is not M-primary but has a unique exceptional prime divisor. A natural candidate for such an ideal is a height one prime of R, as in the following proposition:

PROPOSITION 13. *Suppose that (T, N) is a three-dimensional regular local ring with a regular system of parameters $\{A, B, C\}$, and that p is an element of the ideal $(A, B)T$. Assume that p is in N^2, not in N^3, and that $T/(p) = R$ is a two-dimensional normal local domain. Then the (height one prime) ideal of R generated by the images a and b of A and B has a unique exceptional prime divisor.*

PROOF: Write $p = rA + sB$. Then r and s are in N, but, since R is normal and hence $R_{(a,b)} = T_{(A,B)}/(p)$ is a DVR, r and s are not both in (A, B). Since the height one prime ideal (a, b) of the normal local domain (R, M) is not principal, a and b are analytically independent [**NR**, p.149]. Therefore the R-homomorphism of $R[X]$ (X an indeterminate) onto $R[a/b]$ that maps X to a/b has kernel contained in $MR[X]$. It follows that the natural map $T[A/B] \to R[a/b]$ takes $NT[A/B]$ onto $MR[a/b]$ so it extends to a map $T[A/B]_{NT[A/B]} \to R[a/b]_{MR[a/b]}$. Now the extension of N to $T[A/B]$ is generated by B and C, so $T[A/B]_{NT[A/B]}$ is a two-dimensional regular local ring, and the kernel of this map is generated by the regular parameter $r(A/B)+s$. Thus $R[a/b]_{MR[a/b]}$ is a DVR and hence the unique exceptional prime divisor of (a, b). $\square$

Continuing the discussion (and maintaining the hypotheses and notation) of this proposition for a moment, we would like to see how "close" $X_{(a,b)}$ is to being the normalized blow-up of an M-primary ideal. By [**B**], $R[at, bt]$, for t an indeterminate, is Cohen-Macaulay, so the same is true of the homogeneous localizations $R[a/b]$ and $R[b/a]$; so these rings are the intersections of their localizations at height one primes. Most of these height one primes meet R in height one primes; the height one primes of $R[a/b]$ and $R[b/a]$ that meet R in M are the extensions of M to these rings, and, as we saw in the proof of Proposition 13, the localization $R[a/b]_{MR[a/b]} = R[b/a]_{MR[b/a]}$ is a DVR. Thus, $R[a/b]$ and $R[b/a]$ are normal. So $X_{(a,b)}$ is just the union of the spectra of $R[a/b]$ and $R[b/a]$, and a/b is residually transcendental for the exceptional prime divisor V of (a, b). Since $R/(a, b) = T/(A, B)$ is a DVR, only one of the two-dimensional elements of $X_{(a,b)}$ is contained in the DVR $R_{(a,b)}$. For any *other* two-dimensional element S of $X_{(a,b)}$, the intersection of $(a, b)S$ with R properly contains $(a, b)R$; we can then pick d

in that intersection not in any height one prime containing b and (replacing d with $a+ud, u$ a unit in R, if necessary) we obtain a d for which d/b is residually transcendental for V. Then S is on the normalized blow–up of (d,b). So only one of the two–dimensional elements of $X_{(a,b)}$ (the one contained in the DVR $R_{(a,b)}$) can fail to be in the normalized blow–up of an M–primary ideal.

If S in $\mathbf{s}_2(R)$ is such that $\operatorname{xpd}(S/R) = \{V\}$ and S is on X_I for an M–primary ideal I, then IS is principal and the center of V on S is the only height one prime containing the generator of IS; so that center is the radical of a principal ideal. The next lemma, whose statement and proof are only small modifications of those of [**Go**, Theorem 2.11], is another variation on the theme "unique exceptional prime divisor implies a prime is the radical of a principal". In its proof, we use the following fact: If J is an ideal of R and S in $\mathbf{s}_2(R)$ is not dominated by any exceptional prime divisor of J, then JS is a principal ideal. For, by taking a two–generated reduction of J and applying Proposition 4, we see that the exceptional prime divisors of JS are among those of J. Thus, there is no hidden prime divisor of S that becomes essential in an element of the normalized blow–up of JS over S; and by (a) of Proposition 1 it follows that the normalized blow–up of JS is $\operatorname{Spec}(S)$, i.e., JS is principal.

LEMMA 14. *(Göhner) Let V and W be distinct hidden prime divisors of R, the unique exceptional prime divisors of M–primary ideals I and J respectively. Assume I is normal, and let P be the homogeneous (height two) prime ideal of $R[It]$ for which the homogeneous localization $S = R[It]_{(P)}$ is dominated by W, and let P^* be its image in $A = R[It]/MR[It]$. Then there is a homogeneous element x^* of A for which P^* is the only height one prime containing x^*.*

PROOF: Replacing J with a power, if necessary, we may assume that $v(J) = v(I^n)$, where v denotes the valuation associated to V. Now for any two–dimensional element $\tilde{S}$ of X_I other than $S, \tilde{S}$ is not dominated by (or even contained in) W, so by the discussion above $J\tilde{S}$ is principal. But that implies $\tilde{S}$ dominates an element of X_J, and since $\tilde{S}$ is contained in V, it dominates the same (necessarily two–dimensional) element T of X_J as does V. Let x be an element of J that generates JT. We claim that the image x^* of xt^n in A has the desired property. Note that x also generates $J\tilde{S}$ for every $\tilde{S}$ in X_I other than S. In particular, $v(x) = v(J) = v(I^n)$. If $\tilde{Q}$ is a height one prime of $x\tilde{S}$, then the intersection of $\tilde{Q}$ with R contains J

and so is M; so $\tilde{Q}$ is the center of V on $\tilde{S}$. Since $v(x) = v(I^n), x\tilde{S} = I^n\tilde{S}$.

Now let $\tilde{P}^*$ be any homogeneous height one prime of A other than P^*; let $\tilde{P}$ be its preimage in $R[It]$, and let $\tilde{S} = R[It]_{(\tilde{P})}$. Since $\tilde{P}$ is prime and does not contain all of It, it does not contain all of $I^n t^n$, so there is an element y of I^n for which yt^n is not in $\tilde{P}$. Now since $x\tilde{S} = I^n\tilde{S}$, there are elements u and v of the same power of I, v not in $\tilde{P}$, for which $y = xu/v$; and so $xt^n u = yt^n v$, and hence xt^n, are not in $\tilde{P}$. Thus, x^* is not in $\tilde{P}^*$. Finally, since x^* is homogeneous and is not contained in the height zero prime ideal of A (which is determined by $V\sharp$), x^* is not contained in any nonhomogeneous height one prime of A. □

In preparation for our final example, we recall and extend some ideas from **[Go]**. Consider a faithfully flat extension (R^*, M^*) of R, also an excellent two–dimensional normal local domain, of which R is a dense subspace. (For instance, R^* might be the completion of R.) The hidden prime divisors of R are in one–to–one correspondence with those of R^*, via intersection of the latter with K or extension of the former to valuation domains dominating R^* within its field of fractions; and for an M–primary ideal I, the exceptional prime divisors of I correspond to those of the M^*–primary ideal IR^* [**Go**, Lemma 1.10]. Now we consider two–dimensional spots: For (S, N) in $\mathbf{s}_2(R)$, note that $S[R^*]/N^n S[R^*] = S/N^n$ for each positive integer n [**L2**, page 161], so $NS[R^*]$ is prime. Set $S^* = S[R^*]_{NS[R^*]}$. Then it can be checked that S is a dense subspace of S^*; so S and S^* have the same completion, which, by the excellence of S, is a two–dimensional normal local domain. Thus, S^* is in $\mathbf{s}_2(R^*)$ and meets K in S; and clearly S^* is minimal among the rings with these properties. Conversely, for T^* in $\mathbf{s}_2(R^*)$, set $T^* \cap K = T$; then T is a quasilocal Krull domain dominating R. Also, since $\text{hpd}(T^*)$ is an infinite subset of $\text{hpd}(R^*), T$ is birationally dominated by many elements of $\text{hpd}(R)$, so T has dimension 2; i.e., T is in $\mathbf{s}_2(R)$. Moreover, $\text{xpd}(T/R) \subseteq \{V^* \cap K | V^* \text{ in } \text{xpd}(T^*/R^*)\}$; and if T^* is in the normalized blow–up over R^* of an M^*–primary ideal, then it is the S^* described above, where S is the intersection of T^* and K, and S is on the normalized blow–up over R of an M–primary ideal.

It is natural to ask when the surjection $\mathbf{s}_2(R^*) \to \mathbf{s}_2(R)$ given by intersection with K is a bijection. Since M^*–primary ideals of R^* are the extension of M–primary ideals of R, a sufficient condition that the map $\mathbf{s}_2(R^*) \to \mathbf{s}_2(R)$ is a bijection is that each element of $\mathbf{s}_2(R^*)$ is on the normalized blow–up of an M^*–primary ideal. This condition is also necessary if every element of $\mathbf{s}_2(R)$ is on the normalized blow–up of an M–primary ideal. (For, if (S, N) is on X_I, then the S^* described above is on the normalized blow–up $X^*_{IR^*}$ of IR^* over R^*.) Thus, it is also natural to ask: If R is complete and every element of $\mathbf{s}_2(R)$ is on the normalized blow–up of an M–primary ideal, does it follow that R satisfies condition (N)?

LEMMA 15. *Suppose R has torsion divisor class group and is a dense subspace of the excellent two–dimensional normal local domain (R^*, M^*), which is faithfully flat over R. Suppose also that the ideals I in R and J^* in R^* have unique exceptional prime divisors V and W^* respectively, that I is M–primary, and that W^* is not the unique exceptional prime divisor of an M^*–primary ideal. Finally, let $W = W^* \cap K$, and suppose that the element S_1 of X_I dominated by W has condition (N). Then there is no element S_2 of $\mathbf{s}_2(R)$ for which V dominates S_2 and* $\operatorname{xpd}(S_2/R) = \{W\}$.

PROOF: Assume that such an S_2 exists. We will show that, not only V, but every hidden prime divisor other than W has a similar spot, i.e., for every $\tilde{V}$ in $\operatorname{hpd}(R)$ different from W, we can find an element $\tilde{S}$ in $\mathbf{s}_2(R)$ dominated by $\tilde{V}$ and for which $\operatorname{xpd}\left(\tilde{S}/R\right) = \{W\}$. By Corollary 7, it will follow that there is an M–primary ideal of which W is the only exceptional prime divisor; and so there is an M^*–primary ideal of which W^* is the only exceptional prime divisor, the desired contradiction.

Every hidden prime divisor $\tilde{V}^*$ of R^* other than W^* dominates a two–dimensional element $\tilde{S}^*$ of the normalized blow–up $X^*_{J^*}$ of J^* over R^*. If $\tilde{S}^*$ is not the element S^*_2 of $X^*_{J^*}$ dominated by V^*, then $\tilde{S}^*$ is contained in W^*, so it dominates the same element S^*_1 of $X^*_{IR^*}$ as does W^*. Thus $\tilde{S} = \tilde{S}^* \cap K$ dominates S_1 (which is $S^*_1 \cap K$), and since S_1 has condition (N), $\tilde{S}$ is in the normalized blow–up over S_1 of an ideal primary for the maximal ideal of S_1. Therefore $\tilde{V} = \tilde{V}^* \cap K$ dominates $\tilde{S}$, and $\operatorname{xpd}\left(\tilde{S}/R\right) = \{W\}$, so the proof is complete. □

A final bit of preparation: Let k denote the field of complex numbers and t be an indeterminate. Then any maximal ideal M in $k\left[t^2, t^3\right]$ is the intersection of this ring and the set of multiples in $k[t]$ of $t - c$ for some c in k. Assume M is the radical of a principal ideal, generated by $g(t)$, and $c \neq 0$. Then $g(t)$ divides powers of both $t^2 - c^2$ and $t^3 - c^3$; and since $t + c$ and $t^2 + ct + c^2$ are relatively prime in $k[t], g(t)$ divides a power of $t - c$ in $k[t]$, i.e., $g(t)$ is a power of $t - c$. But no power of $t - c$ lies in $k\left[t^2, t^3\right]$. So $\left(t^2, t^3\right)$ is the only maximal ideal of $k\left[t^2, t^3\right]$ that is the radical of a principal ideal.

Now all the scenery is on stage for the final example. In it, the complete ring $\hat{R}$ is such that (1) there is a hidden prime divisor that is the unique exceptional prime divisor of a height one prime, but not of any $\hat{M}$–primary ideal; in fact, (2) $\hat{R}$ satisfies condition (E) but not condition (N). Moreover, (3) the surjection $s_2\left(\hat{R}\right) \to s_2(R)$ is not a bijection; indeed, R has more than one preimage. Thus, also (4) there is an element of $s_2\left(\hat{R}\right)$ that is not on the normalized blow–up of any $\hat{M}$–primary ideal.

EXAMPLE 16: Let k denote the field of complex numbers and consider $R = k[x,y,z]_{(x,y,z)}$, subject to the relation $z^2 = x^3 + y^7$. (This R, as a UFD, has torsion divisor class group; but its completion has elements of infinite order in its divisor class group, and neither ring satisfies condition (N). Cf. [**L1**, pages 197–198].) We show first that the ideal $I = \left(x, y^2\right)$ has a unique exceptional prime divisor; i.e., that the integral closure $R\left[x/y^2\right]'$ has a unique height one prime lying over M. To see this, we observe that $R\left[x/y^2\right]' = R\left[x/y^2, z/y^3\right]$. The equation $\left(z/y^3\right)^2 = \left(x/y^2\right)^3 + y$ shows that z/y^3 is integral over $R\left[x/y^2\right]$, so it suffices to show that $R\left[x/y^2, z/y^3\right]$ is a (locally) regular ring. Localizing at a prime that lies over a height one prime of R yields a DVR, so we consider a prime P of $R\left[x/y^2, z/y^3\right]$ that lies over M. Since $\left(R\left[x/y^2, z/y^3\right]\right) / \left(MR\left[x/y^2, z/y^3\right]\right)$ is isomorphic to the domain $k\left[t^2, t^3\right]$ (t an indeterminate), the only height one prime of $R\left[x/y^2, z/y^3\right]$ that lies over M is the extension of M, which is contained in the ideal generated by $M, x/y^2$, and z/y^3, so we may assume P has height 2. Since R is the image of a three–dimensional regular local ring T, there is a prime Q of $T[U,V]$ (U and V indeterminates) such that Q contains the maximal ideal (X,Y,Z) of T and $R\left[x/y^2, z/y^3\right]_P$ is the image of $T\left[U,V\right]_Q$, a regular local ring in which $Y^2U - X, Y^3V - Z$ and $V^2 - U^3 - Y$ are part of a regular system of parameters. Since these elements are in the kernel of the map from $T[U,V]_Q$ onto $R\left[x/y^2, z/y^3\right]_P$, the latter is regular. The extension of M to $R\left[x/y^2\right]'$ is the only height one prime lying over M; so, the localization V of $R\left[x/y^2\right]'$ at the extension of M is the unique exceptional prime divisor of I.

Now consider the completion $\hat{R} = k\left[[x,y,z]\right]$ of R. The argument above also shows that the ideal $I\hat{R}$ has a unique exceptional prime divisor V^*; indeed, $V = V^* \cap K$. In $\hat{R}$, the polynomial $x^3 - x^2y^2 + y^7$ factors as two power series. (To see this, give x and y the weights 2 and 1 respectively; then the leading form $x^3 - x^2y^2$ factors into the coprime factors x^2 and $x - y^2$.) Let f denote the factor that begins $x - y^2 + y^3 + \cdots$. Then $P^* = (z - xy, f)$ is a height one prime of $\hat{R}$, and by Proposition 13 (with A, B and C the preimages of $z - xy, f$ and y in the formal power series ring $k[[X,Y,Z]] = k[[A,B,C]]$), P^* has a unique exceptional prime divisor $W^* = \hat{R}[(z-xy)/f]_{\hat{M}\hat{R}[(z-xy)/f]}$, and W^* is not equal to V^*. Since $w^*(z^2 - x^3 - y^7) = \infty$, we see that $w^*(z) > w^*(x)$ and $w^*(z) > w^*(y)$. Thus, $w^*(f) = w^*(z - xy) \geq \min\{w^*(z), w^*(xy)\} > \min\{w^*(x), w^*(y^2)\}$, so we must have $w^*(x) = w^*(y^2)$. Then, since $w^*(x^3) < w^*(y^7)$ and $w^*(z^2 - x^3 - y^7) = \infty$, we see that $w^*(z^2) = w^*(x^3) = w^*(y^6)$. Thus, x/y^2 and z/y^3 are units in W^*. It follows that W^* contains $\hat{R}[x/y^2, z/y^3]$. The center of W^* on $\hat{R}[x/y^2, z/y^3]$ is a maximal ideal different from $(\hat{M}, x/y^2, z/y^3)$, so the image of that center in $\hat{R}[x/y^2, z/y^3]/\hat{M}\hat{R}[x/y^2, z/y^3] = k[t^2, t^3]$ is not the radical of a principal ideal. By Lemma 14, W^* is not the unique exceptional prime divisor of any $\hat{M}$–primary ideal of $\hat{R}$.

Let $W = W^* \cap K$, and let S_1 be the element of X_I dominated by W. Since W contains the regular ring $R[x/y^2, z/y^3]$, S_1 is a localization of $R[x/y^2, z/y^3]$ and hence is regular. Let S_1^* be the "minimal extension" of S_1 to $\mathbf{s}_2(\hat{R})$, in the sense of the discussion above. Since S_1^* is regular and $W^* \in \text{hpd}(S_1^*)$, there is an ideal J^* of S_1^*, primary for the maximal ideal, for which W^* is the only exceptional prime divisor. Let S_2^* and T^* be the centers of V^* on $X_{P^*}^*$ and the normalized blow–up of J^* over S_1^* respectively. Then $\text{xpd}(S_1'/\hat{R}) = \{V^*\}$ and $\text{xpd}(T^*/S_1^*) = \{W^*\}$, while symmetrically $\text{xpd}(S_2^*/\hat{R}) = \{W^*\}$ and $\text{xpd}(T^*/S_2^*) = \{V^*\}$. But when we intersect all these domains with K, the symmetry fails: Lemma 15 implies that there does not exist an element S_2 of $\mathbf{s}_2(R)$ for which $\text{xpd}(S_2/R) = \{W\}$ and V dominates S_2; so $S_2^* \cap K = R$. It follows that for this R and $\hat{R}$ the map $\mathbf{s}_2(\hat{R}) \to \mathbf{s}_2(R)$ is not bijective.

Thus setting $T = T^* \cap K$, we have $\text{xpd}(S_1/R) = \{V\}$ and $\text{xpd}(T/S_1) = \{W\}$, but there is no element S_2 of $\mathbf{s}_2(R)$ for which $\text{xpd}(S_2/R) = \{W\}$ and $\text{xpd}(T/S_2) = \{V\}$.

Acknowledgement.

We would like to thank Professor Craig Huneke for helpful conversations on the material in this paper, and the referee for a careful reading of the paper and several helpful suggestions.

REFERENCES

[A1] S. Abhyankar, "Resolution of Singularities of Embedded Algebraic Surfaces," Academic Press, New York, 1966.

[A2] S. Abhyankar, *Quasi-rational singularities*, Amer. J. Math. **100** (1978), 267–300.

[B] M. Brodmann, *Rees rings and form rings of an almost complete intersection*, Nagoya Math. J. **88** (1982), 1–16.

[Co] I. Cohen, *Lengths of prime ideal chains*, Amer. J. Math. **76** (1954), 654–668.

[Cu] S. Cutkosky, *On unique and almost unique factorization of complete ideals*, (preprint).

[F] R. Fossum, "The Divisor Class Group of a Krull Domain," Springer–Verlag, New York, 1973.

[Gi] R. Gilmer, "Multiplicative Ideal Theory," Marcel Dekker, New York, 1972.

[Go] H. Göhner, *Semifactoriality and Muhly's condition (N) in two-dimensional local rings*, J. Algebra **34** (1975), 403–429.

[EGA] A. Grothendieck and J. Dieudonné, *Eléments de Géométrie Algébrique*, Publ. Math. I. H. E. S. **20** (1964), 101–346.

[Ha] R. Hartshorne, "Algebraic Geometry," Springer–Verlag, New York, 1977.

[He] W. Heinzer, *On Krull overrings of a Noetherian domain*, Proc. Amer. Math. Soc. **22** (1969), 217–222.

[HHS] W. Heinzer, C. Huneke and J. Sally, *A criterion for spots*, J. Math. Kyoto Univ. **26** (1986), 667–671.

[HS] C. Huneke and J. Sally, *Birational extensions in dimension two and integrally closed ideals*, J. Algebra (to appear).

[K] D. Katz, *On the number of minimal prime ideals in the completion of a local domain*, Rocky Mountain J. Math **16** (1986), 575–578.

[L1] J. Lipman, *Introduction to resolution of singularities*, in "Proceedings of Symposia in Pure Mathematics," American Mathematical Society, Providence, Rhode Island, 1975.

[L2] J. Lipman, *Desingularization of two-dimensional schemes*, Ann. of Math. **107** (1978), 151–207.

[L3] J. Lipman, *Equimultiplicity, reduction, and blowing up*, in "Commutative Algebra: Analytic Methods", Lecture Notes in Pure and Applied Mathematics **68**, Marcel Dekker, New York, 1982.

[Ma] H. Matsumura, "Commutative Algebra," second edition, Benjamin/Cummings, Reading, Massachusetts, 1980.

[Mu] H. Muhly, *On the existence of asymptotically irreducible ideals*, J. London Math. Soc. **40** (1965), 99–107.

[MS] H. Muhly and M. Sakuma, *Asymptotic factorization of ideals*, J. London Math. Soc. **38** (1963), 341–350.

[N] M. Nagata, "Local Rings," Interscience, New York, 1962.

[NR] D. Northcott and D. Rees, *Reductions of ideals in local rings*, Proc. Cambridge Phil. Soc. **50** (1954), 145–158.

[O] N. Onoda, *Subrings of finitely generated rings over a pseudo-geometric ring*, Japan. J. Math. **10** (1984), 29–53.

[R] D. Rees, *Izumi's theorem*, Proc. Microprogram on Commutative Algebra (1987) (to appear), M.S.R.I..

[Sy1] J. Sally, *One-fibered ideals*, Proc. Microprogram on Commutative Algebra (1987) (to appear), M. S. R. I..

[Sy2] J. Sally, *Nonsplitting of prime divisors*, Proc. Camb. Phil. Soc. (to appear).

[Sm] P. Samuel, *Some asymptotic properties of powers of ideals*, Ann. of Math. **56** (1952), 11–21.

[**T**] J. Tate, *The arithmetic of an elliptic curve*, Inv. Math. **23** (1974), 179–206.
[**ZS**] O. Zariski and P. Samuel, "Commutative Algebra, vol. II," Van Nostrand, Princeton, New Jersey, 1960.

Purdue University, West Lafayette, IN 47907

Colgate University, Hamilton, NY 13346

Prof. Heinzer gratefully acknowledges the support of NSF Grant DMS-8521767A1, and Prof. Lantz the hospitality of Purdue University while this work was done.

Tight Closure

MELVIN HOCHSTER AND CRAIG HUNEKE

1. Introduction.

Throughout this paper all rings are commutative, with identity, and Noetherian, unless otherwise specified. We will summarize many of the results in [**H-H**] concerning the theory of tight closure and prove several basic theorems using this theory in characteristic p, including the theorem of Briançon-Skoda that the integral closure of the nth power of an n-generator ideal of a regular ring is contained in the ideal, the monomial conjecture, the syzygy theorem, and that summands of regular rings are Cohen-Macaulay (C-M).

We only define the tight closure of an ideal (or submodule) in characteristic p or for algebras which are essentially of finite type over a field of characteristic 0. However many of our results which use tight closure but do not specifically refer to tight closure (e.g., that summands of regular rings are C-M) can be proved for any ring containing a field by using the Artin approximation theorem to reduce to the case of algebras essentially of finite type over a field. It would be of great importance if the notion of tight closure could be extended to mixed characteristic.

We summarize some of the salient features of the tight closure. First of all and most importantly, every ideal of a regular ring is tightly closed. We call the Noetherian rings all of whose localizations have this property *F-regular*. The 'F' stands for Frobenius for reasons which will be obvious when the definition is given. This class of rings is closed under taking direct summands and so includes summands of regular rings. As it turns out, under weak conditions (e.g., being a homomorphic image of a C-M ring) any F-regular ring is C-M. Also F-regular rings are always normal. These rings are closely related to those with rational singularities, at least in the case of an isolated singularity in characteristic 0 (see Theorem 4.3).

Suppose that R is local and that $x_1, \ldots, x_n$ are part of a system of parameters of R. A key property of the tight closure is that the ideal $(x_1, \ldots, x_{n-1}) : x_n$ is contained in the tight closure of $(x_1, \ldots, x_{n-1})$ under mild conditions on R. In general, the tight closure captures the cohomological obstructions to Cohen-Macaulayness. Also the tight closure of an ideal is always contained in the integral closure. Combining this observation with the above remarks yields that certain sequences of operations (such as intersection, colon, sum and product) on ideals generated by monomials in systems of parameters are always contained in the tight closure (where it makes sense to speak of the tight closure) of the 'expected answer' one would get in a C-M ring and, if the ring contains a field, are in any case contained in the integral closure of the expected answer. This gives many powerful constraints on systems of parameters. (We should note that there are some restrictions on the use of colon in iterations: see the discussion following Theorem 8.1.)

Another remarkable consequence of the theory of tight closure is that if R is a locally equidimensional equicharacteristic ring which is a homomorphic image of a C-M ring, $x_1, \dots, x_n$ are elements in R such that $\mathrm{ht}(x_1, \dots, x_n)R = n$, and R is contained in S, where S is regular, then $((x_1, \dots, x_{n-1}) :_R x_n)S = (x_1, \dots, x_{n-1})S$. A closely related result which is more general in some ways is:

THEOREM 1.1 (VANISHING THEOREM). *Let $A \subseteq R \subseteq S$ be excellent equicharacteristic rings such that A and S are regular domains and R is module finite over A. Let M be a finitely generated A-module. Then the map $\mathrm{Tor}_i^A(M, R) \to \mathrm{Tor}_i^A(M, S)$ is 0 for $i \geq 1$.*

Note in this theorem that there is no condition on the map $R \subseteq S$. To prove this result we must consider tight closures of modules: see Section 6. We are also led to introduce the notion of phantom homology: a complex $\cdots \to F_{i+1} \to F_i \xrightarrow{d_i} F_{i-1} \to \cdots$ has phantom homology at the ith spot if $\mathrm{Ker}\, d_i$ is in the tight closure of $\mathrm{Im}(d_{i+1})$ in F_i. Section 6 contains some consequences and discussion of Theorem 1.1 as well as further generalizations, including a parallel of the Buchsbaum-Eisenbud criterion for acyclicity which yields "phantom acyclicity."

2. Tight closures of ideals.

Let $R^0 = R - \cup\{P : P \text{ is a minimal prime of } R\}$. Let I be an ideal of R. If $\mathrm{char}(R) = p > 0$ we say that $x \in R$ is in the tight closure, I^*, of I, if there exists $c \in R^0$ such that for all $e \gg 0$, $cx^q \in I^{[q]}$, where $q = p^e$ and

$I^{[q]} = (i^q : i \in I)$. If R is of finite type over a field K of char 0, we say that $x \in I^*$ if there exist $c \in R^0$, a finitely generated $\mathbb{Z}$-subalgebra D in K, a finitely generated D-flat D-subalgebra R_D of R, and an ideal $I_D \subset R_D$ such that $R \cong K \otimes_D R_D$, $I = I_D R$, and for all maximal ideals M of D, if $\kappa = D/M$ (with the subscript $_\kappa$ denoting images after tensoring with κ), and $p = \mathrm{char}(\kappa)$, then $c_\kappa(x_\kappa)^q \in (I_\kappa)^{[q]}$ for all $q \gg 0$ of the form $q = p^e$. If R is essentially of finite type over a field of characteristic 0 we define I^* as $\cup_B (I \cap B)^*$ as B runs through all subrings of R of finite type over K such that R is a localization of B. (One annoying point is that if we take the latter definition for algebras essentially of finite type over a field of characteristic $p > 0$, then we do not know whether it is the same as our definition in characteristic p. This ignorance is tied up with the fact that we do not yet know whether the tight closure commutes with localization.)

Although these definitions are somewhat intricate they yield a very powerful tool. We note that $I \subseteq I^* \subseteq I^-$, the integral closure of I. Recall that an element $a \in R$ is in I^- if there is an integer k such that a satisfies an equation of the form $a^k + i_1 a^{k-1} + \cdots + i_k = 0$ where $i_j \in I^j$ for $1 \le j \le k$. I^- is an ideal. It is easy to see that if a is in I^- then there exists a k such that for all $n \ge 0$,

$$a^{n+k} \in I^n. \qquad (*)$$

The converse is also true for Noetherian rings. Another useful characterization of the integral closure is that $a \in I^-$ iff for all valuations v (resp. discrete valuations in the Noetherian case) nonnegative on R and infinite only on a minimal prime of R, $v(a) \ge v(I) = \min\{v(i) : i \in I\}$. In characteristic p this last criterion yields that $I^* \subseteq I^-$. For if $ca^q \in I^{[q]}$ for all $q = p^e \gg 0$, then apply such a discrete valuation v and let q go to infinity; we obtain that $v(c)/q + v(a) \ge v(I)$ and so $v(a) \ge v(I)$ unless $v(c) = \infty$. Since $c \in R^0$ $v(c)$ is not ∞, however.

Heuristically, one may think of elements in I^* as being "almost" in I for the following reason: suppose R is reduced of characteristic $p > 0$ and let $R^\infty = \cup_q R^{1/q}$ where q runs through all integers of the form p^e. If $cy^q \in I^{[q]}$ for all sufficiently high $q = p^e$, we may take qth roots to obtain that $c^{1/q} y \in IR^{1/q}$ and hence is in IR^∞ for all $q \gg 0$. For large q the exponent $1/q$ of c is close to 0, and so y is multiplied into IR^∞ by elements arbitrarily "close" to 1.

We say that I is *tightly closed* if $I = I^*$. As we mentioned above, the following proposition is crucial.

PROPOSITION 2.1. *Let R be a regular ring of* char $p > 0$. *Then $I = I^*$ for all ideals I of R.*

PROOF: We may assume that R is local with maximal ideal m and that $y \in I^* - I$. Then for some $c \neq 0$ and all $q = p^e \gg 0$, we have that $cy^q \in I^{[q]}$. Since the Frobenius F is flat [**Ku**], $I^{[q]} : y^q = (I : y)^{[q]}$. It follows that $c \in m^q$, and since this is true for all large q we obtain that $c = 0$, a contradiction. □

Of course, in general, the integral closure of an ideal in a regular ring (or any ring) is much larger than the ideal, so this proposition shows that the tight closure is, usually, much smaller than the integral closure of an ideal. The tight closure of I is a very 'tight' fit for I; hence the name.

DEFINITION 2.2: Let R be a ring. If every ideal in every localization of R is tightly closed, we say that R is *F-regular.* If $I = I^*$ for every ideal of R we say that R is *weakly F-regular.*

It may be that F-regular is the same as weakly F-regular, but we cannot prove this. It is not difficult to see that R is weakly F-regular iff R_m is weakly F-regular for all maximal ideals m of R.

Suppose that $R \subseteq S$ are domains, that I is an ideal of R and S is regular. Then $I^*S = IS$ since IS is tightly closed in S. Hence I^* is contained in all ideals containing I which are contracted from regular rings. The intersection of all such ideals we call the *regular closure* of I. Obviously the regular closure of I is also contained in the integral closure of I. In fact we do not know whether the regular closure and the tight closure coincide. Many of our results which say something about the tight closure, where it is defined, can be rephrased for any ring containing a field by replacing the tight closure by the regular closure. There is also a notion of regular closure for submodules.

3. Subrings of regular rings.

It is fairly easy to show that F-regular rings are normal. Of much greater significance is:

THEOREM 3.1. *Let R be an F-regular ring which is the homomorphic image of a C-M ring. Then R is C-M.*

The proof is given after Lemma 3.2 and Theorem 3.3 below.

Lemma 3.2. *Let $R = S/I$ where S is a C-M local ring and assume that R is equidimensional. Let $\{Q_1, \ldots, Q_n\}$ be the minimal primes over I. Assume that $x_1, \ldots, x_d$ are parameters of R. Then there exist elements $z_1, \ldots, z_h$ in I and $y_1, \ldots, y_d$ such that the y_i lift x_i, the z's and y's together form a regular sequence, and there exist a $c \notin \cup Q_i$ and an integer k such that $cI^k \subseteq (z_1, \ldots, z_h)$.*

We omit the proof, which uses prime avoidance and the fact that, if we let $W = S - \cup Q_i$, then I_W will be, up to radical, generated by $h = \mathrm{ht}(I)$ elements.

Theorem 3.3. *Let $R = S/I$ be equidimensional where S is a C-M ring. Assume that* $\mathrm{char}(R) = p > 0$. *Let $x_1, \ldots, x_n$ be elements of R which are part of a s.o.p. (system of parameters) in R_P for all primes P which contain them. Let $J = (x_1, \ldots, x_{n-1})R$. Then $J :_R x_n \subset J^*$.*

Proof: We may reduce to the case where S is local by localizing at the maximal ideals of S. Let $a' \in J : x_n$. Let c, z_i, y_i, and k be as in Lemma 3.2 and let a be a lifting of a' to S. Since $a'x_n \in J$, $ay_n \in (y_1, \ldots, y_{n-1}) + I$. Then for all $q = p^e$ we obtain that $a^q y_n^q \in (y_1, \ldots, y_{n-1})^{[q]} + I^{[q]}$. Multiply by c and use Lemma 3.2 to get that $ca^q y_n^q \in (y_1^q, \ldots, y_{n-1}^q, z_1, \ldots, z_h)$. As the z's and y's together form a regular sequence in S, this containment forces that $ca^q \in (y_1^q, \ldots, y_{n-1}^q, z_1, \ldots, z_h)$ for all a. Reading modulo I gives that $c'a'^q \in J^{[q]}$ where c' is the image of c in R and is in R^0 by Lemma 3.2. Hence $a' \in J^*$. ☐

Proof of Theorem 3.1: We may assume that R is local. Since R is F-regular it is normal and therefore is a domain. Hence R is equidimensional. Let $x_1, \ldots, x_d$ be a s.o.p. Let $J_i = (x_1, \ldots, x_{i-1})$. It is enough to show that $J_i : x_i = J_i$. By Theorem 3.3, $J_i : x_i$ is contained in $J_i^* = J_i$. ☐

An immediate consequence of Theorem 3.1 is:

Theorem 3.4. *Assume that $R \subseteq S$ are char $p > 0$ and that R is a direct summand of S as an R-module. If S is F-regular (respectively weakly F-regular) then R is F-regular (respectively weakly F-regular). If moreover R is a quotient of a C-M ring and S is weakly F-regular then R is C-M.*

Even when the tight closure is not defined we may use the Artin approximation theorem to deduce:

THEOREM 3.5. *Let R be an equicharacteristic ring which is a direct summand as an R-module (or a pure subring, cf.* [**HR1**]*) of a regular ring S. Then R is C-M.*

A result similar to 3.5 was first proved in [**HR1**] for rings of invariants of linearly reductive groups acting on regular rings. The result was extended to summands of regular rings for algebras of finite type in [**K**], while in [**B**] it is shown that in the affine and analytic cases (where one may apply the Grauert-Riemenschneider theorem) direct summands of rational singularities are rational. None of these earlier arguments permitted extension to the general equicharacteristic case. The results of [**B**] suggest a relationship between rational singularities and F-regularity which we shall discuss below. The difficulty of applying the Artin approximation theorem for this theorem is that it is difficult to capture equationally enough information from the fact that R is a direct summand of a regular ring. In fact the proof of 3.5 rests on the stronger result below which mimics 3.3.

THEOREM 3.6. *Let R be a locally formally equidimensional equicharacteristic 0 Noetherian ring with C-M formal fibers. Let $h : R \to S$ be a homomorphism to a regular Noetherian domain S. Suppose that $\mathrm{ht}(x_1, \ldots, x_n) = n$ and there is a parameter $c \in R$ such that R_c is C-M and $h(c) \neq 0$. Then $((x_1, \ldots, x_{n-1})R :_R x_n)S = (x_1, \ldots, x_{n-1})S$.*

4. Integral closure and the Briançon-Skoda theorem.

As we have seen in the introduction, the tight closure of an ideal is always contained in the integral closure. In fact there is a very tight connection between many proofs involving statements concerning the integral closures of ideals and the theory of tight closures. A good example of this phenomenon is the theorem of Briançon-Skoda [**BS**].

They proved in the regular analytic case that the integral closure of the nth power of an n generator ideal is always contained in the ideal. Later Lipman and Sathaye [**LS**] and Lipman and Teissier [**LT**] gave algebraic proofs of this fact for arbitrary regular local rings. In [**LT**] there is also the realization that the validity of the Briançon-Skoda theorem is closely related to having a rational singularity. In this section we present an extremely simple proof of a stronger statement in $\mathrm{char}\, p > 0$. It is not difficult to use the technique of reduction to characteristic p to obtain the same statement for any ring containing a field.

THEOREM 4.1. *Let R be a Noetherian ring of characteristic p and let I be an ideal of positive height generated by n elements, say $u_1, \ldots, u_n$. Then for every $m \in \mathbb{N}$, $(I^{n+m})^- \subseteq (I^{m+1})^*$. In particular, if R is regular or even weakly F-regular then $(I^{n+m})^- \subseteq I^{m+1}$. If $m = 0$ we obtain that $(I^n)^- \subseteq I$ in this case.*

PROOF: If $(I^{n+m})^-$ is contained in the union of $(I^{m+1})^*$ and the minimal primes of R then it must be contained in one of them, and then it must be contained in $(I^{m+1})^*$, since $\text{ht}(I) > 0$. Hence, we assume to the contrary that it is not contained in this union, and choose $y \in (I^{n+m})^-$ but not in $(I^{m+1})^*$ and not in any minimal prime of R. Set $J = I^{n+m}$. By (*) of Section 2 we can choose a positive integer k such that $y^{k+h} \in J^h$ for all $h \geq 0$. However $J^h = I^{hn+hm}$ is generated by monomials of degree $hn+hm$ in the u_i. Then $J^h \subseteq (u_1^h, \ldots, u_n^h)^{m+1}$. Now take $c = y^k \in R^0$ and when h has the form $q = p^e$ we obtain that $cy^q \in J^q \subseteq (u_1^q, \ldots, u_n^q)^{m+1} = (I^{m+1})^{[q]}$, and so $y \in (I^{m+1})^*$. The rest of the theorem follows directly from the definition of weakly F-regular and Proposition 2.1. □

Notice that if $J \subseteq I \subseteq J^-$ and J has n generators with $\text{ht}(I) > 0$, then $(I^{n+m})^- \subseteq (I^{m+1})^*$ in this case as well, since $(I^{n+m})^- = (J^{n+m})^- \subseteq (J^{m+1})^* \subseteq (I^{m+1})^*$. If R is local with infinite residue field then every ideal is integral over an ideal generated by $\dim(R)$ elements [**NR**]. In general we may take the n in the theorem to be the analytic spread of I rather than the number of generators of I.

Another corollary of this result is that for a principal ideal of height 1, the tight closure and the integral closure coincide. This follows by taking $n = 1$, $m = 0$ in Theorem 4.1. An easy consequence of this remark is that every weakly F-regular ring is normal.

The Briançon-Skoda theorem provides a bridge between F-regularity and rational singularities. This bridge is given by the following result:

PROPOSITION 4.2. *Let R be essentially of finite type over a field of characteristic 0 with an isolated singularity m with $\dim(R_m) = d$. Then R_m has a rational singularity iff $(I^d)^- \subseteq I$ for all m-primary ideals I of R. In fact R has a rational singularity iff $(I^d)^- \subseteq I$ for all m-primary ideals of the form $I = (x_1^k, \ldots, x_d^k)$ for some s.o.p. $x_1, \ldots, x_d$ of R.*

A consequence of this proposition is:

THEOREM 4.3. *Suppose that R is of finite type over a field K of characteristic 0 and that R is weakly F-regular. If either: a) R has isolated* *singularities or b) R is $\mathbb{N}$-graded with $R_0 = K$, $m = \oplus_{i\geq 1} R_i$, and R has rational singularities except possibly at m, then R has rational singularities.*

In particular, weakly F-regular surfaces have rational singularities. We conjecture that all F-regular rings of finite type over a field of characteristic 0 have rational singularities. The converse is not true: an example of [**W**] shows that a surface in characteristic 0 may have rational singularities without even being of F-pure type in the sense of [**HR2**]. Fedder and Watanabe [**FW**] (see this volume) have cleared up some of the relationship between the concepts of F-pure type, weakly F-regular, and rational singularity, at least in the case that R is Gorenstein. It turns out that Gorenstein algebras are much easier to deal with, as we shall discuss in the next section.

5. Gorenstein rings.

In this section we wish to prove that Gorenstein algebras have particularly nice behavior with respect to the notion of F-regularity. All of our results are based on the next proposition.

PROPOSITION 5.1. *Let (R, m) be a Gorenstein local ring. Then the following are equivalent.*

i) *R is weakly F-regular.*
ii) *There exists a s.o.p. $x_1, \ldots, x_d$ such that the ideal J generated by this s.o.p. is tightly closed.*

PROOF: Clearly i) implies ii). Any intersection of tightly closed ideals is tightly closed. Since every ideal is an intersection of m-primary ideals, to see that R is weakly F-regular it suffices to see that every m-primary ideal I is tightly closed.

First we claim that $J_t = (x_1^t, \ldots, x_d^t)$ is tightly closed. Since R is Gorenstein, R/J_1 is also Gorenstein; hence we may choose an element y in R whose image generates the socle of R/J_1. Then $yx_1^t \ldots x_d^t$ generates the socle of R/J_{t+1}. To see that J_{t+1} is tightly closed it suffices to see that $yx_1^t \ldots x_d^t$ is not in $(J_{t+1})^*$. Assume the contrary. Then there is a $c \in R^0$ such that $c(yx_1^t \ldots x_d^t)^q \in (J_{t+1})^{[q]}$ for all $q = p^e \gg 0$. Since the x_i form a regular sequence this implies that $cy^q \in (J_1)^{[q]}$ and hence that $y \in (J_1)^*$ which contradicts the assumption ii).

Now let I be any m-primary ideal. We may choose a t such that $J_t \subset I$. Since R/J_t is Gorenstein, there is an ideal K such that $J_t : K = I$ (in fact take $K = J_t : I$). It is easy to see that whenever A is tightly closed, so is $A : B$. Hence, as J_t is tightly closed, it follows that I is also, which proves i). □

REMARK 5.2: It is also true, although we shall not prove it here, that for a Gorenstein ring weakly F-regular implies F-regular.

COROLLARY 5.3. *Let R be a local Gorenstein ring and suppose that x is not a zero-divisor in R. If $S = R/Rx$ is weakly F-regular, then R is weakly F-regular.*

PROOF: Choose a s.o.p. $x = x_1, \ldots, x_d$. By Proposition 5.1 it is enough to prove that the ideal I generated by the x_i is tightly closed. Suppose that $y \in I^*$. Then there exists a $c \in R^0$ such that $cy^q \in I^{[q]}$ for all large q which are powers of the characteristic p of R. If c is not in Rx then the image of c is in S^0 (S is a normal ring, hence a domain) and therefore the image of y would be in $(IS)^* = IS$, which shows that y is in I. Take the largest power x^h which divides c and write $c = x^h d$. Then $cy^q \in (x^q, x_2^q, \ldots, x_d^q)$ implies that $dy^q \in (x^{q-h}, x_2^q, \ldots, x_d^q)$ since the x_i form a regular sequence. Working modulo Rx we see that $y \in (IS)^*$ and hence, as before, $y \in I$. □

This corollary shows that the property of weak F-regularity behaves well under deformation, at least in the Gorenstein case. We do not know if the corollary is true when R is not Gorenstein. Proposition 5.1 makes it very easy to test whether a hypersurface is weakly F-regular, at least in theory, since one only has to check if a single ideal is tightly closed. We do not know whether there is a finite list of canonically describable ideals in a C-M ring which, if tightly closed, force the ring to be weakly F-regular.

It is very much worth noting that if R is a local ring which is the homomorphic image of a C-M ring, and the ideal generated by a single s.o.p. is tightly closed, then R must be C-M. Thus in the context of Proposition 5.1 in the proof of ii) implies i) it is worth noting that ii) by itself forces the ring to be C-M. This remark is also valuable in considering various determinantal rings in characteristic p. If the ideal generated by a single s.o.p. happens to be contracted from a regular overring, we obtain that the ring is weakly F-regular. The condition that a single s.o.p. be contracted is much weaker than the assumption of being a summand, which forces every ideal to be contracted.

6. Modules, Vanishing Theorems, and the Syzygy Theorem.

If R is a ring of char $p > 0$, we denote the Frobenius map by $\mathbf{F}$, and let $\mathbf{F}^e R$ denote the ring R viewed as a right-module over R via e iterations of the Frobenius map, and as a left R-module in the usual way (see [**PS**, p. 330, Def. (1.2)]).

DEFINITION 6.1: Let R be a ring of char. $p > 0$. Suppose that M is an R-module and N is a submodule of M. Then an element $x \in M$ is said to be in the *tight closure* of N with respect to M (which we denote by $x \in N^*$ if the module M is understood) if there exists an element $c \in R^0$ such that the image of the element $c(1 \otimes x)$ in $\mathbf{F}^e(M/N) = \mathbf{F}^e R \otimes_R (M/N)$ is zero for all large e.

This definition only depends on the quotient module M/N. To say that an element $x \in M$ is in the tight closure of a submodule N is the same as saying that the image of x in M/N is in the tight closure of (0). If we map a free module F onto M/N and let K be the kernel then it is enough to understand the tight closure of K in F. Suppose that $F \cong R^n$ and that K is generated by $u_1 = (a_{11}, \ldots, a_{1n}), \ldots, u_m = (a_{m1}, \ldots, a_{mn})$. Then an element $x = (x_1, \ldots, x_n)$ in F is in the tight closure of K iff there exists an element $c \in R^0$ such that cx^q $(x^q = (x_1^q, \ldots, x_n^q))$ is in the submodule spanned by $(a_{11}^q, \ldots, a_{1n}^q) = u_1^q, \ldots, (a_{m1}^q, \ldots, a_{mn}^q) = u_m^q$ for all $q = p^e \gg 0$. Notice that this definition agrees with our original definition in case F is R and K is an ideal of R. It is true, although not obvious, that if R is weakly F-regular then every submodule N of an R-module M is tightly closed in M.

The vanishing Theorem 1.1 of the introduction is a powerful result; but it is a special case of a more general theorem which also relates to the Buchsbaum-Eisenbud criterion for acyclicity and to the syzygy theorem of Evans and Griffith. Before we state our theorem we need to give some definitions. If $\phi : F \to G$ is a homomorphism of free R-modules, then we say the rank of ϕ is n if $\Lambda^n \phi \neq 0$, but $\Lambda^j \phi = 0$ for $j > n$. If $\operatorname{rank}(\phi) = n$, then we set $I(\phi)$ equal to the ideal generated by all $n \times n$ minors of a matrix which represents ϕ after choosing bases of F and G. This ideal is independent of the bases chosen.

DEFINITION 6.2: Let $G_\bullet$ be a complex of R-modules with cycles $Z_\bullet$, boundaries $B_\bullet$, and homology $H_\bullet$. We say that H_j is *phantom* if Z_j is in the tight closure of B_j in G_j. Equivalently the homology H_j is in the tight closure of (0) in G_j/B_j.

Suppose that $G_\bullet$ is a free complex. To say that $H_j(G_\bullet)$ is phantom means in particular that if R is contained in any regular local ring S with $R^0 \subseteq S^0$, and $z \in Z_j$, then $z \in B_jS$, i.e., all homology coming from R disappears when we extend to any regular local ring containing R.

THEOREM 6.3. *Let R be a Noetherian ring of characteristic $p > 0$ which is locally equidimensional and the homomorphic image of a Gorenstein ring S of finite Krull dimension. Let*

$$G_\bullet = 0 \to G_n \xrightarrow{\phi_n} G_{n-1} \xrightarrow{\phi_{n-1}} \cdots \to G_1 \xrightarrow{\phi_1} G_0$$

be a complex of finitely generated free R-modules. Assume that $\mathrm{rank}(\phi_i) + \mathrm{rank}(\phi_{i-1}) = \mathrm{rank}(F_i)$ *and that* $\mathrm{ht}(I(\phi_i)) \geq i$ *for all i. Then $G_\bullet$ has phantom homology.*

The proof of Theorem 6.3 is based on a "constructive" reformulation of the Buchsbaum-Eisenbud acyclicity theorem. To establish notation, let $G_\bullet =$

$$0 \to G_n \xrightarrow{\phi_n} G_{n-1} \to \cdots \xrightarrow{\phi_1} G_0 \tag{6.4}$$

be a complex of finitely generated free R-modules, where R is a Noetherian ring.

Suppose that the hypothesis of the Buchsbaum-Eisenbud criterion for the complex $G_\bullet$ is satisfied after localizing at some $c \in R$. Then there is a power of c which kills the homology of $G_\bullet$. What we need to do is set up the hypothesis and conclusion in such a way that the power of c in the conclusion is bounded in a manner independent of the maps of the complex, so that, in characteristic p, the same power of c can be used even after applying the Frobenius endomorphism.

DEFINITION 6.5: We shall say an ideal $I \subseteq R$ has c-depth $\geq n$ if for every integer i with $1 \leq i \leq n$ there are elements $x_1, \ldots, x_i \in I$ such that for every integer $j \geq 1$, and for every integer $t \geq 1$, c kills $H_j(x_1^t, \ldots, x_i^t; R)$, the Koszul homology of $x_1^t, \ldots, x_i^t$ on R.

A key point is the following result which guarantees the existence of an element $c \in R^0$ such that every ideal I of R with $\mathrm{ht}(I) \geq n$ has c-$\mathrm{depth}(I) \geq n$.

THEOREM 6.6. *Let R be a Noetherian ring, locally equidimensional, which is a homomorphic image of a Gorenstein ring of finite Krull dimension. Let c' be any element of R^0 such that $R_{c'}$ is C-M. Then c' has a fixed power c such that for every sequence of elements $x_1, \ldots, x_n$ of R such that $\mathrm{ht}(x_1, \ldots, x_n)R \geq n$, c kills $H_i(x_1^t, \ldots, x_n^t; R)$ for all $i \geq 1$, and for all $t \in \mathbb{N}$.*

The importance of Theorem 6.6 is given by the following theorem, a "constructive" acyclicity result, used to prove Theorem 6.3.

THEOREM 6.7. *Let R be an arbitrary Noetherian ring. Let $G_\bullet$ be a complex of finitely generated free R-modules, as in (6.4). Suppose for $2 \leq i \leq n$ that $\mathrm{rk}\,\phi_i + \mathrm{rk}\,\phi_{i-1} = \mathrm{rk}(G_i)$, and let $I_i = I(\phi_i)$. Let $c \in R$ and suppose that $c-\mathrm{depth}(I_i) \geq i$ for $1 \leq i \leq n$. Then $c^{\Box t}$ kills $H_{n-t}(G_\bullet)$, $0 \leq t \leq n-1$, where $\Box\, t = f_{2t+2} - 1$, and f_n denotes the nth Fibonacci number; $f_0 = f_1 = 1$, and $f_{n+1} = f_n + f_{n-1}$.*

REMARK 6.8: If $c = 1$, then the assumptions are basically that $\mathrm{rk}\,\phi_i + \mathrm{rk}\,\phi_{i-1} = \mathrm{rk}\, G_i$, and that $\mathrm{depth}\, I_i \geq i$. The conclusion with $c = 1$ is that $G_\bullet$ is acyclic. Thus, this theorem recovers one direction of the Buchsbaum-Eisenbud acyclicity criterion, under the hypotheses of (6.7). Notice that the exponent $\Box\, t$ is independent of the maps in the complex.

COROLLARY 6.9. *Let R be a local ring of characteristic $p > 0$, and let N be a finitely generated R-module of finite projective dimension. Let $(G_\bullet, \phi_\bullet)$ be a finite free resolution of N. Suppose that R maps to an equidimensional ring S which is a homomorphic image of a Gorenstein ring of finite Krull dimension. Assume that there exists an integer d such that $\mathrm{ht}(IS) \geq \mathrm{ht}(I) - d$ for all ideal I of R. Then $H_i(G_\bullet \otimes_R S)$ is phantom for $i > d$.*

PROOF: The Buchsbaum-Eisenbud criterion shows that $\mathrm{grade}(I(\phi_i)) \geq i$ for all i, so that in particular $\mathrm{ht}(I(\phi_i)) \geq i$ for all i. Hence if we denote the induced maps $\phi_i \otimes 1$ by γ_i, our assumption reads that $\mathrm{ht}(I(\gamma_i)) \geq i - d$ for all i. The ranks of the maps also add correctly to the ranks of the free modules for $i > d$ since $I(\gamma_i) \neq 0$ in this range. Thus we may apply Theorem 6.3 to the truncated complex $\{G_i \otimes_R S\}_{i \geq d}$ to see that it has phantom homology, as required. $\Box$

COROLLARY 6.10. *(Vanishing Theorem 1.1) Let $A \subseteq R \subseteq S$ be excellent equicharacteristic rings such that A and S are regular domains and R is module-finite over A. Let M be a finitely generated A-module with finite free resolution $G_\bullet$. Then the map $\mathrm{Tor}_i^A(M, R) \to \mathrm{Tor}_i^A(M, S)$ is 0 for $i \geq 1$.*

PROOF: The assumptions guarantee that R is equidimensional and the homomorphic image of a Gorenstein ring of finite Krull dimension. Furthermore $\mathrm{ht}(I) = \mathrm{ht}(IR)$ for every ideal I of A so we may apply Corollary 6.9 to conclude that the complex $G_\bullet \otimes_A R$ has phantom homology. The conclusion of the theorem now follows from our discussion above. □

COROLLARY 6.11. *Let S be a local ring of char. p which is a homomorphic image of a Gorenstein local ring, and assume that N is a finite S-module having finite projective dimension, with $G_\bullet$ a finite free resolution of N. Suppose that R is an equidimensional S-algebra, finite as an S-module, and let n be the codimension of R as an S-module. If $j > n$, then $\mathrm{Tor}_j^S(N, R)$ is contained in the tight closure of (0) in $(G_j \otimes_S R)/\mathrm{im}(G_{j+1} \otimes_S R)$.*

PROOF: The assumptions show that $\mathrm{ht}(IR) \geq \mathrm{ht}(I) - n$ for any ideal I of S. The assertion now follows immediately from Corollary 6.9. □

The above corollary is closely related to the syzygy theorem of Evans and Griffith [**EG**]. We give a slight generalization of the syzygy theorem which uses the techniques of this paper. First we recall a definition.

If M is a module and $x \in M$ recall that the *order ideal of* x, denoted by $0_M(x)$, is $\{f(x) : f \in \mathrm{Hom}_R(M, R)\}$.

THEOREM 6.12 (SYZYGY THEOREM). *Let R be a local ring of characteristic $p > 0$ and suppose that M is a finitely generated kth syzygy having finite projective dimension. Let x be any minimal generator of M. Then $\mathrm{grade}(0_M(x)) \geq k$.*

PROOF: Let $G_\bullet$ be a minimal free resolution of a module N whose kth syzygy is M. Let $y \in G_k$ be a minimal generator of G_k which is mapped to x. We assume that $\mathrm{grade}(0_M(x)) = n < k$. Set $I = 0_M(x)$ and observe that $\tilde{y}$ is a cycle in the complex $\tilde{G}_\bullet$ where $\tilde{}$ denotes tensoring with $S = R/I$. Choose a maximal regular sequence inside I and let J be the ideal it generates. Choose a $c \in R$ such that $cI \subset J$ but c is not in J. Write $\tilde{y} = (\tilde{y}_1, \ldots, \tilde{y}_m) \in \tilde{Z}_k \subset \tilde{G}_k$. At least one of the $\tilde{y}_i$ is a unit. If the map from G_k to G_{k-1}, after choosing bases, is given by the matrix (a_{ij}) and y_i lifts $\tilde{y}_i$ then $\sum_j y_j a_{ij} \in I$ for each i. Raise this to the $q = p^e$ power and multiply by c to obtain that $\sum_j c y_j^q a_{ij}^q \in cI^{[q]} \subset J$. Thus we obtain that cy^q is a cycle in the complex $K_\bullet = F^e G_\bullet \otimes_S S/J$, where F^e is the Frobenius functor. Since the Frobenius is exact on modules of finite projective dimension [**PS**], the homology of $K_\bullet$ at the kth spot is exactly

$\mathrm{Tor}_k^S(F^eN, S/J)$, which in turn is zero since J is generated by a regular sequence of length $n < k$. Therefore cy^q is a boundary in this complex. Suppose that the map from G_{k+1} to G_k is given by the matrix (b_{ij}) and let $b_1, \ldots, b_v$ be the rows of this matrix. The fact that cy^q is a boundary implies that this element lies in the submodule of G_k generated by $b_1^q, \ldots, b_v^q$ and JG_k. Since some y_i is a unit it follows that c is in $m^q + J$ for all q, which in turn implies that $c \in J$. This is a contradiction. □

We would like to thank G. Evans for pointing out that we could remove the C-M assumption we originally had without changing the proof.

As usual one can use the Artin approximation theorem to reduce to characteristic p and prove the above theorem for any local ring containing a field. It is still open whether this theorem is valid in mixed characteristic. We next give a theorem which is an easy application of the vanishing Theorem 1.1, and which generalizes results in [**HR1**, Prop. A] and [**K**, Thm. 2.1]. We also note in passing that the vanishing theorem in the case that $M = A/J$ and S is a DVR dominating the local ring R implies the canonical element conjecture [**H**] (of course, only for rings containing fields).

THEOREM 6.13. *Let $A \subseteq R \subseteq S$ be equicharacteristic excellent rings such that A and R are local, A and S are regular domains, and R is finite over A. Let m be the maximal ideal of R. Then the canonical map from the local cohomology $H_m^i(R) \to H_{mS}^i(S)$ is zero for $0 \leq i < d = \dim(R)$.*

PROOF: Choose regular parameters $x_1, \ldots, x_d$ for the maximal ideal of A; then they form a s.o.p. in R. We may calculate the local cohomology as a direct limit of the Koszul cohomology of $x_1^n, \ldots, x_d^n$ and the map from the local cohomology of R to the local cohomology of S is induced by mapping the Koszul cohomology of these elements over R to the Koszul cohomology of them over S. As the elements form a regular sequence in A, the map of $H_m^j(R)$ to $H_{mS}^j(S)$ is the direct limit of the maps:

$$\mathrm{Tor}_{d-j}^A(A/(x_1^n, \ldots, x_d^n), R) \to \mathrm{Tor}_{d-j}^A(A/(x_1^n, \ldots, x_d^n), S),$$

which are all zero for $j < d$ by the vanishing Theorem 1.1. □

REMARK 6.14: With only a little more work, we can also prove a generalization of Theorem 6.13 as follows: use the notation of that theorem, and let I be any ideal of A. Then the canonical map $H_{IR}^i(R) \to H_{IS}^i(S)$ is the zero map for $i < \mathrm{ht}(I)$.

7. Test elements.

In the definition of the tight closure the multiplier c is allowed to vary in R^0 as I changes when one tests whether various elements are in I^*. For many R it turns out that it is not necessary to vary c: a single c can be used as a 'test element' for all tight closure tests. This fact is extremely useful and we briefly explore it in the section.

To be precise, $c \in R^0$ is a *test element* if for all $x \in R$ and I an ideal of R, $x \in I^*$ iff $cx^q \in I^{[q]}$ for all $q = p^e$.

One can also ask that the test element c be a test element in every localization of R and also for the completions of each local ring of R. We call such an element a *completely stable test element.* Our main theorem regarding the existence of such elements is:

THEOREM 7.1. *Let R be module-finite, torsion free and generically smooth over a regular domain A of characteristic $p > 0$. (The last condition means that $L \otimes_A R$ is smooth over the fraction field L of A, which in turn means that $L \otimes_A R$ is a product of fields each of which is a finite, separable extension of L.) Then every element $d \in A^0$ such that R_d is smooth over A_d has a power c which is a completely stable test element in $B \otimes_A R$ for every regular domain $B \supseteq A$. A sufficient condition for c to have this property is that $cR^\infty \subseteq A^\infty[R]$.*

Rather than try to sketch a proof of this theorem here, we will concentrate on how one can find elements c satisfying the last condition of the theorem.

THEOREM 7.2. i) *Let R be a ring of characteristic $p > 0$ which is module-finite, torsion-free, and generically smooth over a regular ring A. Let d be an element of A^0 such that R_d is A_d-smooth. Then d has a power b such that $bR^{1/p} \subseteq A^{1/p}[R] \cong A^{1/p} \otimes_A R$. Let $c = b^2$. Then $cR^\infty \subseteq A^\infty[R] \cong A^\infty \otimes_A R$ as well.*

ii) *Let R be module-finite, torsion-free, and generically smooth over a normal ring A. Let $r_1, \ldots, r_d \in R$ be a vector space basis for $L' = L \otimes_A R$ over the fraction field L of R and let $c = \det(Tr_{L'/L}(r_i r_j))$. Then $c \in A^0$ and $cR^\infty \subseteq A^\infty[R]$.*

PROOF: We first note that when a ring S is smooth and module-finite over a reduced ring B we always have that $S^{1/q} \cong B^{1/q} \otimes_B S$ (the isomorphism is induced in the obvious way). We leave the proof for the reader.

Now we sketch the proof of i). Since the natural map of $A^{1/q} \otimes_A R$ into $R^{1/q}$ has image $A^{1/q}[R]$, this map becomes an isomorphism after localizing at d, and since these modules are torsion-free over A we see that we may identify $A^\infty \otimes_A R$ with $A^\infty[R] \subseteq R^\infty$. Since $R^{1/p} \supseteq A^{1/p}[R]$ are both finitely generated over $A^{1/p}$ and become the same after inverting d, it follows that there is a power b of d which multiplies $R^{1/p}$ into $A^{1/p}[R]$. Let $h = 1 + 1/p + \cdots + 1/p^e$. We claim that $b^h R^{1/pq}$ is contained in $A^{1/pq}[R]$ for all $q = p^e$ by induction on e: we already know the case $e = 0$. Take pth roots repeatedly to obtain that $b^{1/q} R^{1/q}$ is contained in $A^{1/pq}[R^{1/q}]$ for all q, whence with $h' = h - 1/q$, we have

$$b^h R^{1/pq} = b^{h'} b^{1/q} R^{1/pq} \subseteq A^{1/pq}[b^{h'} R^{1/q}] \subseteq A^{1/pq}[A^{1/q}[R]] = A^{1/pq}[R].$$

Since b^h divides b^2 (in $A^{1/q}$) for all h we have that $b^2 R^{1/q} \subset A^{1/q}[R]$ for all q, and the result follows by taking the union.

Next we prove ii). The element $c \neq 0$ since L' is separable over L, and it is clear that $c \in A$. The fraction field of A^∞ may be identified with L^∞. Since L' is separable over L it is linearly disjoint from L^∞ and hence the r_i are also a basis for $L'^\infty \cong L^\infty \otimes_L L'$ over L^∞. Note also that A^∞ is normal. The result now follows by the usual argument after noting that R^∞ is contained in the integral closure of A^∞ in L'^∞. □

Notice that the set of all test elements forms an ideal of R if R is a domain and if we throw in 0. In particular Theorem 7.1 shows that when R is finite, torsion-free, and generically smooth over a regular domain A then this ideal, up to radical, contains all elements $d \in A$ such that R_d is smooth over A_d; this is a very useful fact since it can allow us a good deal of freedom in picking test elements.

REMARK 7.3: If R is reduced and $R^{1/p}$ is module-finite over R, there is always a completely stable test element: in fact, if $c \in R^0$ is any element such that R_c is regular, then c has a power which is a completely stable test element.

8. Constraints on systems of parameters.

In this section we will discuss the constraints satisfied by a s.o.p. $x_1, \ldots, x_n$ in a equidimensional local Noetherian ring of characteristic p which is a homomorphic image of a C-M ring. In particular we will be interested in the result of performing various algebraic operations on monomial ideals in $x_1, \ldots, x_n$, i.e., ideals generated by monomials in the x_i. The prototype of this kind of result is Theorem 3.3. However a more general result holds, and we shall also see that there are comparable results for ideals constructed by iterated use of the addition, multiplication, intersection, and colon operations on monomial ideals in the parameters.

THEOREM 8.1. *Let R be an equidimensional local ring of characteristic p which is a homomorphic image of a C-M local ring S. Let $x_1, \ldots, x_n$ be parameters of R and map $A = C[X_1, \ldots, X_n]$ to R by sending X_i to x_i where $C = \mathbb{Z}/p\mathbb{Z}$. Let I, J be monomials ideals in the X_i in A and let $\#$ be any of the operations $+$, $\times$, $:$, and $\cap$. Then*

$$(I\#J)R \subseteq IR\#JR \subseteq ((I\#J)R)^*.$$

PROOF: If $\#$ is $+$ or $\times$ then $(I\#J)R = IR\#JR$ for any ideals I, J of A for an arbitrary homomorphism from A to R of commutative rings. Henceforth we assume that $\#$ is $:$ or $\cap$. The first inclusion in the statement of the theorem is trivial so we focus on the second inclusion. We will show that there exists a $c \in R^0$ and an integer q' such that for all $q = p^e \geq q'$, $c(IR\#JR)^{[q]} \subset ((I\#J)R)^{[q]}$. This claim for R follows from the same claim for R_{red}, so we may assume that R is reduced. Write $R = S/Q$, where Q is an intersection of primes Q_i which all have the same height, say d, by assumption. By Lemma 3.2 we may find $z_1, \ldots, z_d$ in Q, $y_1, \ldots, y_n$ in S lifting $x_1, \ldots, x_n$ and a c not in any Q_i such that the z_i and y_i together form a regular sequence and $cQ^{q'} \subseteq (z_1, \ldots, z_d) = K$ for some q'.

Let $B = A[Z_1, \ldots, Z_d]$ and map B to S by sending X_i to y_i and Z_j to z_j. The induced map to R agrees with the original map on A and sends all the Z_j to 0. Now suppose that $s \in S$ and its image r in R is in $IR\#JR$. Let $I' = IB + (Z_j)B$ and $J' = JB + (Z_j)B$. It is not difficult to see that $I'\#J' = (I\#J)B + (Z_j)B$ and that $I'S\#J'S = (I'\#J')S$. It follows that $s \in (IS + Q)\#(JS + Q)$, and so for all $q \geq q'$, $cs^q \in (I^{[q]}S + (z_j)S)\#(J^{[q]}S + (z_j)S) = ((I\#J)^{[q]} + (z_j)S)$. Taking images in R yields that $c'r^q \in (I\#J)^{[q]}R = ((I\#J)R)^{[q]}$ for all $q \geq q'$, and the result follows. □

To explain our main result concerning iterated operations, we need some discussion. Let $U_1, \ldots, U_k$ denote variable ideals. We consider ideal-valued functions of these variables which can be constructed recursively from the functions U_i (technically, this is the ith projection), $U_i + U_j$, U_iU_j, $U_i \cap U_j$, and $U_i : U_j$ by allowing the substitution of functions already constructed for any of the variables U_i, U_j in the functions listed above, *subject to the restriction that one may only substitute a function for the numerator U_i of $U_i : U_j$ and not for the denominator U_j.* We call the functions which can be constructed in this way *permissible.* For example, $(U_1 : U_2) : U_3$ is permissible as is $((U_1U_2 + ((U_3U_2) : U_1)) \cap ((U_5 + (U_6 : U_3)) : U_2)$. However $U_1 : (U_2 : U_3)$ is not permissible.

Given a homomorphism $A \to R$ and an ideal I of A,where A, R are Noetherian of characteristic $p > 0$, then we shall say that $I' \subseteq R$ is *trapped* over I if $IR \subseteq I' \subseteq (IR)^*$. Note that Theorem 8.1 can be rephrased in terms of trapping the result of intersection, colon, sum, or product of two monomial ideals over the result of performing the same operation in a polynomial ring with variables in place of the parameters. Also observe that if I' is trapped over I and if we have another homomorphism from R into a regular domain S, then necessarily $IS = I'S$. The following theorem is our main result concerning iterated operations in characteristic p.

THEOREM 8.2. *Let R be a Noetherian ring of characteristic $p > 0$ which is a locally equidimensional homomorphic image of a C-M ring. Let $(x_1, \ldots, x_n)R$ have height n. Set $C = \mathbb{Z}/p\mathbb{Z}$ and map $A = C[X_1, \ldots, X_n]$ to R by sending X_i to x_i. Let Γ be a family of monomial ideals in the X_i in A. Then for every permissible ideal-valued function T of k variable ideals $I_1, \ldots, I_k \in \Gamma$, if $\underline{I} = (I_1, \ldots, I_k)$, and $\underline{I}R = (I_1R, \ldots, I_kR)$, then $T(\underline{I}R)$ is trapped over $T(\underline{I})$, i.e. $T(\underline{I})R \subseteq T(\underline{I}R) \subseteq (T(\underline{I})R)^*$.*

This theorem follows from Theorem 8.1 and induction on the number of operations used in T.

The reason that the colon function $U : V$ requires special treatment is that as the denominator V gets larger the value of $U : V$ becomes smaller. In the case of all the other ideal-valued functions we are considering (as well as for the numerator of $U : V$), enlarging the value of the variable ideal enlarges the value of the variable function. It is because of this that $I' : J'$ need not be trapped over $I : J$ when J' is strictly larger than J.

More specifically, let (R, m) be a two-dimensional equidimensional local ring of characteristic p which is not C-M and let x, y be a s.o.p. Since x, y is not a regular sequence, $xR : yR$ is strictly larger than xR, and so $xR : (xR : yR)$ is contained in m, and is not trapped over $XA : (XA : YA) = XA : XA = A$.

Many well-known constraints on s.o.p.s can be phrased as saying that 1 is not in some colon ideal. For instance the monomial conjecture simply says that $1 \notin (x_1^{t+1}, \ldots, x_d^{t+1}) : (x_1 \ldots x_d)^t$ for a s.o.p. $x_1, \ldots, x_d$ of a d-dimensional local ring R. This is easily deduced from the above result in characteristic p; first complete and kill a minimal prime of maximal dimension. Then Theorem 6.2 says that the colon ideal is actually in $(x_1, \ldots, x_d)^*$ and so in particular is a proper ideal.

If R is simply an equicharacteristic 0 equidimensional local ring which is the homomorphic image of a C-M ring, then even though tight closure is not necessarily defined in R, we still can prove a result similar to Theorem 8.2 by replacing the tight closure by the integral closure or even the regular closure.

References

[**B**] J.-F. Boutot, *Singularités rationelles et quotients par les groups réductifs*, Invent. Math. **88** (1987), 65–68.

[**BS**] J. Briançon and H. Skoda, *Sur la clôture intégrale d'un idéal de germes de fonctions holomorphes en un point de* $\mathbb{C}^n$, C.R. Acad. Sci. Paris Ser. A **278** (1974), 949–951.

[**EG**] G. Evans and P. Griffith, *The syzygy problem*, Annals of Math. **114** (1981), 323–353.

[**FW**] R. Fedder and K. Watanabe, *A characterization of F-regularity in terms of F-purity*, preprint (1987).

[**H**] M. Hochster, *Canonical elements in local cohomology modules and the direct summand conjecture*, J. of Algebra **84** (1983), 503–553.

[**HH**] M. Hochster and C. Huneke, *Tight closure, invariant theory, and the Briançon-Skoda theorem*, in preparation.

[**HR1**] M. Hochster and J. Roberts, *Rings of invariants of reductive groups acting on regular rings are Cohen-Macaulay*, Advances in Math. **13** (1974), 115–175.

[**HR2**] M. Hochster and J. Roberts, *The purity of Frobenius and local cohomology*, Advances in Math. **21** (1976), 117–1722.

[**K**] G. Kempf, *The Hochster-Roberts theorem of invariant theory*, Michigan Math. J. **26** (1979), 19–32.

[**Ku**] E. Kunz, *Characterizations of regular local rings of characteristic p*, Amer. J. Math. **41** (1969), 772–784.

[**LS**] J. Lipman and A. Sathaye, *Jacobian ideals and a theorem of Briançon-Skoda*, Michigan Math. J. **28** (1981), 100–222.

[**LT**] J. Lipman and B. Teissier, *Pseudo-rational local rings and a theorem of Briançon-Skoda about integral closures of ideals*, Michigan Math. J. **28** (1981), 97–116.

[**NR**] D. G. Northtcott and D. Rees, *Reductions of ideals in local rings*, Proc. Cambridge Phil. Soc. **50** (1954), 145–158.

[**PS**] C. Peskine and L. Szpiro, *Dimension projective finie et cohomologie locale*, I.H.E.S. Publ. Math. **42** (1973), 323–395.

[**W**] K. Watanabe, *Study of F-purity in dimension two*, preprint, Tokai Univ. Hiratsuka, 259-12, Japan.

Department of Mathematics, University of Michigan, Ann Arbor MI 48109

Department of Mathematics, Purdue University, West Lafayette IN 47907

Both authors were partially supported by the NSF.

Complete Ideals in Two-Dimensional Regular Local Rings

CRAIG HUNEKE

1. Introduction.

These notes present the main facts concerning the theory of complete ideals in a 2-dimensional regular local ring. Almost all of the theorems in this paper can be found in Appendix 5 of Zariski and Samuel's Commutative Algebra volume II, in Lipman's I.H.E.S. notes on rational singularities, or in some cases the papers [**G**], [**R**], and [**H-S**]. Most of the theorems are due to Zariski or to Lipman. The main point of these notes is to make this beautiful material more easily accessible to a reader who does not wish to invest a large amount of time. The main difference in our treatment from the ones above is using the Hilbert-Burch theorem systematically and inducting upon the multiplicity of the ideal considered.

Throughout these notes (R, m, k) will denote a 2-dimensional regular local ring R with maximal ideal m and infinite residue field k. An ideal I is said to be *complete* if it is integrally closed. We will say that I is *contracted* if there is some x not in m^2 such that $IS \cap R = I$, where S is equal to $R[m/x]$. Since R is regular, the powers of its maximal ideal give rise to a valuation which we denote by o (standing for 'order'). If a is in R, $o(a)$ is the greatest n such that a is in m^n. By $e(I)$ we denote the multiplicity of I and by $\mu(I)$ we denote the least number of generators of I. We say an ideal is *simple* if it is not the product of two proper ideals. We will assume familiarity with the notion of a reduction of an ideal. From the point of view of these notes practically everything we prove follows from two observations:

1. An ideal I is contracted if and only if $\mu(I) = o(I) + 1$.
2. The multiplicity of the transform of an ideal decreases.

We will first prove the two most basic theorems of Zariski; namely that the product of complete ideals in R remains complete and secondly that every complete ideal factors uniquely (up to order) into a product of simple complete ideals. Then we will show how each simple ideal corresponds to a prime divisor of the second kind and prove a basic reciprocity law due to Lipman. Finally we show that each complete ideal has reduction number 1 and compute the Hilbert function of such an ideal. A reader who wishes more information concerning these topics and generalizations to general normal rings should consult the papers in the bibliography – our purpose is to develop these basic facts as simply as possible.

We begin by discussing quadratic transformations. Let $m = (x, y)$, and let $S = R[m/x] = R[y/x]$. Then S is isomorphic to the ring $R[T]/(xT - y)$ since x and y form a regular sequence and so $S/mS = S/xS$ is isomorphic to the ring $k[T]$. In particular the maximal ideals of S which contain m are in $1-1$ correspondence with the irreducible polynomials in $k[T]$. If g is such an irreducible polynomial then we let $N = N_g$ be the corresponding maximal ideal of S and let $T = S_N$. Since $N/(x)$ is generated by g it follows that T is a 2-dimensional regular local ring with maximal ideal N. Furthermore R and T have the same quotient field and N contains m. We say T is a first quadratic transformation of R. The transform of an ideal I in T (or S) is defined as follows: let $o(I) = r$. Then if a is in I, a/x^r is in S so we may write $IS = x^r I'$ for some ideal I' in S. We call I' the *transform* of I in S, and similarly call $(I')_N = I^{\sim}$ the transform of I in T. Notice that $xS = mS$ is a height one prime ideal of S. If P is another height one prime ideal of S then x is not in P. Since $R_x = S_x$ it follows that $S_P = R_{P\cap R}$. Hence if I is any m-primary ideal of R the only height one prime of S which contains I is xS. Since by construction I' is not contained in xS it follows that I' is not contained in any height one ideal of S. As the dimension of S is two there are only finitely many maximal ideals of S which contain I'. Finally notice that S_{xS} is exactly the valuation ring corresponding to the m-adic valuation o of R. We are ready to begin the proofs of our main results.

2. Contracted Ideals.

PROPOSITION 2.1. *An ideal I is contracted from $S = R[m/x]$ if and only if $I : x = I : m$.*

PROOF: First assume I is contracted from S. If a is in R and xa is in I then we wish to show ma is contained in I. If not there is an element y in m such that ya is not in I. In this case, however, $ya = (xa)(y/x)$ is in $IS \cap R$. It follows that ya must be in I contradicting our choice of y. Next suppose that $I : x = I : m$, and let a be in $IS \cap R$. Since a is in IS it follows that $a = r/x^n$ for some n with r in Im^n. Choose n least. Write r equal to $by^n + cx$ for some b in I and c in Im^{n-1}. Since r is equal to $x^n a$ we see that $b = xd$ for some d in R. Then $r = x((yd)y^{n-1} + c)$ and yd is in I since b is in I and $I : x = I : m$. Therefore $a = r'/x^{n-1}$ for some $r' \in Im^{n-1}$ and we are done by induction. □

REMARK 2.2: If I is any m-primary ideal of R then $\mu(I) \leq o(I) + 1$.

PROOF: By the Hilbert-Burch theorem I is minimally generated by the n by n minors of an n by $n+1$ matrix with coefficients in the maximal ideal. The remark follows immediately from this fact.

PROPOSITION 2.3. *I is contracted from $S = R[m/x]$ for some x not in m^2 if and only if $\mu(I) = o(I) + 1$.*

PROOF: Set $o(I) = r$. Choose an x not in the square of m such that the leading form of x in $G = gr_m(R)$ does not divide the leading form of some f in I of order r. Then the length of $R/(I, x)$ is exactly r. On the other hand since the length of R/I is finite the length of $R/(I, x)$ is equal to the length of the module $I : x/I$. Since this module contains $I : m/I$ it follows that r is at least the dimension of $I : m/I$ with equality if and only if I is contracted from S. (Here we use Proposition 2.1). However from the resolution of I it follows that the dimension of $I : m/I =$ the dimension of $\text{Tor}_2(k, R/I) = \mu(I) - 1$. The proposition follows. □

DEFINITION 2.4. *If I is an m-primary ideal of order r, then by $c(I)$ we denote the greatest common divisor of the elements in $(I + m^{r+1})/m^{r+1}$ in $G = gr_m(R)$.*

PROPOSITION 2.5. *If I is contracted then $I = mJ$ for some J if and only if $o(I) \neq \deg(c(I))$.*

PROOF: First observe that $I = mJ$ for some J if and only if $I = m(I : m)$. Hence set $J = I : m$. Consider the exact sequence,

$$0 \to I/mJ \to J/mJ \to J/I \to 0.$$

The dimension of J/I is $\mu(I) - 1$ as in the proof of Proposition 2.3. Since I is contracted this number is $o(I)$. The dimension of the middle term of the exact sequence is $\mu(J)$ which is at most $o(J) + 1$ by Remark 2.2. Since J contains I, $o(J) \leq o(I)$. It follows that the dimension of I/mJ is at most 1 with equality if and only if $o(J) = o(I)$. Hence $I \neq mJ$ if and only if $o(J) = o(I)$ and in this case the dimension of I/mJ is one. Then mJ is contained in m^{r+1} where $r = o(I)$. Thus the dimension of $(I+m^{r+1})/m^{r+1}$ is also 1. Hence the degree of $c(I)$ is $r = o(I)$. Conversely, if $I = mJ$ then since $c(I) = c(m)c(J)$ and since $c(m) = 1$, it follows that $\deg(c(I)) = \deg(c(J)) < o(I)$. □

PROPOSITION 2.6. *The product of two ideals contracted from S is also contracted from S.*

PROOF: Suppose I and J are contracted from S. By proposition 2.1 it suffices to show that $IJ : x = IJ : m$. Let ax be in IJ. Since $R/(x)$ is a DVR we may find b in I and c in J such that $I = (b, x(I : x))$ and $J = (c, x(J : x))$. Then $IJ = (bc, xI(J : x), xJ(I : x))$. Since x, b and x, c form a regular sequence so do x, bc. Therefore ax in IJ implies that a is in the ideal $(bc, I(J : x), J(I : x))$. Hence ma is contained in IJ also since $J : x = J : m$ and $I : x = I : m$. □

COROLLARY 2.7. *If I and J are contracted from S and m divides IJ, then m divides either I or J.*

PROOF: From Proposition 2.6 IJ is contracted and so we may apply Proposition 2.5. Since m divides IJ we must have $o(IJ) > \deg(c(IJ))$. As $o(IJ) = o(I) + o(J)$ and $c(IJ) = c(I)c(J)$ it follows that either $o(I) > \deg(c(I))$ or $o(J) > \deg(c(J))$. Applying Proposition 2.5 gives the required result. □

3. Complete Ideals.

If I is an ideal of R, by $\overline{I}$ we denote the integral closure of I.

PROPOSITION 3.1. *Suppose I is complete. Then there is an x not in m^2 such that $I : x = I : m$.*

PROOF: Since I is complete, I is the intersection of IV over all valuation rings V in the quotient field K of R which contain R. As the length of R/I is finite, it follows that there are finitely many valuation rings $V_1, \ldots, V_n$ such that $I = IV_1 \cap \cdots \cap IV_n \cap R$. In particular if a is in R then a is in I if and only if aV_i is contained in IV_i for each i. Choose an x in R, necessarily not in the square of m, such that $xV_i = mV_i$ for each i. Since k is infinite this choice is possible. Now suppose a is in $I : x$. Then xa is in I so that

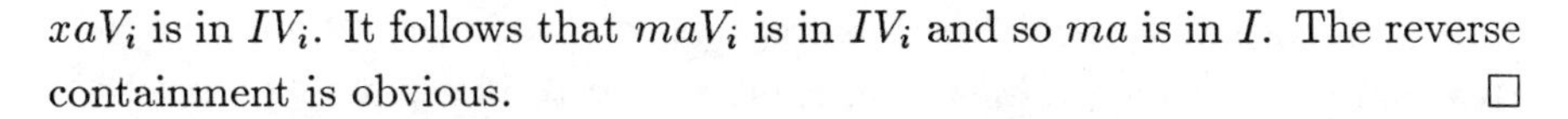
xaV_i is in IV_i. It follows that maV_i is in IV_i and so ma is in I. The reverse containment is obvious. □

COROLLARY 3.2. *Any complete ideal is contracted.*

PROOF: This follows immediately from Propositions 2.1 and 3.1.

LEMMA 3.3. *If I is complete then mI is complete.*

PROOF: Set $J = \overline{mI}$. Choose an x not in the square of the maximal ideal in the same manner as in Proposition 3.1. Since I is complete it follows that $I = J : x$. As R/Rx is a DVR every ideal is complete and so in particular J is contained in (mI, x). Hence $J = mI + x(J : x) = mI$. □

PROPOSITION 3.4. *Suppose I is complete and let $S = R[m/x]$ be such that I is contracted from S. Furthermore let $J = I'$ be the transform of I in S. Then J is complete.*

PROOF: If $r = o(I)$, then $x^r J = IS$. Since S is normal to show that J is complete it suffices to show that IS is complete. Suppose s is in S and is integral over IS. Writing out an equation of integral dependence for s over IS and clearing denominators shows that there is an n such that $t = x^n s$ is in R and is integral over Im^n. From Lemma 3.3 it follows that Im^n is complete and so t is in Im^n. Hence $s = t/x^n$ is in IS. □

We now need to define the inverse transform of an ideal J in S. Since any such J is finitely generated there is an a which is least with respect to the property that $x^a J$ is extended from an ideal in R. We call $I = x^a J \cap R$ the inverse transform of J. Obviously I is contracted.

PROPOSITION 3.5. *Let I be a simple complete m-primary ideal in R which is contracted from $S = R[m/x]$. Then $I'S$ is also simple.*

PROOF: Suppose there are proper ideals J' and K' in S such that $I'S = J'K'$. Let J (respectively K) be the inverse transforms of J' (respectively K'). Let a and b be the integers such that $JS = x^a J'$ and $KS = x^b K'$. Then $x^{a+b} I'S = JKS$, and so $m^{a+b} IS = JKm^r S$ where $r = o(I)$. Since I is contracted from S so is $m^{a+b} I$ by Proposition 2.6. By the same proposition $m^r JK$ is contracted from S since both J and K are contracted from S by construction. Hence $m^r JK = m^{a+b} I$. By Proposition 2.7 we can cancel the powers of the maximal ideal from this equation. Since I is simple it follows that either J or K, say J, is a power of m. This forces J' to be equal to S. □

This essentially is the end of the progress we can make without an inductive tool to reduce questions concerning complete ideals in R to complete ideals in some quadratic transformation T or R. To use induction we need some numerical invariant of a complete ideal which goes down as we pass from I to I'. As it turns out, the multiplicity works. Recall if I is an m-primary ideal with a minimal reduction (u,v) then $e(I) = \ell(R/(u,v))$ where by $\ell()$ we denote the length of an R module. The next proposition was essentially known to Northcott [**N**].

PROPOSITION 3.6. *Let I be an m-primary ideal of R and let J be the transform of I in some quadratic transformation T of R. Then $e(I) > e(J)$.*

PROOF: We assume T is the localization of $S = R[m/x]$ at a maximal ideal N containing mS. Choose a minimal reduction u, v of I in R and set $r = o(I)$. We may assume that $o(u) = r$ and that x, u is an s.o.p. First we claim that the elements $u'' = u/x^r$ and $v'' = v/x^r$ are a minimal reduction of I' in T. It suffices to show that I' is integral over (u'', v''). It is enough to prove $IS = x^r I'$ is integral over $x^r(u'', v'') = (u,v)S$ which is obvious. We wish to show that $\ell_R(R/(u,v)) > \ell_T(T/(u'',v''))$. Since T/N is a finite algebraic extension of k we have the following sequence of inequalities: $\ell_T(T/(u'',v'')) \leq \ell_R(T/(u'',v'')) \leq \ell_R(S/(u'',v'')) < \ell_R(S/(u'',v))$. We claim that $u''S \cap R = uR$. Since S and R are the same after inverting x this equation holds provided x is a non zero-divisor modulo uR. This is true by our choice of u. Hence $R/(u)$ imbeds in $S/(u'')$ and the quotient has finite length as an R-module since $(R/(u))_x = (S/(u''))_x$. It follows that the multiplicity of $R/(u)$ with respect to v is the same as that of $S/(u'')$ with respect to v. Hence $\ell_R(R/(u,v)) = \ell_R(S/(u'',v))$ which proves our claim. □

We are now ready to prove the first major theorem of Zariski concerning products of complete ideals.

THEOREM 3.7. *If I and J are complete ideals of R then IJ is also a complete ideal.*

PROOF: Induct on $t = \min(e(I), e(J))$. If this minimum is equal to 1 then either I or J is the maximal ideal m and in this case Lemma 3.3 gives the theorem. Hence we may assume that $t > 1$. By Corollary 3.2 and the proof of Proposition 3.1 we may assume that I and J are both contracted from $S = R[m/x]$ for some x not in the square of m. By

Proposition 2.6 IJ is also contracted from S so to show IJ is complete it suffices to show that IJS is complete. As S is normal it suffices to show that $I'J'$ is complete. This ideal is contained in only finitely many prime ideals of S, all of them maximal. Hence $I'J'$ is the intersection of its localizations at each of these maximal ideals so to show it is complete it suffices to show that $I'J'T$ is complete where T is the localization of S at a maximal ideal N which contains $I'J'$. By Proposition 3.5 $e(I'T) < e(I)$ and $e(J'T) < e(J)$. Therefore the induction finishes the proof provided these ideals are complete. However Proposition 3.4 tells us that I' and J' are complete. □

We will next show the main result of Zariski, namely that the class of complete ideals in R have unique factorization into simple complete ideals. We begin by making some observations. Firstly from the noetherian property every ideal can be factored into simple ideals. By taking integral closures and continuing this process it is easy to see that every complete ideal is factorable into a product of simple complete ideals. Furthermore the monoid of complete ideals (it is closed under multiplication by Theorem 3.7) has cancellation since if I and J are complete ideals and if M is any finitely generated faithful module such that $IM = JM$ then necessarily $I = J$. Hence to prove unique factorization it suffices to prove that if a complete simple m-primary ideal I divides a product $J_1 \ldots J_n$ of m-primary simple ideals then $I = J_i$ for some i. We have already shown that the maximal ideal has this property. (See Corollary 2.7.)

REMARK 3.8: The transform I' of a simple complete ideal in S is contained in a unique maximal ideal N, and $(I')_N$ is simple.

PROOF: We know that I' is contained in only finitely many maximal ideals. The Chinese Remainder Theorem shows that I' is the product of its primary components. Since I' is simple by Proposition 3.5 it follows that there is a unique maximal ideal N which contains I'. If $(I')_N$ were a product of two proper ideals then it easily follows that I' would be also. □

THEOREM 3.9. *Every m-primary complete ideal of R is uniquely factorable (disregarding the order of the factorization) into simple complete ideals.*

PROOF: The discussion above reduces the theorem to proving that if I is simple and complete and divides a product $J_1 \ldots J_n$ of simple complete ideals then I must equal one of the J's. We prove this by induction on the multiplicity of I. If the multiplicity is one then $I = m$ and we have already shown the result. Assume the multiplicity is greater than 1 and choose an x not in the square of m such that I and all of the J's are contracted from $S = R[m/x]$. Let N be the unique maximal ideal containing I'. Since o is a valuation clearly I' divides $J_1' \ldots J_n'$. Now localize at N. When we localize the J' at N either they become equal to the whole ring or else they remain simple, in which case N is the only maximal ideal containing them. We apply our induction to the ideal $(I')_N$. We may conclude that there is an i such that $(J_i')_N$ is equal to $(I')_N$. However since N is the only prime ideal containing either of these ideals we obtain that in fact $I' = J_i'$. Let $o(I) = r$ and $o(J_i) = s$. Then $m^s IS = x^{r+s} I' = x^{r+s} J_i' = m^r J_i S$. Since products of contracted ideals are contracted we obtain that $m^s I = m^r J_i$. We may cancel common powers of m and since both I and J_i are assumed to be simple it follows that they are equal which proves the theorem. □

4. Simple Complete Ideals.

We are now ready for the second main step of our program. In this section we associate a valuation v to a simple complete ideal I. Throughout this section we will work with $S = R[m/x]$ such that I is contracted from S.

As a result of Remark 3.8 there is a unique maximal ideal N containing I' and $(I')_N$ will also be simple and complete. Therefore we obtain a sequence of quadratic transformations and simple ideals, (R, I), $(R_1, I_1), \ldots, (R_n, I_n), \ldots$. Since the multiplicity of these ideals is strictly decreasing it follows that there is an n such that $I_n = m_n$, the maximal ideal of R_n. Thus we obtain a valuation associated to I, namely the m_n-adic order valuation of R_n. We denote this valuation by v_I, or if there is no confusion by v. Our first objective is to show that the powers of I are valuation ideals for this valuation.

Proposition 4.1. *Let I and v be as above. Then for all $s \geq 1$, I^s is a valuation ideal for v, i.e. if V is the valuation ring of v then $I^s V \cap R = I^s$.*

Proof: We induct on n, where n is chosen as above. If $n = 1$ then $I = m$ and we know that the powers of m are valuation ideals for the order valuation. Now set $J = I_1$. By induction we know that J^s are valuation ideals for v for all s. Let $T = R_1$. If $o(I) = r$ by definition $x^r J = IT$. Set $q_s = I^s V \cap R$. We first claim that $q_s \cap m^{rs}$ is equal to I^s. Clearly I^s is contained in the intersection. To see the converse let $u \in q_s \cap m^{rs}$. Since

$v(u) \geq v(I^s) = v(x^{rs}J^s)$ we see that $u' = u/x^{rs}$ has the property that u' is in T (since u is in m^{rs}) and furthermore $v(u') \geq v(J^s)$. Hence by induction u' is in J^s. Then $u = x^{rs}u'$ is in $I^sT \cap R = I^s$ since I and therefore I^s is contracted from $R[m/x]$ and hence from T.

To finish the proof it suffices to show that q_s is contained in m^{rs}. Assume that this is false. Set $b = o(q_s)$. We see that $m^{rs-b}q_s$ is contained in the intersection of m^{rs} with q_s and hence lies in I^s. Since the order of I^s is equal to the order of $m^{rs-b}q_s$ it follows that $c(I^s)$ must divide $c(m^{rs-b}q_s) = c(q_s)$. In this case $\deg(c(I^s)) \leq \deg(c(q_s)) \leq o(q_s) < o(I^s)$. By Proposition 2.5 m divides I^s and hence by Corollary 2.7 m must divide I which is a contradiction. □

THEOREM 4.2. *Let I be a simple complete m-primary ideal of R and let v be the associated valuation with valuation ring V. Choose an element f in I of minimal value with respect to the valuation v and set $B = R[I/f]$. Let p be the center of v on B. Then B is normal, p has height one, p is the only minimal prime containing mB, and B_p is V.*

PROOF: First we claim that $B = V \cap R_f$. One containment is clear while if r/f^n is in V then $v(r) > v(I^n)$. By the above proposition, r is in I^n and so r/f^n is in B. Consequently B is normal. We next claim p is the nilradical of fB. For if t is in p then for some q, $v(t^q) > v(f)$. It follows that t^q/f is in B, and hence t is in the nilradical of p. Therefore p has height one and as B is normal it follows that B_p is a DVR contained in V, and thus equal to V. Since p is the only prime containing f, it is the only prime containing mB. □

Now we will prove a beautiful formula due to Lipman [**L**, Prop. 21.4]. Let I and J be two m-primary simple ideals with associated DVR's v and w. In the case that k is algebraically closed Lipman's formula gives that $w(I) = v(J)$. Our proof shows that this number is essentially the length of R modulo the ideal generated by a general element of I and a general element of J. Before we begin the proof we observe some preliminary facts.

Let I be an m-primary simple ideal and let a, b be a minimal reduction of I. By A denote the ring $R[a/b]$ and by B denote the ring $R[I/b]$. Since I is integral over (a, b) it is easily seen that B is integral over A. On the other hand as we saw in the proof of Theorem 4.2, B is integrally closed and is therefore the integral closure of A. Since a, b form a regular sequence the ring A is isomorphic to $R[x]/(bx - a)$ where x is a variable.

Hence A/mA is isomorphic to a polynomial ring $k[a/b]$. In Theorem 4.2 we proved that there is a unique minimal prime p containing mB which must then necessarily contract to mA. Furthermore $B_p = V$, the valuation ring of v. Let $W = A - mA$ and put $C = A_W$ and $D = B_W$. Then C is a 1-dimensional local domain and D is integral over it. Hence D is 1-dimensional and every maximal ideal of D contracts to the maximal ideal of C. Since p is the only such ideal in B we see that $D = V$. In particular the residue field of V is finite over the residue field of C which is $k(a/b)$. We denote the degree of this extension by $\Delta(v)$. We can now state our main result.

THEOREM 4.3. *Let I and J be two m-primary simple complete ideals with associated valuations v and w respectively. Then,*

$$w(I)\Delta(w) = v(J)\Delta(v).$$

PROOF: Let a, b be a reduction of I and c, d be a reduction of J. We prove the theorem by showing that both $v(J)\Delta(v)$ and $w(I)\Delta(w)$ are the length of the ring $H = R(x,y)/(b - ax, c - dy)$ where x and y are variables and $R(x,y) = R[x,y]_{m[x,y]}$. To achieve this it is enough to show that this length is equal to $v(J)\Delta(v)$ since both are symmetric. Let $S = R(y) = R[y]_{m[y]}$. We extend v to a valuation $\mathbf{v}$ of $R(y)$ by defining $\mathbf{v}(a_0 + a_1 y + \cdots + a_n y^n) = \min\{v(a_0), \ldots, v(a_n)\}$. This defines the valuation on $R[y]$ which one extends in the usual way to the quotient field of $R[y]$. Let A and B be defined as above. Then $S[I/b]$ is equal to $B(y)$ and $S[a/b]$ is equal to $A(y)$. Set $S[I/b] = G$ and $S[a/b] = L$. Then G is still integrally closed and nL is prime where $n = m(y)$ is the maximal ideal of S. Also pG must still be the unique minimal prime of nG. The residue field of $E = L_{nL}$ is $k(y, a/b)$. This shows that the valuation determined by the DVR $\mathbf{V} = G_{pG}$ has the property that the image of y in its residue field is purely transcendental and this implies the valuation determined by this valuation ring is exactly $\mathbf{v}$. Let $f = c - dy$ in S. Then $\ell_E(E/fE) = \ell_E(V/fV) = \ell_V(V/fV)[k(V) : k(E)] = \mathbf{v}(f)[k(V)(y) : k(a/b, y)] = \min\{v(c), v(d)\}[k(V) : k(a/b)] = v(J)\Delta(v)$. On the other hand consider what E/fE is isomorphic to. E is equal to L_{nL} which is isomorphic to $(R(y)[x]/(a - bx))_{n[x]}$ which is isomorphic to $R(x,y)/(a-bx)$ so that E/fE is isomorphic to $R(x,y)/(a-bx, c-dy) = H$. This equality finishes the proof of the theorem. □

COROLLARY 4.4. *Let I, J, v, w be as in Theorem 4.4. Assume further that k is algebraically closed. Then $v(J) = w(I)$.*

PROOF: It clearly suffices to show that $\Delta(v) = 1$ since by symmetry we will obtain that $\Delta(w) = 1$. Recall that $\Delta(v)$ is the degree of the field extension $k(v)$ over $k(a/b)$ where a, b is a reduction of I. Consider the quadratic sequence determined by $I : (R, I), \ldots, (R_n, I_n)$, where v is the $I_n = m_n$-adic order valuation of R_n. If $r = o(I)$ then we let $a_1 = a/x^r$ and $b_1 = b/x^r$. We have seen, I_1 is still integral over (a_1, b_1) and so a_1, b_1 is a reduction of I_1. By induction we define a_i, b_i in a similar manner. We eventually obtain that $t = a_n$ and $s = b_n$ must generate the maximal ideal m_n of R_n. Since the valuation ring V of v is just $R_n[t/s]_{sR_n[t/s]}$, it follows that $k(v)$ is just $k(R_n)(t/s)$ and in addition t/s is purely transcendental over $k(R_n)$. On the other hand by construction $a/b = a_1/b_1 = \cdots = t/s$. Hence we obtain that $\Delta(v) = [k(v) : k(a/b)] = [k(R_n)(a/b) : k(a/b)] = [k(R_n) : k]$. However R_n is a finitely generated algebra over R localized at a maximal ideal contracting to m so that $k(R_n)$ is algebraic over k. Since k is algebraically closed this degree is just 1. ☐

Observe that this proof shows that $\Delta(v)$ is just the degree of the algebraic closure of k in $k(v)$ over k.

5. Hilbert Functions.

In this section we give the Hilbert function of an m-primary complete ideal I. By definition the Hilbert function is the function, $H_I(n) = \ell(R/I^n)$. It is well known this function is a polynomial of degree two in n for large n. In fact for complete ideals it turns out to be a polynomial for all n. This is due to the next theorem which is due to Lipman and Teissier [**L-T**], but the proof is from [**H-S**].

THEOREM 5.1. *Let I be an m-primary complete ideal, and let a, b be a minimal reduction of I. Then $I^2 = (a, b)I$.*

PROOF: We prove this theorem by induction on $e(I)$. If $e(I) = 1$ then I is m and the theorem is true since in this case $m = (a, b)$. So assume $e(I) > 1$. We first claim that the ideal $J = (a, b)I$ is contracted. By Proposition 2.3 it suffices to show that $o(J) + 1 = \mu(J)$. Since I is contracted, $\mu(I) = o(I) + 1$. Set $\mu(I) = r + 2$ and choose $c_1, \ldots, c_r$ which together with a and b form a minimal generating set of I. Then I claim that $ac_1, \ldots, ac_r, bc_1, \ldots, bc_r, ab$ form a minimal generating set of J. Clearly these generate J. Suppose there is a relation,

$$\sum_{i=1}^{r} r_i a c_i + \sum_{i=1}^{r} s_i b c_i + uab = 0.$$

Since a and b form a regular sequence we obtain there exists an element v such that,

$$av = \sum_{i=1}^{r} s_i b_i + au, \text{ and}$$

$$-bv = \sum_{i=1}^{r} r_i c_i.$$

Since a, b and the c's form a minimal generating set of I these equations force that u, v, r_i, and s_i are all in m. This proves our claim. Hence $\mu(J) = 2r + 1 = o(J) + 1$ so that J is contracted. Since k is infinite we may assume that I and J are contracted from the same $S = R[m/x]$. To show that $J = I^2$ it suffices to show that $JS = I^2 S$ since both ideals are contracted from S. Notice that both of these ideals have the same order. In fact if $o(I) = r$, then $JS = x^{2r}(a', b')I'$ where $a' = a/x^r$ and $b' = b/x^r$. Similarly $I^2 S = x^{2r}(I')^2$. We also have seen that a' and b' form a reduction of I'. Obviously it is enough to prove that $(a', b')I' = (I')^2$. This question is local on the primes which contain $(a', b')I'$. There are only finitely many such ideals, all of them maximal. If we localize at one of these primes then the multiplicity of I' has decreased from that of I by proposition 3.5 and the induction then implies the needed equality. □

We now give the formula for the Hilbert function.

THEOREM 5.2. *Let I be an m-primary complete ideal. Then for all $n \geq 1$, $\ell(R/I^n) = en^2/2 + (2\ell - e)n/2$, where $e = e(I)$ and $\ell = \ell(R/I)$.*

PROOF: Let a, b be a reduction of I. Then using Theorem 5.1 we see there are exact sequences for all n,

$$0 \to (a,b)^{n-1}/I^n \to R/I^n \to R/(a,b)^{n-1} \to 0.$$

However $(a,b)^{n-1}/I^n = (a,b)^{n-1}/(a,b)^n \otimes R/I$ and so it is a free R/I-module of rank n. Since $\ell(R/(a,b)^{n-1}) = (1 + 2 + \cdots + (n-1))\ell(R/(a,b))$ and as $\ell(R/(a,b)) = e$, the formula follows easily. □

REMARK 5.3: The fact that the constant term of the Hilbert polynomial given in (5.2) is zero is quite important. See **[R]** for a substantial development of this.

ACKNOWLEDGEMENT: I'd like to thank Bill Heinzer, Joe Lipman, and Judith Sally for many valuable conversations concerning the material in this paper.

References

[C] S. Cutkosky, *On unique factorization of complete ideals*, preprint.

[G] H. Göhner, *Semifactoriality and Muhly's condition* (N) *in two dimensional local rings*, J. Algebra **34** (1975), 403–429.

[H-S] C. Huneke and J. Sally, *Birational extensions in dimension two and integrally closed ideals*, J. Algebra **115** (1988), 481–500.

[L] J. Lipman, *Rational singularities with application to algebraic surfaces and unique factorization*, I.H.E.S. **36** (1969), 195–279.

[L2] J. Lipman, *On complete ideals in regular local rings*, preprint.

[L-T] J. Lipman and B. Teissier, *Pseudo-rational local rings and a theorem of Briançon-Skoda about integral closures of ideals*, Mich. Math. J. **28** (1981), 97–116.

[N] D. G. Northcott, *Abstract dilatations and infinitely near points*, Proc. Cambridge Philos. Soc. **52** (1956), 178–197.

[R] D. Rees, *Hilbert functions and pseudo-rational local rings of dimension 2*, J. London Math. Soc. (2) **24** (1981), 467–479.

[Z-S] O. Zariski and P. Samuel, "Commutative Algebra, Vol. 2," Van Nostrand, Princeton, 1960.

Department of Mathematics, Purdue University, West Lafayette IN 47907

The author is partially supported by the NSF.

Powers of Licci Ideals

C. HUNEKE AND B. ULRICH

§1. Introduction and definitions.

In this paper we are concerned with the question whether or not the powers of a licci ideal I define Cohen-Macaulay (C-M) rings, and if the higher conormal modules I^n/I^{n+1} are C-M. Two ideals I and J in a C-M local ring R are said to be *linked* if there is a regular sequence $\underline{\alpha} = \alpha_1, \ldots, \alpha_g$ such that $(\underline{\alpha}) : I = J$ and $(\underline{\alpha}) : J = I$ (we write $I \sim J$). Two ideals I and J in R are said to be in the same linkage class if there exist $I_1, \ldots, I_n$ such that $I \sim I_1 \sim \cdots \sim I_n \sim J$. Finally I is *licci* if it is in the linkage class of an ideal generated by a regular sequence. A licci ideal I is known to be strongly Cohen-Macaulay (SCM) [**H1**] which means that if one chooses any generating set $f_1, \ldots, f_n$ of I then the Koszul homology modules $H_i(f_1, \ldots, f_n; R)$ are C-M modules. As it turns out one can control the powers of an SCM ideal sufficiently to allow us to prove our main Theorem 2.8.

Theorem 2.8 gives an affirmative answer in the special case of licci ideals to a question posed by W. Vasconcelos: Let I be a perfect generically complete intersection ideal in a regular local ring R such that the conormal module I/I^2 is C-M, then is I a Gorenstein ideal (i.e., is R/I Gorenstein)? Theorem 2.8 also generalizes results of [**H3**] to arbitrary dimension: let R be a regular local ring and let p be a licci prime ideal such that $\dim(R/p) = 1$. If R/p is not Gorenstein, then $p^{(n)} \neq p^n$ for all $n \geq 2$ where $p^{(n)}$ is the nth symbolic power of p. If R/p is Gorenstein but not a complete intersection then $p^{(2)} = p^2$ but $p^{(n)} \neq p^n$ for all $n \geq 3$.

We need a few more definitions before we begin the proofs. Let R be a Noetherian local ring and let I be an R-ideal. Then $\mu(I)$ denotes the minimal number of generators of I, $\mathrm{ht}(I)$ is the height of I, $d(I) = \mu(I) - \mathrm{grade}(I)$ is the deviation of I, and I is called generically a complete intersection if I_p is generated by an R_p-regular sequence for all $p \in \mathrm{Ass}(R/I)$. We say that I satisfies G_∞ if $\mu(I_p) \leq \dim R_p$ for all prime ideals p containing I. A pair (S, J) consisting of a local ring S and an ideal J, is a deformation of (R, I) if there exists a sequence $\underline{a} = a_1, \ldots, a_n$ in S which is regular on S and S/J such that $(S/(\underline{a}), (J, \underline{a})/(\underline{a})) = (R, I)$ (we also say that S/J is a deformation of R/I). Finally, (S,J) is essentially a deformation of (R,I) if (S,J) is obtained from (R,I) by a sequence of deformations, localizations, and purely transcendental changes of the residue class field (cf. [**HU**], 2.2).

The length of a finitely generated R-module M will be denoted by $\ell(M)$. We say that M has a rank and write $\operatorname{rank}_R M = n$ if for all $p \in \operatorname{Ass}(R)$, M_p is a free R_p-module of constant rank n. Moreover, for a system of parameters $\underline{x}$ of M, $e(\underline{x}; M)$ will denote the multiplicity symbol from [**N**], or equivalently, the multiplicity of M with respect to the ideal $(\underline{x})R$.

§2. Proof of the main result.

We first need several lemmas. The statements of the lemmas are in some cases stronger than we need, but we give them for future reference.

LEMMA 2.1. *Let R be a local Cohen-Macaulay ring, let I be an R-ideal which is SCM and G_∞, and let (S,J) be a deformation of (R,I). Then $S[Jt]$, $S[Jt,t^{-1}]$, and $\operatorname{gr}_J(S)$ are deformations of $R[It] \cong S[Jt] \otimes_S R$, $R[It,t^{-1}] \cong S[Jt,t^{-1}] \otimes_S R$, and $\operatorname{gr}_I(R) \cong \operatorname{gr}_J(S) \otimes_S R$ respectively.*

PROOF: We only show the claim for the Rees algebra. First we claim that since I satisfies G_∞, J also satisfies G_∞. Let q be a prime in S such that $q \supset J$. Set $\operatorname{ht}(q) = t$, and let $x_1, \ldots, x_d$ ($d = \dim S - \dim R$) be a regular sequence on S and S/J such that $(R,I) \cong (S/(\underline{x}), (J,\underline{x})/(\underline{x}))$. There must be a prime Q in S of height at most $t+d$ which contains $q + (\underline{x})$ (S is C-M and hence equidimensional and catenary). Then $\mu(J_q) \leq \mu(J_Q) = \mu(((J,\underline{x})/(\underline{x}))_Q)$ (since $x_1, \ldots, x_d$ is a regular S/J-sequence) $= \mu(I_{\overline{Q}})$ (where the bar denotes the image of Q in R) $\leq \operatorname{ht}(\overline{Q})$ since I satisfies G_∞. Hence $\mu(J_q) \leq \operatorname{ht}(\overline{Q}) \leq \operatorname{ht}(Q) - d \leq (t+d) - d = t$. Thus J satisfies G_∞.

Next we claim that since I is SCM and I and J are generically complete intersections, then J is also SCM. Let $f_1, \ldots, f_n$ generate J. Then the images $f_1', \ldots, f_n'$ of $f_1, \ldots, f_n$ in R generate I. By induction on i we claim that $H_i(f_1, \ldots, f_n; S)$ is C-M. If $i = 0$, $H_0(f_1, \ldots, f_n; S) \cong S/J$ is C-M since R/I is C-M and $x_1, \ldots, x_d$ are a regular sequence on S/J. Suppose that $H_0(\underline{f}; S), \ldots, H_{i-1}(\underline{f}; S)$ are C-M and $i \leq n - \operatorname{ht}(J)$. Then Lemma 2.15 of [**H1**] shows that $H_i(\underline{f}; S) \otimes R \cong H_i(\underline{f}; R) = H_i(\underline{f}'; R)$. This latter module is C-M of dimension $\dim(R/I)$ (see 2.4 of [**H1**]). We want to show that

$H_i(\underline{f};S)$ is C-M. This is a module of dimension $\dim(S/J) = \dim(R/I) + d$, and it has a well-defined rank since J is generically a complete intersection. If $\mathrm{ht}(J) = g$ then (see 2.5 of [**H1**]), $\mathrm{rank}_{S/J} H_i(\underline{f};S) = \binom{n-g}{i}$. Let $\underline{y}$ be a system of parameters for $A = R/I$, which is lifted to $B = S/J$. Then,

$$\begin{aligned} e(\underline{x},\underline{y};H_i(\underline{f};S)) &= \binom{n-g}{i} e(\underline{x},\underline{y};B) \\ &= \binom{n-g}{i} \ell(B/(\underline{x},\underline{y})) \quad \text{since } B \text{ is C-M,} \\ &= \binom{n-g}{i} \ell(A/(\underline{y})) = \binom{n-g}{i} e(\underline{y};A) \\ &= e(\underline{y};H_i(\underline{f}';R)), \end{aligned}$$

since I is generically a complete intersection and $\mathrm{ht}(I) = g$. On the other hand, since I is SCM, $e(\underline{y};H_i(\underline{f}';R)) = \ell(H_i(\underline{f}';R) \otimes A/(\underline{y})) = \ell(H_i(\underline{f};S) \otimes_B B/(\underline{x}) \otimes_A A/(\underline{y}))$ by above, $= \ell(H_i(\underline{f};S) \otimes_B B/(\underline{x},\underline{y}))$. Therefore $e(\underline{x},\underline{y};H_i(\underline{f};S)) = \ell(H_i(\underline{f};S) \otimes B/(\underline{x},\underline{y}))$, and it follows from [**N**], that $H_i(\underline{f};S)$ is C-M.

Because I and J are strongly Cohen-Macaulay and G_∞, it follows that $\mathrm{Sym}(I) \cong R[It]$, and $\mathrm{Sym}(J) \cong S[Jt]$, and that these rings are Cohen-Macaulay [**HSV**]. Also $I \cong J/(\underline{x}) \cap J = J/(\underline{x})J \cong J \otimes_S R$, therefore

$$\begin{aligned} R[It] \cong \mathrm{Sym}(I) \cong \mathrm{Sym}(J \otimes_S R) &\cong \mathrm{Sym}(J) \otimes_S R \\ &\cong S[Jt] \otimes_S R \cong S[Jt]/(\underline{x})S[Jt]. \end{aligned}$$

Since moreover $S[Jt]$ is a Cohen-Macaulay ring, and $d = \dim S - \dim R = \dim S[Jt] - \dim R[It]$, we conclude that $\underline{x} = x_1, \ldots, x_d$ is a regular sequence on $S[Jt]$.

□

LEMMA 2.2. *Let R be a local Cohen-Macaulay ring, let I be an R-ideal which is generically a complete intersection, let (S,J) be a deformation of (R,I), and let $n \geq 1$ be a fixed integer. Further assume that either I^i/I^{i+1} are Cohen-Macaulay modules for all $0 \leq i \leq n-1$, or else that I is strongly Cohen-Macaulay and G_∞. Then I^n/I^{n+1} is a Cohen-Macaulay module if and only if J^n/J^{n+1} is a Cohen-Macaulay module.*

PROOF: The ideals I and J are both generically complete intersections [**HU**, 2.3], Cohen-Macaulay, and they have the same grade g. Write $A = R/I$, $B = S/J$, and let $\underline{x}$ be a regular sequence on S and B with $(R,I) = (S/(\underline{x}), (J,\underline{x})/(\underline{x}))$.

Let us first assume that I^i/I^{i+1} are Cohen-Macaulay for $0 \leq i \leq n-1$. Using induction on n, we may assume that J^i/J^{i+1} are Cohen-Macaulay for $0 \leq i \leq n-1$, and therefore S/J^n is a Cohen-Macaulay ring. In particular, $\underline{x}$ is a regular sequence on S/J^n, and then tensoring the exact sequence

$$0 \to J^n/J^{n+1} \to S/J^{n+1} \to S/J^n \to 0$$

by $\otimes_S R$, it follows that

$$I^n/I^{n+1} \cong J^n/J^{n+1} \otimes_S R \cong J^n/J^{n+1} \otimes_B A.$$

If however I is strongly Cohen-Macaulay and G_∞, then the same isomorphism $I^n/I^{n+1} \cong J^n/J^{n+1} \otimes_B A$ follows from Lemma 2.1.

In either case, $I^n/I^{n+1} \cong J^n/J^{n+1} \otimes_B A$, where $A = B/(\underline{x})$, B is a local Cohen-Macaulay ring and $\underline{x}$ is a B-regular sequence. Moreover, since I and J are generically complete intersections of the same grade g, $\operatorname{rank}_A I^n/I^{n+1} = \binom{n+g-1}{n} = \operatorname{rank}_B J^n/J^{n+1}$. But then it follows from [**He2**, 1.2], or [**N**], or an argument as in the proof of 2.1 that I^n/I^{n+1} is Cohen-Macaulay if and only if J^n/J^{n+1} is Cohen-Macaulay. □

The next lemma is a slight generalization of [**H3**, 2.1].

LEMMA 2.3. *Let R be a local Cohen-Macaulay ring, let I be a Cohen-Macaulay R-ideal which is generically a complete intersection, let (S, J) be a deformation of (R, I), and let $n \geq 1$ be a fixed integer. Then R/I^n is a Cohen-Macaulay ring if and only if S/J^n is a Cohen-Macaulay ring.*

PROOF: Again, I and J are both generically complete intersections and Cohen-Macaulay ideals of the same grade g. Let $\underline{x}$ be a regular sequence on S and S/J with $(R, I) = (S/(\underline{x}), (J, \underline{x})/(\underline{x}))$.

Now let $\underline{y}$ be a sequence in S such that the images of $\underline{x}, \underline{y}$ form a system of parameters in the ring S/J. Then $\underline{x}, \underline{y}$ (or rather their images) also form a system of parameters in S/J^n, $\underline{y}$ form a system of parameters in R/I, and $\underline{y}$ form a system of parameters in R/I^n. Also note that $S/(J^n, \underline{x}, \underline{y}) \cong R/(I^n, \underline{y})$.

Now by [**N**], S/J^n is a Cohen-Macaulay ring if and only if

$$e(\underline{x}, \underline{y}; S/J^n) = \ell(S/(J^n, \underline{x}, \underline{y})),$$

and likewise, R/I^n is a Cohen-Macaulay ring if and only if

$$e(\underline{y}; R/I^n) = \ell(R/(I^n, \underline{y})).$$

Since $S/(J^n, \underline{x}, \underline{y}) \cong R/(I^n, \underline{y})$, it therefore suffices to prove that

$$e(\underline{x}, \underline{y}; S/J^n) = e(\underline{y}; R/I^n).$$

Now let $P = \{p \in V(J^n) | \dim(S/p) = \dim(S/J^n)\}$. It turns out that $P = \{p \in V(J^n) | \dim(S/p) = \dim(S/J)\} = \operatorname{Ass}(S/J)$. For all $p \in P$, $\ell((S/J^n)_p) = \binom{n-1+g}{n-1}\ell(S_p/J_p)$, since J is generically a complete intersection of grade g. Now the associativity formula for multiplicities ([**Na**], 24.7) yields

$$\begin{aligned} e(\underline{x}, \underline{y}; S/J^n) &= \sum_{p \in P} \ell((S/J^n)_p) e(\underline{x}, \underline{y}; S/p) \\ &= \binom{n-1+g}{n-1} \sum_{p \in P} e(\underline{x}, \underline{y}; S/p) \ell(S_p/J_p). \end{aligned} \tag{2.4}$$

In the special case $n = 1$, (2.4) reads as

$$e(\underline{x}, \underline{y}; S/J) = \sum_{p \in P} e(\underline{x}, \underline{y}; S/p) \ell(S_p/J_p). \tag{2.5}$$

Inserting (2.5) into (2.4) we obtain

$$e(\underline{x}, \underline{y}; S/J^n) = \binom{n-1+g}{n-1} e(\underline{x}, \underline{y}; S/J). \tag{2.6}$$

On the other hand, S/J is Cohen-Macaulay, and therefore $e(\underline{x}, \underline{y}; S/J) = \ell(S/(J, \underline{x}, \underline{y}))$. Thus (2.6) implies that

$$e(\underline{x}, \underline{y}; S/J^n) = \binom{n-1+g}{n-1} \ell(S/(J, \underline{x}, \underline{y})).$$

By the same arguments it follows that

$$e(\underline{y}; R/I^n) = \binom{n-1+g}{n-1} \ell(R/(I, \underline{y})).$$

But then since $S/(J, \underline{x}, \underline{y}) \cong R/(I, \underline{y})$, we conclude that $e(\underline{x}, \underline{y}; S/J^n) = e(\underline{y}; R/I^n)$.

□

Aspects of the next lemma are also contained in [**HSV**], [**H2**], [**SV**]. Recall that an ideal I is said to be *perfect* if $\operatorname{grade}(I) = \operatorname{pd}(R/I)$.

LEMMA 2.7. *Let R be a local Gorenstein ring, let I be a perfect R-ideal of deviation d which is strongly Cohen-Macaulay and G_∞, and write $D = \dim(R/I)$. Then for all $n \geq d$, $\operatorname{depth}(I^n/I^{n+1}) = D - d$, and $\operatorname{depth}(R/I^{n+1}) = D - d$.*

PROOF: Some of the arguments in this proof are due to J. Herzog (cf. also [**SV**, proof of 4.2]).

Let H_i be the ith Koszul homology of a minimal generating set of I, set $A = R/I$, $F = A^{\mu(I)}$, and let $S_i(F)$ denote the ith symmetric power of F. Since I is strongly Cohen-Macaulay and G_∞, the graded pieces of the approximation complex [**HSV**] yield exact sequences for all $n \geq 0$,

$$0 \to H_n \otimes S_0(F) \to H_{n-1} \otimes S_1(F) \to \dots$$
$$\to H_0 \otimes S_n(F) \to I^n/I^{n+1} \to 0.$$

Since $H_i = 0$ for all $i > d$, and $\operatorname{depth}(H_i) = D$ for $d \geq i \geq 0$, the above sequences imply that for all $n \geq 0$, $\operatorname{depth}(I^n/I^{n+1}) \geq D - d$ [**HSV**, **H2**]. But then the short exact sequences

$$0 \to I^n/I^{n+1} \to R/I^{n+1} \to R/I^n \to 0$$

yield inductively that for all $n \geq 0$, $\operatorname{depth}(R/I^{n+1}) \geq D - d$.

Now suppose that for some $n \geq d$, $\operatorname{depth}(I^n/I^{n+1}) > D - d$. Then $d > 0$ and $\operatorname{Ext}^d_A(I^n/I^{n+1}, H_d) = 0$ since H_d is the canonical module for A. But then the above exact sequence (note that $H_i = 0$ for $i > d$)

$$0 \to H_d \otimes S_{n-d}(F) \to H_{d-1} \otimes S_{n-d+1}(F) \to \dots$$
$$\to H_0 \otimes S_n(F) \to I^n/I^{n+1} \to 0$$

yields an exact sequence

$$\operatorname{Hom}_A(H_{d-1}, H_d) \otimes S_{n-d+1}(F) \to \operatorname{Hom}_A(H_d, H_d) \otimes S_{n-d}(F) \to 0.$$

Notice that $\operatorname{Hom}_A(H_{d-1}, H_d) \cong H_1$ [**He1**] and $\operatorname{Hom}_A(H_d, H_d) \cong A$. Hence H_1 would have a free direct summand, which by [**GL**] would imply that I is a complete intersection (we may apply [**GL**] since I, as a perfect ideal, has finite projective dimension).

This contradiction shows that for all $n \geq d$, $\operatorname{depth}(I^n/I^{n+1}) = D - d$. Since moreover $\operatorname{depth}(R/I^n) \geq D - d$, the exact sequence

$$0 \to I^n/I^{n+1} \to R/I^{n+1} \to R/I^n \to 0$$

now also implies that $\operatorname{depth}(R/I^{n+1}) = D - d$.

□

THEOREM 2.8. *Let R be a regular local ring, let I be a licci R-ideal which is generically a complete intersection, but not a complete intersection.*
a) *Assume that I is not Gorenstein, then*

i) *I/I^2 is not Cohen-Macaulay,*
ii) *if I satisfies G_∞, then for all $n \geq 1$, I^n/I^{n+1} is not Cohen-Macaulay,*
iii) *for all $n \geq 2$, R/I^n is not Cohen-Macaulay.*

b) *Assume that I is Gorenstein, then*

i) *I/I^2 is Cohen-Macaulay, but I^2/I^3 is not Cohen-Macaulay,*
ii) *if I satisfies G_∞, then for all $n \geq 2$, I^n/I^{n+1} is not Cohen-Macaulay,*
iii) *for all $n \geq 3$, R/I^n is not Cohen-Macaulay.*

PROOF: We first prove part a). Since I is licci, but not Gorenstein, it follows from [**U**, 2.6.b], that (R, I) has essentially a deformation (S, J) such that J is perfect, S/J is normal, and $d(J) = 1$. Hence Lemma 2.7 implies that for all $n \geq 1$, J^n/J^{n+1} is not Cohen-Macaulay and S/J^{n+1} is not Cohen-Macaulay. Now suppose that i), ii), or iii) are false, then by Lemmas 2.2 and 2.3, for some $n \geq 1$, J^n/J^{n+1} or S/J^{n+1} would be Cohen-Macaulay. But this is impossible by the above.

To prove part b), we first note that by [**B**], I/I^2 is Cohen-Macaulay (at least if R contains a field, for the proof in the general case see [**BU**]). Since I is licci, Gorenstein, but not a complete intersection, it follows from [**HU**, 4.2], that (R, I) has essentially a deformation (S, J) such that J is perfect, S/J is normal, and $d(J) = 2$. Now Lemma 2.7 implies that for all $n \geq 2$, J^n/J^{n+1} and S/J^{n+1} are not Cohen-Macaulay. Then our claim follows again from Lemmas 2.2 and 2.3. □

We conclude this paper with a result concerning symbolic powers of prime ideals that generalizes results from [**H3**]. Remember, that the symbolic power $p^{(n)}$ of a prime ideal p in a domain R is defined to be the ideal $(p^n R_p) \cap R$.

COROLLARY 2.9. *Let R be a regular local ring, let p be a licci prime ideal in R which is not a complete intersection, and assume that $\dim(R/p) = 1$.*

a) *If R/p is not Gorenstein, then for all $n \geq 2$, $p^{(n)} \neq p^n$.*
b) *If R/p is Gorenstein, then $p^{(2)} = p^2$, and for all $n \geq 3$, $p^{(n)} \neq p^n$.*

PROOF: As $\dim(R/p) = 1$, it follows that $p^{(n)} = p^n$ if and only if R/p^n is Cohen-Macaulay. Now the claim follows from Theorem 2.8. □

References

[B] R. O. Buchweitz, *Contributions à la théorie des singularités*, Thesis, l'Université Paris VII (1981).

[BU] R. O. Buchweitz and B. Ulrich, *Homological properties which are invariant under linkage*, preprint.

[GL] T. Gulliksen and G. Levin, *Homology of local rings*, Queens Papers in Pure and Applied Mathematics, no. 20, (1969), Kingston.

[He1] J. Herzog, *Komplexe, Auflösungen und Dualität in der lokalen Algebra*, Habilitationsschrift, Regensburg (1974).

[He2] J. Herzog, *Ein Cohen-Macaulay-Kriterium mit Anwendungen auf den Konormalenmodul und den Differentialmodul*, Math. Z. **163** (1978), 149–162.

[HSV] J. Herzog, A. Simis and W. Vasconcelos, *Approximation complexes and blowing-up rings*, J. Algebra **74** (1982), 466–493.

[H1] C. Huneke, *Numerical invariants of liaison*, Inventiones Math. **75** (1984), 301–325.

[H2] C. Huneke, *The theory of d-sequences and powers of ideals*, Adv. Math. **46** (1982), 249–279.

[H3] C. Huneke, *The primary components of and integral closures of ideals in 3-dimensional regular local rings*, Math. Ann. **275** (1986), 617–635.

[HU] C. Huneke and B. Ulrich, *The structure of linkage*, Annals of Math. **126** (1987), 277–334.

[Na] M. Nagata, "Local rings," Krieger, Huntington, 1975.

[N] D. G. Northcott, "Lessons on rings, modules, and multiplicities," Cambridge Univ. Press, London, 1968.

[SV] A. Simis and W. Vasconcelos, *The syzygies of the conormal module*, Amer. J. Math. **103** (1981), 203–224.

[U] B. Ulrich, *Theory and applications of universal linkage*, in "Commutative Algebra and Combinatorics," Advanced Studies in Pure Mathematics 11, North Holland, Amsterdam, 1987, pp. 285–301.

Department of Mathematics, Purdue University, West Lafayette IN 47907

Department of Mathematics, Michigan State University, East Lansing MI 48824

Both authors were partially supported by the NSF.

The Hilbert Function of a Gorenstein Artin Algebra

A. IARROBINO

Summary.

The self-duality of a Gorenstein Artin algebra A with maximal ideal m over a field $k = A/m$ carries over in a weaker form to the associated graded algebra $A^* = \mathrm{Gr}_m A = \oplus A_i$, where $A_i = m^i/m^{i+1}$. If $j = \max\{i \mid A_i \neq 0\}$ is the socle degree of A, then A^* has a canonically defined decreasing sequence of ideals $A^* = C(0) \supset C(1) \supset \cdots \supset C(j+1) = 0$, whose successive quotients $Q(a) = C(a)/C(a+1)$ are reflexive A^*-modules. For $a = 0, \ldots, j$, $Q(a)_i = \mathrm{Hom}(Q(a)_{j-a-i}, k)$, and up to a shift in grading $Q(a) = \mathrm{Hom}_k(Q(a), k)$. We define the i-th graded piece $C(a)_i \subset A_i$ of $C(a)$ by

$$C(a)_i = ((0 : m^{j+1-a-i}) \cap m^i)/((0 : m^{j+1-a-i}) \cap m^{i+1}).$$

Thus, the Hilbert function $H(A) \underset{\mathrm{def}}{=} \sum \ell(A_i) Z^i$, with $\ell(A_i)$ denoting k-length, satisfies

$$H(A) = \sum H(Q(a)), \text{ with the } a\text{-th summand symmetric about } (j-a)/2.$$

Here we show that the A^*-modules $Q(a)$ are reflexive. We determine them for complete intersection (C.I.) quotients of the power series ring $k[[x,y]]$, where they are isomorphic to graded Gorenstein algebras; and we give some quite different examples in higher codimension. We suggest a way to use these results in finding examples of Gorenstein Artin (GA) quotients of a power series ring R, having certain specified Hilbert functions. The results also lead to further inequalities for the Hilbert function of Gorenstein algebra quotients of R.

1. Magic Rectangles.

The hills of Berkeley are certainly a marvelous, mysterious place to consider the behavior of certain "magic rectangles" of integers, that occur in the study of Gorenstein Artin algebras, under the benevolent direction of a magician's Kap. We'll say that a sequence of non-negative integers $H = (t_0, t_1, \ldots, t_j, 0, 0, \ldots)$ has a "magic rectangle" decomposition as a sum of sequences $H = H(0) + \cdots + H(s)$ iff each $H(a) = (t_{a0}, t_{a1}, \ldots, t_{a,j-a}, 0, 0, \ldots)$ is a non-negative sequence of integers symmetric about $(j-a)/2$. For example the sequence $H = (1,3,6,8,6,4,2,1)$ has the two decompositions

$$\begin{array}{rl} H = & (1,3,6,8,6,4,2,1) \\ \hline H(0) = & (1,2,3,4,4,3,2,1) \\ H(1) = & (0,1,2,3,2,1) \\ H(2) = & (0,0,1,1) \\ \hline D = & (1,2,4,6,8,6,3,1) \end{array} \quad \text{and} \quad \begin{array}{rl} H = & (1,3,6,8,6,4,2,1) \\ \hline H(0) = & (1,2,3,3,3,3,2,1) \\ H(1) = & (0,1,3,5,3,1) \\ H(2) = & (0) \\ \hline D = & (1,2,4,6,8,6,3,1) \end{array}$$

Here the term t_{ab} of $H(a)$ is zero if it is not listed. A property of the rectangles with rows $H(a)$, $a = 0, \ldots, [j/2]$ is that the diagonal sums $D = (d_0, \ldots, d_i, \ldots, d_j)$ where $d_i = \sum t_{a,i-a}$ satisfy $D = H^\vee = (t_j, \ldots, t_0)$; the symmetry implies $d_i = h_{j-i}$. For the left rectangle the diagonal sums are $d_0 = 1$, $d_1 = 2+0$, $d_2 = 3+1+0$, $d_3 = 4+2+0$, $d_4 = 4+3+1$, etc. Our rectangles, arising like the two above (see §4, Example 5) from Gorenstein Artin algebras, have some further properties: namely, $H(0)$ is the Hilbert function of a graded GA algebra, and $t_{a0} = 0$ for $a > 0$. For the hypothetical lay person imagined by an earlier speaker as a participant in this meeting, this is a talk about magic rectangles, occuring in Nature.

(1.1) Introduction and Notation: Throughout A will be an equicharacteristic Gorenstein Artin local (GA) algebra with maximal ideal m over an algebraically closed field $k = A/m$. The associated graded algebra $A^* = \mathrm{Gr}_m A = \oplus A_i$, where $A_i = m^i/m^{i+1}$. The integer $j = \max\{i \mid A_i \neq 0\}$ is the *socle degree* of A; the socle $(0 : m) = \{f \in A \mid mf = 0\}$ is the minimal ideal m^j of A, and has length one as a k-vector space. Suppose β is a k-linear homomorphism $\beta : A \to k$, whose restriction to $(0 : m)$ is surjective; then the inner product $\langle \cdot, \cdot \rangle_\beta$ defined by $\langle a, b \rangle_\beta = \beta(ab)$ is an exact pairing $\langle \cdot, \cdot \rangle_\beta : A \times A \to k$ such that m^i has annihilator $0 : m^i$.

Thus $A = \mathrm{Hom}_k(A,k)$ is self-dual under $\langle\cdot,\cdot\rangle$, where we suppose β chosen. (See [**E-L**], [**KAP**, §4-6], [**MAC-2**, §73], or [**I-4**] for discussions.)

We let $\ell(A)$ denote the length of the k-module A. The function $H(A) = \sum \ell(A_i)Z^i$ is the Hilbert *function* of A; we will usually write $H(A) = (t_0, t_1, \ldots, t_j)$ in sequence form, in place of $t_0 + \cdots + t_j\, Z^j$.

In Section 2 we introduce the reflexive "factors" $Q(a)$ of the associated graded algebra of a GA algebra. Section 3 determines $Q(a)$ for a GA quotient A of $k[[x,y]]$. Section 4 gives some examples when A is a quotient of $k[[x,y,z]]$, and suggests a procedure using Macaulay's inverse systems for finding examples with $H(a) = H(Q(a))$ specified. Section 5 discusses consequences for the Hilbert function of Gorenstein algebras.

2. Associated graded algebra of a Gorenstein Artin algebra: the reflexive factors $Q(a)$.

When the GA algebra A is itself graded Gorenstein, F. H. S. Macaulay showed that $\langle\cdot,\cdot\rangle$ restricts to an exact pairing $A_i \times A_{j-i} \to k$; hence the Hilbert function $H(A) = (t_0, \ldots, t_j)$ satisfies $h_i = h_{j-i}$ (See [**MAC-2**, §70]).

The associated graded algebra of the complete intersection $A = k[[x,y]]/(xy, x^2 + y^3)$ is not itself Gorenstein, a fact reflected in the nonsymmetry of its Hilbert function $H(A) = (1,2,1,1)$; for $A^* = k[x,y]/(xy, x^2, y^4)$ with basis the classes $\bar{1}, \bar{x}, \bar{y}, \bar{y}^2, \bar{y}^3$. That A here is Gorenstein is a well-known result of Macaulay [**MAC-2**, §71]. As we shall see, A^* has reflexive subquotient "factors" $Q(0) = A^*/(\bar{x}) = k[x,y]/(x, y^4)$ with basis $\bar{1}, \bar{y}, \bar{y}^2, \bar{y}^3$, and $Q(1) = (\bar{x})$, leading to the Hilbert function decomposition

$$\begin{array}{rl} \underline{H(A) = (1,2,1,1)} & \\ H(Q(0)) = (1,1,1,1) & \quad Q(0) = \langle \bar{1}, \bar{y}, \bar{y}^2, \bar{y}^3 \rangle = k[x,y]/(x, y^4). \\ H(Q(1)) = (0,1,0) & \quad Q(1) = \langle \bar{x} \rangle = xA^*. \end{array}$$

Figure 1. Simple structure of A^* for a GA algebra in codimension two.

The reflexive modules $Q(a)$ are defined from the intersection of the usual m-adic and Loëwy filtrations on A, as successive quotients of ideals $C(a)$ in A^*; each ideal $C(a)$ is defined piecewise. The action of the maximal ideal of A^* on $Q(a)$ is in general non-trivial.

DEFINITION: If A is an equicharacteristic Gorenstein Artin local (GA) k-algebra of socle degree j, we define a descending sequence of ideals of $A^* = \mathrm{Gr}_m A$, namely $A^* = C(0) \supset C(1) \supset \cdots \supset C(j+1) = 0$, by (letting b denote $j+1-a$)

$$(1) \quad C(a)_i = (0 : m^{b-i}) \cap m^i / (0 : m^{b-i}) \cap m^{i+1}$$
$$= i\text{-th graded piece of } \mathrm{Gr}_m(0 : m^{b-i}) \text{ in } A^*.$$

The "factor" $Q(a)$ is the quotient $C(a)/C(a+1)$. Thus

$$(2) \qquad Q(a)_i = (0 : m^{b-i}) \cap m^i / \left((0 : m^{b-i}) \cap m^{i+1} + (0 : m^{b-i-1}) \cap m^i \right).$$

The A^*-module $Q(0) = A^*/C(1)$ is the unique socle-degree-j graded Gorenstein quotient of A^*.

THEOREM 1. *The A^*-module $Q(a)$ is reflexive, up to a shift in degree: $\mathrm{Hom}_k(Q(a), k) = Q(a)$ as A^*-module, and in particular*

$$(3) \qquad \mathrm{Hom}_k(Q(a)_i, k) = Q(a)_{j-a-i}.$$

The Hilbert function $H(a) = H(Q(a))$ is symmetric about $(j-a)/2$, and $H(A) = \sum H(a)$.

PROOF OF DEFINITION AND THEOREM: That $C(a)$ is an ideal is immediate from $m(0 : m^{b-i}) \subset (0 : m^{b-(i+1)})$. That $Q(0)$ is a graded Gorenstein algebra will follow from its being a k-algebra with generator 1, and the reflexivity of $Q(0)$ as A^*-module: since $Q(0)$ has a single generator (so the length $\ell(Q(0)/mQ(0)) = 1$), its socle $(0 : m)$ has length one, implying $Q(0)$ is Gorenstein. The uniqueness is well-known (see [**EM**] or [**EM-I**]).

We now show that $Q(a)$ is reflexive. Recall that under the pairing $\langle \cdot, \cdot \rangle$, the annihilator of $m^i \cap (0 : m^{b-i})$ satisfies

$$(4) \qquad \left(m^i \cap (0 : m^{b-i}) \right)^{\perp} = m^{b-i} + (0 : m^i).$$

Note that $Q(a)_i$ is isomorphic to a quotient D/E, namely

$$(5) \quad Q(a)_i = \left((m^i \cap (0 : m^{b-i})) + m^{i+1} \right.$$
$$\left. + (0 : m^{b-i-1}) \right) / \left(m^{i+1} + (0 : m^{b-i-1}) \right) = D/E$$

where D denotes numerator, E denominator. The inner product $\langle\cdot,\cdot\rangle$ induces an exact pairing $(D/E)\times(E^\perp/D^\perp)\to k$. Thus the dual module $Q(a)_i^\vee$ under $\langle\cdot,\cdot\rangle$ satisfies

$$\begin{aligned}Q(a)_i^\vee = E^\perp/D^\perp &= \left(m^{b-i-1}\cap(0:m^{i+1})\right)/\\ &\qquad\qquad\left(((0:m^i)+m^{b-i})\cap(0:m^{i+1})\cap m^{b-i-1}\right)\\ &= \left(m^{b-i-1}\cap(0:m^{i+1})\right)/\left(m^{b-i}\cap(0:m^{i+1})+m^{b-i-1}\cap(0:m^i)\right)\\ &= Q(a)_{j-a-i}.\end{aligned}$$

We now show that the action of A^* on $Q(a)$ respects the duality we have just defined piecewise. Suppose that $\bar{c}$, $\bar{t}$, and $\bar{u}$ are elements, respectively of R_s, $Q(a)_{j-a-i}$ and $Q(a)_{i-s}$. It suffices to show that $(\bar{c}\bar{t})\circ\bar{u}=\bar{t}\circ(\bar{c}\bar{u})$, where $(\bar{c}\bar{t})\circ\bar{u}$ in k arises from the pairing $Q(a)_{j-a-i+s}\times Q(a)_{i-s}$ to k, and $\bar{t}\circ(\bar{c}\bar{u})$ from the pairing $Q(a)_{j-a-i}\times Q(a)_i$ to k. Let $c\in m^s$, $t\in m^{b-i-1}\cap(0:m^{i+1})$, and $u\in m^{i-s}\cap(0:m^{b-i-s})$ induce respectively $\bar{c}$, $\bar{t}$, and $\bar{u}$. Then $(\bar{c}\bar{t})\circ\bar{u}=\langle ct,u\rangle=\beta(ctu)=\langle cu,t\rangle=\bar{t}\circ(\bar{c}\bar{u})$. This completes the proof. □

For examples, see §3 and §4.

3. Intersection of two plane curves.

F. H. S. Macaulay and C. A. Scott studied the Hilbert function of complete intersection quotients $A=R/I=k[[x,y]]/(f,g)$ of the power series ring R in two variables [**SC,MAC-1**]. Macaulay concluded[1], using his inverse systems and some ideas of C. A. Scott that the Hilbert function $H(A)$ satisfied

(6) $\quad H(A)=(1,2,\ldots,d,t_d,\ldots,t_j,0),$

$$\text{where } d\geqq t_d\geqq t_{d+1}\geqq\cdots\geqq t_j=1,\text{ and } |t_i-t_{i+1}|\leqq 1 \text{ for all } i.$$

---

[1]Macaulay shows in [**MAC-1**, §7-13 ff.] the "one set theorem" that the ideal $I=(f,g)$ has an inverse system that is principal: a complete intersection quotient of R is Gorenstein. He quotes in §15 a key diagram from [**SC**] that depicts a basis of the inverse system, arranged according to degree; his proof of the "one set theorem" involves a discussion of the shape of the diagram, which satisfies (6). He presents his proof as an emendation closing gaps he finds in the proof of [**SC**]. That the shape of the inverse system is also the shape of a complementary basis to I in R — arranged according to order — is basic; see [**MAC-2**, §70] for an application. Thus, Macaulay has shown (6). In [**MAC-1**, §22–23] he proves a "t-set theorem" for $t+1$-generated ideals I in R: his proof involves the generalization of (6), also studied in [**BRI**] and [**I-1**].

More generally, if I has $t+1$ generators, he showed $H(R/I)$ satisfies $|t_i - t_{i+1}| \leqq t$. Subsequently, J. Briançon and the author studied these ideals using standard bases, with a view to parametrizing them; in the process they reproved (6), and showed that any sequence $(t_0, \ldots)$ of integers satisfying (6) occurs as the Hilbert function of some C.I. quotient of R, where order $I = d$. The bar graph of the Hilbert function can thus be viewed as a series of steps, of height one and length varying; in degrees greater than d we find platforms where $t_u = t_{u+1} = \cdots = t_v$ with $t_{u-1} \neq t_u$, and $t_{v+1} < t_v$ (except for the 0-th platform). We label the steps — extended left — strictly above this platform, and below or equal in height to the next higher platform with the integer $a = (j+1-u) - t_u$, which is the difference between the platform's expected height were the steps to the right each of length one, and the platform's actual height. The extended steps labelled "a" comprise the bar graph of the summand $H(a)$ of the Hilbert function H (see Figure 2).

```
  * *
 ______
* * * * *                 H(3) = (0,0,0,1,1,0),        a = (11 - 5) - 3
                          H(2) = (0,1,2,2,2,2,2,1),    a = (11 - 8) - 1
 * * * * * * *
______________            H(0) = (1,1,...,1).          a = (11 - 11) - 0
* * * * * * * * * * *
```

Figure 2. Hilbert function decomposition $H = \sum H(a)$,
for $H = (1,2,3,4,4,3,3,2,1,1,1)$ in codimension two ($j = 10$).

In the following Theorem we use the previous work to show that in codimension two, A^* already contains the information concerning the ideals $C(a)$, and the "factors" $Q(a)$; the Hilbert decomposition $H = \sum H(Q(a))$ is the decomposition depicted in Figure 2. We view this as a deeper understanding of the Macaulay-Scott characterization of the Hilbert function, for the intersection of two plane curves. A key step in the somewhat technical proof is that when A^* is not Gorenstein, $C(1)$ is the principal ideal (h) of A^*, and "$C(1)/h$" turns out to be an algebra B^* associated to a simpler complete intersection $B = R/(f', g')$. This permits the inductive conclusion that $Q(a)$ is a graded complete intersection, up to a shift in degree. I'd like to thank J. Briançon for supplying a candidate algebra B early on in this work. Recall that here $A = R/I$ Gorenstein $\iff$ A is a C.I.

THEOREM 2. *Suppose A is a C.I. quotient $A = R/(f,g)$, and suppose $H = \sum H(a)$ is the decomposition of the Hilbert function $H = H(A)$ into* *nonzero parts, as described above. For each integer a that appears, there is a form $h(a)$ in R of degree t_u (u depends on a) such that $C(a) = h(a)A^*$. If a' is the label on the next higher platform then the A^*-module $Q(a) = (h(a)A^*/(h(a')A^*)$, and it has Hilbert function $H(a)$. After a degree-shift by t_u, $Q(a)$ is isomorphic to a graded complete intersection quotient of R^*.*

PROOF: A graded Gorenstein quotient $R/(F,G)$ of R has the Hilbert function of $R/(x^s, y^{s'})$, where $s' = \deg G$, $s = \deg F$. We first use this knowledge about graded Gorenstein quotients, and Theorem 1 to describe $Q(0)$ and determine $H(Q(0))$ from $H(A)$. We then use the work on standard generators to produce a GA algebra B such that B^* behaves like the A^*-module $C(1)$, leading to an inductive characterization of $Q(a)$.

Since A is GA, the length $\ell(A_j) = t_j = 1$. Theorem 1 and the first, easy formula of (6) — that $H(A) = (1, 2, \ldots, d, t_d, \ldots, t_j, 0)$ where $d \geqq t_d \geqq t_{d+1} \geqq \cdots \geqq t_j = 1$ — show that for any integer $s \geqq 0$,

$$t_j = 1,\ t_{j-1} = 2, \ldots, t_{j-s+1} = s \implies t_{j-s} \leqq s + 1.$$

Otherwise, $\sum_{j-s}^{j} t_i \geqq \sum_0^s t_i$, contradicting the formula of Theorem 1 that $H(A)$ is the sum of symmetric sequences $H(Q(a))$ with centers each $(j - a)/2$, less than or equal to $j/2$. Thus, there is a minimal integer $v = j - s + 1$, and (given v) a minimal $u \leq v$ such that

$$(7) \qquad t_u = \cdots = t_v = s, \text{ and } (t_v, t_{v+1}, \ldots, t_j) = (s, s-1, \ldots, 1).$$

Furthermore, the Hilbert function $H(0) = H(Q(0)) = (t_{00}, t_{01}, \ldots)$ must satisfy $t_{0i} = t_i$ for $i \geqq v - 1$, for the same reason, that $H(A) = H(0) + H'$ where H' is the sum of symmetric sequences with centers less than $j/2$. It follows from our first remark that $H(0)$ is the Hilbert function of $R/(x^s, y^{v+1})$, namely $H(0) = (1, 2, \ldots, s, \ldots, t_v = s, s-1, \ldots, 2, 1)$ of socle degree j. Thus $Q(0)$ is a graded Gorenstein quotient $Q(0) = R/(F, G)$ where degree $F = s$ and degree $G = v + 1$.

Should $u = v$, then letting $T(j) = (1, 2, \ldots, s, s-1, \ldots, 1)$ we find $j = 2s - 2$ and $H(0) = T(j)$; since $H(A) \leq (1, 2, 3, \ldots)$ it follows that $H' = 0$ and $H(A) = H(0) = T(j)$, whence A is extremal Gorenstein in the sense of P. Schenzel (see [**I-2**]) and also $A^* = Q(0)$. Likewise, if $H(A)$ is symmetric, then $C(1) = 0$ and $A^* = Q(0)$. Henceforth we assume $u < v$ and $C(1) \neq 0$.

We now recall the well-known and elementary fact, that for quotients A

of $R = k[[x, y]]$

(8) $t_{v-1} = t_v \Longrightarrow$ there is a form $h \in R$ of degree $s = t_v$, such that

$$I_v = (h) \cap R_v,\ I_{v-1} = (h) \cap R_{v-1},\ \text{and } h \mid I_i \text{ for } i \leqq v.$$

(See, for example, [**I-1**, Prop. 5.1 and (4.4)], or [**I-3**, Lemma 2.2.1].) The quotient $Q(0) = A^*/C(1)$ agrees with A^* in degrees at least u, and satisfies $Q(0)_i = R_i/(h) \cap R_i$ in degrees smaller than u; it follows from this and (7) that $C(1) = hA^*$. Furthermore, $h = F$, and G may be taken any element of I_{v+1} not in (h) — for example a standard generator (see below) of I^* in degree $v + 1$. Thus, we have found

(9)
$$C(1) = hA^* \text{ with degree } h = s = \text{height of first platform.}$$
$$Q(0) = A^*/C(1) = R^*/(h, G) \text{ where } G \in I_{v+1} \text{ with } (h, G) = 1.$$

That $(h, G) = 1$ follows from $Q(0)$ being a complete intersection that is Artin.

We now prepare an induction step. We suppose, possibly after a change of basis x, y that $I^* = (F_d, \dots, F_0)$ where F_d has degree $d = \operatorname{order} I$, $F_d = y^d + \dots$, and $h = y^s + \dots$ have pure y terms. Recall that $I = (f, g)$, with $\operatorname{in}(f) = F_d$ and $\operatorname{in}(g) = F_{d-1}$, and write (in unique fashion)

(10) $$f = f'h + f'', \quad g = g'h + g'' \text{ with } \operatorname{order}_y f'' < s,\ \operatorname{order}_y g'' < s.$$

Let J denote the ideal (f', g'), and let $B = R/J$. We will show

CLAIM. *The associated graded algebra $B^* = C(1) : h$, as graded R^*-module. Thus, $B_i^* = C(1)_{i+s} : h$, and the Hilbert function $H(B) = H'$ satisfies $H' = S^s(H - H(0))$, where S denotes a left shift: $t'_i = t_{i+s} - t_{0,i+s}$. Furthermore, the R^*-modules $C'(b)$ and $Q'(b)$ for the algebra B are related to those for A by $C'(b) = C(a) : h$, and $Q'(b) = Q(a) : h$ where $a = b + (j + 1 - u - s)$.*

From the claim, the rest of Theorem 2 is immediate, for B is a complete intersection, and Artin, hence $Q'(0)$ is a graded Gorenstein algebra by the first part of the proof, and one may continue inductively to find graded Gorenstein algebras $Q''(0), \dots$, isomorphic to the factors $Q(a)$ of A.

PROOF OF CLAIM WHEN $\operatorname{char} k = 0$: We use the standard basis $I = (f_d, \dots, f_0)$, $I^* = (F_d, \dots, F_0)$ where $F_i =$ initial form (f_i) and the orders satisfy $v(f_d) \leqq \dots \leqq v(f_0) = j + 1$ having standard relations E_i:

(11) $$x^{w_i} f_i = y f_{i-1} + \sum \alpha_{ik} f_k = 0, \text{ with } i = 1, \dots, d$$
$$\text{and } k = 0, \dots, i - 1, \text{ and } \alpha_{ik} \in k[x].$$

In addition, for each $i \leqq d$, $(f_i, f_{i-1}) = (f_i, \ldots, f_0)$, implying that H satisfies (6) (see [**BRI**] or [**I-1**, Thm. 4.3]). Certain of the terms in α_{ik} are key in determining $(F_d, \ldots, F_s = x^{u-s}h)$. We show these terms are the same as corresponding terms for the ideal $hJ = (f'h, g'h)$, hence for J; the f'' and g'' are out of the picture. The key terms are the non-zero initial terms $\alpha_{i,i-2}(0)$ of $\alpha_{i,i-2}$ for $i = d, \ldots, c+1$. If we run up the ideal I, using the relations $E_{s+1}, \ldots, E_d$ to define $f_{s+1}, \ldots, f_d$, and the equations $f_i = f_i'h + f_i''$ (with y-degree $f_i'' < s$), we find that the initial forms $F_i'h$ of $f_i'h$ are the same as F_i. Consider E_{s+1}: the terms $(\sum_0^{s-1} \alpha_{ik} f_k)$ are in f_{s+1}'' and have order greater than (order f_s). While it is possible that in the portion yf_{s+1} of E_{s+2} these terms may enter via $f_{s+2} = x^{-w_{s+2}}(yf_{s+1} + \sum \alpha_{ik} f_k)$ in f_{s+2}', they have order too high to affect the initial form of $x^{w_{s+2}} f_{s+2}' - yf_{s+1}' + \sum \alpha_{s+2,s+1} f_{s+1}'$, which is $\alpha_{s+2,s}(0)F_s : h$. Hence, the ideal $(f_{s+2}', f_{s+1}')^*$ contains $F_s : h$. Similarly, $(f', g') = (f_d', f_{d-1}')^*$ contains $F_d : h, \ldots, F_s : h)$, and a converse argument shows equality. This completes the proof of the claim. The translation $a = b + (j+1-u-s)$ arises from $Q'(0) = Q(j+1-u-s)$, as $(j+1-u-s)$ is the a-label of the first platform $(t_u = \cdots = t_v = s)$.

Furthermore, we have shown that $Q(a) = h(a)Q'(b)$, where $Q'(b)$ is the graded complete intersection $R^*/(h(a')/h(a),\ F(a)/h(a))$, where a' is the label of the next higher platform, and $F(a)$ is the standard generator of I^* just following the a'-labelled platform (of y-degree the height t_u' of that platform). When a labels the top platform, then $Q'(b) = R^*/(F_d/h(a), F_{d-1}/h(a))$.

The proof of the claim when char $k = p$ involves the weak normal patterns and standard bases for I that are described in [**I-1**, §3B]. Although the proof is conceptually similar to that above, the order $v(f)$ of a standard generator f no longer corresponds to its y-degree. Thus one needs, differently from [**I-1**], to write the basis $I = (f_d, \ldots, f_0)$ such that $v(f_d) \leqq v(f_{d-1}) \leqq \cdots \leqq v(f_0) = j+1$.

For examples see the beginning of §2, or [**I-3**].

REMARK: A different line of proof occurs in [**I-3**]: it relies on showing that a graded algebra B^* of Hilbert function $H(B)$ satisfying (6), satisfies $B^* = A^*$ for a C.I. iff it is defined by an ideal I^* with the minimum possible number of generators, $v(H) =$ (2+ the number of platforms of height between 1 and $d-1$). A simpler proof, not using standard bases, would be closer to the spirit of Theorem 1.

4. Inverse system, and the factors $Q(a)$.

We now use Macaulay's principal systems to give examples of Gorenstein Artin quotients A of the power series ring $R = k[[x,y,z]]$ having certain Hilbert function decompositions $H(A) = \sum H(a)$, including the two decompositions of $H = (1,3,6,8,6,4,2,1)$ specified in §1.0. Let D denote the divided power ring $D = K[X,Y,Z]$, upon which R acts as higher partial differential operators without coefficients: thus, $x^s \circ X^t = X^{t-s}$, understood as zero if $t-s$ is negative, and the action of monomials in R on monomials in D extends bilinearly to an action of R on D. We'll denote by $[X+Z]^n = X^n + X^{n-1}Z + \cdots + Z^n$ the power in D. Macaulay showed

LEMMA. *There is a one-to-one correspondance of sets sending A to its embedded dual $\hat{A}$ in D*

$$\left\{\begin{matrix}\text{Socle-degree } j \text{ Gorenstein Artin} \\ \text{quotients } A = R/I\end{matrix}\right\} \longleftrightarrow \left\{\begin{matrix}R\text{-cyclic submodules } \hat{A} = R\circ F \\ \text{of } D \text{ with degree } F = j\end{matrix}\right\}$$

given by

$$\begin{aligned} A = R/I \quad &\longrightarrow \hat{A} = \{f \in D \text{ such that } I \circ f = 0\} \\ R/\operatorname{Ann} F \quad &\longleftarrow \hat{A} \text{ with generator } F. \end{aligned}$$

Thus $\hat{A} \equiv \operatorname{Hom}_k(A,k)$ is the Matlis dual of A, and $\hat{A}_i = \operatorname{Hom}(A_i,k)$ where M_i for a submodule M of D is $(M \cap D_{\leq i} + D_{<i})/(D_{<i})$. The Hilbert function $H(A) = H(\hat{A}) \underset{\text{def}}{=} (\ldots, \ell(\hat{A}_i), \ldots)$. Key for us is the fact, whose proof is beyond our scope now

THEOREM 3. *The R^*-modules $A^*/C(a+1)$ and $Q(a)$ depend only on $(\hat{A} + D_{<(j-1-a)}/D_{<(j-1-a)}$, hence only on the terms $F_{j-a} + \ldots F_j$ of a generator F of $\hat{A}$.*

Thus, F_j determines $Q(0)$, $F_j + F_{j-1}$ determines $Q(0)$ and $Q(1)$ and how they fit together in $A^*/C(2)$, and so on.

Below we give examples, constructed using Theorem 1 and 2, of GA quotients A of $R = k[[x,y,z]]$ (Examples 1,2,5) or of $R = k[[x,y,z,w]]$ (Examples 3,4) with striking properties. $Q(j-2)$ may have a length-two socle; $Q(1)$ may have a length-two socle though A is a complete intersection. There may be two different structures of $Q(a)$ decomposition for the same graded algebra $A^* = B^*$, for which the Hilbert function decompositions are distinct (Example 4). Example 5 yields the "magic rectangle" of §1.0, and Example 6 warns that $Q(2)$ may be nonzero, even though $F = F_j + F_{j-1}$ has zero term F_{j-2}. Examples 2,5 and 6 are from [**I-3**].

In the case of codimension three, the Hilbert functions possible for $H(Q(0))$ are known to be those whose first differences $\Delta = (\delta_0 = t_0 - 0, \delta_1 = t_1 - t_0, \ldots, \delta_{[j/2]})$ up to degree $[j/2]$ are a 0-sequence: these differences must be the Hilbert function of some quotient of $k[[x,y]]$, so must satisfy $\Delta = (1, 2, \ldots, d, \delta_d, \ldots, \delta_{[j/2]})$ with $d \geqq \delta_d \geqq \delta_{d+1} \geqq \cdots \geqq \delta_{[j/2]}$, where d is the initial degree of the ideal in R defining A. (See Prop. 3.3 of [**B-E**] and the explication, Thm. 4.2 of [**ST**].) Despite the Pfaffian structure theorem of [**B-E**], the Hilbert functions of $Q(a) \mid a > 0$ are not well understood (Examples 1,2 and 5 below are in codimension three). We suppose the algebra $Q(0) = R/I(0)$.

Example 1. $Q(j-2)$ with length two socle. Take $F = X^j + YZ$, and then $I = \operatorname{Ann} F = (xy, xz, y^2, z^2, yz - x^j)$, and $I(0) = \operatorname{Ann} X^j = (y, z, x^{j+1})$ defining $Q(0)$ with Hilbert function $(1, \ldots, 1)$. The dual-module $\hat{A} \subset D$ is $\hat{A} = \langle 1; X, Y, Z; X^2, \ldots, X^{j-1}, F\rangle$ of Hilbert function $H(A) = H(\hat{A}) = (1, 3, 1, \ldots, 1, t_j = 1)$, thus $\hat{Q}(j-2) = \langle Y, Z\rangle$ and $Q(j-2) = (\bar{y}, \bar{z})$ of Hilbert function $(0, 2, 0)$, while $Q(a) = 0$ for $a \neq 0, j-2$, by Theorem 3. □

Example 2. Complete intersection where $H(Q(1)) = (0, 1, 0, 1)$. Take $F = X^2Y^3 - X^3Z$. Then $I = \operatorname{Ann} F = (xz + y^3, yz, z^2)$, and $I(0) = \operatorname{Ann} X^2Y^3 = (z, x^3, y^4)$ defining $Q(0)$ of Hilbert function $H(0) = (1, 2, 3, 3, 2, 1)$. The R-module $\hat{A} = \langle 1; X, Y, Z; X^2, XY, Y^2; X^3, X^2Y, XY^2, Y^3; X^2Y^2, XY^3 - X^2Z; F\rangle$ of Hilbert function $H(A) = (1, 3, 3, 4, 2, 1)$, so $Q(1) = \langle \bar{z}, \bar{x}^3\rangle$ and $H(1) = (0, 1, 0, 1)$ while $Q(a) = 0$ for $a > 2$. Again, $Q(1)$ has length two socle. □

Example 3. Hilbert function with two decompositions, in four variables. Take $F = X^2Y^3 + X^3W + ZW^2$. Then $\hat{A} = \langle 1; X, Y, Z, W; X^2, XY, Y^2, W^2; X^3 + ZW, X^2Y, XY^2, Y^3 + XW; X^2Y^2, XY^3 + X^2W; F\rangle$, the associated graded dual module $\hat{A}^*$ has generators X^2Y^3, X^3, W^2, Z, and the ideal $I = (xz, xw - y^3, yz, yw, z^2, zw - x^3, w^3)$. The dual $\hat{Q}(0)$ is generated by X^2Y^3, so with $R = k[[x, y, z, w]]$ the graded GA algebra $Q(0) = R/(z, w, x^3, y^4)$ of Hilbert function $H(0) = (1, 2, 3, 3, 2, 1)$; the dual $\hat{Q}(1) = \langle Z, X^3\rangle$, and $Q(1) = \langle \bar{z}, \bar{x}^3\rangle$ with $H(1) = (0, 1, 0, 1)$;

$\hat{Q}(2) = \langle W, W^2 \rangle$, so $Q(2) = \langle \bar{w}, \bar{w}^2 \rangle$ with $H(2) = (0,1,1,0)$; and $Q(a) = 0$ for $a > 2$. Thus $H(A) = (1,4,4,4,2,1)$ has the decomposition of Figure 2a.

Take $G = X^2Y^3 + X^2W^2 + Z^2$. Then $\hat{B} = R \circ G = \langle 1; X, Y, Z, W; X^2, XY, Y^2, XW; X^2Y, XY^2, X^2W, Y^3 + W^2; X^2Y^2, XY^3 + XW^2; G \rangle$, $\hat{B}^*$ has generators X^2Y^3, X^2W, Z, and the ideal J defining $B = R/J$ is $J = (xz, yz, yw, z^2 - x^2w^2, zw, w^2 - y^3, x^3)$. The graded GA algebra $Q(0)$ is that of A above; the dual $\hat{Q}(1) = \langle X^2W, XW, W \rangle$ of Hilbert function $H(1) = (0,1,1,1)$; and $\hat{Q}(3) = \langle Z \rangle$ of Hilbert function $H(3) = (0,1,0)$. Thus $H(B) = H(A)$ has the decomposition of Figure 2b.

$H(A) = (1,4,4,4,2,1)$	$H(B) = (1,4,4,4,2,1)$
$H(0) = (1,2,3,3,2,1)$	$H(0) = (1,2,3,3,2,1)$
$H(1) = (0,1,0,1)$	$H(1) = (0,1,1,1)$
$H(2) = (0,1,1)$	$H(2) = (0,1,0)$

Figure 2a, 2b. Two decompositions of $H(A) = H(B)$ for GA quotients of R (Example 3).

Example 4. Graded algebra A^* with two different $Q(a)$ structures, in four variables. The algebra $A^* = k[x,y,z,w]/(x^3V, y^2V, z^4V, w^3V)$ where $V = \langle x,y,z,w \rangle = R_1$ has two different $Q(a)$ structures, with different Hilbert function decompositions. Notice that the dual module $\hat{A}^*$ has generators $X^2YZ^3W^2, Z^4, X^3, W^3, Y^2$, and that the defining ideal I^* is related to that of the complete intersection $Q(0) = R/(x^3, y^2, z^4, w^3)$ defined by $X^2YZ^3W^2$, of Hilbert function $H(0) = (1,4,9,14,16,14,9,4,1)$, while $H(A^*) = H(0) + (0,0,1,2,1) = (1,4,10,16,17,14,9,4,1)$ of length 76. The GA algebra A defined by $F = X^2YZ^3W^2 + Y^2Z^4 + X^3W^3$ has $Q(2) = \langle \bar{y}^2, \bar{x}^3, \bar{w}^3, \bar{z}^4 \rangle$, so $H(2) = (0,0,1,2,1)$; it is defined by the ideal $I = (x^3w - x^2yz^3, x^4, x^3z, x^3w; y^2z - x^2yw^2, y^3, y^2x, y^2w; z^4y - x^2z^3w^2, z^5, z^4x, z^4w; w^3x - yz^3w^2, w^4, w^3y, w^3z)$ and has associated graded algebra A^*.

The GA algebra B defined by $G = X^2YZ^3W^2 + X^3Z^4 + Y^2W^3$ has $Q(1) = \langle \bar{x}^3, \bar{z}^4 \rangle$ and $Q(3) = \langle \bar{y}^2, \bar{w}^3 \rangle$ of Hilbert functions $H(1) = (0,0,0,1,1)$ and $H(3) = (0,0,1,1)$, respectively; B is defined by the ideal $J = (x^3z - x^2yw^2, x^4, x^3y, x^3w;\ z^4x - yz^3w^2, z^5, z^4y, z^4w;\ y^2w - x^2yz^3, y^3, y^2x, y^2z;\ w^3y - x^2w^2z^3, w^4, w^3x, w^3z)$, and also has associated graded algebra $B^* = A^*$. Key here is the action of the generators of the complete intersection (x^3, y^2, z^4, w^3) defined by the leading term $X^2YZ^3W^2$ on the two other terms of F, G. The decomposition is checked by finding $H(R \circ X^2YZ^3W^2)$,

$H(R \circ (X^2YZ^3W^2 + X^3Z^4))$ and $H(R \circ G)$; Theorem 3 above implies the term Y^2W^3 can affect only $Q(a)$ for $a \geq 3$; the symmetry properties of $H(a)$ from Theorem 1 thus show H has the specified decomposition. Note that $Q(0)$ is an invariant of A^* (provided A is GA), thus any such examples with $B^* = A^*$ have identical $Q(0)$, so are defined by principal systems with $\langle F_j \rangle = \langle G_j \rangle$.

Example 5. Power sums and finding an algebra with $H(A) = \sum H(a)$ specified. We do not know how to construct such examples with precision, but there is a procedure that often helps. Consider the first of the decompositions of §1.0, where $H(0) = (1,2,3,4,4,3,2,1)$, $H(1) = (0,1,2,3,2,1)$, and $H(2) = (0,0,1,1)$, of $H = (1,3,6,8,6,4,2,1)$ of socle degree $j = 7$ and length 31. That $t_{01} = 2$ implies that the top degree (leading) term F_7 of F be written using two variables (after a change of basis in D). In two variables the relation between F_7 and $H(0)$ is well understood; that the maximum term of $H(0)$ is 4 implies F_7 is a generalized sum of 4 seventh powers of linear forms (see [**EM-I**] and [**I-4**]). We choose $F_7 = X^7 + Y^7 + [X+Y]^7 + [X-Y]^7$. That $h_{11} = 1$ implies that F_6 involves another variable, and we try a sum of three powers, $F_6 = Z^6 + [Z+X]^6 + [Z-X]^6$. We add the term $F_5 = Y^2Z^3$ so that the yz and yz^2 derivates of F will yield YZ^2 and YZ in A independent of $R \circ (F_7 + F_6)$. Thus, we try $F = F_7 + F_6 + F_5$, which yields $A = R/I$, with $R = k[x,y,z]$ and $I = \big(zxy, zy^2 - z^4 + z^2x^2, 3z^3y - 3y^6 + 2x^6 - 2y^7, xz(x^2 - z^2), xy(x^2 - y^2), 6z^4 - 9z^2x^2 + 3x^5 - 3x^3y^2 - 2x^6 + 4y^7, 6(x^5 + y^5 - x^3y^2) - 3x^4y - 4x^6 + 8y^7\big)$. The Hilbert function decomposition is checked in $\hat{A}^*$ by verifying successively that $H(R \circ F_j) = H(0)$, $H(R \circ (F_j + F_{j-1})) = H(0) + H(1)$ with $H(1)$ symmetric around $(j-1)/2$, etc. Finding the ideal I is the hardest step here.

Consider now the second decomposition of §1.0, where $H(0) = (1,2,3,3,3,3,2,1)$, $H(1) = (0,1,3,5,3,1)$, and $H(a) = 0$ for $a \geq 2$. The power sum approach suggests writing G_7 as a sum of three powers of linear forms in two variables, and G_6 as a sum of five powers of linear forms involving the third variable — because of the maximum value of 5 for terms of $H(1)$. Thus, we try $G = X^7 + Y^7 + [X+Y]^7 + Z^6 + [Z+X]^6 + [Z-X]^6 + [Z+Y]^6 + [Z-Y]^6$. The ideal $J = \operatorname{Ann} G$ is $J = \big(zxy, xy(x-y), z^2y^2 - z^2x^2 + 2x^5 - 2y^5, xy^3 - z^3y, zx^3 - z^3x, 4z^4 - 5z^2y^2 + 5z^2x^2 + 2x^5 + 2y^5 - 4x^4y - 4x^5y, x^6 + y^6 - 3x^5y - 2x^4z^2\big)$. For further discussion see [**I-3**, Examples 4.2.1 and 4.2.2].

Example 6. The factor $Q(2)$ can be nonzero when $F_{j-2} = 0$. Let $R = k[[x,y]]$, and $F = Y^6 + Y^4X$. Then $H(Q(2)) = (0,1,1,1)$ is nonzero, although $F_{j-2} = F_4 = 0$. Here, $Q(0) = R/(x, y^7)$ and $H(0) = (1,1,1,1,1,1,1)$, while $Q(a) = 0$ if $a \neq 0, 2$. The ideal $I = (x^2, xy^3 - y^5)$, and $Q(2) = xA^* = \langle \bar{x}, \bar{x}\bar{y}, \bar{x}\bar{y}^2 \rangle$. Thus, the simplest choice of F_{j-2} is not $F_{j-2} = 0$, but rather $F_{j-2} =$ generic. Here, X^4 is generic enough, in which case $Q(0)$ is that above, $Q(1) = 0$, but $Q(2) = \langle \bar{x}, \bar{x}\bar{y}, \bar{x}^2, \bar{x}\bar{y}^2 \rangle$ of Hilbert function $H(2) = (0,1,2,1)$ for B^* determined by $G = Y^6 + Y^4X + X^4$. See §5 for a discussion.

5. The Hilbert function of a Gorenstein algebra.

The "magic rectangle" property of the Hilbert function of a Gorenstein Artin algebra leads to inequalities satisfied by the Hilbert function $H(A)$ of any Gorenstein quotient A of a Gorenstein ring R having a nonzero divisor of order one. We suppose for simplicity that R is a power series ring $R = k[[x_1, \ldots, x_r]]$ in r-variables, where $r \geq 2$. We will state these results, whose proof is beyond the scope of the paper. But first, we illustrate the principle, that given $H(0), H(1), \ldots, H(a-1)$ and r, there is a maximum possible $H(a)$.

Example 7. If $r = 3$, and $H(0) = (1,3,4,5,6,5,4,3,1)$, then since $H(R) = (1,3,6,10,\ldots)$ and $\sum H(a) = H(A) \leq H(R)$, the Hilbert function $H(1) \leq H(R) - H(0) = (0,0,2,5,9,\ldots)$. The symmetry of $H(1)$ around $(j-1)/2 = 7/2$ implies that $H(1) \leq (0,0,2,5,5,2,0)$, termwise. We call this sequence $M(1)$, for maximum, given $H(0)$. Similarly, if $H(1)$ is in fact $(0,0,1,3,3,1)$, then $H(2) \leq H(R) - H(0) - H(1) = (0,0,1,2,6,\ldots)$, and symmetry of $H(2)$ around $(j-2)/2 = 3$ implies $H(2) \leq M(2) = (0,0,1,2,1,0)$, termwise.

The following definition of $M(a)$ generalizes Example 7. When a is fixed, let $j' = (j-a)/2$ and $H_{\leq j'}$ denote the initial sequence of H.

Definition and Lemma: Upper bound for $H(a)$. *Given $H(0), \ldots, H(a-1)$, and r, we define a sequence $M(a) = (m_{a0}, \ldots, m_{ak}, \ldots)$ by*

$$M(a)_{\leq j'} = \left(H(R) - \sum_{0}^{a-1} H(i) \right)_{\leq j'}, \text{ and} \tag{12}$$

$$m_{ak} = m_{a,j-a-k} \text{ for } k \geq j'.$$

The Hilbert function $H(a) = H(Q(a))$ for a GA quotient of R satisfies

$$H(a) \leqq M(a) \text{ termwise.} \tag{13}$$

The proof of the Lemma is immediate from the symmetry of $H(a)$ and the inequality $\sum H(i) \leqq H(R)$.

We now state the key inequalities for the Hilbert function of GA quotients of R.

THEOREM 4A: UPPER BOUND FOR $H(a)$. *If A is a GA quotient of R, then for each $a \leqq j$,*

$$H(A) \leqq \sum_{0}^{a-1} H(i) + M(a), \text{ termwise.} \tag{14}$$

THEOREM 4B: MAXIMUM FOR $H(A)$ (with J. Emsalem). *If the sequences $T(0), T(1), \ldots, T(a-1)$ occur as $H(0), H(1), \ldots, H(a-1)$ for a GA quotient $A = R/I$ defined by the principal system $R \circ F \subset D$, then the principal system $R \circ (F + g_{j-a})$ for a generic choice of the degree-$(j-a)$ form $g_{j-a} \in D$ defines a GA quotient $A' = R/I' = R/\operatorname{Ann}(F + g_{j-a})$ of R having the Hilbert function decomposition*

$$H(i) = \begin{cases} T(i) & \text{for } i < a, \\ M(a) & \text{for } i = a, \\ 0 & \text{for } i > a. \end{cases} \tag{15}$$

Thus A' has the maximal Hilbert function (given $H(0), \ldots, H(a-1)$ and r) of (14). Furthermore $Q'(i)$ for A' are related to $Q(i)$ for A by

$$Q'(i) = Q(i) \text{ as } R^*\text{-module for } i \leqq a-1.$$

The proof of Theorem 4A is combinatorial, and uses the fact $H(A/C(a)) = \sum_0^{a=1} H(i)$ is the Hilbert function of a quotient of R, as well as the symmetry of $H(a)$. The extant proof of Theorem 4B is difficult, and uses a result of A. Miri [**MI**, Prop. III.1.2], improved by J. Emsalem, as well as a detailed study of the ideals $(I : m^i) \cap m^k$ of R (the ideal tables of [**I-4**]). The algebra A' is a maximal length Gorenstein quotient of R such that $I' \cap m^{j+1-a} = I \cap m^{j+1-a}$, so A is the same as A' in degrees at least $j+1-a$; we call such algebras A' *relatively compressed* quotients of R (relative to A in degrees at least $j+1-a$).

COROLLARY: Stretched and nearly stretched GA algebras. *If A is a GA quotient of R for which there is an integer $a > 0$ such that $H(A)$ satisfies $t_{j-a} = \cdots = t_j = 1$, then $H(0) = (1, 1, \ldots, 1)$, and $H(i) = 0$ for $1 \leqq i \leqq a-1$. Let $M(a)$ be the sequence of (12), given these values of $H(0), \ldots, H(a-1)$, and r. If $t = [(j-a)/2]$, and $\#R_i$ denotes $\ell(R_i)$, then*

$$(16) \quad H(A) \leqq H(0) + M(a) = \\ (1, r, \#R_2, \ldots, \#R_t, \#R_{t-1}, \ldots, \#R_2, r, 1, \ldots, 1),$$

*where r occurs in degree $j - a - 1$ and there is a repetition of $\#R_t$ if $(j-a)$ is odd. The maximum possible Hilbert function here is just that of a compressed Gorenstein quotient of R with socle degree $j-a$, followed by a ones. When $a = j-2$, the relatively compressed algebras A' of Hilbert function $H(0)+M(j-2) = (1, r, 1, \ldots, 1)$ are the Artin case of the stretched Gorenstein algebras defined by J. Sally [**SA-1**].*

PROOF: It is easy to see that $F = F_j + F_{j-1} + \ldots$ must satisfy, after a change of variable, $F = X^j + b_1 X_1^{j-1} + \cdots + b_{j+1-a} X_1^{j+1-a} + \ldots$, in order for $t_{j-a}, \ldots, t_j$ to be 1; thus after change by a unit F may be chosen $F = X_1^j + g_{j-a}$ + lower terms. It follows from Theorem 3 that $Q(0) = R/(x_2, \ldots, x_r, x_1^{j+1})$ of Hilbert function $H(0) = (1, \ldots, 1)$, and that $Q(i) = 0$ for $1 \leq i \leq a-1$. Thus (16) is a consequence of (12) and (14); that the bound can be attained is a consequence of Theorem 4B.

Example 8. Suppose $r = 3$, $j = 7$, and $H(A)$ ends in $(\ldots, 4, 3, 1)$. Then $H(A) \leq (1, 3, 6, 10, 7, 4, 3, 1)$, termwise. To see this, we first notice that $H(0)$ must also end in $(\ldots, 4, 3, 1)$ as $t_{1,j-1} = 0$ implies $t_{0,j-1} = 3$, $r = 3$ so $t_{01} = 3$ and $t_{i1} = 0$ for $i \geq 1$: thus $t_{1,j-2} = 0$ and $t_{0,j-2} = 4$. Thus $H(0) = (1, 3, 4, 4, 4, 4, 3, 1)$ or $(1, 3, 4, 5, 5, 4, 3, 1)$, by the result of [**B-E**] (see the remark preceding Example 1). In the former case, $M(1) = (0, 0, 2, 6, 2)$, giving a maximum Hilbert function of $(1, 3, 5, 10, 6, 4, 3, 1)$; in the latter case $M(1) = (0, 0, 2, 5, 2)$, giving $H(A) \leq (1, 3, 6, 10, 7, 4, 3, 1)$. Either is obtained by adding to F_j a general enough g_{j-1}, forming the principal system $R \circ (F_j + g_{j-1})$.

Examples 7 and 8 show that information about the tail $t_{j-s}, \ldots, t_j$ of $H(A)$ implies there are inequalities satisfied by $(t_0, \ldots, t_{j-s-1})$. To find these inequalities in Example 8, we needed to go through the possible tails of the sequences $H(a)$ that are consistent with the specification of $H(A)_{\geq (j-s)}$, then apply Theorem 4A.

What can we say about the Hilbert function of Gorenstein algebras A having dimension greater than zero? A result joint with L. Avramov shows that the lengths of the terms in the intersection of the Loëwy $(0:\bar{m}^s)$ and $\bar{m}$-adic $(\bar{m}^i)$ filtrations on a general enough minimal reduction $\bar{A} = A/L$ of A, are constants $c_A(i,s)$. (Here L is an A-sequence, see [**E-I**, Lemma 1.1] for a discussion.) It follows that the sequences $H(a)$ for $\bar{A}$ are invariant for such reductions. It follows from a result of Singh ([**SI**], quoted in [**SA-2**, §3, 3.1]; see [**E-I**, (2)]) that

$$H(A) \leqq H^t(\bar{A}) \underset{\text{def}}{=} H(\bar{A})/(1-Z)^t, \text{ the } t\text{-th sum function of } H(\bar{A}).$$

Consequently, information about the tail of the Hilbert function $H(\bar{A})$ of a general minimal reduction of A implies a termwise upper bound for the Hilbert function $H(A)$ of the original Gorenstein algebra A. The case only $t_j = 1$ specified, in other words only the socle degree j fixed, is discussed in [**E-I**] as the Gorenstein case of Theorem 1 there.

The inequalities (14) bounding $H(A)$ for GA algebras A are I believe the "stronger inequalities" nascent in a remark at the end of [**E-I**, §1]. Example 8 above pertains to that remark, and the Gorenstein structure of $Q(a)$ in codimension two gives an algebraic explanation of the Hilbert function (6) of the intersection of two plane curves. There are similar inequalities for the Hilbert functions of Artin algebras that are not Gorenstein, resting on a similar notion of relatively compressed (see [**E-I**] for the case where only the socle-type of A is specified).

There is much that is mysterious about the reflexive factors $Q(a)$, when A is Gorenstein Artin. May the magicians delight!

REFERENCES

[A-A] R. Achilles and L. Avramov, *Relations between properties of a ring and of its associated graded ring*, in "Seminar D. Eisenbud/B. Singh/W. Vogel," vol. 2, Teubnertexte, Band 48, Teubner, Leipzig, 1982, pp. 5–29.

[BRI] J. Briançon, *Description de* Hilb^n $C\{x, y\}$, Invent. Math. **41** (1977), 45–89.

[B-E] D. Buchsbaum and D. Eisenbud, *Algebra structures for finite free resolutions, and some structure theorems for codimension three*, Amer. J. Math. **99** (1977), 447–485.

[E-I] J. Elias and A. Iarrobino, *The Hilbert function of a Cohen-Macaulay local algebra: extremal Gorenstein algebras*, J. Algebra **110** (1987), 344–356.

[E-L] D. Eisenbud and H. Levine, *An algebraic formula for degree of a map germ*, Annals of Math. **106** (1977), 19–38.

[EM] J. Emsalem, *Géomètrie des points épais*, Bull. Soc. Math. France **106** (1978), 399–416.

[EM-I] J. Emsalem and A. Iarrobino, *Some zero-dimensional generic singularities: finite algebras having small tangent space*, Compositio Math. **36** (1978), 145–188.

[I-1] A. Iarrobino, "Punctual Hilbert Schemes," AMS Memoir #188, 1977.

[I-2] A. Iarrobino, *Compressed algebras*, Trans. AMS **285** (1984), 337–378.

[I-3] A. Iarrobino, *Tangent cone of a Gorenstein singularity*, in "Proceedings of the Conference on Algebraic Geometry, Berlin 1985," H. Kurke and M. Roczen, eds., Teubnertexte zur Math. vol. 92, Teubner, Leipzig, 1986, pp. 163–176.

[I-4] A. Iarrobino, *Gorenstein Artin algebras and their associated graded algebras*, preprint.

[KAP] I. Kaplansky, "Commutative Rings," Allyn and Bacon, Boston, 1970.

[MAC-1] F. H. S. Macaulay, *On a method for dealing with the intersections of two plane curves*, Trans. AMS **5** (1904), 385–410.

[MAC-2] F. H. S. Macaulay, "The Algebraic Theory of Modular Systems," Cambridge University Press, London, New York, 1916.

[MI] A. Miri, *Artin modules having extremal Hilbert series: compressed modules*, Ph.D. thesis, Northeastern University (1985).

[NO] D. G. Northcott, *Injective envelopes and inverse polynomials*, J. London Math. Soc. (2) **8** (1974), 290–296.

[SA-1] J. Sally, *Stretched Gorenstein rings*, J. London Math. Soc. (2) **20** (1979), 19–26.

[SA-2] J. Sally, "Numbers of generators of ideals in local rings," Lecture Notes in Pure and Applied Math, vol. 35, Dekker, New York, 1978.

[SC] C. A. Scott, *On a recent method for dealing with the intersections of plane curves*, Trans. AMS **3** (1902), 216–263.

[SI] B. Singh, *Effect of a permissible blowing up on the local Hilbert function*, Invent. Math. **26** (1974), 201–212.

[ST] R. Stanley, *Hilbert functions of graded algebras*, Advances in Math. **28** (1978), 57–83.

Department of Mathematics, Northeastern University, Boston MA 02115

On Unramifying Transcendence Base

Wei-Eihn Kuan and Christel Rotthaus

In this article we prove the following theorem: Every t-adic complete factorial domain containing a field k of characteristic 0 contains an unramified transcendence base over k for all height 1 prime ideals.

0. Motivation.

We start with a special case of Artin's Conjecture [**1**]:

(0.1) Let k be a field of characteristic 0; $R = k[x_1, \ldots, x_N]$; $\hat{R} = k[[x_1, \ldots, x_N]]$ and $\varphi : R \to \hat{R}$ the canonical morphism.

Question: Is $\hat{R}$ a direct limit of smooth R-algebras of finite type? I.e. can every commutative diagram

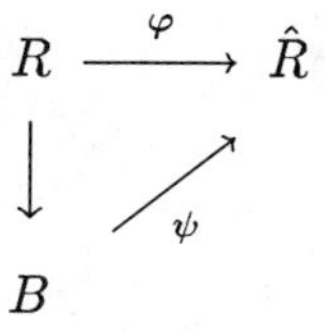

$$\begin{array}{ccc} R & \xrightarrow{\varphi} & \hat{R} \\ \downarrow & \nearrow_{\psi} & \\ B & & \end{array}$$

where B is a finite type R-algebra, be embedded in a commutative diagram:

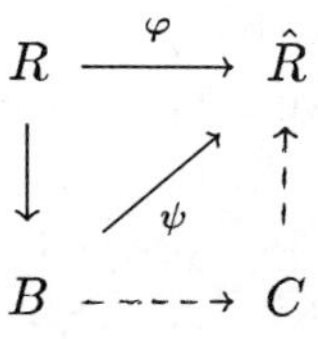

$$\begin{array}{ccc} R & \xrightarrow{\varphi} & \hat{R} \\ \downarrow & \nearrow_{\psi} & \uparrow \\ B & \dashrightarrow & C \end{array}$$

where C is of finite type and smooth over R?

In [**2**] and [**4**] this special case was solved by methods "similar" to those in Artin's proof of the approximation property for $R^h_{m_R}$. But the algebras C constructed in [**4**] are not any more subalgebras of $\hat{R}$.

In this paper we would like to discuss a result which might be useful in trying to get $\hat{R}$ as a direct limit of smooth finite type R-algebras. To begin with we look at a different approach of Néron's desingularization [**1**]:

Case: $N = 1$. We consider the canonical morphism $\varphi : R = k[x] \to \hat{R} = k[[x]]$; and pick a transcendence basis $\{u_i\}_{i\in I} \subseteq k[[x]]$ of $\hat{R}$ over R, and consider the diagram

$$\begin{array}{ccc} R & \xrightarrow{\varphi} & \hat{R} \\ \downarrow & \nearrow_{\nu} & \\ D = Q(R[\{u_i\}_{i\in I}]) \cap \hat{R} & & \end{array}$$

where $Q(*)$ stands for the quotient field of $*$. Then

(a) D is a discrete valuation ring

(b) ν is étale

(c) D is a direct limit of finite type R-algebras of the form: blow-ups of polynomial rings $k[x, u_{i_1}, \ldots, u_{i_r}]$ with center x and localization.

In (c) one can always arrange that the finite type algebras are regular rings, hence smooth over $k[x]$.

As for Case $N > 1$, one can show the following: suppose there is a transcendence basis $\{u_i\}_{i\in I} \subseteq \hat{R}$ over R such that

$$\begin{array}{ccc} R & \xrightarrow{\varphi} & \hat{R} \\ \downarrow & \nearrow_{\nu:\text{étale}} & \\ D = Q(R[\{u_i\}_{i\in I}]) \cap \hat{R} & & \end{array}$$

then D is a direct limit of regular blow-ups of polynomial rings over R, and $D = \varinjlim(R[u_{i_1}, \ldots, u_{i_j}, g_{j/f}]_{\mathfrak{n}})$. $R[u_{i_1}, \ldots, u_{i_j}, g_{j/f}]_{\mathfrak{n}}$ is regular and hence smooth over R. The method of proof is again similar to Artin's proof of the approximation property for $R^h_{m_R}$, [1].

A natural question: Does such a transcendence basis exist making $\hat{R}$ étale over R?

We do not have an answer but can prove that there is always a transcendence basis such that the ht 1 prime ideals are unramified, which is the main object of this paper.

1. Definitions.

Let R be a factorial domain containing field k of characteristic 0. Let $\{u_i\}_{i\in I} \subseteq R$ be a transcendence basis of the quotient field $Q(R)$ of R over k. Then

(1.1) the Krull domain $R_{\{u\}} = R \cap k(\{u_i\}_{i \in I})$ is called the ring of the transcendence basis $\{u_i\}_{i \in I}$ or, shortly, ring of $\{u_i\}_{i \in I}$.
(1.2) Let $P \in R$ be a prime element, then P is called unramified with respect to $\{u_i\}_{i \in I}$ if the prime ideal $\mathcal{P}_{(P)} = (P) \cap R_{\{u\}}$ satisfies:

$$\mathcal{P}_{(P)} R_{(P)} = P R_{(P)}.$$

Otherwise P is called ramified.

REMARK: If $P \in R$ is unramified then the extension of the discrete valuation rings

$$R_{(P)} \cap k(\{u_i\}_{i \in I}) \to R_{(P)}$$

is unramified. The converse is not true in general unless ht $(\mathcal{P}_{(P)}) = 1$.

(1.3) The transcendence basis $\{u_i\}_{i \in I} \subseteq R$ is called *unramifying* if all prime elements $P \in R$ are unramified with respect to $\{u_i\}_{i \in I}$.

2. Main Theorem.

LEMMA 2.1. *Let R be a factorial domain containing a field k, and let t be a prime element of R. Then R contains a transcendence basis containing t and consisting of prime elements of R.*

PROOF: Let U be the set of all prime elements of R. Considering subsets of U consisting of algebraically independent prime elements of R over k, and the partial order relation $\subseteq$, we have that every chain of such subsets $\{U_i\}$ has an upper bound. Thus by Zorn's lemma U has a maximal subset U_0 consisting of algebraically indepdendent prime elements of R over k. We claim that U_0 is a transcendence basis of R over k. Indeed, suppose that $Z \in R$ is algebraically independent over $k(U_0)$, and $Z = \varepsilon P_1^{e_1} \dots P_n^{e_n}$ where $P_1, \dots, P_n$ are prime elements of R and ε a unit in R and $e_1, \dots, e_n$ are positive integers. Then we have

Case 1: At least one of the P_i's is algebraically independent over $k(U_0)$, say P_1. Then $U_0 \cup \{P_1\}$ is a larger subset then U_0 consisting of algebraically independent prime elements over k and contradicts the maximality of U_0; and

Case 2: None of the P_i's is algebraically independent over $k(U_0)$. Then $\varepsilon \neq 1$ and each element of $\{\varepsilon, \varepsilon P_1, \dots, \varepsilon P_n\}$ is algebraically independent over $k(U_0)$. Again $U_0 \cup \{\varepsilon P_1\}$ is a larger set than U_0, and contradicts that U_0 is maximal. Thus U_0 is a transcendence basis consisting of prime elements of R. Now if the chain $\{U_i\}$ starts with $\{t\}$ then U_0 contains t, and is a desired transcendence basis.

In the following we assume that R is a factorial domain containing a field of characteristic 0. Let $t \in R$ be a prime element such that R is complete and separated with respect to the (t)-adic topology. Moreover, we suppose $\operatorname{tr}\deg_k R = \infty$. Then we have

PROPOSITION 2.2. *There is an unramifying transcendence basis of R over k.*

PROOF (2.2.1): By Lemma 2.1, we start with an arbitrary transcendence basis

$$\underline{w}_0 = \{t\} \cup \{w_i^{(0)}\}_{i \in I_0} \subset R$$

of R over k containing prime elements $w_i^{(0)}$'s of R.

(2.2.2) We construct recursively, for each $n \in N$, a transcendence basis

$$\underline{w}_n = \{t\} \cup \{w_i^{(n)}\}_{i \in I_n} \subseteq R, \text{ where } I_n \text{ is an index set,}$$

and an element

$$z_n \in R$$

such that $\underline{w}_n$ and $\underline{w}_{n+1}$ differ by a single element and $z_{n+1} - z_n \in (t^n)$.

Let $R_n = k(\underline{w}_n) \cap R$. Suppose the construction does not stop for any n, then we shall end up with a transcendence basis

$$\underline{w}_\infty \subseteq R$$

and the element

$$z = \lim z_n$$

such that z is algebraically independent over $k(\underline{w}_\infty)$. This is a contradiction and leads to the existence of an unramifying basis.

(2.2.3) If $\underline{w}_0$ is unramifying, then we are done. Otherwise there is a ramifying prime element $P_1 \in R$, and a finite subset $\mathcal{F}_0 \subset I_0$ such that P_1 is algebraic over the field $k(t, \{w_j^{(0)}\}_{j \in \mathcal{F}_0}$. We pick an element $\tilde{z}_1 \in \{w_i^{(0)}\}_{i \in I_0} \setminus (\{w_j^{(0)}\}_{j \in \mathcal{F}_0} \cup \{t\})$ and replace $\tilde{z}_1$ by

$$z_1 = P_1 \tilde{z}_1 t.$$

Then the set

$$\underline{w}_1 = \{t\} \cup \left[\{w_i^{(0)}\}_{i\in I_0} \setminus \{\tilde{z}_1\}\right] \cup \{z_1\}$$

is again a transcendence basis of R over k, which is constructed from $\underline{w}_0$ and P_1. We note that P_1 is ramifying over R_0 and unramifying over R_1.

(2.2.4) **Recursive step.**

(A) Suppose we have constructed transcendence bases:

$$\begin{aligned} \underline{w}_0 &= \{t\} \cup \{w_i^{(0)}\}_{i\in I_0} \subseteq R, \\ &\vdots \\ \underline{w}_n &= \{t\} \cup \{w_i^{(n)}\}_{i\in I_n} \subseteq R, \end{aligned}$$

where almost all elements of $\underline{w}_j$ are prime elements of R;

(B) Suppose furthermore $P_1, \ldots, P_n$ are prime elements such that, for all $\ell \in \{1, \ldots, n\}$, P_ℓ is ramified over $R_{\ell-1}$. Now we describe the procedure of the construction of $\underline{w}_\ell$ from $\underline{w}_0, \ldots, \underline{w}_{\ell-1}$ and $P_1, \ldots, P_\ell$ as follows:

(C) For each $\ell \in \{0, 1, \ldots, n-1\}$ there is a finite subset $\mathcal{F}_\ell \subseteq I_\ell$ such that

(C.1) $P_{\ell+1}$ is algebraic over $k(t, \{w_j^{(\ell)}\}_{j\in\mathcal{F}_\ell})$; and
(C.2) For each $\nu \in \mathcal{F}_{\ell-1}$ there is a $\mu \in \mathcal{F}_\ell$ such that

$$w_\nu^{(\ell-1)} = w_\mu^{(\ell)},$$

i.e., $\{w_\nu^{(\ell-1)}\}_{\nu\in\mathcal{F}_{\ell-1}} \subseteq \{w_\mu^{(\ell)}\}_{\mu\in\mathcal{F}_\ell}$.

(D) For every $\ell \in \{0, \ldots, n-1\}$, we pick an element

$$\tilde{z}_{\ell+1} \in \underline{w}_\ell \setminus \left[\{t\} \cup \{w_j^{(\ell)}\}_{j\in\mathcal{F}_\ell}\right]$$

and construct an element

$$z_{\ell+1} \in \underline{w}_{\ell+1}$$

such that

(D.1) $\underline{w}_\ell \setminus \{\tilde{z}_{\ell+1}\} = \underline{w}_{\ell+1} \setminus \{\tilde{z}_{\ell+1}\}$,
(D.2) $z_{\ell+1} = Q_{\ell+1}\, \tilde{z}_{\ell+1}\, t^{\ell+1} + z_\ell$, where $Q_{\ell+1} = P_{\ell+1}(P_1 \cdots P_\ell)^2$, and
(D.3) There is $j \in \mathcal{F}_{\ell+1}$ such that $z_{\ell+1} = w_j^{(\ell+1)}$ for $\ell \in \{1, \ldots, n-1\}$.

If $\underline{w}_n$ is an unramifying transcendence basis, we are done. Otherwise we construct $\underline{w}_{n+1}$ as follows:

(E) We pick a ramified prime element P_{n+1} of R over R_n, and take a finite subset $\mathcal{F}_n \subseteq I_n$ with the following properties:

(E.1) For each $\nu \in \mathcal{F}_{n-1}$, there is $\mu \in \mathcal{F}_n$ such that

$$w_\nu^{(n-1)} = w_\mu^{(n)},$$

(E.2) There is $j \in \mathcal{F}_n$ such that $z_n = w_j^{(n)}$,

(E.3) P_{n+1} is algebraic over $k(t, w_j^{(n)})_{j\in\mathcal{F}_n}$.

Note that (E.1) and (E.2) imply $\{t\} \cup \{z_1, \ldots, z_{n-1}\} \subseteq \{w_\nu^{(n-1)}\}_{\nu\in\mathcal{F}_{n-1}} \cup \{z_n\} \subseteq \{w_\mu^{(n)}\}_{\nu\in\mathcal{F}_n}$.

Now pick an element

$$\tilde{z}_{n+1} \in \{w_i^{(n)}\}_{i\in I_n} \setminus \{w_j^{(n)}\}_{j\in\mathcal{F}_n},$$

and define $z_{n+1} = Q_{n+1}\,\tilde{z}_{n+1}\,t^{n+1} + z_n$, where

$$Q_{n+1} = P_{n+1}(P_1 \cdots P_n)^2.$$

REMARK: $\tilde{z}_{n+1}$ is a prime element of R, because all non-prime elements of $\underline{w}_n$ are in $\{w_\nu^{(n-1)}\}_{\nu\in\mathcal{F}_{n-1}} \cup \{z_n\}$.

Defining $\underline{w}_{n+1} = \left[\underline{w}_n \setminus \{\tilde{z}_{n+1}\}\right] \cup \{z_{n+1}\} \subseteq R$, we make

(2.2.6) CLAIM. *$\underline{w}_{n+1}$ is a transcendence basis of R over k.*

PROOF OF (2.2.6): By choice of $\mathcal{F}_n$ we get

(2.2.6.1) every P_ℓ, for $\ell = 1, \ldots, n+1$, is algebraic over $k(t, \{w_j^{(n)}\}_{j\in\mathcal{F}_n})$. Therefore Q_{n+1} is algebraic over $k(\underline{w}_n \setminus \{\tilde{z}_{n+1}\})$. Since z_n and t are in $\underline{w}_n \setminus \{\tilde{z}_{n+1}\}$, z_n and t are both algebraic over $k(\underline{w}_n \setminus \tilde{z}_{n+1})$. Therefore we get

(2.2.6.2) $\underline{w}_{n+1}$ is an algebraically independent set.

Thus $\underline{w}_{n+1}$ is a transcendence basis of R over k.

(2.2.7) CLAIM. *For all $k \in \{1, \ldots, n+1\}$, $z_k - z_{k-1} \in R_{n+1}$ is divisible in R by P_k but not P_k^2. In particular $P_1, \ldots, P_{n+1}$ are unramifying prime elements over R_{n+1} and it also follows that P_i and P_j are non-associated prime elements of R, for $i \neq j$ and $i, j \in \{1, \ldots, n+1\}$.*

PROOF OF (2.2.7): Induction on n. By construction of $\underline{w}_{n+1}$, $z_k \in \underline{w}_{n+1}$ for $k = 1, \ldots, n+1$, and $z_{n+1} - z_n = Q_{n+1}\,\tilde{z}_{n+1}\,t^{n+1}$. We observe (1) $\tilde{z}_{n+1}$ and t are prime elements in R and are elements of R_n, and (2) $Q_{n+1} =$

$P_{n+1}(P_1 \cdots P_n)^2$ and $P_1, \ldots, P_n$ are unramified prime elements of R over R_n by induction hypothesis. Since P_{n+1} is a ramified prime element over R_n, therefore P_{n+1} is relatively prime to each of $\tilde{z}$, t, $P_1, \ldots, P_n$. Therefore $P_{n+1} \mid z_{n+1} - z_n$ and $P_{n+1}^2 \nmid z_{n+1} - z_n$.

(2.2.8) Suppose the successive construction of transcendence bases does not stop, then we consider

$$\underline{w}_\infty = \left[\underline{w}_n \setminus \bigcup_{\ell > n} \{\tilde{z}_\ell\}\right] \cup \left[\bigcup_{\ell > n} \{z_\ell\}\right] \subseteq R.$$

In view of (2.2.4)(D.1) $\underline{w}_\infty$ is independent of n.

We shall show the following contradicting statements:

(2.2.9) $\underline{w}_\infty$ is a transcendence basis of R over k and

(2.2.10) $z = \lim z_n$ is algebraically independent over $k(\underline{w}_\infty)$.

PROOF OF (2.2.9): It suffices to show

(2.2.9.1) $\tilde{z}_{\ell+1}$ is algebraic over $k(w_\infty)$ for all $\ell \in \mathbb{N}$, and

(2.2.10.1) $\underline{w}_\infty$ is a set of algebraically independent elements over k.

PROOF OF (2.2.9.1): By the construction of $\underline{w}_{\ell+1}$ and $\underline{w}_\ell$, $Q_{\ell+1}$, t, and z_ℓ are algebraic over $k(\underline{w}_\ell \setminus \{\tilde{z}_{\ell+1}\})$. By the construction of $\mathcal{F}_n$ and $\underline{w}_n$ for $n \geq \ell$, $Q_{\ell+1}$, t, z_ℓ are also algebraic over $k\big(\underline{w}_\ell \setminus \bigcup_{n>\ell}\{\tilde{z}_n\}\big)$. On the other hand, $\tilde{z}_{\ell+1} = \frac{1}{Q_{\ell+1}\, t^{\ell+1}}\, z_{\ell+1} + z_\ell$, hence $\tilde{z}_{\ell+1}$ is algebraic over $k\big([\underline{w}_\ell \setminus \bigcup_{n>\ell}\{\tilde{z}_n\}] \cup \{z_{\ell+1}\}\big) \subseteq k(\underline{w}_\infty)$.

PROOF OF (2.2.9.2): Suppose $\underline{w}_\infty$ is not an algebraically independent set. Then a finite subset F of $\underline{w}_\infty$ is an algebraically dependent set. By the construction of $\underline{w}_\infty$, there is $\underline{w}_n$ for some n such that $\underline{w}_n \supset F$. Since $\underline{w}_n$ is a transcendence basis of R over k, $F = \varphi$.

PROOF OF (2.2.10): Since R is (t)-adically complete and separated, then $z = \lim z_k = \sum_{k=1}^{\infty} Q_k\, \tilde{z}_k\, t^k$ is in R. Suppose z is algebraic over $k(\underline{w}_\infty)$. Then there is a non-zero polynomial $H(x)$ of minimal degree in x in $k(\underline{w}_\infty)[x]$ such that $H(z) = 0$. Let $H(x) = \sum_{i=0}^{m} h_i\, x^i$, where $h_m \neq 0$ and $h_i \in k[\underline{w}_\infty]$, the polynomial ring in transcendence elements in $\underline{w}_\infty$. Since $\underline{w}_\infty = [\underline{w}_0 \setminus \bigcup_{\ell>0}\{\tilde{z}_\ell\}] \cup [\bigcup_{\ell>0}\{z_\ell\}]$, then there is $r \in \mathbb{N}$ such that $h_i \in k\big[(w_0 \setminus \{\tilde{z}_\ell\}) \cup \bigcup_{\ell=1}^{r}\{z_\ell\})\big]$ for $i = 0, 1, 2, \ldots, m$. It follows that $h_i \in k[\underline{w}_n]$ for $i = 0, 1, \ldots, m$ and for all $n > r$. Let $n \in \mathbb{N}$ and $n > r$, and write $H(z)$ in Taylor expansion around z_n. We have

$$0 = \sum_{i=0}^{m} h_i\, z^i = H(z) = H(z_n) + (z - z_n)\, H'(z_n) + (z - z_n)^2\, G(z, z_n),$$

where $G(z, z_n) \in k[\underline{w}_n][z, z_n]$. It follows that
(2.2.10.1) $H(z_n) = -\big[(z - z_n)\, H'(z_n) + (z - z_n)^2\, G(z, z_n)\big]$. Since
(2.2.10.2) $z - z_n = \sum_{k=n+1}^{\infty} Q_k\, \tilde{z}_k\, t^k$, we have $P_{n+1} \mid (z - z_n)$ and $P_{n+1}^2 \nmid (z - z_n)$ in R by (2.2.7).
(2.2.10.3) Since P_{n+1} is ramified over $R_n \supseteq k[\underline{w}_n]$, every element of R_n which is divisible in R by P_{n+1} is also divisible by P_{n+1}^2. Thus from (2.2.10.1) and that $H(z_n) \in R_n$, we get $P_{n+1}^2 \mid H(z_n)$ and $P_{n+1} \mid H'(z_n)$ in R.
(2.2.10.4) Applying Taylor expansion to $H'(z)$, we have

$$H'(z) = H'(z_n) + \sum_{\nu=2}^{m} \frac{H^{(\nu)}(z_n)}{(\nu - 1)!}\, (z - z_n)^{\nu - 1},$$

and $P_{n+1} \mid H'(z)$ in R. Consequently $P_{n+1} \mid H'(z)$ in R for all $n + 1 \geq r$. But (2.2.7) implies that $P_{n+1}, P_{n+2}, \ldots$ are mutually relatively prime in R. Thus $H'(z)$ has infinitely many prime factors in the factorial domain R, hence $H'(z) = 0$. Since k is of characteristic 0 and $H'(x)$ is a polynomial of degree $m - 1 (< m)$ having z as a root, we thus have a contradiction to the minimality of m. z is therefore algebraically independent over $k(\underline{w}_\infty)$.

References

1. M. Artin, *Algebraic approximation of structures over complete local rings*, Publ. Math. Inst. Hautes Étud. Sci. **36** (1969), 23–58.
2. M. Artin, *Algebraic structure of power series rings*, Comtemp. Math. **13** (1982), 223–227.
3. M. Artin and J. Denef, *Smoothing of a ring morphism along a section*, in "Arithmetic and Geometry," Vol. II, Birkhäuser, Boston, 1983.
4. M. Artin and C. Rotthaus, *A structure theorem for power series rings*, to appear in a special volume dedicated to M. Nagata's sixtieth birthday.
5. M. Nagata, "Local rings," Interscience, 1962.

Department of Mathematics, Michigan State University, East Lansing MI 48824-1027

The second author is partially supported by the National Science Foundation.

A Survey of Problems and Results on the Number of Defining Equations

GENNADY LYUBEZNIK

§0. Introduction.

How many equations are necessary to define a given algebraic set in n-space set-theoretically? Despite its simple formulation this question is highly non-trivial.

We can ask the same question in a more general setting. Namely, let B be a commutative Noetherian ring, $X = \operatorname{Spec} B$ the associated affine scheme, $V \subset X$ a closed set defined by an ideal $I = I(V) \subset B$ and $f_1, \dots, f_s \in I$ some elements. Or let k be a field, $B = k \oplus B_1 \oplus \dots \oplus B_n \oplus \dots$ a graded commutative Noetherian ring which is generated by B_1 as a k-algebra, $X = ProjB, V \subset X$ a closed subset defined by a *homogeneous* ideal $I = I(V) \subset B$ and $f_1, \dots, f_s \in B$ some *homogeneous* elements. We say that V is defined set-theoretically by s equations $f_1 = \dots = f_s = 0$ if $f_1, \dots, f_s$ generate I up to radical, i.e. $\sqrt{(f_1, \dots, f_s)} = \sqrt{I}$. This definition is motivated by Hilbert's Nullstellensatz.

Right at the very start we run into an unsolved problem.

PROBLEM 0.1: Let B be a graded ring as above, $M = B_1 \oplus B_2 \oplus \dots \subset B$ the irrelevant maximal ideal and $I \subset B$ a homogeneous ideal. If $I_M \subset B_M$ can be generated up to radical by s elements, can $I \subset B$ also be generated up to radical by s elements? If $I \subset B$ can be generated up to radical by s elements, can it be generated up to radical by s *homogeneous* elements?

Due to space constraints we cannot make this survey exhaustive and, in particular, we do not discuss the analytic aspects. For this see [**Fo2, Schn**].

§1. n Equations.

The first result in the field apparently belongs to L. Kronecker [**Kr1**], who stated without a detailed proof that every algebraic set in n-space could be defined set-theoretically by $(n + 1)$ equations. Proofs were given in [**Kö**], [**Pe2**] and [**Wa**]. O. Forster generalized this as follows.

THEOREM 1.1 [**Fo1**]. *Let B be a commutative Noetherian ring of dimension n and $X = \operatorname{Spec} B$ the associated affine scheme. Then every closed subset $V \subset X$ can be defined set-theoretically by $(n+1)$ equations.*

As the next step one would like to know when n equations are enough.

THEOREM 1.2 [**FW**]. *Let I be a finitely generated ideal of a commutative ring B such that $\operatorname{Supp}(I/I^2)$ is Noetherian. Suppose there exists an integer m such that I_P can be generated by $\min\{m - \dim(B/P), m-1\}$ elements for every prime ideal $P \supset I$. Then there exists an m-generated ideal $J \subset I$ such that $\sqrt{J} = \sqrt{I}$ and $\dim(\operatorname{Supp}(I/J)) \leq 0$.*

The following theorem gives a complete solution of the "n equations problem" for regular affine algebras over algebraically closed fields. The symbol $A_o(*)$ stands for the group of zero-cycles modulo rational equivalence.

THEOREM 1.3 [**Ly7**]. *Let k be an algebraically closed field, X a smooth affine n-dimensional variety over k with coordinate ring B, and $V = V' \cup P_1 \cup P_2 \cdots \cup P_r$ an algebraic subset of $X = \operatorname{Spec} B$, where V' is the union of irreducible components of positive dimensions and $P_1, P_2, \ldots, P_r$ some isolated closed points (which do not belong to V'). Then V can be set-theoretically defined by n equations if and only if one of the following conditions holds.*

(i) *$r = 0$, i.e. V consists only of irreducible components of positive dimensions.*
(ii) *V' is empty, i.e. V consists only of closed points and there exist positive integers $n_1, n_2, \ldots, n_r$ such that $n_1P_1 + n_2P_2 + \cdots + n_rP_r = 0$ in $A_0(X)$.*
(iii) *V' is nonempty, $r \geq 1$ and there exist positive integers $n_1, n_2, \ldots, n_r$ such that $n_1P_1 + n_2P_2 + \ldots, n_2P_r$ belongs to the image of the natural map $A_0(V') \to A_0(X)$ induced by the inclusion $V' \hookrightarrow X$.*

For every $n \geq 1$ there exists an irreducible affine n-dimensional variety over a suitable algebraically closed field such that its singular locus consists of just one closed point and for every d between 1 and $(n-1)$ it contains an irreducible d-dimensional subvariety which cannot be defined set-theoretically by n equations [Ly7]. Yet the ambient varieties in those examples are not normal.

PROBLEM 1.4: Let B be a commutative Noetherian normal n-dimensional domain which is finitely generated as an algebra over a field and $V \subset X =$

Spec B a closed set which consists of irreducible components of positive dimensions. Can it be defined set-theoretically by n equations?

K. Th. Vahlen [**Vah**] claimed to have produced an example of a curve in 3-space, which could not be defined set-theoretically by 3 equations. Yet O. Perron[**Pe1**] refuted Vahlen's example and M. Kneser [**Kn**] proved that all algebraic subsets of $\mathbb{A}^3$ can be defined set-theoretically by 3 equations. This has been generalized as follows.

THEOREM 1.5 [**Sto, EE**]. *Let $B = A[T]$, where A is a commutative Noetherian ring of dimension $(n-1)$. Then every closed set $V \subset X =$ Spec B can be defined set-theoretically by n equations. In particular, every algebraic subset of $\mathbb{A}^n_k$, where k is any field, can be defined set-theoretically by n equations.*

THEOREM 1.6 [**EE**]. *Let $A = A_0 \oplus A_1 \oplus \cdots \oplus A_i \oplus \ldots$ be a graded n-dimensional commutative Noetherian ring, where $A_0 = k$ is a field. We regard $B = A[T]$ as a graded ring with T homogeneous of positive degree. Let $I \subset B$ be a homogeneous ideal with $I \subset MB$, where $M = A_1 \oplus A_2 \oplus \ldots$. Then I can be generated up to radical by n homogeneous elements of B. In particular, every algebraic subset $V \subset \mathbb{P}^n_k$, which has at least one k-rational point, can be defined set-theoretically by n equations.*

Thus, in particular, if k is algebraically closed, every $V \subset \mathbb{P}^n_k$ can be defined set-theoretically by n equations.

THEOREM 1.7 [**Ly7**]. *Every algebraic subset $V \subset \mathbb{P}^n_k$, where k is any field of positive characteristic, can be defined set-theoretically by n equations.*

PROBLEM 1.8: Let k be any field of characteristic 0. Is every algebraic set $V \subset \mathbb{P}^n_k$ definable set-theoretically by n equations?

It even seems to be unknown whether or not the answer is positive for every closed point in $\mathbb{P}^2_Q$, where Q is the field of rational numbers.

§2. $(n-1)$ Equations.

PROBLEM 2.1: Can every algebraic subset $V \subset \mathbb{A}^n_k$ consisting of irreducible components of positive dimensions be defined set-theoretically by $(n-1)$ equations?

THEOREM 2.2 [**Fe1, Sz, Bo1, MK1**]. *Every locally complete intersection ideal $I \subset k[X_1, \ldots, X_n]$ of pure dimension 1 can be generated up to radical by $(n-1)$ elements.*

The following two results were used in the proof of Theorem 2.2

THEOREM 2.3 [**Bo2**]. *Let $I \subset k[X_1, \ldots, X_n]$ be such that I/I^2 is μ-generated. Then I can be generated up to radical by μ elements.*

THEOREM 2.4 [**MK1**]. *Let $I \subset k[X_1, \ldots, X_n]$ be such that I/I^2 is μ-generated and $\mu \geq 2 + \dim I$. Then I is μ-generated.*

The following theorem is still another step toward Problem 2.1.

THEOREM 2.5 [**CN1**]. *Every curve in $\mathbb{A}_k^n$, where k is any field of characteristic $p > 0$, is a set-theoretic complete intersection.*

We include a short proof of this result based on [**Moh4**]. Let $I \subset B = k[X_1, \ldots, X_n]$ be the radical defining ideal of the curve in question.

If the extension of I in $B \otimes_k k^{1/p^\infty}$ can be generated up to radical by $(n-1)$ elements, then, for some N, their p^N-th powers will produce $(n-1)$ elements of B which generate I up to radical. Thus we may assume $k = k^{1/p^\infty}$, i.e. k is perfect .

There exists a change of variables such that $\varphi : k[X_1, X_2]/(k[X_1, X_2] \cap I) \hookrightarrow B/I$ is integral and birational. If k is infinite, a generic linear change will do. We are leaving the general case to the reader.

Let C be the conductor of φ. Since B/I is one-dimensional and C contains non-zero divisors, $(B/I)/C$ is Artinian, and therefore it is a finite-dimensional vector-space over k. Let x_i be the image of X_i in $(B/I)/C$ and let $L_{i,r}$ be the k-linear span of $x_i^{p^r}, x_i^{p^{r+1}}, \ldots$ The sequence $L_{i,0} \supset L_{i,1} \supset \cdots \supset L_{i,r} \supset \ldots$ eventually stabilizes, since $(B/I)/C$ is finite-dimensional. Thus there exists r, such that $x_i^{p^r} \subset L_{i,r+1}$ for all i, that is $x_i^{p^r} - s_{i,1} x_i^{p^{r+1}} - s_{i,2} x_i^{p^{r+2}} - \cdots - s_{i,t_i} x_i^{p^{r+t_i}} = 0$, i.e. $X_i^{p^r} - s_{i,1} X_i^{p^{r+1}} - \cdots - s_{i,t_i} X_i^{p^{r+t_i}} - c_i = f_i \in I$, where $s_{i,j} \in k$ and $c_i \in k[X_1, X_2]$. Let $Y_i = X_i^{p^r}$, $B' = k[X_1, X_2, Y_3, Y_4, \ldots, Y_n]$ and $I' = I \cap B'$. Since $I'B \subset I$ and every element of I raised to the p^r-th power belongs to I', we see that $\sqrt{I'B} = I$. Since $\frac{\partial f_i}{\partial Y_j} = \delta_{ij} (i, j \geq 3)$, the Jacobian of $f_3, \ldots, f_n$ with respect to $Y_3, \ldots, Y_n$ is the identity matrix. Thus $f_3, \ldots, f_n$ define a non-singular surface and $B'' = B'/(f_3, \ldots, f_n)$ is regular. Since I' has height 1 in B'', it is locally 1-generated because B'' is regular. Therefore $I' \subset B'$ is locally

$(n-1)$ generated (by $f_3, \ldots, f_n$ and a preimage of a local generator of the image of I' in B''). Now we are done by Theorem 2.2.

THEOREM 2.6 [**Ly7**]. *Every algebraic set $V \subset \mathbb{A}_k^n$, (where k is any field of positive characteristic) consisting of irreducible components of positive dimensions can be defined set-theoretically by $(n-1)$ equations.*

This gives an affirmative answer to Problem 2.1 in finite characteristic. Same argument reduces Problem 2.1 in characteristic 0 to the case of an irreducible curve.

At present it is unknown, whether or not every irreducible curve in $\mathbb{A}_{\mathbb{C}}^3$ is even locally a set-theoretic complete intersection. For example, it is unknown whether or not Moh's curve defined parametrically by $X = t^6 + t^{31}, Y = t^8, Z = t^{10}$ is a set-theoretic complete intersection locally at the origin [**Moh1**]. For $n \geq 4$ it is unknown whether or not every monomial curve $X_1 = t^{a_1}, X_2 = t^{a_2}, \ldots, X_n = t^{a_n}$ is a set-theoretic complete intersection, even locally. For $n = 3$ the answer is positive [**He, Br1**].

Let B be a regular local Noetherian ring, $I \subset B$ a prime ideal of dimension 1, $I^{(n)}$ the n-th symbolic power of I, and $R = B \oplus I \oplus I^{(2)} \oplus \cdots \oplus I^{(n)} \oplus \ldots$ the symbolic Rees algebra of I. Cowsik [**C**] has observed that if R is Noetherian, then I is a set-theoretic complete intersection, and he conjectured that R is always Noetherian. P. Roberts [**Ro1**] gave a counterexample to this by exhibiting a prime ideal $I \subset B = \mathbb{C}[X, Y, Z]_{(X,Y,Z)}$ whose symbolic Rees algebra is not Noetherian. Yet his prime ideal splits as an intersection of several different primes in the completion of B with respect to the maximal ideal (X, Y, Z). In connection with this we state

PROBLEM 2.7: Is the symbolic Rees algebra of every prime ideal of dimension 1 in $\mathbb{C}[[X_1, \ldots, X_n]]$ Noetherian?

A positive answer would imply that every 1-dimensional prime in $\mathbb{C}[[X_1 \ldots, X_n]]$ is a set-theoretic complete intersection, and then, by an argument of Mohan Kumar, so is every ideal of pure dimension 1. P. Roberts [**Ro2**] found a prime ideal of height 3 in $\mathbb{C}[[X_1, X_2, X_3, X_4, X_5, X_6, X_7]]$ whose symbolic Rees algebra is not Noetherian.

THEOREM 2.8 [**Hu2**]. *Let (B, m) be a 3-dimensional regular local ring with infinite residue field; let $P \subset B$ be a prime ideal of height 2. Then the following are equivalent.*

(i) $\oplus_{n \geq 0} P^{(n)}$ *is Noetherian.*

(ii) *There exist k, ℓ, two elements $f \in P^{(k)}, g \in P^{(\ell)}$ and an $x \notin P$ such that $\lambda(B/(f,g,x)) = \lambda(B/(P,x))\ell k$, where $\lambda(*)$ denotes the length.*

M. Morales has extended this result to regular rings of any dimension > 3 [**Mor**].

S. Eliahou [**El2**] has written a computer program which finds generators of the Noetherian symbolic Rees algebras of monomial curves. It is unknown whether or not the symbolic Rees algebra of every monomial curve is Noetherian.

More results along these lines are found in [**Hu1**] [**Hu2**] [**El1**] [**El2**] [**El3**] [**Moh3**] [**Oo**] [**Sche1**] [**Sche2**] [**Br1**] [**Br2**] [**C**] [**Fe2**] [**MK2**].

Very little information is available about the projective space. It is known that a geometrically disconnected algebraic subset of $\mathbb{P}_k^n$ cannot be defined set-theoretically by $< n$ equations (Theorem 3.3 below).

PROBLEM 2.9: Can every geometrically connected algebraic subset of $\mathbb{P}_k^n$ of dimension > 0 be defined set-theoretically by $(n-1)$ equations (where k is any field)?

At present the answer is unknown even for smooth curves over algebraically closed fields.

THEOREM 2.10 [**Fe2**]. *Let C be a smooth curve in $\mathbb{P}_k^3$, where k is algebraically closed and* char $k = p > 0$. *Assume that C has a linear projection birational to it, and with only cusps as singularities. Then C is a set-theoretic complete intersection.*

This has been generalized to $\mathbb{P}_k^n$ by [**Moh2**].

Curves with cuspidal birational projections have been studied by R. Piene [**Pi**].

An interesting example is the rational monomial curve $C_d \subset \mathbb{P}_k^3$ defined parametrically by $(u^d, u^{d-1}v, uv^{d-1}, v^d)$. The hypothesis of Theorem 2.10 is satisfied, since C_d is of degree d and has a tangent line with a contact of order $d-1$. Thus, C_d is a set-theoretic complete intersection for every algebraically closed field k of positive characteristic - a result originally proven by R. Hartshorne [**Ha4**]. More generally, if $d > b > a \geq 1$, and k is a field of finite characteristic then the curve in $\mathbb{P}_k^3$, given parametrically by $(u^d, u^b v^{d-b}, u^a v^{d-a}, v^d)$ is a set-theoretic complete intersection [**BrSR**]. The following is a very old open problem.

PROBLEM 2.11: Is C_4 a set-theoretic complete intersection in characteristic 0?

THEOREM 2.12 [**RoV**, **SV**]. *Every arithmetically Cohen-Macaulay monomial curve in* $\mathbb{P}^3$ *is a set-theoretic complete intersection.*

C_4 is not arithmetically Cohen-Macaulay in any characteristic. This suggests the following problem.

PROBLEM 2.13: Let k be an algebraically closed field of characteristic 0 and $C \subset \mathbb{P}^3_k$ an irreducible curve of degree ≤ 7, which is a set-theoretic complete intersection. Is C necessarily arithmetically Cohen-Macaulay?

If C is not irreducible, the answer is negative [**GMV**, **PS2**]. There exists a smooth curve of degree 8 and genus 5 in P^3 which is not arithmetically Cohen-Macaulay, yet it is the set-theoretic intersection of two Kummer surfaces [**GP**]. There also exist 2-dimensional local Cohen-Macaulay rings of characteristic 0, such that B_{red} is a domain but not Cohen-Macaulay [**CN2**].

THEOREM 2.14 [**Fe2**]. *Let k be an algebraically closed field of characteristic $p > 0$ and C a reduced curve in $\mathbb{P}^3_k$, with any singularities. Then there exists a curve D in $\mathbb{P}^3_k$, which is locally a complete intersection, such that $D_{\text{red}} = C$.*

THEOREM 2.15 (Mohan Kumar). *If a curve $C \subset \mathbb{P}^n_k$ (where k is any algebraically closed field) is a set-theoretic complete intersection, then so is a connected union of C with any line. In particular, any connected union of lines in $\mathbb{P}^n_k$ is a set-theoretic complete intersection.*

For more information see [**BrSR**] [**BrSV**] [**BrR**] [**Cr**] [**Fe2**] [**GMV**] [**Gey**] [**Ha4**] [**Pi**] [**RoV**] [**SV**] [**Sta**].

§3. Local Cohomological Dimension.

Let R be the coordinate ring of $\mathbb{A}^n_k$ (resp. the homogeneous coordinate ring of $\mathbb{P}^n_k$), $V \subset \mathbb{A}^n_k$ (resp. $V \subset \mathbb{P}^n_k$) an algebraic subset and $I \subset R$ the defining ideal (resp. the homogeneous defining ideal) of V. We denote by $\operatorname{lcd}(V, \mathbb{A}^n_k)$ (resp. $\operatorname{lcd}(V, \mathbb{P}^n_k)$), or simply by $\operatorname{lcd} V$, if no confusion can arise, the local cohomological dimension of V in $\mathbb{A}^n_k$ (resp. $\mathbb{P}^n_k$), i.e. the biggest integer i such that $H^i_I(R) \neq 0$, where $H^i_I(R)$ is the local cohomology module of R with support in I. As is explained in [**Ha2**], $\operatorname{lcd} V$ provides a lower bound on the number of equations needed to define V set-theoretically.

THEOREM 3.1 (Implicit in [**Ha2**]). *$\operatorname{lcd}(V, \mathbb{A}_k^n) < n$ if and only if V consists of irreducible components of positive dimensions.*

THEOREM 3.2 (Implicit in [**O1, PS1**]). *$\operatorname{lcd}(V, \mathbb{A}_k^n) < n-1$ if and only if all irreducible components of V have dimension ≥ 2 and $V - P$ is formally geometrically connected at P, for every closed point $P \in V$.*

Thus, for example a union of two n-dimensional linear subspaces meeting at the origin in $\mathbb{A}_k^{2n}$ cannot be defined set-theoretically by $< 2n-1$ equations.

R. Hartshorne [**Ha1**] also proved that an algebraic set which is disconnected in codimension 2 is not a set-theoretic complete intersection.

THEOREM 3.3 [**Ha2**]. *$\operatorname{lcd}(V, P^n) < n$ if and only if V is geometrically connected.*

Thus a disconnected algebraic set in $\mathbb{P}_k^n$ cannot be defined set-theoretically by $< n$ equations.

THEOREM 3.4 (implicit in [**Ha2**]). *$\operatorname{lcd}(V, \mathbb{A}_k^n) = h$ for every locally complete intersection $V \subset \mathbb{A}_k^n$ of pure codimension h.*

THEOREM 3.5 [**O1, HS, Fal**]. *$\operatorname{lcd}(V, \mathbb{P}^n) \leq 2h-1$ for every locally complete intersection $V \subset \mathbb{P}_k^n$ of pure codimension h.*

THEOREM 3.6 [**HuL**]. *If V is geometrically irreducible and normal of codimension h, then $\operatorname{lcd}(V, \mathbb{A}_k^n) \leq n - [n/(h+1)] - [(n-1)/(h+1)]$ and $\operatorname{lcd}(V, \mathbb{P}_k^n) \leq n+1-[(n+1)/(h+1)] - [n/(h+1)]$.*

THEOREM 3.7 (Implicit in [**Fal**]). *If V consists of irreducible components of codimensions $\leq h$, then $\operatorname{lcd}(V, \mathbb{A}_k^n) \leq n - [\frac{n-1}{h}]$ and $\operatorname{lcd}(V, \mathbb{P}_k^n) \leq n+1-[\frac{n}{h}]$.*

THEOREM 3.8 [**HuL**]. *If $V \subset \mathbb{P}_k^n$ is geometrically irreducible of codimension h, then $\operatorname{lcd} V \leq n - [\frac{n-1}{h}]$.*

The results of 3.7 and 3.8 are exact for every n and h [**HuL, Ly4**].

These results lead to the following problems.

PROBLEM 3.10: Can every $V \subset \mathbb{A}_k^n$ such that all irreducible components of V have dimensions ≥ 2 and $V - P$ is formally geometrically connected at P, for every closed point $P \in V$, be defined set-theoretically by $(n-2)$ equations?

PROBLEM 3.11 [**Mu1**]: Is every locally complete intersection $V \subset \mathbb{A}_k^n$ a set-theoretic complete intersection?

PROBLEM 3.12: Can every locally complete intersection subscheme of $\mathbb{P}_k^n$ of pure codimension h be defined set-theoretically by $(2h-1)$ equations?

PROBLEM 3.13: Can every normal geometrically irreducible $V \subset \mathbb{A}_k^n$ (resp. $V \subset \mathbb{P}_k^n$) of codimension h be defined set-theoretically by $n - [n/(h+1)] - [(n-1)/(h+1)]$ (resp. $n+1-[(n+1)/(h+1)] - [n/(h+1)]$) equations?

PROBLEM 3.14: Can every $V \subset \mathbb{A}_k^n$ (resp. $V \subset \mathbb{P}_k^n$), consisting of irreducible components of codimensions $\leq h$ be defined set-theoretically by $n - [\frac{n-1}{h}]$ (resp. $n+1-[\frac{n}{h}]$) equations?

PROBLEM 3.15: Can every geometrically irreducible subvariety $V \subset \mathbb{P}_k^n$ of codimension h be defined set-theoretically by $n - [\frac{n-1}{h}]$ equations?

Monomial algebraic subsets $V \subset \mathbb{P}_k^n$ are those defined by ideals $I = I(V) \subset k[X_0, \ldots, X_n]$ generated by monomials in $X_0, \ldots, X_n$. They are unions of coordinate subspaces of $\mathbb{P}_k^n$. They form an interesting special case in which lcd can be easily computed. Namely, $\operatorname{lcd} V$, for a monomial $V \subset \mathbb{P}_k^n$, equals the projective dimension of $k[X_0, \ldots, X_n]/\sqrt{I(V)}$ [**Ly3**].

THEOREM 3.16 [**Ly6**]. *Let $I \subset k[X_0, \ldots, X_n]$ be a monomial ideal, such that every minimal prime of I has height $\leq h$. Then I can be generated up to radical by $n+1-[\frac{n}{h}]$ elements in $k[X_0, \ldots, X_n]_{(X_0, \ldots, X_n)}$.*

This provides additional supporting evidence for a positive answer to Problem 3.14.

An interesting example are ideals

$$\begin{aligned} \mathfrak{A}_n = (&X_0X_1 \ldots X_n, X_1X_2 \ldots X_{n+1}, \ldots, \\ &X_nX_{n+1} \ldots X_{2n}, X_{n+1}X_{n+2} \ldots X_{2n}X_0) \subset k[X_0, \ldots X_{2n}] \end{aligned}$$

They have pure height 2 and their local cohomological dimensions also are equal to 2, yet it is unknown whether or not they are set-theoretic complete intersections for $n \geq 4$. For $n = 2, 3$ they are [**NV**].

More information on monomial ideals is found in [**Ly3**] [**Ly6**] [**Grä**] [**Ly1**] [**Ly2**] [**ScheV**] [**SchmV**] [**NV**].

There also exist necessary conditions on the Betti numbers of $V \subset \mathbb{P}_{\mathbb{C}}^n$ for V to be set-theoretically definable by s equations [**Ha3**] [**O1**] [**N**]. Namely, $b_i(V)$ must be equal to $b_i(\mathbb{P}_{\mathbb{C}}^n)$ for $i \leq n-s-1$.

For more information on local cohomology see [**Ha3**, Ch. III] [**HaS**] [**O1**] [**O2**] [**Ba1**] [**Ba2**] [**Sp**] [**N**] [**Fa1**] [**Gr1**] [**Gr2**] [**Ha2**] [**Ly3**] [**Ly4**] [**PS1**].

On the number of defining equations of determinantal varieties see [**Va2**] [**Bru**] [**BV**].

§4. Locally Complete Intersections.

THEOREM 4.1 [**Ly7**]. *Every locally complete intersection subscheme of $\mathbb{A}_k^n$ of pure positive dimension can be defined set-theoretically by $(n-1)$ equations (where k is any field.)*

This has been generalized as follows.

THEOREM 4.2 [**Ma2, Ly7**]. *Let $R = A[X]$ be a polynomial ring over a commutative Noetherian ring A and I a locally complete intersection ideal of R of pure height $\leq \dim A$. Suppose I contains a monic polynomial. Then I can be generated up to radical by $\dim A$ elements.*

This result uses the following generalizations of Theorems 2.3 and 2.4.

THEOREM 4.3 [**Ma1**]. *Let $R = A[X]$ be a polynomial ring over a commutative Noetherian ring A and I an ideal of R. Suppose I contains a monic polynomial and I/I^2 is generated by $\mu \geq \dim(R/I) + 2$ elements. Then I can be generated by μ elements.*

THEOREM 4.4 [**MR**]. *Let $R = A[X]$ be a polynomial ring over a commutative Noetherian ring A and I an ideal of R. Suppose I contains a monic polynomial and I/I^2 is generated by μ elements. Then I can be generated up to radical by μ elements.*

THEOREM 4.5 [**Ly7**]. *Let k be an algebraically closed field of characteristic $p > 0$ and $V \subset \mathbb{A}_k^n$ a locally complete intersection subscheme of pure dimension d such that $2 \leq d \leq n-4$. Then V can be defined set-theoretically by $(n-2)$ equations. In particular, every locally complete intersection surface in $\mathbb{A}_k^n$, where $n \geq 6$, is a set-theoretic complete intersection.*

If k is the algebraic closure of a finite field the restriction $d \leq n-4$ can be removed.

THEOREM 4.6 [**BMS**]. *Let $V \subset \mathbb{A}_k^4$ be a smooth affine surface, where k is any algebraically closed field. Then V is a set-theoretic complete intersection in the following cases.*

(i) *V is* not *birationally equivalent to a surface of general type.*

(ii) *V is birationally equivalent to a smooth projective surface in $\mathbb{P}_k^3$.*
(iii) *V is birationally equivalent to a product of two curves.*

N. Mohan Kumar has extended 4.6(iii) to $\mathbb{A}_k^n, n \geq 5$. He has also proven that a smooth surface $X \subset \mathbb{A}_k^n (n \geq 5)$ is a set-theoretic complete intersection if it is birational to an abelian surface [**MK2**].

THEOREM 4.8 [**Ly7**]. *Let k be any algebraically closed field of characteristic 2 and $V \subset \mathbb{A}_k^n$ a locally complete intersection subscheme of pure dimension d, such that $3 \leq d \leq n-6$. Then V can be defined set-theoretically by $(n-3)$ equations. In particular, every locally complete intersection three-fold in $\mathbb{A}_k^n$, where $n \geq 9$, is a set-theoretic complete intersection.*

More results on the number of equations defining locally complete intersections in affine space are found in [**BMS**] [**Bo1**] [**Bo2**] [**Bo4**] [**Bo5**] [**Bo6**] [**Fo2**] [**Ly7**] [**Ma2**] [**Mu1**] [**Mu2**] [**Va1**] [**MK2**].

In projective space there exist smooth subvarieties which are not set-theoretic complete intersections, for example a quintic elliptic scroll, embedded as a smooth surface in $\mathbb{P}_{\mathbb{C}}^4$ [**Ha2**]. This follows from the next result.

THEOREM 4.9 [**Ha2**]. *Let V be a non-singular closed subscheme of $\mathbb{P}_{\mathbb{C}}^n$. If* $\operatorname{lcd} X < n-1$, *then* $\operatorname{Pic} X$ *is finitely generated.*

PROBLEM 4.10 [**Ha6**]: If V is a non-singular subvariety of $\mathbb{P}^n$ of dimension bigger than $\frac{2}{3}n$, then V is an ideal-theoretic complete intersection.

Work on this problem seems to be well-known and we do not discuss it here. See [**Ba1**] [**Ba2**] [**Ba3**] [**BaL**] [**BaV**] [**Fa2**] [**Hu3**] [**FuL**] [**Ha6**] [**Ha7**] [La] [**Ra**] [**EG**].

REFERENCES

[**Ba1**] W. Barth, *Transplanting cohomology classes in complex projective space*, Amer. J. Math. **92** (1970), 951–967.

[**Ba2**] W. Barth, *Locale Cohomologie bei isolierten Singularitäten analytischer Mengen*, Schrift. Math. Inst. Univ. Münster **(2) 5** (1971).

[**Ba3**] W. Barth, *Submanifolds of low codimension in projective space,*, Proc. Intern. Congr. Math., Vancouver, British Columbia, Canada (1975), 409–413.

[**BaL**] W. Barth and M. F. Larsen, *On the homotopy groups of complex projective algebraic manifolds*, Math. Scand. **30** (1972), 88–94.

[**BaV**] W. Barth and A. Van de Ven, *A decomposability criterion for algebraic 2-bundles on projective spaces*, Inv. Math. **25** (1974), 91–106.

[**BMS**] S. Bloch, P. Murthy and L. Szpiro, *Zero-cycles and the number of generators of an ideal*, in "Pierre Samuel Colloquium Proceedings" (to appear).

[**Bo1**] M. Boratynski, *A note on set-theoretic complete intersection ideals*, J. Algebra **54** (1978), 1–5.

[**Bo2**] M. Boratynski, *Generating ideals up to radical and systems of parameters of graded rings*, J. Algebra **78** (1982), 20–24.

[**Bo3**] M. Boratynski, *Every curve on a nonsingular surface can be defined by two equations*, Proc. AMS **96** (1986), 391–393.

[**Bo4**] M. Boratynski, *On the conormal module of smooth set-theoretic complete intersection*, Trans. Amer. Math. Soc. **296** (1986), 291–300.

[**Bo5**] M. Boratynski, *A remark on locally complete intersection surfaces in A^4*, Queen's papers in pure and applied math. **76**.

[**Bo6**] M. Boratynski, *Poincare forms, Gorenstein algebras and set-theoretic complete intersections*, in "Complete Intersections," S. Greco and R. Strano, eds., Lecture Notes in Math., vol. 1092, Springer-Verlag, 1984, pp. 270–290.

[**Br1**] H. Bresinsky, *Monomial space curves in A^3 as set-theoretic complete intersection*, Proc. Amer. Math. Soc. **75** (1979), 23–24.

[**Br2**] H. Bresinsky, *Monomial Gorenstein curves in A^4 as set-theoretic complete intersections*, Manuscripta Math. **27** (1979), 353–358.

[**BrR**] H. Bresinsky and B. Renschuch, *Basisbestimmung Veronesescher Projektionsideale mit Allgemeiner Nullstelle $(t_o^m, t_o^{m-r}t_1^r, t_o^{m-s}t_1^s, t_1^m)$*, Math. Nachr. **96** (1980), 257–269.

[**BrV**] H. Bresinsky and W. Vogel, *Some remarks on a paper by L. Kronecker*, in "Seminar D. Eisenbud/B. Singh/W. Vogel, vol. 2," Teubner-Texte zur Mathematik, Band 48, Leipzig, 1982.

[**BrSR**] H. Bresinsky, J. Stückrad and B. Renschuch, *Mengentheoretisch vollständige Durchschnitte verschiedener rationaler Raumkurven im $\mathbb{P}^3$.*, Math. Nachr. **104** (1981), 147–169.

[**BrSV**] H. Bresinsky, P. Schenzel and W. Vogel, *On liaison, arithmetically Buchsbaum curves and monomial curves in $\mathbb{P}^3$.*, J. of Algebra **86** (1984), 283–301.

[**Bru**] W. Bruns, *Additions to the theory of algebras with straightening law*, in "Commutative Algebra," Proceedings of the Microprogram held June 1987, MSRI Publications, Springer-Verlag, New York, 1989.

[**BV**] W. Bruns and U. Vetter, "Determinantal rings," Lecture Notes in Math., vol. 1327, Springer-Verlag, New York, 1988.

[**C**] R. Cowsik, *Symbolic powers and the number of defining equations*, in "Algebra and its applications, New Delhi 1981," Lecture Notes in Pure and Appl. Math., vol. 91, Dekker, New York, 1984.

[**CN1**] R. Cowsik and M. Nori, *Affine curves in characteristic p are set-theoretic complete intersections*, Inv. Math. **45** (1978), 111–114.

[CN2] R. Cowski and M. Nori, *On Cohen-Macaulay rings*, J. Algebra **38** (1976), 536–538.

[Cr] P. Craighero, *Una asservazione sulla curva di Cremona di* $\mathbb{P}^3_k$, Rend. Seminario Mat. Padova **65** (1981), 177–190.

[EE] D. Eisenbud and E. G. Evans, *Every algebraic set in n-space is the intersection of n hypersurfaces*, Inv. Math. **19** (1973), 107–112.

[El1] S. Eliahou, *Idéaux de définition des courbes monomiales*, in "Complete Intersections," S. Greco and R. Strano, eds., Lecture Notes in Math., vol. 1092, Springer-Verlag, 1984, pp. 229–240.

[El2] S. Eliahou, *Symbolic powers of monomial curves*, J. Algebra **117** (1988), 437–456.

[El3] S. Eliahou, *A problem about polynomial ideals*, in "The Lefschetz Centennial Conference," Contemporary Math., AMS vol. 58, part I, pp. 107–120.

[EG] E. G. Evans and P. A. Griffith, *The syzygy problem*, Annals of Math. **114** (1981), 323–353.

[Fa1] G. Faltings, *Über locale Kohomologiegruppen hoher Ordnung*, J. Reine Angew Math. (1980), 43–51.

[Fa2] G. Faltings, *Ein Kriterium für vollständige Durchschnitte*, Inv. Math. **62** (1981), 393–401.

[Fe1] D. Ferrand, *Courbes gauches et fibres de rang 2*, C.R. Acad. Sc. Paris **281** (1975), 345–347.

[Fe2] D. Ferrand, *Set-theoretical complete intersections in characteristic* $p > 0$, Springer Lecture Notes in Math., vol. 732.

[Fo1] O. Forster, *Über die Anzahl der Erzeugenden eines Ideals in einem Noetherschen Ring*, Math. Z. **84** (1964), 80–87.

[Fo2] O. Forster, *Complete intersections in affine algebraic varieties and Stein spaces*, in "Complete Intersections," S. Greco and R. Strano, eds., Lecture Notes in Math., vol. 1092, Springer-Verlag, 1984.

[FuL] W. Fulton and R. Lazarsfeld, *Connectivity and its applications in algebraic geometry*, in "Algebraic Geometry Proceedings," Lecture Notes in Math., vol. 862, Springer-Verlag, 1981, pp. 26–92.

[FW] O. Forster and K. Wolfhardt, *A theorem on zero schemes of sections in two-bundels over affine schemes with applications to set-theoretic intersections*, Lecture Notes in Math., vol. 1273, Springer-Verlag, New York (1987), 372–383.

[GMV] A. V. Geramita, P. Maroscia and W. Vogel, *A note on artihmetically Buchsbaum curves in* $\mathbb{P}^3$, Le Mathematiche, vol. XL, FASC. I-II (1985), 21–28.

[Gey] W. D. Geyer, *On the number of equations which are necessary to describe an alagebraic set in n-space*, Atlas 3ª Escola de Algebra, Brasília (1976), 183–317.

[Gr1] A. Grothendieck, *Cohomologie locale des faiseaux et théorèmes de Lefschetz locaux et globaux*, (SGA2) North-Holland (1968).

[Gr2] A. Grothendieck, *Local Cohomology*, notes by R. Hartshorne, Springer Lecture Notes in Math., vol. 41.

[Grä] H.-G. Gräbe, *On the arithmetical rank of square-free monomial ideals*, Math. Nachr. **120** (1985), 217–227.

[GP] L. Gruson and C. Peskine, *Genre des courbes de l'espace projectif*, Springer Lecture Notes in Math., vol. 687.

[Ha1] R. Hartshorne, *Complete intersections and connectedness*, Amer. J. Math. **84** (1962), 497–508.

[Ha2] R. Hartshorne, *Cohomological dimension of algebraic varieties*, Ann. of Math. **88** (1968), 403–450.

[Ha3] R. Hartshorne, *Ample subvarieties of algebraic varieties*, Springer Lecture Notes in Math., vol. 156.

[Ha4] R. Hartshorne, *Complete intersections in characteristic $p > 0$*, Amer. J. Math. **101** (1979), 380–383.

[Ha5] R. Hartshorne, *Topological conditions for smoothing algebraic singularities*, Topology **13** (1974), 241–253.

[Ha6] R. Hartshorne, *Varieties of small codimension in projective space*, Bull. Amer. Math. Soc. **80** (1974), 1017–1032.

[Ha7] R. Hartshorne, *Algebraic vector bundles on projective spaces: a problem list*, Topology **18**, 117–128.

[HaS] R. Hartshorne and R. Speiser, *Local cohomological dimension in characteristic p*, Ann. Math. **105** (1977), 45–79.

[He] J. Herzog, *Generators and relations of abelian semigroups and semigroup rings*, Manuscripta Math. **3** (1970), 153–193.

[Hu1] C. Huneke, *On the finite generation of symbolic blow-ups*, Math Z. **179** (1982), 465–472.

[Hu2] C. Huneke, *Hilbert functions and symbolic powers*, Michigan Math. J. **34** (1987), 293–318.

[Hu3] C. Huneke, *Criteria for complete intersections*, J. London Math. Soc. **(2) 32** (1985), 19–30.

[HuL] C. Huneke and G. Lyubeznik, *On the vanishing of local cohomology modules*, (to appear).

[Kn] M. Kneser, *Über die Darstellung algebraischer Raumkurven als Durchschnitte von Flächen*, Arch. Math. (Basel) **11** (1960), 157–158.

[Kö] J. König, "Einleitung in die algemeine Theorie der algebraischen Grössen," Leipzig, 1903.

[Kr1] L. Kronecker, *Grundzüge einer arithmetischen Theorie der algebraischen Grössen*, J. Reine Angew Math. **92** (1882), 1–123.

[Kr2] L. Kronecker, *Zur Theorie der Formen höherer Stufen*, in "Sitz.-Bericht Königlichen Preussischen Akad. Wiss. und Nature. Mitteilungen," Jahrgang, 1883, pp. 521–524.

[Ku1] E. Kunz, *Über die Anzahl der Gleichungen, die zur Beschreibung einer algebraischen Varietät nötig sind*, Jahresber. Deutsch. Math.-Verein **81** (1979), 97–108.

[Ku2] E. Kunz, "Introduction to commutative algebra and algebraic geometery," Birkhäuser, 1985.

[La] R. Lazarsfeld, *Some applications of the theory of positive vector bundles*, in "Complete Intersections," S. Greco and R. Strano, eds., Lecture Notes in Math., vol. 1092, Springer-Verlag, 1984, pp. 29–61.

[Ly1] G. Lyubeznik, *Set-theoretic intersections and monomial ideals*, Thesis, Columbia University (1984).

[Ly2] G. Lyubeznik, *On set-theoretic intersections*, J. Algebra **87** (1984), 105–112.

[Ly3] Lyubeznik, G., *On the local cohomology modules $H^i_{\mathfrak{a}}(R)$ for ideals $\mathfrak{a}$ generated by monomials in an R-sequence.*, in "Complete Intersections," S. Greco and R. Strano, eds., Lecture Notes in Math., vol. 1092, Springer-Verlag, 1984, pp. 214–220.

[Ly4] G. Lyubeznik, *Some algebraic sets of high local cohomological dimension in $\mathbb{P}^n_k$*, Proc. Amer. Math. Soc. **95** (1985), 9–10.

[Ly5] G. Lyubeznik, *A property of ideals in polynomial rings*, Proc. Amer. Math. Soc. **98** (1986), 399–400.

[Ly6] G. Lyubeznik, *On the arithmetical rank of monomial ideals*, J. Algebra **112** (1988), 86–89.

[Ly7] G. Lyubeznik, *The number of defining equations of affine algebraic sets*, (to appear).

[Ma1] S. Mandal, *On efficient generation of ideals*, Inv. Math. **75** (1984), 59–67.

[**Ma2**] J. Mandal, *On set-theoretic intersection in affine spaces*, J. Pure Appl. Alg. **51** (1988), 267–275.

[**MaR**] S. Mandal and A. Roy, *Generating ideals in polynomial rings*, Math. Z. **195** (1987), 315–320.

[**MK1**] N. Mohan Kumar, *On two conjectures about polynomial rings*, Invent. Math. **46** (1978), 225–236.

[**MK2**] N. Mohan Kumar, *Set-theoretic generation of ideals*, in "Pierre Samuel Colloquium Proceedings" (to appear).

[**MK3**] N. Mohan Kumar, these Proceedings (to appear).

[**Moh1**] T. T. Moh, *On the unboundedness of generators of prime ideals in powerseries rings of three variables*, J. Math. Soc. Japan **26, No. 4** (1974), 722–734.

[**Moh2**] T. T. Moh, *Set-theoretic complete intersections*, Proc. Amer. Math. Soc. **94** (1985), 217–220.

[**Moh3**] T. T. Moh, *A result on the set-theoretic complete intersection problem*, Proc. Amer. Math. Soc. **86**, 19–20.

[**Moh4**] T. T. Moh, *On the set-theoretic complete intersection problem*, unpublished.

[**Mor**] M. Morales, *Noetherian symbolic blow-up and examples in any dimensions*, preprint.

[**Mu1**] P. Murthy, *Complete intersections*, in "Conference on Commutative Algebra," Queen's papers Pure Appl. Math., vol. 42, 1975, pp. 196–211.

[**Mu2**] P. Murthy, *Affine varieties as complete intersections*, in "Intl. Symp. on Algebraic Geometerey Kyoto," 1977, pp. 231–236.

[**Mu3**] P. Murthy, *Zero-cycles, splitting of projective modules and number of generators of a module*, Bulletin AMS **19** (1988), 315–317.

[**N**] P. E. Newstead, *Some subvarieties of the Grassmanians of codimension 3*, Bull. London Math. Soc. **12** (1980), 176–182.

[**NV**] U. Nagel and W. Vogel, *Über mengentheoretische Durchschnitte und Zusammenhang algebraischer Mannigfaltigkeiten* im $\mathbb{P}^n$, Arch. Math. **49** (1987), 414–419.

[**O1**] A. Ogus, *Local cohomological dimension of algebraic varieties*, Ann. Math. **98** (1973), 327–365.

[**O2**] A. Ogus, *On the formal neighborhood of a subvariety of projective space*, Amer. J. Math. **97** (1976), 1085–1107.

[**Oo**] A. Ooishi, *Noetherian property of symbolic Rees algebras*, Hiroshima Math. J. **15** (1985), 581–584.

[**Pe1**] O. Perron, *Über das Vahlensche Beischpiel zu einem Satz von Kronecker*, Math. Z. **47** (1942), 318–324.

[**Pe2**] O. Perron, *Beweis und Verschärfung eines Satzes von Kronecker*, Math. Ann. **118** (1941/1943), 441–448.

[**PS1**] C. Peskine and L. Szpiro, *Dimension projective finie et cohomologie locale*, Publ. Math. IHES **42** (1973), 47–119.

[**PS2**] C. Peskine and L. Szpiro, *Liaison des variétés algébriques*, Invent. Math. **26** (1974), 271–302.

[**Pi**] R. Piene, *Cuspidal projections of space curves*, Math. Ann. **256** (1981), 95–119.

[**Ra**] Z. Ran, *On projective varieties of codimension 2*, Inv. Math. **73** (1983), 333–336.

[**RoV**] L. Robbiano and G. Valla, *Some curves in* $\mathbb{P}^3$ *are set-theoretic complete intersections*, Springer Lecture Notes in Math., vol. 997.

[**Ro1**] P. Roberts, *A prime ideal in a polynomial ring whose symbolic blow-up is not Noetherian*, Proc. Amer. Math. Soc. **94** (1985), 589–592.

[**Ro2**] P. Roberts, *An infinitely generated symbolic blow-up in a power series ring and a new counterexample to Hilbert's fourteen's problem*, preprint.

[**Sche1**] P. Schenzel, *Filtrations and Noetherian symbolic blow-up rings*, preprint.

[**Sche2**] P. Schenzel, *Examples of Noetherian symbolic blow-up rings*, preprint.

[ScheV] P. Schenzel and W. Vogel, *On set-theoretic intersections*, J. Algebra (1977), 401–408.

[SchmV] T. Schmitt and W. Vogel, *Note on set-theoretic intersections of subvarieties of projective space*, Math. Ann. **245** (1979), 247–253.

[Schn] M. Schneider, *On the number of equations needed to describe a variety*, in "Complex Analysis of Several Variables," AMS Proceedings of Symposia in Pure Math., vol. 41, 1984.

[Sp] R. Speiser, *Cohomological dimension and abelian varieties*, Amer. J. Math. **95** (1973), 1–34.

[Sta] E. Stagnaro, *Su quintiche di Vahlen ed altre curve razionali ohe sono sottoinsieme intersezione completa di P_k^3 in caratteristica positiva*, Boll. Un. Mat. Ital. **B(5) 17** (1980), 278–285.

[Sto] U. Storch, *Bemerkungen zu einem Satz von M. Kneser*, Arch. Math. (Basel) **23** (1972), 403–404.

[Sz] L. Szpiro, "Lectures on equations defining space curves," Tata Institute of Fundamental Research, Springer-Verlag, Bombay, 1979.

[SV] J. Stückrad and W. Vogel, *On the number of equations defining an algebraic set of zeros in n-space*, in "Seminar D. Eisenbud/B. Singh/W. Vogel, vol. 2," Teubner-Texte zur Mathematik, Band 48, Leipzig, 1982, pp. 88–107.

[Vah] K. Th. Vahlen, *Bemerkung zur vollstandigen Darstellung algebraischer Raumkurven*, J. f. d. Reine Angew Math. **108** (1891), 346–347.

[Va1] Valla, G., *On set-theoretic complete intersections,*, in "Complete Intersections," S. Greco and R. Strano, eds., Lecture Notes in Math., vol. 1092, Springer-Verlag, 1984, pp. 85–101.

[Va2] G. Valla, *On determinantal ideals, which are set-theoretic complete intersections*, Comp. Math **42** (1981), 3–11.

[Vo] W. Vogel, *Eine Bemerkung über die Anzahl von Hyperflächen zur Darstellung algebraischer Varietäten vol 13*, Monatsber. Dtsch. Akad. Wiss. Berlin (1971), 629–633.

[Wa] B. L. van der Waerden, *Review of Perron's article*, Zbl. Math. **24** (1941), p. 276.

Department of Mathematics, University of Chicago, 5734 University Avenue, Chicago IL 60637

Integrally Closed Projectively Equivalent Ideals

STEPHEN McADAM, L. J. RATLIFF, JR., AND JUDITH D. SALLY

To David Rees, in honor of
his seventieth birthday.

1. Introduction and preliminaries.

In this paper, I will be a regular ideal in a Noetherian ring R, and $\bar{I}$ will denote the integral closure of I. If J is another ideal of R, I and J are said to be projectively equivalent if there are positive integers n and m with $\overline{I^n} = \overline{J^m}$. Projective equivalence was introduced by Samuel in [**S**], and further developed by Nagata in [**N**]. The concept of projective equivalence is rather interesting, but there appears to be much about it which is not known. We hope to dispel some of the darkness. We study the set of integrally closed ideals which are projectively equivalent to I, and show that this set is rather well behaved. For instance, it is linearly ordered by inclusion, and "eventually periodic", a phrase we will make more precise in the next paragraph. An interesting corollary to our work is that there is a fixed positive integer d such that for any ideal J projectively equivalent to I, there is a positive integer n with $\overline{I^n} = \overline{J^d}$.

Our work will build upon that of Rees in [**R**], which in turn extends the work of Samuel and Nagata. Let us recall some relevant results. Define $V_I(x)$ to be the largest power of I which contains the element x of R. ($V_I(x) = \infty$ if x is in all powers of I.) Next, define $\bar{V}_I(x)$ to be the limit of $V_I(x^k)/k$ as k goes to infinity. $\bar{V}_I(x)$ is well defined, and for n a positive integer, $\bar{V}_I(x) \geq n$ if and only if $x \in \overline{I^n}$. If $W = \{\bar{V}_I(x) \mid x \in R$ and $0 < \bar{V}_I(x) < \infty\}$, then W is a discrete subset of the positive rationals. In this paper, for $\alpha > 0$ a real number, we let $I_\alpha = \{x \in R \mid \bar{V}_I(x) \geq \alpha\}$, and show that I_α is an integrally closed ideal. Now let $U = \{\alpha \in W \mid I_\alpha$ is projectively equivalent to $I\}$. We show that $\{I_\alpha \mid \alpha \in U\} = \{J \mid J$ is an integrally closed ideal projectively equivalent to $I\}$. Thus, this set inherits a natural ordering from U. Concerning our earlier statement that this set is "eventually periodic", we mean that the set U is "eventually periodic". Specifically, we show that there are positive integers N and d such that

$\{\alpha \in U \mid \alpha \geq N\} = \{N + (h/d) \mid h \geq 0 \text{ is an integer}\}$. In fact, the d here is the same d mentioned at the close of the previous paragraph. Furthermore, we show that d must be a common factor of the integers $e_1, \ldots, e_r$ discussed in (1.1) below.

Our work will require a more intimate knowledge of the results in [**R**] than is outlined above. Therefore, after setting notation, we will give a technical description of Rees' work.

NOTATION: R will be a Noetherian ring with integral closure R'. I will be a regular ideal of R with integral closure $\bar{I}$. By $R(I)$, we will mean the Rees ring of R with respect to I. Thus $R(I) = R[u, It]$ with t an indeterminate and $u = t^{-1}$. $\bar{V}_I(x)$, I_α, W, and U will be as defined above. By (2.1)(c), I_α is an ideal. Thus one can discuss $\bar{V}_{I_\alpha}$. We will denote this by $\bar{V}_\alpha$. (Remark: Lejeune and Teissier first introduced the ideals I_α in the setting of analytic geometry, [**LT**]. Their notation for our I_α is $\overline{I^\alpha}$.)

(1.1) REMARK: [**R**] shows that there are finitely many valuations $v_1, \ldots, v_r$ defined on R, (with values in the positive integers union $\{\infty\}$), and positive integers $e_1, \ldots, e_r$, such that for $x \in R$, $\bar{V}_I(x) = \min\{v_i(x)/e_i \mid i = 1, \ldots, r\}$. In more detail, let q be a minimal prime of R with $I + q \neq R$, let $T = R/q$ and $H = IT$. Let P be a prime in $T(H)'$, the integral closure of the Rees ring $T(H)$, with P minimal over $uT(H)'$. As $T(H)'$ is a Krull domain, height $P = 1$, and $T(H)'_P$ is a discrete valuation domain. Let w be the associated discrete valuation. Now define v on R as follows. For $x \in R$, let $v(x) = w(x + q/q)$. Then v is a valuation on R, and $v(x) = \infty$ exactly when $x \in q$. The set $v_1, \ldots, v_r$ mentioned above consists of all the v which arise in this way, as q varies over all minimal primes in R with $I + q \neq R$, and P varies over all primes minimal over $uT(H)'$. If $v_i = v$ arises from w, as above, then $e_i = w(u)$. See [**R**], or [**M**, Chapter XI].

2. The integrally closed ideals projectively equivalent to I.

(2.1) LEMMA. *Let $0 < \alpha < \infty$. Let x and y be elements of R, and let J be an ideal of R.*

a) *If n and m are positive integers, $\overline{I^n} = \overline{J^m}$ if and only if $mV_I(x) = nV_J(x)$ for all $x \in R$.*
b) *If h, k, n, and m are positive integers with $h/k = n/m$, then $\overline{I^h} = \overline{J^k}$ if and only if $\overline{I^n} = \overline{J^m}$.*

c) *For n a positive integer, $I_n = \overline{I^n}$.*
d) $\bar{V}_I(xy) \geq \bar{V}_I(x) + \bar{V}_I(y)$, *and* $\bar{V}_I(x+y) \geq \min\{\bar{V}_I(x), \bar{V}_I(y)\}$. *Also, for n a positive integer, $\bar{V}_I(x^n) = n\bar{V}_I(x)$.*
e) *For m a positive integer, $\overline{I_\alpha{}^m} \subseteq I_{m\alpha}$.*
f) $\bar{V}_I \geq \alpha \bar{V}_\alpha$.
g) *I_α is an integrally closed ideal.*

PROOF: a) [**M**, Corollary 11.9(ii)].

b) [**M**, Lemma 11.24].
c) [**M**, Corollary 11.8].
d) This follows easily from the characterization of $\bar{V}_I$ mentioned in (1.1).
e) Let $x \in \overline{I_\alpha{}^m}$. We must show that $\bar{V}_I(x) \geq m\alpha$. Suppose to the contrary that $\bar{V}_I(x) = \beta < m\alpha$. Say $x^k = a_1 x^{k-1} + \cdots + a_k$, with $a_i \in I_\alpha{}^{mi}$. By (d), we get $\bar{V}_I(a_i) \geq im\alpha$, and $k\beta = \bar{V}_I(x^k) = \bar{V}_I(a_1 x^{k-1} + \cdots + a_k) \geq \min\{im\alpha + (k-i)\beta \mid 1 \leq i \leq k\} = m\alpha + (k-1)\beta$. However, $k\beta < m\alpha + (k-1)\beta$, a contradiction.
f) We take an $x \in R$, and consider two cases. First, suppose that $\bar{V}_\alpha(x) = \beta < \infty$. By (1.1), β is rational. Pick $n \geq 1$ an integer with $n\beta$ integral. By (d) applied to the ideal I_α, we see that $\bar{V}_\alpha(x^n) = n\beta$. By (c) applied to I_α, $x^n \in \overline{I_\alpha{}^{n\beta}}$. By (e), $x^n \in I_{n\beta\alpha}$, so that $\bar{V}_I(x^n) \geq n\beta\alpha$. By (d), $\bar{V}_I(x) \geq \beta\alpha = \alpha\bar{V}_\alpha(x)$. Next, we consider the case that $\bar{V}_\alpha(x) = \infty$. We must show that $\bar{V}_I(x) = \infty$. It follows from (1.1) (or from (c) together with [**M**, (3.11)]) that for any ideal J, $\bar{V}_J(x) = \infty$ if and only if x is in the intersection of those minimal primes q for which $J + q \neq R$. However, it is easily seen that $\operatorname{Rad} I_\alpha = \operatorname{Rad} I_1 = \operatorname{Rad} \bar{I} = \operatorname{Rad} I$. Our result quickly follows.
g) That I_α is an ideal is trivial by (d). That it is integrally closed follows from (e), taking $m = 1$.

Recall that W and U were defined in Section 1. By (1.1), W is a discrete subset of the positive rationals.

(2.2) LEMMA. *Let $0 < \alpha < \infty$. If either $\alpha\bar{V}_\alpha = \bar{V}_I$, or $\overline{I_\alpha{}^m} = I_{m\alpha}$ for all integers $m \geq 1$, then $\alpha \in W$.*

PROOF: Suppose $\alpha \notin W$. Let β be the smallest member of W with $\beta > \alpha$. The definitions clearly show $I_\alpha = I_\beta$, and $\bar{V}_\alpha = \bar{V}_\beta$. Suppose first that $\alpha\bar{V}_\alpha = \bar{V}_I$. Let x be a regular element in I. By (2.1)(c), $\bar{V}_I(x) \geq 1$. Also, as argued in the proof of (2.1)(f), $\bar{V}_I(x) < \infty$. As $\alpha\bar{V}_\alpha = \bar{V}_I$, we see that $0 < \bar{V}_\alpha(x) < \infty$. Now $\beta\bar{V}_\beta(x) = \beta\bar{V}_\alpha(x) > \alpha\bar{V}_\alpha(x) = \bar{V}_I(x)$. This contradicts (2.1)(f) applied to $\bar{V}_\beta$.

Now suppose $\overline{I_\alpha{}^m} = I_{m\alpha}$ for all integers $m \geq 1$. Let m be large enough that there is an integer k with $m\alpha \leq k < k+1 \leq m\beta$. Using (2.1)(e), we see that $\overline{I_\alpha{}^m} = \overline{I_\beta{}^m} \subseteq I_{m\beta} \subseteq I_{k+1} \subseteq I_k \subseteq I_{m\alpha}$. As $\overline{I_\alpha{}^m} = I_{m\alpha}$, we get $I_k = I_{k+1}$. This is impossible, by (2.1)(c) and [**M**, 11.27].

We come to a main result.

(2.3) PROPOSITION. *Let J be an ideal, and let n and m be positive integers.*

i) *If $\overline{J^m} = \overline{I^n}$, then $\bar{J} = I_{n/m}$ and $n/m \in U$.*
ii) *If $n/m \in U$, then $\overline{(I_{n/m})^m} = \overline{I^n}$.*

PROOF: i) Say $\overline{J^m} = \overline{I^n}$. Now $x \in \bar{J} \iff x^m \in \overline{J^m}$, since the monic polynomial which shows the second inclusion, also shows the first inclusion. Thus $x \in \bar{J} \iff x^m \in \overline{J^m} = \overline{I^n} \iff \bar{V}_I(x^m) \geq n \iff \bar{V}_I(x) \geq n/m \iff x \in I_{n/m}$. That is, $\bar{J} = I_{n/m}$, as desired. Since $I_{n/m} = \bar{J}$ is projectively equivalent to I, in order to show that $n/m \in U$, we must show that $n/m \in W$. Let α be the smallest element in W with $\alpha \geq n/m$. Clearly $\bar{J} = I_{n/m} = I_\alpha$. By (2.1)(c) and (a), $1\bar{V}_1 = \bar{V}_1 = \bar{V}_{\bar{I}} = \bar{V}_I$. Thus by (2.2), $1 \in W$, and so we may pick $y \in R$ with $\bar{V}_I(y) = 1$. If $x = y^n$, then $\bar{V}_I(x) = n$. By (2.1)(c) and (e), $x \in \overline{I^n} = \overline{J^m} = \overline{I_\alpha{}^m} \subseteq I_{m\alpha}$. Thus, $n = \bar{V}_I(x) \geq m\alpha$, so $n/m \geq \alpha$, showing that $n/m = \alpha \in W$.

ii) If $n/m \in U$, then $I_{n/m}$ is projectively equivalent to I, and so there are positive integers h and k with $\overline{I^h} = \overline{(I_{n/m})^k}$. By part (i), $I_{n/m} = \overline{(I_{n/m})} = I_{h/k}$, and $h/k \in U \subseteq W$. Since n/m and h/k are both in W, and since $I_{n/m} = I_{h/k}$, the various definitions easily show that $n/m = h/k$. By (2.1)(b), we have $\overline{(I_{n/m})^m} = \overline{I^n}$.

(2.4) COROLLARY. *$\{I_\alpha \mid \alpha \in U\} = \{J \mid J$ is an integrally closed ideal projectively equivalent to $I\}$, and this set is linearly ordered by inclusion.*

PROOF: One containment is by (2.1)(g) and the definition of U. The other follows easily from (2.3)(i). The linear ordering is obvious.

We strengthen (2.2).

(2.5) PROPOSITION. *Let $0 < \alpha < \infty$. The following are equivalent.*

i) *$\overline{I_\alpha{}^m} = I_{m\alpha}$ for all integers $m \geq 1$.*
ii) *$\alpha\bar{V}_\alpha = \bar{V}_I$.*
iii) *$\alpha \in U$.*

PROOF: i) $\Rightarrow$ iii). By (2.2), if (i) holds, then $\alpha \in W$. In particular, α is rational, and so we may choose an integer $m \geq 1$ with $m\alpha = n$ an integer. By (i), $\overline{I_\alpha{}^m} = I_n = \overline{I^n}$, so I_α is projectively equivalent to I, showing $\alpha \in U$.

iii) $\Rightarrow$ ii). If $\alpha \in U$, write $\alpha = n/m$. By (2.3)(ii), $\overline{I_\alpha{}^m} = \overline{I^n}$. By (2.1)(f), we have $\bar{V}_I \geq \alpha \bar{V}_\alpha$. Suppose for some $x \in R$ that $\bar{V}_I(x) > \alpha \bar{V}_\alpha(x) = (n/m)\bar{V}_\alpha(x)$. Then $m\bar{V}_I(x) > n\bar{V}_\alpha(x)$. For a large integer c, the gap between $cm\bar{V}_I(x)$ and $cn\bar{V}_\alpha(x)$ exceeds 1, and so there is an integer k with $\bar{V}_I(x^{cm}) \geq k > \bar{V}_\alpha(x^{cn})$. Therefore, (2.1)(c) shows $x^{cm} \in \overline{I^k}$, but $x^{cn} \notin \overline{I_\alpha{}^k}$. On the other hand, since we have $\overline{I_\alpha{}^m} = \overline{I^n}$, and $x^{cm} \in \overline{I^k}$, we see $x^{cmn} \in \overline{I^{kn}} = \overline{I_\alpha{}^{mk}}$. This implies $x^{cn} \in \overline{I_\alpha{}^k}$ (as argued in the proof of (2.3)(i)), which gives a contradiction. Thus $\bar{V}_I = \alpha \bar{V}_\alpha$.

ii) $\Rightarrow$ i). By (2.1)(e), we always have $\overline{I_\alpha{}^m} \subseteq I_{m\alpha}$. Suppose (ii) holds. Then $x \in I_{m\alpha} \Rightarrow \bar{V}_I(x) \geq m\alpha \Rightarrow \alpha\bar{V}_\alpha(x) \geq m\alpha \Rightarrow \bar{V}_\alpha(x) \geq m \Rightarrow x \in \overline{I_\alpha{}^m}$, the last implication by (2.1)(c) applied to I_α.

(2.6) PROPOSITION. *If $\alpha \in U$ and $\beta \in U$, then $\alpha + \beta \in U$, and $\overline{I_\alpha I_\beta} = I_{\alpha+\beta}$. Also, $1 \in U$.*

PROOF: In the proof of (2.3)(i), we showed that $1 \in W$. Since $I_1 = \bar{I}$, $1 \in U$. Let α and β be in U. Write $\alpha = n/m$ and $\beta = h/k$ with n, m, h, and k positive integers. Now (2.3)(ii) shows that $\overline{I_\alpha{}^m} = \overline{I^n}$, and $\overline{I_\beta{}^k} = \overline{I^h}$. It is an easy exercise to see that $\overline{(I_\alpha I_\beta)^{mk}} = \overline{I^{nk+mh}}$. By (2.3)(i), (taking $J = I_\alpha I_\beta$), we see that $\alpha + \beta = (nk + mh)/mk \in U$, and $\overline{I_\alpha I_\beta} = I_{\alpha+\beta}$.

(2.7) LEMMA. *Let V be a set of positive rational numbers which is closed under addition, with $1 \in V$. Suppose that the set $\{\beta - \gamma \mid \beta > \gamma$ are both in $V\}$ has a smallest element. Then there are positive integers d (unique) and N such that $\{\alpha \in V \mid \alpha \geq N\} = \{N + (h/d) \mid h \geq 0$ is an integer$\}$. Furthermore, $d\alpha$ is an integer for all $\alpha \in V$.*

PROOF: The least element of $\{\beta - \gamma \mid \beta > \gamma$ are both in $V\}$ must be a rational number, which we may write as c/d with c and d relatively prime positive integers. We claim that $c = 1$. There are integers u and v with $cu + dv = 1$. We may take u to be positive, so that $v \leq 0$. Suppose that $c/d = \beta - \gamma$ with β and γ in V. Thus, $\beta = \gamma + c/d$. As V is closed under addition, V contains $u\beta = u\gamma + uc/d = u\gamma + (1 - dv)/d = u\gamma - v + 1/d$. However, since γ and 1 are in V, and since $-v \geq 0$, we see that that $u\gamma - v$ is in V. Thus $1/d = (u\beta) - (u\gamma - v)$ is in the set whose minimum was said to be c/d. Therefore, $c = 1$, and that minimum is $1/d$.

We now have γ and $\beta = \gamma + 1/d$ both in V. As V is closed under addition, we may add γ to both of these, and get that 2γ and $2\gamma + 1/d$ are both in V. However, we also have $2(\gamma+1/d) = 2\gamma+2/d$ in V. Thus, each of 2γ, $2\gamma+1/d$, and $2\gamma + 2/d$ are in V. Adding γ gives that 3γ, $3\gamma + 1/d$, and $3\gamma + 2/d$ are in V. However, we also have that $3(\gamma+1/d) = 3\gamma+3/d$ is in V. Continuing in this fashion, we see that $d\gamma$, $d\gamma + 1/d$, $d\gamma + 2/d, \ldots, d\gamma + d/d = d\gamma + 1$ are all in V. For any integer $k \geq 0$, we may add k to all of these, and so we see that $d\gamma + h/d \in V$ for all integers $h \geq 0$. Obviously, for any integer $m \geq d$, we have $m\gamma + h/d \in V$ for all integers $h \geq 0$. As γ is rational, for some such m, $m\gamma$ is a positive integer, which we take for our N. Thus $\{N + (h/d) \mid h \geq 0 \text{ is an integer}\} \subseteq \{\alpha \in V \mid \alpha \geq N\}$. For the reverse inclusion, say $\alpha \in V$ with $\alpha \geq N$. For some integer $h \geq 0$, we have $N + h/d \leq \alpha < N + (h+1)/d$. Since $N + h/d$ is in V, the minimal property of $1/d$ shows that α must equal $N + h/d$. Clearly d is unique. As for the final statment in the theorem, for any $\alpha \in V$, since $N \in V$, we have $N + \alpha \in V$, and so $N + \alpha = N + h/d$ for some integer $h \geq 0$, by what we have just shown. Thus $\alpha = h/d$, and $d\alpha = h$ is an integer.

(2.8) THEOREM. *a) $n \in U$ for all positive integers n.*

Also, there are positive integers d (unique) and N such that

- b) $\{\alpha \in U \mid \alpha \geq N\} = \{N + (h/d) \mid h \geq 0$ *is an integer*$\}$.
- c) *$d\alpha$ is an integer for all $\alpha \in U$.*
- d) *d is a common divisor of $e_1, \ldots, e_r$, with $e_1, \ldots, e_r$ as in (1.1).*

PROOF: (a) is immediate from (2.6). The definition of W and the characterization of $\bar{V}_I$ in (1.1), shows that every $\alpha \in U \subseteq W$ has the form n/e_i for some e_i and some integer n. From this, we quickly see that $\{\beta - \gamma \mid \beta > \gamma$ are members of $U\}$ has a smallest element. Thus (2.6) and (2.7) give parts (b) and (c). The proof of (d) will be given after (2.10).

(2.9) COROLLARY. *Let d be as in (2.8). Then for any ideal J projectively equivalent to I, there is a positive integer n such that $\overline{I^n} = \overline{J^d}$.*

PROOF: For some positive integers h and k, we have $\overline{I^h} = \overline{J^k}$. By (2.3), $h/k \in U$, and so by (2.8), h/k can be written as n/d. By (2.1)(b), $\overline{I^n} = \overline{J^d}$.

Recall that (1.1) shows that $\bar{V}_I(x) = \min\{v_i(x)/e_i \mid i = 1, \ldots, r\}$. Suppose that J is an ideal projectively equivalent to I, with $\overline{I^n} = \overline{J^m}$. If we apply (1.1) to the ideal J, we see that there are valuations $v'_1, \ldots, v'_s$, and positive integers $e'_1, \ldots, e'_s$, such that $\bar{V}_J(x) = \min\{v'_j(x)/e'_j \mid j = 1, \ldots, s\}$.

We will now show that $s = r$, and when appropriately ordered, $v_i(x) = v'_i(x)$ for all $x \in R$, and $e'_i = (n/m)e_i$, for $i = 1, \ldots, r$. (Remark: It is an easy exercise to see that this implies that for $\alpha \in U$, $\alpha \bar{V}_\alpha = \bar{V}_I$, a fact we already know from (2.5).)

(2.10) PROPOSITION. *Let J be an ideal with $\overline{J^m} = \overline{I^n}$. Let $v_1, \ldots, v_r$ and $e_1, \ldots, e_r$ be as in (1.1) applied to I, and analogously, let $v'_1, \ldots, v'_s$, and $e'_1, \ldots, e'_s$ be as in (1.1) applied to the ideal J. Then $r = s$, and when appropriately ordered, $v'_i = v_i$, and $e'_i = (n/m)e_i$.*

PROOF: We first do the case that $J = I^n$, (so that $m = 1$). As is explained in (1.1), to produce the various v_i (alternately, v'_j), we first go to R/q for q some minimal prime of R with $I + q \neq R$ (alternately, $J + q \neq R$). Since I and J have the same radical, we see that it does no harm to assume R is a domain. Let $\mathcal{R} = R(I)$ be the Rees ring of R with respect to I. Note that the integral closure $\mathcal{R}'$ has the form $\cdots + R'u^2 + R'u + R' + H_1 t + H_2 t^2 + \ldots$ where $H_i = \overline{I^i R'}$. Let $\mathcal{R}'_n$ be the subring of this having the form $\cdots + R'u^{2n} + R'u^n + R' + H_n t^n + H_{2n} t^{2n} + \ldots$. There is an obvious isomorphism between $\mathcal{R}'_n$, and the integral closure of the Rees ring of R with respect to $J = I^n$, carrying u^n to u and elements in R' to themselves. Now $\mathcal{R}'$ is an integral extension of $\mathcal{R}'_n$, and the n-th power of any homogeneous element of $\mathcal{R}'$ is in $\mathcal{R}'_n$. It is an easy exercise to see that there is a one-to-one correspondence between the sets $\{P \in \operatorname{Spec} \mathcal{R}' \mid P$ is minimal over $u\mathcal{R}'$ (equivalently, over $u^n \mathcal{R}'$)$\}$ and $\{Q \in \operatorname{Spec} \mathcal{R}'_n \mid Q$ is minimal over $u^n \mathcal{R}'_n\}$, P corresponding to Q when $Q = P \cap \mathcal{R}'_n$. Let P correspond to Q in this way. Let L and F be the quotient fields of $\mathcal{R}'$ and $\mathcal{R}'_n$, respectively. Let w be the discrete valuation that $(\mathcal{R}')_P$ determines on L and let w' be the discrete valuation that $(\mathcal{R}'_n)_Q$ determines on F. Now if we define v on R by $v(x) = w(x)$ for all $x \in R$, then v is one of the valuations v_i discussed in (1.1). The associated positive integer $e = e_i$ was defined to be $e = w(u)$. Furthermore, utilizing the isomorphism between $\mathcal{R}'_n$ and the integral closure of the Rees ring of R with respect to $J = I^n$ mentioned above, since that isomorphism fixes $x \in R$, we see that defining $v'(x) = w'(x)$ makes v' one of the v'_j in the statement of this proposition, and the associated $e' = e'_j$ is $w'(u^n)$ (since our isomorphism carries u^n to u). To complete the proof (in the case that $J = I^n$), it will suffice to show that w extends w', so that $v(x) = w(x) = w'(x) = v'(x)$ for all $x \in R$, and $e' = w'(u^n) = w(u^n) = nw(u) = ne = (n/m)e$. Since $(\mathcal{R}')_P \cap F$ is a valuation ring between $\mathcal{R}'_n$ and F whose maximal ideal intersects $\mathcal{R}'_n$ at Q, clearly $(\mathcal{R}')_P \cap F = (\mathcal{R}'_n)_Q$. Let π be an element of P which generates $P(\mathcal{R}')_P$. We may assume that π is homogeneous (since in the DVR $(\mathcal{R}')_P$, one component of π divides all the other components). We claim that there is an $x \in I$, with $xt \in \mathcal{R}' - P$. This claim follows from the argument in the proof of [M, Lemma 3.3]. (Note: the reference just cited has a misprint. In the fourth line of that proof, delete the appearance of height $Q \cap \bar{R}[t^{-1}]$.) Now let xt be as in our claim, and let h be a positive integer such that $h + \text{degree}\, \pi$ is a multiple of n. Let $\delta = (xt)^h \pi$. Since $(xt)^h \in \mathcal{R}' - P$, δ is a generator of $P(\mathcal{R}')_P$. However, since δ is homogeneous and has degree a multiple of n, $\delta \in P \cap \mathcal{R}'_n = Q$. We easily see that δ is also a generator of $Q(\mathcal{R}'_n)_Q$. Now for any $\beta \in (\mathcal{R}'_n)_Q$, we may write $\beta = \delta^r \kappa$, with $r \geq 0$ an integer, and κ a unit in $(\mathcal{R}'_n)_Q$ and so also a unit in $(\mathcal{R}')_P$. By definition of the associated valuations, $w'(\beta) = r = w(\beta)$. Thus w extends w', completing the case that $J = I^n$.

We now consider the case that $J = \bar{I}$, (so that $n = m = 1$). Again we may take R to be a domain. Since the Rees ring of R with respect to I and the Rees ring of R with respect to $\bar{I}$ have the same integral closure, this case is straightforward.

The general case is now easy, passing from I to I^n, hence to $\overline{I^n} = \overline{J^m}$, then to J^m, and finally to J.

PROOF OF (2.8)(d): Let d be as in (2.8). By (2.8)(b), we may choose a positive integer n large enough that $n/d \in U$. We may also assume that n is relatively prime to d. Let $J = I_{n/d}$. By (2.3)(ii), $\overline{J^d} = \overline{I^n}$. Let $e_1, \ldots, e_r$, and $e'_1, \ldots, e'_s$ be as in the statement of (2.10). Then $s = r$, and for $i = 1, \ldots, r$, $e'_i = (n/d)e_i$. However, as e'_i is an integer, and as n is relatively prime to d, we see d divides e_i. This proves (2.8)(d).

We are studying the set of integrally closed ideals projectively equivalent to I. Clearly that set is identical to the set of integrally closed ideals projectively equivalent to I_α for any $\alpha \in U$. Thus we could have replaced I by I_α at the start. We now show how such a replacement would affect U.

(2.11) LEMMA. *Let $\alpha \in U$. For any $0 < \beta < \infty$, $I_\beta = (I_\alpha)_{\beta/\alpha}$. Also, if U_α is the set corresponding to U when these ideas are applied to the ideal I_α, then $U_\alpha = \{\gamma/\alpha \mid \gamma \in U\}$.*

PROOF: By (2.5), we have $\alpha \bar{V}_\alpha = \bar{V}_I$. Now $I_\beta = \{x \in R \mid \bar{V}_I(x) \geq \beta\} = \{x \in R \mid \bar{V}_\alpha(x) \geq \beta/\alpha\} = (I_\alpha)_{\beta/\alpha}$. This proves the first part. Now for $\gamma \in U$, since $\gamma \in W$, there is an $x \in R$ with $0 < \bar{V}_I(x) = \gamma < \infty$. As

$\alpha\bar{V}_\alpha = \bar{V}_I$, $\bar{V}_\alpha(x) = \gamma/\alpha$, and so $\gamma/\alpha \in W_\alpha$. To show that $\gamma/\alpha \in U_\alpha$, we now must show that $(I_\alpha)_{\gamma/\alpha}$ is projectively equivalent to I_α. This is easy, since $(I_\alpha)_{\gamma/\alpha} = I_\gamma$, which is projectively equivalent to I, which in turn is projectively equivalent to I_α. Thus $\{\gamma/\alpha \mid \gamma \in U\} \subseteq U_\alpha$. For the reverse containment, say $\delta \in U_\alpha$. Then (by the various definitions applied to I_α) there is a $y \in R$ with $0 < \bar{V}_\alpha(y) = \delta < \infty$, and $(I_\alpha)_\delta$ is projectively equivalent to I_α. We need $\alpha\delta \in U$, which will suffice to show $U_\alpha \subseteq \{\gamma/\alpha \mid \gamma \in U\}$. Now $\alpha\delta = \alpha\bar{V}_\alpha(y) = \bar{V}_I(y) \in W$, and so we only need $I_{\alpha\delta}$ projectively equivalent to I. By the first part of this result (using $\beta = \alpha\delta$), $I_{\alpha\delta} = (I_\alpha)_\delta$ is projectively equivalent to I_α, which in turn is projectively equivalent to I, and we are done.

(2.12) REMARKS: (a) Let N be as in (2.8)(b), and let $\{\alpha \in U \mid \alpha < N+1\} = \{\alpha_1, \ldots, \alpha_m\}$. It is easily seen that any $\alpha \in U$ can be written $\alpha = c_1\,\alpha_1 + \cdots + c_m\,\alpha_m$, for appropriate nonnegative integers $c_1, \ldots, c_m$. By (2.6) and (2.4), if J is an integrally closed ideal projectively equivalent to I, then J is the integral closure of $(I_{\alpha_1})^{c_1}\,(I_{\alpha_2})^{c_2}...(I_{\alpha_m})^{c_m}$ for appropriate nonnegative integers $c_1, \ldots, c_m$. In this sense, we may think of $\{J \mid J$ is an integrally closed ideal projectively equivalent to $I\}$ as being finitely generated.
(b) In the special case that $U = \{n/d \mid n \geq 1$ is an integer$\}$, with d as in (2.8)(b), part (a) can be stregthened to say that any integrally closed ideal J projectively equivalent to I can be written as $\overline{(I_{1/d})^n}$, for an appropriate integer $n \geq 1$.
(c) As is standard, for some ideal J, let $\bar{V}_I(J)$ denote $\min\{\bar{V}_I(x) \mid x \in J\}$. We claim there is a fixed $v_i \in \{v_1, \ldots, v_r\}$ and its associated e_i in $\{e_1, \ldots, e_r\}$ (as in (1.1)), such that for any ideal J projectively equivalent to I, $\bar{V}_I(J) = v_i(J)/e_i$. For some i and j we have $\bar{V}_I(I) = v_i(I)/e_i$ and $\bar{V}_I(J) = v_j(J)/e_j$. Of course $v_i(I)/e_i \leq v_j(I)/e_j$, and $v_j(J)/e_j \leq v_i(J)/e_i$. There are positive integers n and m with $\overline{I^n} = \overline{J^m}$, and so $v_i(J^m)/e_i = v_i(\overline{J^m})/e_i = v_i(\overline{I^n})/e_i = v_i(I^n)/e_i \leq v_j(I^n)/e_j = v_j(J^m)/e_j \leq v_i(J^m)/e_i$. Equality holds throughout, and we see that $v_j(J^m)/e_j = v_i(J^m)/e_i$. Thus $v_i(J)/e_i = v_j(J)/e_j = V_I(J)$.
(d) We mention some facts, without proof.

(i) If $\beta \in W$ and $\alpha > 0$, then $I_\beta{}^n \subseteq I_\alpha \iff n \geq \alpha/\beta$.
(ii) For $n \geq 1$ an integer, $\operatorname{Rad} I = I_{1/n} \iff (\operatorname{Rad} I)^n \subseteq \bar{I}$.
(iii) $\operatorname{Rad} I = I_\beta$ where β is the smallest element in W.
(iv) If β is the smallest element in W, and $n \geq 1$ is an integer, let $r(n)$

be the least integer with $(\text{Rad}\, I)^{r(n)} \subseteq \overline{I^n}$. Then $r(n)$ is the least integer such that $r(n) \geq n/\beta$, so that $r(n)/n \to 1/\beta$ as $n \to \infty$.

3. Examples.

There is a second way of finding the v_i and e_i, $i = 1, \ldots, r$, discussed in (1.1), which is useful in considering examples.

(3.1) PROPOSITION. *Let R be a Noetherian domain, and let $I = (a_1, \ldots, a_n)$ be an ideal. For a $j = 1, \ldots, n$, let $D = R[a_1/a_j, \ldots, a_n/a_j]$, and let p be a height one prime of the integral closure D' with $a_j \in p$. Let v be the discrete valuation associated with the DVR D'_p. Then v restricted to R is one of the v_i of (1.1), and the associated e_i is $v(a_j) = v(I)$ (the smallest value v assumes on I). Furthermore, every v_i and e_i of (1.1) arises in this way.*

PROOF: Let a_j, D, D', and p be as in the statement, and for convenience, let $a = a_j$. Let $(V, N) = (D'_p, p_p)$, and let v be the discrete valuation associated with V. If t is an indeterminate over D', then so is at. Thus $pD'[at]$ is a height 1 prime in the integrally closed ring $D'[at]$. Let (V^*, N^*) be the discrete valuation ring obtained by localizing $D'[at]$ at $pD'[at]$, and let w be the discrete valuation associated with V^*. Now $V \subseteq V^*$, and since $NV^* = N^*$, we see that w extends v. Let $\mathcal{R} = R[u, It]$ be the Rees ring of R with respect to I. We claim that $D'[at] \subseteq \mathcal{R}'[1/at] \subseteq V^*$. Since $D = R[It/at] \subseteq \mathcal{R}[1/at]$, and since $\mathcal{R}'[1/at]$ is integrally closed (being a localization of $\mathcal{R}'$), the first inclusion is obvious. As for the second inclusion, since at $\in D'[at] - pD'[at]$, at is a unit in V^*. Thus $1/at$ and $u = a/at$ are in V^*. We easily see that $It \subseteq D[at] \subseteq V^*$. The second inclusion follows, and our claim is proved. We now see that $\mathcal{R}'[1/at]$ localized at $N^* \cap \mathcal{R}'[1/at]$ must equal V^*. In particular, height $N^* \cap \mathcal{R}'[1/at] = 1$. It follows that $N^* \cap \mathcal{R}'$ is a height 1 prime, which we denote by P. Since $a \in p \subseteq N^*$, we have $u = a/at \in N^* \cap \mathcal{R}' = P$. That is, P is a height 1 prime of $\mathcal{R}'$ containing u, as is discussed in (1.1). Thus, if w' is the discrete valuation associated with $\mathcal{R}'_P$, then w' restricted to R is one of the v_i of (1.1), and the associated e_1 is $w'(u)$. However, since $\mathcal{R}'_P \subseteq V^*$, and these are both discrete valuation rings with the same quotient field, they must be equal. Therefore, $w' = w$. As w extends v, and as $w = w'$ restricted to R is one of the v_i of (1.1), we see that v restricted to R is one of the v_i of (1.1), as desired. Also, the associated $e_i = w'(u) = w(u) = w(a)$ (since at is a unit in V^*) $= v(a) = v(I)$ (since $aV = IV$). This proves half the result.

It remains to show that all of the v_i and e_i discussed in (1.1) arise in this way, for an appropriate choice of $a = a_j$. Let v_i be as in (1.1). Then there is a height 1 prime P in $\mathcal{R}'$ with $u \in P$, such that if w is the discrete valuation associated with $\mathcal{R}'_P$, then w restricted to R equals v_i, and $e_i = w(u)$. Let $(V^*, N^*) = (\mathcal{R}'_P, P\mathcal{R}'_P)$. (Remark: notation in this paragraph will be considered new. For instance, this V^* is technically not the same as used in the last paragraph. However, we will reproduce the situation of the last paragraph, and the notation here will be analogous to what it was there.) By [M, Lemma 3.3], for some $a = a_j$, $j = 1, \ldots, r$, we must have $at \notin P$. Let $D = R[I/a] = R[It/at]$. It is not hard to see that $D'[at, 1/at] = \mathcal{R}'[1/at]$. Now $\mathcal{R}' \subseteq \mathcal{R}'[1/at] \subseteq V^*$ (since at is a unit in V^*), and so $N^* \cap \mathcal{R}'[1/at] = N^* \cap D'[at, 1/at]$ is a height 1 prime. As $D'[at, 1/at]$ is just a localization of $D'[at]$, we have that $N^* \cap D'[at]$ has height 1. Also, $D'[at]$ localized at $N^* \cap D'[at]$ is easily seen to equal V^*. Let $p = N^* \cap D' = (N^* \cap D'[at]) \cap D'$. As at is an indeterminate over D', height $p \leq 1$. However, since $u \in P \subseteq N^*$, $a = (at)u \in N^* \cap D' = p$. Thus $p \neq 0$, and so height $p = 1$. Let $(V, N) = (D'_p, p_p)$, and let v be the discrete valuation associated with V. It is easily seen that $V \subseteq V^*$, and $NV^* = N^*$. Thus w extends v. Since w restricted to R equals v_i, v restricted to R equals v_i, as desired. Also, $e_i = w(u) = w(a)$ (since at is a unit in V^*) $= v(a) = v(I)$ (since $aV = IV$). This completes the proof.

(3.2) Corollary. *Let $I = aR$ be a principal ideal in a Noetherian domain R. In the integral closure R', let $p_1, \ldots, p_r$ be the height 1 primes minimal over aR'. Let w_i be the discrete valuation associated with R'_p, let v_i be w_i restricted to R, and let $e_i = w_i(a)$, for $i = 1, \ldots, r$. Then these v_i and e_i are the same as given in (1.1).*

Proof: Since I is principal, the construction in (3.1) can only lead to $D = R$. Thus the corollary is immediate.

(3.3) Example: Let $R = K[X^3, X^5]$, and $I = X^3R$, with K a field and X an indeterminate. Using (3.2) it is not hard to see that $W = \{3/3, 5/3, 6/3\} \cup \{n/3 \mid n \geq 8 \text{ is an integer}\}$. For $n/3 \in W$, it is easily seen that $(I_{n/3})^3 = I_n = I^n$, so that $U = W$.

(3.4) Example: We will show that in (2.8)(d), d might not be the greatest common divisor of $e_1, \ldots, e_r$. Let $R' = K[X, Y]$, while R equals

$K[X^2, XY, Y]$. It is convenient to note that R consists of those polynomials $f \in R'$ such that each monomial X^aY^b in f has either a even or $b \geq 1$. Let $I = X^2R$. By (3.2), we see that with notation as in (1.1), we have $r = 1$, $e_1 = 2$, and for $f \in R$, $v_1(f)$ is the smallest power of X appearing in f (with nonzero coefficient).

We next show that in (2.8), $d = 1$. By (2.8), we have that either $d = 1$ or $d = 2$. Suppose that $d = 2$. Then (2.8)(b) shows that for a sufficiently large odd integer n, we have $n/2 \in U$. By (2.3)(ii), $\overline{I_{n/2}{}^2} = \overline{I^n}$. Since $X^{2n} \in I^n$, we get $X^{2n} \in \overline{I_{n/2}{}^2} \subset \overline{(I_{n/2}{}^2)R'}$. Now $I_{n/2} = \{f \in R \mid \bar{V}_I(f) \geq n/2\}$. However, by (1.1) and the first paragraph, $\bar{V}_I(f) = v_1(f)/2$, and so $I_{n/2} = \{f \in R \mid$ the smallest power of X appearing in f is at least $n\} = (X^nY, X^{n+1})R$, using that n is odd. Thus, we get X^{2n} is in the integral closure of $(X^{2n}Y^2, X^{2n+1}Y, X^{2n+2})R'$. However, it is not hard to see that this is not true. Thus $d \neq 2$, so $d = 1$.

REMARK: In the previous example, $\overline{I^n} = \overline{X^{2n}R} = X^{2n}R' \cap R =$ (easily seen) $X^{2n}R = I^n$, for all $n \geq 1$.

(3.5) PROPOSITION. *Let $I \neq 0$ be an ideal in a Dedekind domain R. Let $I = P_1{}^{e_1} \cap \ldots P_r{}^{e_r}$ be a primary decomposition of I. Let w_i be the discrete valuation associated to R_P, and let v_i be the restriction of w_i to R. Let d be the greatest common divisor of $e_1, \ldots, e_r$. Then*

a) *The $v_1, \ldots, v_r$ and $e_1, \ldots, e_r$ appearing here are identical to the $v_1, \ldots, v_r$ and $e_1, \ldots, e_r$ discussed in (1.1).*
b) *$W = \{n/e_i \mid i = 1, \ldots, r$, and $n \geq 1$ is an integer$\}$.*
c) *$U = \{n/d \mid n \geq 1$ is an integer$\}$.*
d) *If $f_i = e_i/d$, $i = 1, \ldots, r$, then $I_{1/d} = P_1{}^{f_1} \cap \cdots \cap P_r{}^{f_r}$, and $\{J \mid J$ is an ideal projectively equivalent to $I\} = \{(I_{1/d})^n \mid n \geq 1$ is an integer$\}$.*

PROOF: a) Fix an $i = 1, \ldots, r$. As R is a Dedekind domain, I is generated by two elements. Say $I = (a, b)R$. Without loss, we may assume that $b/a \in R_{P_i}$. Let $D = R[b/a] \subseteq R_{P_i}$. As any overring of a Dedekind domain is integrally closed, $D = D'$. Let $p_i = P_i R_P \cap D$. Clearly $a \in p_i$. Since $R \subseteq D \subseteq R_{P_i}$, clearly $D_{p_i} = R_{P_i}$. Thus w_i is the valuation associated to D'_{p_i}, and so by (3.1), w_i restricted to R is one of the v_i discussed in (1.1), as claimed. We also know that the associated e_i (as in (1.1) and (3.1)) is $v_i(a) = v_i(I)$. Therefore, we must show that $v_i(I) = e_i$ (this e_i as in the statement of this result). However, this is clear from our primary decomposition. Thus each pair v_i, e_i in the statement of this result is among the v_i, e_i discussed in (1.1). The converse, that each pair v_i, e_i appearing in (1.1) also appears in the statement of the present result, similarly follows from (3.1). We leave the details to the reader.

b) By (a) and (1.1), we know for $x \in R$, $\bar{V}_I(x) = \min\{v_i(x)/e_i \mid i = 1, \ldots, r\}$. Thus clearly $W \subseteq \{n/e_i \mid i = 1, \ldots, r$, and $n \geq 1$ is an integer$\}$. For the reverse containment, pick an $i = 1, \ldots, r$, and an $n \geq 1$. We must show that there is an $x \in R$ with $\bar{V}_I(x) = n/e_i$. For this, it will suffice to have $v_i(x)/e_i = n/e_i$, and $v_j(x)/e_j > n/e_i$ for all $j \neq i$, $1 \leq j \leq r$. If k is an integer with $k > \max\{ne_j/e_i \mid j \neq i, 1 \leq j \leq r\}$, it suffices to have $v_i(x) = n$ and $v_j(x) \geq k$ for all $j \neq i$. Standard arguments show that we can find an x in $((\cap P_j{}^k) \cap P_i{}^n) - P_i{}^{n+1}$ $(j \neq i)$, and that such an x satisfies our needs. Thus W has the form stated.

c) Using that $\bar{V}_I(x) = \min\{v_i(x)/e_i \mid i = 1, \ldots, r\}$, and that $f_i/e_i = 1/d$, it is straightforward to see that $I_{1/d} = P_1{}^{f_1} \cap \cdots \cap P_r{}^{f_r}$. As R is a Dedekind domain, $P_1{}^{f_1} \cap \cdots \cap P_r{}^{f_r} = (P_1{}^{f_1}) \ldots (P_r{}^{f_r})$. Thus $(I_{1/d})^d = (P_1{}^{f_1})^d \ldots (P_r{}^{f_r})^d = (P_1{}^{e_1}) \ldots (P_r{}^{e_r}) = I$. Thus, $I_{1/d}$ is projectively equivalent to I. Since $1/d \in W$ (easily seen using (b)), we have that $1/d \in U$. By (2.6), $\{n/d \mid n \geq 1$ is an integer$\} \subseteq U$. However, (2.8) shows that we must have equality here.

d) The first part is proved in the proof of (c). The last part is by (2.12)(b), and the fact in a Dedekind domain, every ideal is integrally closed.

(3.6) Remark: If (R, M) is a 2-dimensional regular local ring, then Zariski's theory of unique factorization for integrally closed ideals, [**ZS**, Appendix 5], allows a determination of the integrally closed ideals projectively equivalent to a fixed integrally closed M-primary ideal I in a manner analogous to the determination in (3.5), cf. also the recent [**HS**]. Zariski proves that any such ideal I has a unique factorization $I = I_1{}^{e_1} \ldots I_m{}^{e_m}$ as a product of simple integrally closed ideals. (An ideal is simple if it is not a product of two proper ideals.) If d is the greatest common divisor of $e_1, \ldots, e_m$, and if $f_i = e_i/d$, $i = 1, \ldots, m$, then $I_{1/d} = I_1{}^{f_1} \ldots I_m{}^{f_m}$, and the integrally closed ideals projectively equivalent to I are exactly the ideals $(I_{1/d})^n$, with $n \geq 1$ an integer. In particular, if I is a simple integrally closed M-primary ideal, then the set of integrally closed ideals projectively equivalent to I is just the set of powers of I.

(3.7) EXAMPLE: We give an example of a 2-dimensional local domain which is not regular, for which the final sentence of (3.6) fails. Let $R = K[[X,Y,Z]]/(X^2+Y^5+Z^7)$, and let $M = (\tilde{X}, \tilde{Y}, \tilde{Z})$ and $I = (\tilde{X}, \tilde{Y}^2, \widetilde{YZ}, \tilde{Z}^2)$, the tilde denoting images. Then $\bar{I} = M^2$, and $M = I_{1/2}$. M and I are both simple.

References

[**HS**] C. Huneke and J. Sally, *Birational extensions in dimension two and integrally closed ideals*, J. Algebra (to appear).

[**LT**] M. Lejuene-Jalabert and B. Teissier, "Cloture intégrale des ideaux et équisingularité," Seminaire Lejuene-Teissier, Centre de Mathématiques École Polytechnique, 1974.

[**M**] S. McAdam, "Asymptotic Prime Divisors," Lecture Notes in Mathematics No. 1023, Springer-Verlag, New York, 1983.

[**N**] M. Nagata, *Note on a paper of Samuel concerning asymptotic properties of ideals*, Mem. Coll. Sci. Univ. Kyoto, Ser. A Math. **30** (1957), 165–175.

[**R**] D. Rees, *Valuations associated with ideals (II)*, J. London Math. Soc. **31** (1956), 221–228.

[**S**] P. Samuel, *Some asymptotic properties of powers of ideals*, Annals of Math. **56** (1952), 11–21.

[**ZS**] O. Zariski and P. Samuel, "Commutative Algebra, Vol. II," D. Van Nostrand, Amsterdam, 1960.

Department of Mathematics, University of Texas, Austin TX 78712

Department of Mathematics, University of California, Riverside CA 92521

Department of Mathematics, Northwestern University, Evanston IL 60201

Izumi's Theorem

DAVID REES

In this paper I want to discuss a recent paper by Shozo Izumi [**I1**] which solved an old problem, that of characterizing analytically irreducible local domains (i.e., local domains whose completion is a domain) in a way which does not involve the idea of completion. I shall go a little beyond the original paper, which was concerned with local analytic algebras, and therefore was restricted to local rings containing a field, but the modifications needed to cover the general case are minor, and the paper contained all the essentials of the solution.

Could I first explain, in my own language, the contents of Izumi's theorem. I will be concerned throughout with a local ring (Q, m, k, d), this signifying that Q has maximal ideal m, residue field k, and that the Krull dimension of Q is d. Now suppose that I is an m-primary ideal of Q. We can associate with I an integer-valued function $I(x)$ on Q by $I(x) = n$ if $x \in I^n - I^{n+1}$, and $I(x) = \infty$ if $x = 0$. (∞ is assumed to be a positive integer.) We have two other functions which can be constructed from I. The first, which I will denote by $I^\wedge(x)$ is defined to be $\lim(I(x^n)/n)$ as $n \to \infty$, this limit always existing, if the value ∞ is allowed. It was, essentially, first introduced into commutative algebra in the paper [**S**] of Samuel given in the references and was further discussed in the papers [**R1**–**R4**] of the present author. In particular we have the Valuation Theorem, proved in general in [**R3**], and which runs as follows:

VALUATION THEOREM. *There exists a finite set of integer-valued valuations $v_1, \ldots, v_s$ on Q, which take non-negative integer values on Q and positive values on m, such that*

$$I^\wedge(x) = \min(v_i(x)/v_i(I))$$

where $v_i(I) = \{\min v_i(z) \mid z \in I\}$. If this representation is irredundant, the valuations v_i are uniquely determined, and v_i takes the value ∞ on a minimal prime ideal $p(v_i)$ of Q (so that v_i determines a valuation on the *field of fractions of $Q/p(v_i)$). The residue field of v_i is a finitely generated extension of k of transcendence degree at most $d-1$ over k.*

The valuations v_i will be termed the valuations associated with I.

The other function that will concern me is $I^*(x)$. We define $I^*(x)$ by the condition that $I^*(x) \geq n$ if x satisfies an equation

$$x^r + a_1 x^r + \cdots + a_r = 0 \qquad a_j \in I^{jn}$$

that is, if x belongs to the integral closure $(I^n)^*$ of I^n. This function is a half-way house to $I^\wedge(x)$ since one can prove that $I^*(x) = [I^\wedge(x)]$. Note that $I^\wedge(x)$ is rational valued.

Note that $I(x), I^\wedge(x), I^*(x)$ are all filtrations ($f(x)$ is a filtration if it satisfies the conditions

i) $f(1) = 0, f(0) = \infty$,
ii) $f(x-y) \leq \min(f(x), f(y))$,
iii) $f(xy) \geq f(x) + f(y)$).

Now I want to say a little about $I(x)$. First the Artin-Rees lemma implies the following;

(A). *If x is a non-zero divisor of Q, there is an integer $\mu(I,x)$ depending only on I, x such that*

$$I(x) \leq I(xy) - I(y) \leq \mu(I,x) \qquad \textit{for all } y \neq 0.$$

The first inequality is immediate, while for the second, if $I(xy) = n$, $y \in I^n : x$, and, by the Artin-Rees lemma, there exists an integer $\mu = \mu(I,x)$ such that $I^n : x \subseteq I^{n-\mu}$ for all n.

Next we need a characterization of the condition that Q be a.u. (= "analytically unramified", i.e. that the completion $Q^\wedge$ of Q has no nilpotent elements). The result following was proved in [**R4**], a simpler proof being implicit in [**R5**].

(B). *Q is a.u. if for at least one I and only if for all I, there exists a constant $K(I)$ depending only on I such that, for all $x \neq 0$,*

$$I^\wedge(x) - I(x) \leq K(I).$$

Now we come to Izumi's Theorem. We say that Q is analytically irreducible, (abbreviated to a.i.), if $Q^\wedge$ is a domain.

(C) IZUMI'S THEOREM. *Q is a.i. if for at least one I, and only if, for all I, there exists constants $C(I)\,C'(I)$ depending only on I such that*

$$I(xy) - I(y) \leq C(I)\,I(x) + C'(I), \qquad \textit{for all } x, y \neq 0 \textit{ in } Q.$$

The "if" part of both (B) and (C) is comparatively easy. The really interesting part of (C), as is the case for (B), lies in the "only if." It is with a sketch of the proof of this part that I will concern myself for the rest of this paper. For fuller details, the reader is referred to [**I1**].

From now on, therefore, I will assume that Q is an a.i. domain. In fact I can assume that Q is a compete domain. For the function $IQ\hat{}(x)$ restricts to $I(x)$ on Q, and hence it is sufficient to prove the result for the function $IQ\hat{}(x)$ to prove it for $I(x)$. Now we take a step further. It is enough to prove, instead of (C), the similar inequality

(D). *There exists a constant $C\hat{}(I)$ such that*

$$I\hat{}(xy) - I\hat{}(y) \leq C\hat{}(I)\,I\hat{}(x) \qquad \textit{for all } x, y \neq 0 \textit{ in } Q$$

For $I(xy) - I(y) \leq I\hat{}(xy) - I\hat{}(y) + K(I) \leq C\hat{}(I)\,(I(x) + K(I)) + K(I)$.

Now we transform (D). First of all I want to use that fact that Q is a complete domain. In this case, the valuations associated with I are all valuations on the field of fractions F of Q. Secondly, the residue field of v_i has transcendence degree exactly $d-1$ over k. Call a variation v on F an m-valuation if $v(x) \geq 0$ on $Q_i > 0$ on m and has residue field a finitely generated extension of k of transcendence degree $d-1$ over Q. Then the valutions associated with I are all m-valuations, and, if S is a finite set of m-valuations, we can choose I so that the valuations of S are included in the set of valuations associated with I. Now we come to my final form of Izumi's Theorem.

(E). *If v, w are two m-valuations of the compete local domain Q, then there exists an integer $C(v, w)$ such that*

$$v(x) \leq C(v, w)\,w(x) \qquad \textit{for all } x \neq 0 \textit{ in } Q.$$

Let us see that this implies (D), and hence Izumi's Theorem. Let $v_1, \ldots, v_s$ be the valuations associated with I and let $C = \max(v_i(I)\,C(v_i, v_j)/v_j(I))$. C clearly depends only on I. Choose i, j such that

$$I\hat{}(x) = v_i(x)/v_i(I); \qquad I\hat{}(y) = v_j(y)/v_j(I).$$

Then

$$I^\wedge(xy) - I^\wedge(y) \leq (v_j(xy) - v_j(y))/v_j(I)$$
$$= v_j(x)/v_j(I) \leq C(v_i, v_j)\, v_i(x)/v_j(I)$$
$$\leq CI^\wedge(x).$$

One last remark. We can assume that Q is a complete normal local domain, simply by replacing Q with its integral closure Q^* which will be a local domain maximal ideal m^* and noting that the m-valuations and m^*-valuations are the same.

Now we come to the sketch of the proof. First we don't need to consider $d = 1$ since a complete normal local domain of dimension 1 is a DVR and so has only one m-valuation. Hence we start with $d = 2$. The proof falls into two parts, the first being the case $d = 2$, and the second being an induction starting with $d = 2$.

Let us first consider the case $d = 2$. Let $I = (a_1, \ldots, a_m)$ be an m-primary ideal of the complete normal local domain Q of dimension 2. Let t be an indeterminate over Q and write u for t^{-1}. Then we define $R(I)$ to be the sub-ring $Q[ta_1, \ldots, ta_m, u]$ of $Q[t, u]$. Let p be a graded prime ideal of $R(I)$. Then we define $Q(p)$ to be the set of fractions at^r/bt^r, at^r, bt^r in $R(I)$ and bt^r not in p, so that $F \supseteq Q(p) \supseteq Q$ and $\dim Q(p) \leq 2$. We now term $R(I)$ a desingularization of Q if all the rings $Q(p)$ are regular local rings. This implies that, if p has height 1, then $Q(p)$ is a DVR and so has a normalized associated valuation v. We consider the set $\sum(I)$ of valuations obtained in this way. First suppose u does not belong to p. Then p meets Q in a height 1 prime ideal P and $Q(p) = Q_P$. In this way we obtain the set of valuations on F whose center on Q has height 1, and we refer to these as the proper prime divisors associated with $R(I)$. Next we consider those arising from p such that $u \in p$. There are only a finite number, and, in fact, they are the valuations associated with I. These are the exceptional prime divisors of $R(I)$. To complete the picture, if v is any other m-valuation of Q, then O_v contains only one of the local rings $Q(p)$ which has dimension 2 and v is a $pQ(p)$-valuation of that ring. I want one further property of the desingularization $R(I)$, and that is that it can be chosen so that the exceptional divisors of $R(I)$ should contain two given m-valuations v, w. Let me explain how this can be ensured. If v is not exceptional, O_v contains exactly one of the local rings $Q(p)$, say $Q(p_0)$. We blow up the maximal ideal of $Q(p_0)$ and obtain a new set of local rings

$Q(p)$, in which $Q(p_0)$ is replaced by a set of regular local rings, one being of dimension 1, the associated valuation being that associated with the maximal ideal of $Q(p)$. Hence the set of exceptional valuations is increased by adding this valuation. This blowing-up can also be obtained by replacing I with a new ideal I' as follows. Let $I_n(p_0)$ be the ideal of Q consisting of all x such that $xt^n \in p_0$. Then $R(I_n(p_0))$ produces the same set of local rings $Q(p)$ as the blowing process just defined providing that n is large enough. Repeating the blowing-up process a sufficient number of times will eventually give us a desingularization in which v is exceptional, and the same process can be used to add w to the set of exceptional valuations.

The next stage depends on an intersection theory. By this I mean the following. With every pair of prime divisors v, w of $R(I)$, we associate an integer $[v,w]$ $(= [w,v])$ subject to the following conditions.

i) if either v or w is proper, or if $v \neq w$, then

$$[v,w] \geq 0$$

ii) if v is exceptional and $x \neq 0$ belongs to Q, then

$$\sum [v,w]\, w(x) = 0$$

iii) if v_0, v_{r+1} are prime divisors then they can be joined by a chain of exceptional prime divisors $v_1, \ldots, v_r$ such that

$$[v_i, v_{i+1}] > 0 \quad i = 0, \ldots, r$$

Note that by choosing x in m, so that $v(x) > 0$ if v is exceptional, ii) and iii) imply that $[v,v] < 0$ if v is exceptional. Now suppose that v, w are two exceptional prime divisors of $R(I)$ such that $[v,w] > 0$ then ii) implies that

$$v(x) \leq C(v,w)\, w(x) \text{ for all } x \neq 0 \text{ on } Q, \text{ with } C = -[w,w]/[v,w]$$

while, if $[v,w] = 0$, we can use iii) to find a suitable $C(v,w)$. Since we can choose a desingularization $R(I)$ with any given pair v, w of m-valuations as exceptional prime divisors this completes the proof of the case $d = 2$. This is only marginally different from Izumi's proof. The new ingredients are the existence of desingularizations of a complete normal domain of dimension 2 and the existence of an intersection theory satisfying i), ii), iii). For this I refer to the two paper of Lipman [**L1, L2**].

Now we turn to the induction taking us from $d = 2$ to all $d > 2$. Here I will need two results which I will quote without proof. The first of these is the following.

(F). *If Q is an a.i. local domain of dimension ≥ 3, $a \neq 0$ belongs to m, and I, J are ideals of Q containing a which are not m-primary and satisfy $IJ \subseteq \mathrm{rad}(aQ)$, then $I + J$ is not m-primary.*

This was stated in a slightly different form by Grothendieck in [**G2**]. A related result was proved by Faltings in [**Fal**]. (The author is grateful to the referee for drawing his attention to this reference.) Finally, there is a more general result, with a particularly nice proof, given by Brodman and Rung in [**B-R**] which is worth stating.

First we require two definitions. If Q is a local ring, $c(Q)$ is the minimum of the dimension of the rings $Q/(I + J)$, where I, J are two radical ideals of Q such that $I \cap J$ is the radical of Q and neither I nor J is contained in $\mathrm{rad}(Q)$. If K is an ideal of Q, $r(K)$ is the minimum number of elements generating an ideal with the same radical as K. Then

(F$'$). *Let (Q, m, k, d) be a complete local ring. Then, if I, J are two ideals of Q such that* $\dim(Q/I+J) < \min(\dim(Q/I), \dim(Q/J))$, $\dim(Q/I+J) \geq c(Q) - r(I \cap J) - 1$.

The second is a criterion for an m-valuation v on a local ring Q to be associated with an m-primary ideal I, of which a proof is given in an appendix.

(G). *Let (Q, m, k, d) be a local ring, $I = (a_1, \ldots, a_m)$ be an m-primary ideal of Q and v be an m-valuation of Q with residue field K. Suppose that $v(I) = v(a_m)$ and let x_i be the image in K of a_i/a_m $(i = 1, \ldots, m-1)$. Then v is associated with I if and only if K is an algebraic extension of $k(x_1, \ldots, x_{m-1})$.*

Now we come to the proof of the inductive stage. We suppose that (Q, m, k, d) is a complete normal local domain with field of fractions F, and that v, w are m-valuations of Q. The idea behind the proof is the construction of an a.i. local ring $(L, M, K, d-1)$ containing Q and M-valuations V, W of L extending v, w. Once this is done, we can apply our inductive hypothesis to L, V, W to deduce (E) for V, W and thereby prove (E) for v, w. The construction of L, V, W is fairly easy. The crux of the proof is the proof that L is a.i.

Now for some notation. Let $(X_1, \ldots, X_r, \ldots)$ be a countable set of indeterminates. Then $Q[r]$ will denote the ring $Q[X_1, \ldots, X_r]$, $Q(r)$ will denote the localization of $Q[r]$ at $mQ[r]$, $Q[r, i]$, where $i \leq r$, will denote the ring $Q[X_1, \ldots, X_{i-1}, X_{i+1}, \ldots, X_r]$ and $Q(r,i)$ the localization of $Q[r,i]$ at $mQ[r,i]$.

Now Q is an excellent normal local domain. Hence $Q[r]$ is normal, implying that $Q(r)$ is normal, and further $Q(r)$ is excellent and therefore analytically normal for all r. We now come to the definition of L, V, W. First we define L. Let $I = (x_1, \ldots, x_d)$ be an m-primary ideal of Q with which both v and w are associated. Since v, w are m-valuations, it follows that $v(I) = v(x_1) = \cdots = v(x_d)$ and $w(I) = w(x_1) = \cdots = w(x_d)$. Now let $x = \sum x_i X_i$ and $y_i = x - X_i x_i$. We define L to be the ring $Q(d)/xQ(d)$, so that $M = mQ(d)/xQ(d)$. Now since $Q[d,i]$ is normal, the ideal (x_i, y_i) has grade 2, and from this it follows that the ring $Q[d,i][-y_i/x_i]$ is isomorphic to $Q[d]/xQ[d]$ for any i, and L is the localization of $Q(d,i)[-y_i/x_i]$ and so is an excellent local domain. While L is not normal, it is analytically unramified, being the localization of the finitely generated extension $Q(d,i)[y_i/x_i]$ of $Q(d,i)$, which is a.u. Now it is clear that the ring of fractions $y/x_i{}^s$ $(y \in L)$ is normal, and hence the conductor $C = C(L^*/L)$ of the integral closure L^* of L in its field of fractions contains a power of x_i for each i, and so is M-primary. Finally we can define the extensions V, W of v, w as follows. We consider L as a localization of $Q(d-1)[y_d/x_d]$, so that the field of fractions of L is $F(X_1, \ldots, X_{d-1})$. Define V, W to be the trivial extensions of v, w from F to $F(X_1, \ldots, X_{d-1})$. Then by (G), V, W are associated with IL and so are M-valuations.

All that remains is the proof that L is a.i., i.e., that $L^\wedge$ is a domain. We already know that it is reduced. Now consider the conductor of the integral closure of $L^\wedge$ in its complete ring of fractions. This will be $CL^\wedge$ since L is excellent. Now suppose $L^\wedge$ is not a domain. Then the complete ring of fractions of $L^\wedge$ will contain idempotents e, f such that $e + f = 1$ and hence $ef = 0$. Since e, f are integrally dependent on $L^\wedge$, $C' = eC$, $C'' = fC$ are ideals of $L^\wedge$ such that $C'C'' = (0)$ and $C = C' + C''$. Hence in the normal complete local domain $Q(d)^\wedge$, we have two ideals whose union is $mQ(d)^\wedge$-primary and whose intersection is contained in $\mathrm{rad}(xQ(d)^\wedge)$. Hence we have a contradiction to (F) and the proof is complete.

This last part of the proof differs from that given by Izumi, which used a Bertini-like theorem due to Flenner.

Appendix.

We conclude with an appendix giving a proof of (G) in the case where Q is an analytically irreducible local domain, the general case being a fairly simple consequence of this case. The proof includes a proof of the valuation theorem for this case.

THEOREM. *Let (Q, m, k, d) be an a.i. local domain and let $I = (a_1, \ldots, a_m = a)$ be an m-primary ideal of Q. Let v be an m-valuation of Q with residue field K, and suppose $v(I) = v(a)$. Let $K(I)$ be the sub-field $k(a_1/a, \ldots, a_{m-1}/a)$ of K. Then v is associated with I if and only if K is algebraic over $K(I)$.*

Let $R = R(I) = Q[ta_1, \ldots, ta_m, u]$, where $u = t^{-1}$. Since Q is analytically unramified, it follows from (B) that the integral closure R^* of R in $Q[t, u]$ (which consists of all finite sums $\sum c_r t^r$ with $c_r \in (I^r)^*$) is a finite R-module, so is finitely generated over Q and is a graded noetherian ring. But $Q[t, u] = R_s$, where S is the set of powers of u. Hence the ideals $u^n R^*$ are unmixed of height 1 for all n, and if $P(1), \ldots P(s)$ are the height 1 graded prime ideals of R^* minimal over uR^*, the local rings $R_{P(i)}$ are discrete valuation rings. Let V_i be the valuation corresponding to $Q_{P(i)}$. Then V_i is a graded valuation. Let v_i be its restriction to the field of fractions F of Q. Then

$$V_i(ct^r) = v_i(c) - rV_i(u).$$

Next we observe that Q is quasi-unmixed. Hence if P is a prime ideal of R^* meeting Q in m, then, by the altitude inequality, if $K(P) = R^*{}_P/PR^*{}_P$,

$$d + 1 = \text{Trans. deg}_k\, K(P) + \text{ht}\, P.$$

Now let V be the graded valuation on R^* defined by $V(ct^r) = v(c) - rv(I)$, so that $V(u) = v(I)$. Let P be the center of V. P contains m and has height 1 if and only if trans. $\deg_k K(P) = d$. But v is an m-valuation, and so its residue field K has transcendence degree $d - 1$ over k, whence that of V has transcendence degree d over k. It follows that P has height 1 if and only if K is algebraic over $K(I)$, that is V is one of the valuations $V_1, \ldots, V_s$ if and only if K is algebraic over $K(I)$. To complete the proof, all we need show is that $v_1, \ldots, v_s$ are the valuations associated with I. First it is clear that x in Q belongs to $(I^n)^*$ if and only if $xt^n \in R^*$, i.e.

if and only if $v_i(x) \geq nV_i(u)$, and this proves that $V_i(u) = v_i(I)$ and this implies that

$$I^\wedge(x) = \min(V_i(x)/v_i(I)).$$

Finally it is not difficult to see that this representation is irredundant.

Notes on the references.

Apart from **[I1]**, Izumi has written a second paper **[I2]**, which contains further applications of this result in the context of analytic algebras.

The results on the existence of desingularizations of complete normal local domains of dimension 2 and intersection theories are taken from **[L1,L2]**.

The references to the theory of excellent rings are taken from **[G1]**, and are all from Section 7.8, particularly Scholie (7.8.3) i)–ix).

[M] is a useful reference, particularly Chapter 13.

The basic result (F) first appeared in the following set of seminar notes **[G2]**. The theorem in question is given on page 5 of Expose XIII at the beginning of Section 2.

The proof of (F′) appeared in **[B-R]**.

REFERENCES

[B-R] M. Brodman and J. Rung, *Local cohomology and the connectedness dimension in algebraic varieties*, Comm. Math. Helv. **61** (1986), 481–490.

[Fal] G. Faltings, *A contribution to the theory of formal meromorphic functions*, Nagoya Math. Jour. **77** (1980), 99–106.

[G1] A. Grothendieck, "Éléments de Géométrie Algébrique: IV Étude Locale des Schémas et des Morphismes des Schémas (Seconde Partie)," Publ. Math. IHES, vol. 24, 1965.

[G2] A. Grothendieck, "Cohomologie Locale des Faisceaux Cohérents et Théorèmes des Lefschetz Locaux et Globaux," SGA II, IHES, 1962.

[I1] S. Izumi, *A measure of integrity for local analytic algebras*, Publications of the Research Institute for Mathematical Sciences, Kyoto University **21** (1985), 719–735.

[I2] S. Izumi, *Gabrielov's rank condition is equivalent to an inequality of reduced orders*, Math. Ann. **276** (1986), 81–89.

[L1] J. Lipman, *Rational singularities, with applications to algebraic surfaces and unique factorization*, Publ. Math. IHES **36** (1969), 195–279.

[L2] J. Lipman, *Desingularisation of two dimensional schemes*, Annals of Math **107** (1978), 151–207.

[M] H. Matsumura, "Commutative Algebra," 2nd edition, Benjamin-Cummings, 1980.

[R1] D. Rees, *Valuations associated with local rings [1]*, Proc. London Math. Soc. (3) **5** (1955), 107–128.

[R2] D. Rees, *Valuations associated with ideals [1]*, Proc. London Math. Soc. (3) **6** (1956), 161–174.

[R3] D. Rees, *Valuations associated with ideals [2]*, Jour. London Math. Soc. **31** (1956), 221–228.

[R4] D. Rees, *Valuations associated with local rings [2]*, Jour. London Math. Soc. **31** (1956), 228–235.

[R5] D. Rees, *A note on analytically unramified local rings*, Jour. London Math. Soc. **36** (1961), 24–28.

[S] P. Samuel, *Some asymptotic properties of powers of ideals*, Annals of Math. **56** (1952), 11–21.

6 Hillcrest Park, Exeter, EX4 4SH, UNITED KINGDOM

Intersection Theorems

PAUL ROBERTS

1. Introduction.

The "intersections" referred to in the title of this paper are intersections of the kind which came into Commutative Algebra from Intersection Theory in Algebraic Geometry, and the "theorems" are descendants of the Intersection Theorem of Peskine and Szpiro [**15,16**]. There are other kinds of theorems which go under this name, notably Krull's Intersection Theorem on the intersection of the powers of an ideal; we will not be discussing these here.

The original form of the theorem is the following:

THEOREM 1 (Peskine-Szpiro [**15**]). *Let M be a finitely generated module of finite projective dimension over a local ring A, and let N be a finitely generated A-module such that $M \otimes_A N$ is a module finite length. Then, if A has positive characteristic or if the completion of A is the completion of a local ring essentially of finite type over a field, we have*

$$\text{projective dimension}(M) \geq \text{Krull dimension}(N).$$

The sense in which this is a theorem about intersections comes from the fact that the support of the tensor product of two modules is the intersection of the supports of the modules. Thus this theorem says something about the codimension of the intersection of two closed subsets of Spec(A); namely, it says that if one of the two closed subsets is the support of a module of finite projective dimension M, then the codimension of the intersection in the other closed subset is bounded by the projective dimension of M. It is perhaps not easy to see the reason behind this kind of inequality, with projective dimension on one side and Krull dimension on the other, but the importance of the theorem was demonstrated by the fact that it implies two earlier conjectures (Peskine-Szpiro [**16**]):

CONJECTURE 1 (Auslander). *If M is a module of finite projective dimension over a local ring A, and if an element x in A is not a zero-divisor on M, then x is not a zero-divisor on A.*

CONJECTURE 2 (Bass). *If a local ring A has a non-zero finitely generated module of finite injective dimension, then A is Cohen-Macaulay.*

Thus these conjectures became theorems for rings satisfying the conditions of Theorem 1. In proving these theorems, Peskine and Szpiro introduced the method of reduction to positive characteristic and the use of the Frobenius morphism to solve homological problems, and they showed that by using Artin's Approximation Theorem one could extend these results to rings whose completions are completions of localizations of rings of finite type over a field. Hochster [**10,11**] extended these methods and proved these results (and others) for arbitrary local rings which contain a field. Meanwhile, Theorem 1 was generalized (Peskine-Szpiro [**17**], Roberts [**18**]) to give

THEOREM 2 (New Intersection Theorem). *Let $F_* =$*

$$0 \to F_k \to \cdots \to F_1 \to F_0 \to 0$$

be a complex of finitely generated free A-modules such that the homology of F_* has finite length but F_* is not exact. Then, if A contains a field, we have

$$k \geq \text{dimension}(A).$$

It is not difficult to see that Theorem 2 implies Theorem 1; if M is a module of finite projective dimension, then it has a resolution of the form

$$0 \to F_k \to \cdots \to F_1 \to F_0 \to M \to 0$$

where k is the projective dimension of M. It suffices to prove Theorem 1 when N is of the form A/P for a prime ideal P of A. The hypothesis that $M \otimes_A (A/P)$ is a module of finite length implies that the complex of free A/P-modules obtained by tensoring F_* with A/P satisfies the hypotheses of Theorem 2, and the conclusion of Theorem 2, that $k \geq \dim(A/P)$, gives the conclusion of Theorem 1, since k is the projective dimension of M.

The New Intersection Theorem was proven by essentially the same methods as the original one; hence the hypothesis that A contain a field in its statement. This left open one remaining case, that of mixed characteristic. We discuss in this paper the recent proof of the theorem in this case (Roberts [**22**]). At the end of the paper we shall mention more recent versions of the theorem which are still conjectures.

We close this section with an outline of a false proof of the New Intersection Theorem, but one which contains part of the idea for a correct one. We first note that it suffices to prove the theorem if A is a complete local integral domain. Let A be a local ring of dimension d, and suppose that $F_* =$

$$0 \to F_{d-1} \to \cdots \to F_1 \to F_0 \to 0$$

is a complex of free modules with homology of finite length and not everywhere zero as in the hypothesis of the theorem; note that the length of the complex is exactly one less than the dimension of A. We wish to show that this is impossible. Choose an element x in A such that A/xA has dimension $d-1$; since A is a domain, x is not a zero-divisor in A. We thus have a short exact sequence of complexes

$$0 \to F_* \xrightarrow{x} F_* \to F_*/xF_* \to 0.$$

Recall that if a bounded complex G_* has homology of finite length, we can define its Euler characteristic by the formula

$$\chi(G_*) = \sum (-1)^i (\text{length}(H_i(G_*))).$$

The long exact sequence associated to the above short exact sequence of complexes and the additivity of Euler characteristics give us:

$$\chi(F_*/xF_*) = \chi(F_*) - \chi(F_*) = 0.$$

Notice that F_*/xF_* is a non-trivial (i.e. non-exact) complex of free A/xA-modules with homology of finite length and with length exactly equal to the dimension of A/xA. If we could show that these conditions implied that the Euler characteristic had to be positive, we would be done. However, this is false in general, although the only known counterexamples are quite deep; they will be discussed later in the paper. The remedy for this can be found in the theory of local Chern characters as introduced by Baum, Fulton, and MacPherson [**1**] and further developed by Fulton [**9**]. If we replace the Euler characteristic by the Chern character of the complex applied to the fundamental class of Spec(A) (this will be explained in Section 3), we can duplicate the first part of the proof and show that we get zero; representing the Chern character as a limit of Euler characteristics under powers of the Frobenius morphism when A/xA has positive characteristic, we can show that, in the above situation, the value of the Chern character is positive. Thus, if A is a domain of mixed characteristic, we can apply this with $x =$ the characteristic of the residue field and prove the result.

2. Local Chern characters — Background.

The first appearance of a form of the theory of Chern characters to prove a result on intersection multiplicities for modules of finite projective dimension was in a result of Peskine and Szpiro [**17**], where they proved some conjectures about multiplicities in the case of a graded module over a graded ring. They used the fact that for a graded module M of finite projective dimension over a graded ring A, where the component of degree zero of A is Artinian, M has a resolution by graded free modules; that is by direct sums of copies of A with grading shifted. Thus there is a resolution

$$0 \to \oplus A[n_{kj}] \to \cdots \to \oplus A[n_{1j}] \to \oplus A[n_{0j}] \to M \to 0,$$

where $A[n]$ denotes the module A with grading shifted by n ($A[n]_k = A_{n+k}$). We recall that the Hilbert polynomial of a graded module M is the polynomial $P_M(x)$ such that for large n, $P_M(n)$ is the length of the component of M of degree n as an A_0-module; the degree of $P_M(x)$ is equal to the Krull dimension of M. Since the above resolution of M is exact in each degree, it gives an expression for the Hilbert polynomial of M as an alternating sum of Hilbert polynomials of the $A[n_{ij}]$. Furthermore, we have $P_{A[n_{ij}]}(x) = P_A(x + n_{ij})$ and, expanding this by Taylor's formula and putting everything together one gets:

$$P_M(x) = \sum \rho_k P_A^{(k)}(x),$$

where, for each k, $P_A^{(k)}(x)$ denotes the k-th derivative of $P_A(x)$, and ρ_k is defined to be $(1/k!)\sum_{i,j}(-1)^i(n_{ij})^k$. There are two things to notice about this formula. First, since taking a derivative reduces the degree of a polynomial by one, knowing the ρ_k's makes it possible to see immediately the degree of $P_M(x)$, and hence the dimension of M; namely, one has $\dim(A) - \dim(M)$ = the lowest integer k for which ρ_k is not zero. Secondly, these numbers remain the same when the resolution is tensored with another graded module, thus giving relations between the dimension of M and Euler characteristics of the sort $\sum(-1)^i \operatorname{length}(\operatorname{Tor}_i(M, N))$; these were used by Peskine and Szpiro in [**17**] to prove a number of conjectures in the graded case. They also recognized that this was a kind of "Riemann-Roch Theorem" and asked whether it was possible to define similar invariants if the ring was not graded and thus generalize these results.

An earlier version of this, in a different special case, can be found in a paper of MacRae [**14**]. He defined a principal ideal associated to a module of finite projective dimension and showed that if the codimension of M (i.e. $\dim(A) - \dim(M)$) was at least two, this had to be the trivial ideal equal to A itself. Foxby [**8**] has used this to prove some of the conjectures referred to above in low dimension.

We should also mention the simplest of these invariants — the alternating sum of the ranks of the free modules in a resolution of M. This has the analogous property that if the codimension of M is at least one this number must be zero.

To summarize, useful invariants had been defined in codimensions zero and one for all Noetherian rings and in arbitrary codimension for graded rings, and it was asked whether they could be generalized to arbitrary codimension in the general case. These invariants should have good functorial properties and have the property that the first one which does not vanish determines the codimension of the module. In the next section, we describe an answer to this question which comes as close as possible to fulfilling all these requirements.

3. Local Chern characters — Outline of the Theory.

The invariants we describe in this section were introduced by Baum, Fulton, and MacPherson [**1**], and a good description of them, together with further results, can be found in Fulton [**9**, Chapter 18]. We present here an outline of what kind of invariants they are and what properties they satisfy; we neither give a complete definition nor define them in the greatest possible generality, and we refer to Fulton's book for this. These invariants are somewhat more complicated than those of the last section, in that they are defined as operators on the Chow group of the ring, so we now define the Chow group.

Let A be a local ring; for technical reasons we assume that A is a homomorphic image of a regular local ring. Let $X = \mathrm{Spec}(A)$. For each integer i, let $Z_i(X)$ denote the free Abelian group on the set of reduced and irreducible closed subschemes of X of dimension i. An alternative description of the generators is the set of prime ideals P of A such that $\dim(A/P) = i$; this generator will be denoted $[A/P]$. Let Q be a prime ideal with $\dim(A/Q) = i + 1$, and let x be a non-zero element of A/Q.

We define the divisor of x, denoted $\operatorname{div}(x)$, to be the following element of $Z_i(X)$:

$$\operatorname{div}(x) = \sum(\operatorname{length}((A/(Q + xA))_P)([A/P]),$$

where the sum is over all P with $\dim(A/P) = i$; note that this sum is finite. We now define the i-th component of the Chow group, denoted $A_i(X)$, to be $Z_i(X)$ modulo the subgroup generated by all elements of the form $\operatorname{div}(x)$ for x and Q as above. This equivalence relation is known as "rational equivalence". The formulas used in the definition require denominators (this is the $k!$ of the ρ_k of Peskine and Szpiro defined in the last section), so we shall use $A_i(X)$ to denote the group defined above tensored with the rational numbers. We let $A_*(X)$ denote the direct sum of $A_i(X)$ for all i between 0 and the dimension of A.

This definition can be carried out, with some modifications, for any scheme of finite type over a regular scheme (see Fulton [**9**, Chapter 20]).

Now let F_* be a bounded complex of free modules. We are most interested in the case where the homology of F_* has finite length, but it is convenient to give the definition in a little more generality. Let Z be the support of F_*; this is the closed subset of $\operatorname{Spec}(A)$ consisting of those prime ideals P for which the localization of F_* at P is not exact. We denote the local Chern character of F_* by $\operatorname{ch}(F_*)$. This is a sum of components:

$$\operatorname{ch}(F_*) = \operatorname{ch}_0(F_*) + \operatorname{ch}_1(F_*) + \cdots + \operatorname{ch}_d(F_*),$$

where d is the dimension of A. For each integer i, $\operatorname{ch}_i(X)$ defines, for each integer k, a homomorphism of $\mathbb{Q}$-modules from $A_k(X)$ to $A_{k-i}(Z)$, where Z is the support of F_*. These operators have the following properties:

(1) They are compatible with the basic operations on $A_*(X)$; in particular, with proper push-forward and intersection with Cartier divisors (in ring-theoretic terms, this last operation means dividing by a non-zero-divisor in A).
(2) They are additive: if $0 \to F_* \to G_* \to H_* \to 0$ is a short exact sequence of complexes, then $\operatorname{ch}(G_*) = \operatorname{ch}(F_*) + \operatorname{ch}(H_*)$.
(3) They are multiplicative: If F_* and G_* are bounded complexes of free modules as above, then $\operatorname{ch}(F_* \otimes G_*) = \operatorname{ch}(F_*)\operatorname{ch}(G_*)$, where multiplication in the right hand side is composition.
(4) They commute with one another: for all i and j, we have

$$\operatorname{ch}_i(F_*)\operatorname{ch}_j(G_*) = \operatorname{ch}_j(G_*)\operatorname{ch}_i(F_*).$$

(5) If ${F_*}^\vee = \mathrm{Hom}(F_*, A)$ is the dual of F_*, then $\mathrm{ch}_i({F_*}^\vee) = (-1)^i \mathrm{ch}_i(F_*)$.

(6) (The local Riemann-Roch formula). Suppose that F_* has homology of finite length (for a general version of this formula, see Fulton [**9**, Example 18.3.12]). Then $Z = x$, the closed point of $\mathrm{Spec}(A)$, and $A_*(x) = A_0(x) \cong \mathbf{Q}$. This formula states that for every finitely generated module M, there is an element $\tau(M)$ in $A_*(\mathrm{Supp}(M))$ such that

$$\chi(F_* \otimes M) = \mathrm{ch}(F_*)(\tau(M)).$$

It is this last property which provides the connection between these invariants and Euler characteristics. We say a little more about $\tau(M)$. Note that, as an element of $A_*(\mathrm{Supp}(M))$, it is a formal sum of prime ideals in the support of M with rational coefficients. There is a possible candidate for such an element; one could take all prime ideals minimal in the support of M, and, for each such prime ideal P, let the coefficient of $[A/P]$ be $\mathrm{length}(M_P)$. Denote this element of $A_*(\mathrm{Supp}(M))$ by $[M]$. It turns out that this is not quite equal to $\tau(M)$, but that their components in $A_i(\mathrm{Supp}(M))$, for $i =$ the dimension of M, are the same. The actual definition of $\tau(M)$ is as follows: let R be a regular local ring of which A is a homomorphic image, and let H_* be a free resolution of M over R. Then $\tau(M) = \mathrm{ch}(H_*)([R])$, where $[R]$ is defined as above.

Now let d be the dimension of A, and let $[A]_d$ denote the component of $[A]$ in $A_d(X)$. If F_* is a complex of free modules with homology of finite length, the local Riemann-Roch formula gives the following expression for the Euler characteristic:

$$\chi(F_*) = \sum \mathrm{ch}_i(F_*)(\tau_i(A)).$$

Note that the term of degree d in this expression is $\mathrm{ch}_d(F_*)([A]_d)$. It turns out, and this is one of the main points of this paper, that this number behaves much better than the Euler characteristic. For example, using this instead of the Euler characteristic in the false proof of the New Intersection Theorem in mixed characteristic (with a little more work — we return to this in Section 5) turns it into a correct proof. Another example is Serre's vanishing conjecture for intersection multiplicities; if the definition of intersection multiplicities in terms of Euler characteristics is replaced by one in terms of the local Chern character applied to $[A]_d$, one gets a vanishing theorem for general local rings, and in the case when $\tau(A) =$

$[A]_d$, which includes regular local rings and complete intersections, one can deduce from this the original vanishing conjecture for Euler characteristics (see Roberts [**19**]).

As far as the actual construction of local Chern characters is concerned, we mention only that it is carried out by a process known as the "graph construction", which uses the graphs of the boundary maps of F_*; these are free submodules of $F_i \oplus F_{i-1}$ with free quotients and define sections of Grassmannians of appropriate rank. The Chern classes of the tautological bundles on the Grassmannians are used to define the local Chern characters. A similar construction was given by Iversen [**13**], but with values elements of cohomology groups rather than operators on Chow groups, so they are not defined over rings of mixed characteristic.

We return now to the question raised in Section 2 as to whether the invariants defined above satisfy all the conditions it was hoped they would. There is one missing — that if F_* is a resolution of a module M and the codimension of M is at least n, then $\mathrm{ch}_i(F_*) = 0$ for $i < n$. Szpiro [**23**] discusses this and various consequences it would have; in particular, it would follow from the Riemann-Roch formula that if F_* were, as usual, a bounded complex of free modules with homology of finite length, and if N were a module of dimension less than the dimension of A, we would have $\chi(F_* \otimes N) = 0$. The hope that this would hold was shattered when Dutta, Hochster, and McLaughlin [**5**] produced an example of a ring of dimension 3, a module of finite length and finite projective dimension with resolution F_*, and a module N of dimension 2 with $\chi(F_* \otimes N) = -1$. We discuss this example and various other examples which can be derived from it in the next section.

4. Some examples of Euler characteristics of complexes.

We begin this section by describing in more detail the example of negative Euler characteristic mentioned above. Let A be the ring $K[[X, Y, Z, W]]/(XY - ZW)$, where K is a field; for later purposes we assume that K is infinite of characteristic not equal to 2. Then A is a local domain of dimension 3. Let P be the prime ideal generated by X and W and let $N = A/P$; the dimension of N is 2. Dutta, Hochster, and McLaughlin [**5**] construct a module M of length 15 and of finite projective dimension such that, if F_* is a resolution of M, then $\chi(F_* \otimes N) = -1$. We shall not go into the details of this construction, which is quite complicated, but describe instead the implications of this for local Chern characters and some other examples based on this one.

First, expressing the above Euler characteristic in terms of Chern characters by means of the local Riemann-Roch formula gives:

$$-1 = \chi(F_* \otimes N) = \mathrm{ch}_0(F_*)(\tau_0(N)) + \mathrm{ch}_1(F_*)(\tau_1(N)) + \mathrm{ch}_2(F_*)(\tau_2(N)).$$

The first two terms of this sum are zero. The first is multiplication by the alternating sum of the ranks of the F_i, which is zero. The second involves the operator $\mathrm{ch}_1(F_*)$, which can be identified with the invariant defined by MacRae discussed in section one, and this is known to vanish since the codimension of M is greater than one (see Roberts [**20**]). Thus $\mathrm{ch}_2(F_*)$ is not zero, even though the codimension of the support of M is 3.

From this example it is not difficult to derive others, contradicting other properties which were once conjectured to be true. First, we note that since $N = A/P$ and $\chi(N \otimes F_*) = -1$, if we let $N_k = A/P^k$, since the localization at P is a discrete valuation ring, we have $\chi(N_k \otimes F_*) = -k$. Second, the short exact sequence $0 \to P^k \to A \to N_k \to 0$ shows that we have $\chi(P^k \otimes F_*) = \chi(F_*) - \chi(N_k \otimes F_*) = 15 + k$. Finally, if we replace P by $Q = (X, Z)$ throughout, we obtain similar formulas but with a change of sign; for example, we have $\chi(Q^k \otimes F_*) = 15 - k$. This last statement follows from the fact that there is a short exact sequence

$$0 \to A/P \to A/XA \to A/Q \to 0;$$

it is easy to show that $\chi((A/XA) \otimes F_*) = 0$.

It is now a simple matter to construct an example of a ring of dimension 3 and a bounded complex of free modules of length 3 with homology of finite length and with negative Euler characteristic. Let $B = A \oplus Q^k$, where k is greater than 15, and let $G_* = F_* \otimes_A B$. It is clear that G_* has negative Euler characteristic, and all that has to be done is to make B into a ring. One way to do this is to let everything in Q^k have square zero, but with only a little more work it can be made into an integrally closed domain, so we describe how to do this. We let $a_1, a_2, \ldots, a_{2k}$ be distinct elements of the field K, and then we let $\eta = \Pi_i(X - a_i W)$. Let $C = A \oplus A[\sqrt{\eta}]$; since η is not a square in A, C is an integral domain. Since the characteristic of K is not 2 and A is an integrally closed domain, it is not difficult to find the integral closure of C in its quotient field. For each i, the principal ideal generated by $X - a_iW$ is the intersection of the height one prime ideals (X, W) and $(X - a_iW, Z - a_iY)$. These latter ideals are distinct, so the order of η at the valuation corresponding to (X, W) is 2k and the order of η at the other height one primes containing η is one. An element $r + s\sqrt{\eta}$, with r and s in the quotient field of A, is integral over $A \oplus A[\sqrt{\eta}]$ if and only if r is in A and the order of s is greater than or equal to zero at all height one primes except (X, W) and is greater than or equal to $-k$ at (X, W). Any such s can be written as t/X^k with t in $(X, Z)^k$, and conversely every such element t/X^k satisfies this condition, so the integral closure is isomorphic to B as an A-module as desired.

Now let B and G_* be as in the above example but with $k = 1$. The exact sequence $0 \to Q \to A \to A/Q \to 0$ together with the fact that A/Q has depth 2 implies that Q has depth 3 and B is a Cohen-Macaulay ring. On the other hand, we have

$$\operatorname{ch}_2(G_*)(\tau_2(B)) = \operatorname{ch}_2(F_*)(\tau_2(Q)) = -1.$$

Thus this is an example of a Cohen-Macaulay ring of dimension 3 with $\tau(B)$ not equal to $[B]_3$, and, what is more significant, with a bounded complex of free modules G_* with homology of finite length with $\operatorname{ch}_2(G_*)(\tau_2(B)) \neq 0$. The fact that B is Cohen-Macaulay implies that G_* is a free resolution of a module of finite length (with length $= 29$); denote this module N'. It had been asked (Szpiro [**23**]) whether in this case, when N' is a module of finite length and finite projective dimension over a Cohen-Macaulay ring of dimension d, the length of $\operatorname{Ext}_B^d(N', B)$ was equal to the length of N'. In this example, however, using the duality property of local Chern characters, denoting the dual of G_* by $G_*{}^\vee$, we have

$$\begin{aligned}\operatorname{length}(\operatorname{Ext}_B^3(N', B)) &= -\chi(G_*{}^\vee) = -\operatorname{ch}_2(G_*{}^\vee)(\tau_2(B)) - \operatorname{ch}_3(G_*{}^\vee)(\tau_3(B)) \\ &= -\operatorname{ch}_2(G_*)(\tau_2(B)) + \operatorname{ch}_3(G_*)(\tau_3(B)) = 1 + 30 = 31.\end{aligned}$$

Thus the length of $\operatorname{Ext}_B^3(N', B)$ is not equal to the length of N'.

As a final example, we show how to prove a result of Dutta [**3**] using this theory. He showed that if A is a Gorenstein ring of dimension at most 5, and if M and N are modules of finite projective dimension such that $M \otimes N$ is a module of finite length and such that $\dim(M) + \dim(N) < \dim(A)$, then

$$\sum(-1)^i \left(\operatorname{length}(\operatorname{Tor}_i(M, N))\right) = 0.$$

We assume that the dimension is equal to 5. To show that this alternating sum of lengths of Tors is zero, we again use the duality property of local Chern characters. Since A is a Gorenstein ring, if we represent A as a homomorphic image of a regular local ring R and take a free resolution H_* of A as an R-module, this resolution will be self dual. More precisely, denoting the dual of H_* by $H_*{}^\vee$, we have that H_* is isomorphic to $H_*{}^\vee$ shifted by $n-d$, where n is the dimension of R and d is the dimension of A (which is 5 in our case). Thus, by the duality property, we have

$$\begin{aligned}\tau_i(A) &= \mathrm{ch}_{n-i}(H_*)([R]) = (-1)^{n-i}\ \mathrm{ch}_{n-i}(H_*{}^\vee)([R]) \\ &= (-1)^{n-i}\,(-1)^{n-d}(\mathrm{ch}_{n-i}(H_*)([R]) = (-1)^{d-i}\tau_i(A).\end{aligned}$$

Thus $\tau_i(A) = 0$ if $d - i$ is odd, and the only possible values of $\tau_i(A)$ which could possibly not be zero are for $i = 3$ and $i = 5$ (note that $A_0(\mathrm{Spec}(A)) = 0$ for any local ring of dimension > 0). Now let F_* be a free resolution of M and G_* a free resolution of N. The alternating sum of lengths of Tors above is

$$\chi(F_* \otimes G_*) = \mathrm{ch}(F_*)\ \mathrm{ch}(G_*)(\tau(A)) = \sum \mathrm{ch}_i(F_*)\ \mathrm{ch}_j(G_*)(\tau_k(B)),$$

the sum being taken over all non-negative integers i, j, and k with $i+j = k$ and, from the above discussion, with $k = 3$ or $k = 5$. Assume first that $k = 5$. If $j < 5-\dim(N)$, then $\mathrm{ch}_j(G_*)(\tau_5(B))$ is an element of $A_{5-j}(\mathrm{Supp}(N))$, which is zero, since $\mathrm{Supp}(N)$ has dimension less than $5 - j$ so can have no closed subschemes of dimension $5 - j$. The hypothesis that $\dim(M) + \dim(N) < 5$ implies that either $j < 5-\dim(N)$ or $i < 5-\dim(M)$ whenever $i + j = 5$, so, using the commutativity of the local Chern characters, we see that all of these terms are zero. Now assume that $k = 3$. If $i + j = 3$, one of i or j must be zero or one, and these are exactly the cases where the local Chern character vanishes if the codimension of the support is greater than zero or one respectively (using the identification of ch_0 with the alternating sum of the ranks of the free modules of a resolution and ch_1 with the MacRae invariant as discussed earlier). Using this, it can be verified that all terms in the sum are zero; we leave the details to the reader.

5. The New Intersection Theorem.

We now return to the proof of the New Intersection Theorem as discussed in Section 1. Recall that we must show that a bounded complex of free modules over a local ring A with homology of finite length, if it is not exact, must have length at least equal to the dimension of A. Since this has been proven for rings containing a field, we can assume that A is a ring of mixed characteristic. By completing and dividing by an appropriate prime ideal, we can assume that, in addition, A is complete (and hence a homomorphic image of a regular local ring) and an integral domain. By taking a suitable flat extension we can also assume that the residue field is perfect. Let $F_* =$

$$0 \to F_k \to F_{k-1} \to \cdots \to F_1 \to F_0 \to 0$$

be a complex as above. We may assume that $H_0(F_*)$ is not zero. Furthermore, if p is the characteristic of the residue field of A, reducing modulo p produces the same type of complex over A/pA, and the dimension is reduced by one. Hence, by the theorem in positive characteristic (Peskine-Szpiro [**17**], Roberts [**18**]), we may assume that $k = d - 1$, where d is the dimension of A.

As outlined in Section 1, the idea of the proof is to reduce modulo p, and to show that the local Chern character of the reduced complex applied to the fundamental class $[A/pA]$ in $A_{d-1}(\mathrm{Spec}(A/pA))$ is both zero (since the complex is the reduction modulo a non-zero divisor of a complex on A), and positive (since the complex has length equal to the dimension of (A/pA)). Let G_* denote the complex F_* reduced modulo p; that is $G_* = F_* \otimes (A/pA)$.

The first of these assertions follows immediately from the compatibility of local Chern characters with intersection with divisors. In fact, we have a commutative diagram:

$$\begin{array}{ccc} A_d(\mathrm{Spec}(A)) & \xrightarrow{\mathrm{ch}_{d-1}(F_*)} & A_1(x) \\ \downarrow & & \downarrow \\ A_{d-1}(\mathrm{Spec}(A/pA)) & \xrightarrow{\mathrm{ch}_{d-1}(G_*)} & A_0(x), \end{array}$$

where the vertical arrows are defined by intersecting with the Cartier divisor defined by the non-zero-divisor p and x denotes the closed point of $\mathrm{Spec}(A)$. Starting in the upper left with the class $[A]$, intersecting with the divisor defined by p gives $[A/pA]$, so we get $\mathrm{ch}_{d-1}(G_*)([A/pA])$ as the result in the lower right. On the other hand, $A_1(x) = 0$ since x, being of dimension zero, can have no subschemes of dimension one, so going around the other way in the diagram gives zero. Thus $\mathrm{ch}_{d-1}(G_*)([A/pA]) = 0$. (This is the analogue of the proof in Section 1 that the Euler characteristic of G_* is zero, which is also true.)

We now prove the other half of this theorem, that $\mathrm{ch}_{d-1}(G_*)([A/pA])$ is positive. To accomplish this we use the asymptotic Euler characteristic, denoted χ_∞, which was introduced by Dutta [2] for rings of positive characteristic. Szpiro [**23**] pointed out that this invariant is the same as the one we are discussing in terms of local Chern characters and that it thus satisfied a number of properties which were conjectured for the ordinary Euler characteristic, but which, as shown in the last section, are often not true. In what follows, if A is a ring of characteristic $p > 0$, we let $f : A \to A$ denote the Frobenius morphism defined by $f(x) = x^p$, and if F_* is a complex of free modules, we let ${F_*}^n$ denote F_* tensored with this map n times; the result of this is to replace the matrices defining the maps of F_* with matrices whose entries are (p^n)-th powers of the original ones.

DEFINITION: Let A be a local ring of characteristic $p > 0$ and dimension d, and let F_* be a bounded complex of free modules with homology of finite length. Then $\chi_\infty(F_*) = \lim_{n\to\infty}(\chi({F_*}^n))/p^{nd}$.

The fact that this limit exists has been shown by Seibert [**22**]. That it can be defined in terms of local Chern characters follows from the local Riemann-Roch formula and the compatibility of local Chern characters with finite maps. This compatibility says the following in our case: let g denote the n-th power of the Frobenius morphism, considered as a map from Spec(A) to Spec(A). Then, if F_* is a complex of free modules over A as usual, and if α is any element of the Chow group $A_*(\mathrm{Spec}(A))$, what we get by applying the local Chern character of the pullback of F_* along g and then pushing down by g is the same as what we get by first pushing down by g and then applying the local Chern character of F_*. We denote α pushed down by g by $g_*(\alpha)$. In this case all of these processes can be described explicitly.

The pullback of F_* along g is the complex ${F_*}^n$ defined above. If α is an element of $A_i(\mathrm{Spec}(A))$, then pushing down by g gives $p^{ni}\alpha$. The reason for this is that if A has perfect residue field (which we are assuming here), and P is a prime ideal such that the dimension of A/P is i, so $[A/P]$ is a generator of $A_i(\mathrm{Spec}(A))$, and we look at $g : A/P \to A/P$, then the rank of A/P over $g(A/P)$ is p^{ni}. Putting all this together and applying the local

Riemann-Roch formula, we obtain:

$$\chi(F_*{}^n) = \mathrm{ch}(F_*{}^n)(\tau(A)) = \sum_i \mathrm{ch}_i(F_*{}^n)(\tau_i(A))$$

$$= g_*\left(\sum_i \mathrm{ch}_i(F_*{}^n)(\tau_i(A))\right) \text{ since this element is in } A_0(\mathrm{Spec}(A))$$

$$= \sum_i \mathrm{ch}_i(F_*)(g_*(\tau_i(A)) \text{ from the compatibility with the Frobenius outlined above}$$

$$= \sum_i p^{ni}\,\mathrm{ch}_i(F_*)(\tau_i(A)) \text{ since } \tau_i(A) \text{ is in } A_i(\mathrm{Spec}(A)).$$

If we now divide by p^{nd} and take the limit as $n \to \infty$, all the terms in the last sum go to zero except the one where $i = d$, so we have

$$\chi_\infty(F_*) = \mathrm{ch}_d(F_*)(\tau_d(A)) = \mathrm{ch}_d(F_*)([A]).$$

It now remains to show that, in the case we are considering here, the element $\mathrm{ch}_{d-1}(G_*)([A/pA])$ of $A_0(x)$ is positive. This will follow from the following theorem:

THEOREM. *Let A be a local ring of characteristic $p > 0$ and dimension d. Let $F_* = 0 \to F_d \to F_{d-1} \to \cdots \to F_1 \to F_0 \to 0$ be a complex of finitely generated free modules of length d with homology of finite length and with $H_0(F_*)$ not equal to 0. Then $\chi_\infty(F_*) > 0$.*

PROOF: The proof uses a dualizing complex for A. Let D^* be a dualizing complex, so that for each i between 0 and d, we have

$$D^i = \oplus E(A/P),$$

where the sum is taken over all prime ideals P with dimension $(A/P) = d - i$, and $E(A/P)$ denotes the injective hull of A/P. The other property of D^* we need is that the homology $H_i(D^*)$ is finitely generated. It then follows from the description of D^i that its dimension is at most $d - i$. We now consider, for each n, the double complex $\mathrm{Hom}(F_*{}^n, D^*)$ and the two associated spectral sequences whose abutments are its total complex. The double complex looks like this:

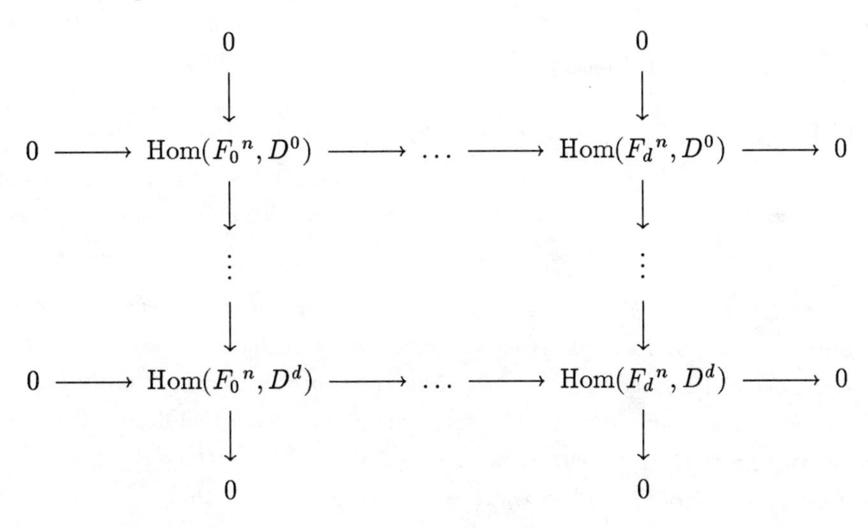

We look first at the rows of this double complex. Since, for $i < d$, D_i is a direct sum of modules of the type E(A/P), which are modules over the localizations A_P for P different from the maximal ideal of A, and since F_* has homology of finite length, every row except the last one is exact. In the last row, since $D_d = E(k)$, we have $H_i(\mathrm{Hom}(F_*{}^n, D^d)) \cong \mathrm{Hom}(H_i(F_*{}^n), E(k))$, which is a module of the same length as $H_i(F_*{}^n)$.

We now look at the columns of the double complex. If we take the homology of the columns and look at the complex induced on the j-th row, we get

$$0 \to \mathrm{Hom}(F_0{}^n, H_j(D^*)) \to \cdots \to \mathrm{Hom}((F_d{}^n, H_j(D^*)) \to 0.$$

We claim that if $j > 0$, since the dimension of $H_j(D^*)$ is at most $d - j \leq d - 1$, as n goes to infinity, the length of the homology of this complex is bounded by a constant times $p^{n(d-1)}$. Assuming this, we now consider the abutment of this spectral sequence. We have seen from the other spectral sequence that the abutment is $\mathrm{Hom}(H_i(F_*{}^n), D^d) \cong \mathrm{Hom}(H_i(F_*{}^n), E(k))$ for i between 0 and d and is zero everywhere else. The terms which contribute to $\mathrm{Hom}(H_i(F_*{}^n), E(k))$ for a given $i > 0$ come from the terms $\mathrm{Hom}(F_k{}^n, D^j)$ with $k + j = i + d$. Since the length of F_* is exactly d (this is where this hypothesis is used), any of these terms which are not zero must have $k \leq d$ and hence $j \geq i > 0$. Since there are only a finite number of terms contributing to $\text{Hom}(H_i(F_*{}^n, E(k))$ for each $i > 0$, and this module has the same length as $H_i(F_*{}^n)$, we can conclude that there is a constant C such that $\text{length}(H_i(F_*{}^n)) \cong Cp^{n(d-1)}$ for all n, and hence

$$\lim_{n\to\infty} \text{length}(H_i(F_*{}^n))/p^{nd} = 0.$$

We prove the claim in a separate Lemma (proofs of this fact, or something essentially the same, can be found also in Dutta [**2**] and Seibert [**22**]).

LEMMA. *Let $G_* = F \to G \to H$ be a complex of finitely generated free A modules, where A has characteristic $p > 0$. Let N be a finitely generated A-module of dimension r. Assume that every proper localization of G_* is split exact at G; this means that there are maps, both denoted s, from G to F and from H to G, such that $ds + sd = 1_G$, where the d's denote the boundary maps of G_*. Then there is a constant C such that*

$$\text{length}(H(\text{Hom}(G_*{}^n, N))) < Cp^{nr}$$

for all $n > 0$, where H denotes the homology of the complex.

PROOF OF THE LEMMA: We prove this by induction on r, the dimension of N. By taking a filtration of N whose quotients are of the form A/P for various prime ideals P, we can assume that N is itself of this form. If $r = 0$, then $N = k$ and the length of $H(\text{Hom}(G_*{}^n, k))$ is bounded by the rank of G, so the result is true in this case. If $r > 0$, let x be an element of the maximal ideal of A which is not in P. When G_* is localized at x (i.e. x is inverted), it becomes split exact at G, and hence there is a power of x, say x^m, such that we can find maps s from H to G and from G to F such that $sd + ds =$ multiplication by x^m on G, where, as above, d denotes the boundary maps of G_*. Replacing G_* by $G_*{}^n$, we have the same situation with m replaced by mp^n. This implies that $y = x^{mp^n}$ annihilates $H(\text{Hom}(G_*{}^n, M))$ for any module M. Since y is a non-zero-divisor on N, we have a short exact sequence

$$0 \to \text{Hom}(G_*{}^n, N) \xrightarrow{y} \text{Hom}(G_*{}^n, N) \to \text{Hom}(G_*{}^n, (N/yN)) \to 0,$$

which induces an exact sequence:

$$H(\text{Hom}(G_*{}^n, N)) \xrightarrow{y} H(\text{Hom}(G_*{}^n, N)) \to H(\text{Hom}(G_*{}^n, (N/yN))).$$

Since multiplication by y is zero, this implies that the length of $H(\mathrm{Hom}(G_*{}^n, N))$ is less than or equal to the length of $H(\mathrm{Hom}(G_*{}^n, (N/yN)))$. There is a filtration of N/yN whose quotients are p^n copies of N/x^mN, so that

$$\mathrm{length}(H(\mathrm{Hom}(G_*{}^n, N))) \leq p^n(\mathrm{length}(H(\mathrm{Hom}(G_*{}^n, (N/x^mN))))$$

We now apply the induction hypothesis to N/x^mN to obtain a constant C such that $\mathrm{length}(H(\mathrm{Hom}(G_*{}^n, (N/x^mN)))) \leq Cp^{n(r-1)}$ for all n. We then have

$$\mathrm{length}(H(\mathrm{Hom}(G_*{}^n, N))) \leq p^n(Cp^{n(r-1)}) = Cp^{nr}$$

for all n, as was to be shown.

To conclude the proof of the theorem, we must show that the limit of $\mathrm{length}(H_0(F_*{}^n))/p^{nd}$ is positive. By assumption, $H_0(F_*)$ is non-zero, so we may assume that the map from F_1 to F_0 is defined by a matrix with coefficients in the maximal ideal m of A. Then the map from $F_1{}^n$ to $F_0{}^n$ will be defined by a matrix with coefficients in m^{p^n}, so the length of $H_0(F_*{}^n)$ is at least equal to the length of A/m^{p^n}. For large n, this number is given by the value of the Hilbert-Samuel polynomial of m at p^n; since A has dimension d, this is a polynomial of degree d with positive leading coefficient. Thus this contribution to $\chi_\infty(F_*)$ is positive, so, since we have shown that the other contributions are zero, $\chi_\infty(F_*)$ is positive, and this concludes the proof.

6. The Improved New Intersection Conjecture.

In the first section of this paper, we presented the original version of the Intersection Theorem of Peskine and Szpiro and some later theorems which developed from it. This development, however, did not stop at the point we left it. In proving a theorem on the ranks of syzygies of a module known as the "Syzygy Theorem", Evans and Griffith [**6**,**7**] noticed that a newer version of the New Intersection Theorem would imply the theorem they were interested in. The theorem states the following:

THEOREM (Improved New Intersection Theorem). *Let* $F_* =$

$$0 \to F_k \to \cdots \to F_1 \to F_0 \to 0$$

be a complex of finitely generated modules over a local ring A. If $H_i(F_*)$ has finite length for $i > 0$ and $H_0(F_*)$ has a minimal generator annihilated by a power of the maximal ideal of A, then $k \geq \text{dimension}(A)$.

This is actually a theorem, and not a conjecture, for rings containing a field; in fact, as was pointed out by Evans and Griffith, it follows from the existence of big Cohen-Macaulay modules, which Hochster [**11**] has proven to exist for these rings. It is still a conjecture in mixed characteristic.

The name "improved" was given to this version of the theorem by M. Hochster, and the reason that it is an improvement over the earlier ("new") version is that it is a stronger result and implies more other conjectures. The first example of this was the Syzygy Theorem mentioned above, but recently Dutta [4] has shown that it implies the "Canonical Element" conjecture and thus several others of Hochster's homological conjectures. We state two of these conjectures here, both of which have been shown by Hochster [**12**] to be equivalent and to imply the Improved New Intersection Conjecture; hence all three of these conjectures are equivalent (in mixed characteristic; if A contains a field, they are equivalent since they are all true).

CANONICAL ELEMENT CONJECTURE. *Let $x_1, \ldots, x_d$ be a system of parameters for A, let K_* be the Koszul complex on $x_1, \ldots, x_d$, and let F_* be a free resolution of A/m. Let $f_* : K_* \to F_*$ be induced by the map $A/(x_1, \ldots, x_d) \to A/m$. Then f_d is not zero.*

DIRECT SUMMAND CONJECTURE. *If R is a regular local ring and S is a module-finite ring extension of R, then R is a direct summand of S as an R-module.*

In the course of Dutta's proof of the implication mentioned above, he shows also that an apparently weaker version of the theorem, in which one assumes in addition that any proper localization of $H_0(F_*)$ is free, is actually equivalent to the version stated here. This is useful since the extra hypothesis implies that the condition that the homology in degrees greater than zero has finite length remains true after tensoring with any finitely generated module.

We conclude with a brief discussion (without proofs) of what happens if one tries to imitate the proof of the "unimproved" intersection theorem in this case. The obvious problem is that one does not have an Euler characteristic defined here, since the zeroth homology does not have finite length, or, from the point of view of local Chern characters, the support of the complex could be all of Spec(A), so the Chern character does not give any information. This can be remedied by taking a subcomplex as follows: Let F_0' denote a free direct summand of F_0 of rank one generated by an element whose image in $H_0(F_*)$ is the minimal generator annihilated by a power of the maximal ideal. Let $F_0'' = F_0/F_0'$, and let M be the submodule of F_1 consisting of all elements whose image in F_0'' is zero. It is then easy to verify that we have a complex

$$0 \to F_k \to \cdots \to F_2 \to M \to F_0' \to 0$$

with homology everywhere of finite length. But now M is not free. One can blow up an appropriate determinantal ideal to make it locally free, and apply the theory of local Chern characters there then push back down. Most of the proof goes through as before (using the hypothesis on the generator of F_0'), but a new term enters into the formula. Let Y denote the scheme obtained by blowing up Spec(A) at this determinantal ideal and reducing modulo p, and let f denote the map from Y to Spec(A/pA). Let $\mathcal{M}$ denote the (locally free) pullback of M to Y. Let $\mathcal{M}_n$ denote the pullback of $\mathcal{M}$ by the n-th power of the Frobenius map on Y. Let $\chi_1(\mathcal{M})$ denote

$$\text{length}(R_1 f_*(\mathcal{M})) - \text{length}(R_2 f_*(\mathcal{M})) + \text{length}(R_3 f_*(M)) - \ldots$$

Now let $\chi_1{}^\infty(\mathcal{M})$ denote the limit of $\chi_1(\mathcal{M}_n)/p^{nd}$ as $n \to \infty$, where d is the dimension of A/pA. If it could be shown that $\chi_1{}^\infty(\mathcal{M}) \geq 0$, the proof could be carried through. However, it is not known whether $\mathcal{M}$ has this property. It can be shown that in the analogous situation in which one divides by a non-zero-divisor in a ring which already had positive characteristic, every locally free sheaf which is a reduction of a locally free sheaf from the larger ring has $\chi_1{}^\infty \geq 0$. The same is true for any direct sum of invertible sheaves in our situation, when one reduces from a ring of mixed characteristic to one of positive characteristic. However the sheaf $\mathcal{M}$ we defined above fits into neither of these categories, so it is not known whether it satisfies this inequality or not, and the improvement to the New Intersection Theorem remains a conjecture in mixed characteristic.

References

1. P. Baum, W. Fulton and R. MacPherson, *Riemann-Roch for singular varieties*, Publ. Math. IHES **45** (1975), 101–145.
2. S. P. Dutta, *Frobenius and Multiplicities*, J. of Algebra **85** (1983), 424–448.
3. S. P. Dutta, *Generalized intersection multiplicities of modules II*, Proc. Amer. Math. Soc. **93** (1985), 203–204.
4. S. P. Dutta, *On the canonical element conjecture*, Trans. Amer. Math. Soc. **299** (1987), 803-811.
5. S. P. Dutta, M. Hochster, and J. E. McLaughlin, *Modules of finite projective dimension with negative intersection multiplicities*, Invent. Math. **79** (1985), 253–291.
6. E. G. Evans and P. Griffith, *The syzygy problem*, Ann. of Math. **114** (1981), 323–333.
7. E. G. Evans and P. Griffith, *The syzygy problem: a new proof and historical perspective*, in "Commutative Algebra," (Durham 1981), London Math Soc. Lecture Note Series, vol. 72, 1982, pp. 2–11.
8. H.-B. Foxby, *The MacRae invariant*, in "Commutative Algebra," (Durham 1981), London Math Soc. Lecture Note Series, vol. 72, 1982, pp. 121–128.
9. W. Fulton, "Intersection Theory," Springer-Verlag, Berlin, 1984.
10. M. Hochster, *The equicharacteristic case of some homological conjectures on local rings*, Bull. Amer. Math. Soc. **80** (1974), 683–686.
11. M. Hochster, "Topics in the homological theory of modules over commutative rings," Regional Conference Series in Mathematics, vol. 24, 1975.
12. M. Hochster, *Canonical elements in local cohomology modules and the direct summand conjecture*, Journal of Algebra **84** (1983), 503–553.
13. B. Iversen, *Local Chern classes*, Ann. Scient. Éc. Norm. Sup. **9** (1976), 155–169.
14. R. E. MacRae, *On an application of the Fitting invariants*, J. of Algebra **2** (1965), 153–169.
15. C. Peskine and L. Szpiro, *Sur la topologie des sous-schémas fermés d'un schéma localement noethérien, définis comme support d'un faisceau cohérent localement de dimension projective finie*, C. R. Acad. Sci. Paris Sér. A **269** (1969), 49–51.
16. C. Peskine and L. Szpiro, *Dimension projective finie et cohomologie locale*, Publ. Math. IHES **42** (1973), 47–119.
17. C. Peskine and L. Szpiro, *Syzygies et Multiplicités*, C. R. Acad. Sci. Paris Sr. A **278** (1974), 1421–1424.
18. P. Roberts, *Two applications of dualizing complexes over local rings*, Ann. Sci. Éc. Norm. Sup. **9** (1976), 103-106.
19. P. Roberts, *The vanishing of intersection multiplicities of perfect complexes*, Bull. Amer. Math. Soc. **13** (1985), 127–130.
20. P. Roberts, *The MacRae invariant and the first local Chern character,*, Trans. Amer. Math. Soc. **300** (1987), 583–591.
21. P. Roberts, *Le théorème d'intersection*, C. R. Acad. Sc. Paris Sér. I no. 7, 304 (1987), 177–180.
22. G. Seibert, *Complexes with homology of finite length and Frobenius functors*, to appear.
23. L. Szpiro, *Sur la théorie des complexes parfaits*, in "Commutative Algebra," (Durham 1981), London Math Soc. Lecture Note Series, vol. 72, 1982, pp. 83–90.

Department of Mathematics, University of Utah, Salt Lake City UT 84112

This research was supported in part by a grant from the National Science Foundation.

One-Fibered Ideals

JUDITH D. SALLY

This is essentially an expository paper about $\underline{m}$-primary ideals I in a local ring $(R, \underline{m})$ which have a single exceptional prime, i.e., whose closed fiber in the normalization $\overline{R(I)}$ of the blow-up $R(I)$ is irreducible. Stated slightly differently, it is about $\underline{m}$-primary ideals which have a single associated Rees valuation. If R is analytically unramified then the existence of such an ideal in R implies that R is analytically irreducible. It is conjectured that, conversely, every analytically irreducible local domain has such an ideal.

We begin by setting up notation and recalling facts about prime divisors and Rees valuations. Henceforth $(R, \underline{m})$ is a d-dimensional normal local domain with quotient field K and I is an $\underline{m}$-primary ideal. Let $R(I)$ denote the Rees ring $R \oplus It \oplus I^2t^2 + \dots$ and $\overline{R(I)}$ its normalization:

$$\overline{R(I)} = R \oplus \overline{I}t \oplus \overline{I^2}t^2 + \dots$$

(We will consistently use $^{-}$ to denote integral closure both of rings and of ideals.) If P is a prime ideal in $\overline{R(I)}$ minimal over $I\overline{R(I)}$, i.e., an exceptional prime of I, then $\overline{R(I)}_P \cap K$ is a discrete valuation ring. Let $T(I)$ denote the set of all discrete valuations obtained in this way. These valuations and their corresponding rings are called the Rees valuations and the Rees valuation rings corresponding to I. Note that since $\overline{R(I)}$ is a Krull domain, $T(I)$ is a finite set. It follows that $T(I) = T(\overline{I}) = T(I^n) = T(J)$ for any J projectively equivalent to I, i.e., any J satisfying $\overline{J^k} = \overline{I^n}$ for some positive integers k, n. Thus, if $R/\underline{m}$ is infinite and $\underline{x} = x_1, \dots, x_d$ is any minimal reduction of I, then $T(I) = T(\underline{x}R)$. For convenience, we will assume that I has such a minimal reduction.

When working with $\overline{R(I)}$ and Rees valuations, it is often advantageous to work with the affine pieces of $\overline{R(I)}$ and it will be useful to note when we can be assured that *all* the Rees valuations will "show up" on one such piece. To see this we will need to distinguish those elements of $T(I)$ which are prime divisors. (Recall that a discrete valuation v birationally dominating R is a prime divisor of R if the transcendence degree of the residue field of v over that of R is as large as possible, i.e., $d - 1$.)

Suppose then that $\underline{x} = x_1, \ldots, x_d$ is a minimal reduction of I. Then $\overline{R[I/x_i]} = \overline{R[\underline{x}/x_i]}$ and Proj $\overline{R(I)}$ is covered by Spec $\overline{R[\underline{x}/x_i]}$ for $i = 1, \ldots, d$. Thus $T(I)$ is the finite collection of discrete valuations whose rings have the form $\overline{R[\underline{x}/x_i]}_P$, where P is minimal over $x_i\overline{R[\underline{x}/x_i]}$. If R is quasi-unmixed or if $T(I) = \{v\}$, then all the valuations in $T(I)$ are prime divisors of R. For $x_1, \ldots, x_d$ is analytically independent so $R[\underline{x}/x_i]/\underline{m}R[\underline{x}/x_i]$ is isomorphic to a polynomial ring in $d-1$ variables over $R/\underline{m}$. Thus every prime P minimal over $x_i\overline{R[\underline{x}/x_i]}$ which contracts to $\underline{m}R[\underline{x}/x_i]$ satisfies tr. d. $\overline{R[\underline{x}/x_i]}/P : R/\underline{m} = d-1$. Since $\overline{R[\underline{x}/x_i]}$ is an integral extension of $R[\underline{x}/x_i]$, there exists at least one such. If R is assumed to be quasi-unmixed, i.e., every prime p in the $\underline{m}$-adic completion $R^\wedge$ satisfies $\dim R^\wedge/p = d$, then the dimension formula holds, so all such P contract to $\underline{m}R[\underline{x}/x_i]$.

We note also that the prime divisors in $T(I)$ have center in *every* $\overline{R[\underline{x}/x_i]}$. For $x_i/x_j \notin \underline{m}R[\underline{x}/x_i]$ so $R[\underline{x}/x_i]_{mR[x/x_i]} = R[\underline{x}/x_j]_{mR[\underline{x}/x_j]}$ and, since localization commutes with integral closure, every prime divisor v in $T(I)$ has center in $\overline{R[\underline{x}/x_i]}$.

DEFINITION: I is said to be one-fibered (abbreviated: 1-fibered) if $T(I) = \{v\}$, i.e., the radical of $I\overline{R(I)}$ is prime.

Thus, the maximal ideal of a regular local ring is 1-fibered. Here is another example. Let R be the power series ring $k[[x,y]]$ with k a field. Let $I = (y^2, x^3, x^2y)$. Then $I^2 = (y^2, x^3)I$ and $\overline{R[I/y^2]} = R[I/y^2] \cong R[u,v]/I_2(A)$, where $I_2(A)$ is the ideal generated in $R[u,v]$ by the 2×2 minors of the matrix

$$A = \begin{pmatrix} x & u & v \\ y & v & x \end{pmatrix}$$

If P is a prime ideal minimal over $IR[I/y^2]$, then the preimage of P in $R[u,v]$ contains x, y and v so it *is* $(x,y,v)\,R[u,v]$, since ht $P = 1$. Thus I is 1-fibered.

There is another way to characterize 1-fibered ideals. This is in terms of Samuel's limit order function [**S**]. Recall that if we let v_I denote the order function associated to I, i.e., $v_I(x) = n$ if $x \in I^n \backslash I^{n+1}$, Samuel defined the function

$$\overline{v}_I(x) = \lim_{n \to \infty} v_I(x^n)/n.$$

Rees [**R1**] proved that

$$\overline{v}_I(x) = \inf_{1 \leq i \leq s} v_i(x)/v_i(I),$$

where $T(I) = \{v_1, \ldots, v_s\}$, and that the representation is irredundant and unique up to equivalence. Thus, I is 1-fibered if and only if $\overline{v}_I$ is a valuation. Now, $\overline{I^j} = \{x \in R \mid \overline{v}_I(x) \geq j\}$, so if I is 1-fibered, then there exists a prime divisor v such that the ideals $\overline{I^j}$ are valuation ideals for v.

If we assume that R is analytically unramified, then $\overline{R(I)}$ is a finitely generated $R(I)$-module. This means that some power of I, say I^r, has normal integral closure, i.e., $(\overline{I^r})^n = \overline{(\overline{I^r})^n}$. Thus, if R is analytically unramified, then I is 1-fibered if and only if there is a prime divisor v of R such that the powers of the integral closure of some power of I are valuation ideals for v. The necessity follows from the above remarks. For the sufficiency, suppose that $\overline{I^r} = J$ is normal and its powers are valuation ideals for a prime divisor v. Let V with maximal ideal generated by t be the valuation ring for v and suppose that $v(J) = e$. Then $J^n V \cap R = t^{en} V \cap R = J^n$ implies that $\overline{v}_I(x) = r\overline{v}_J(x) = r\overline{v}_{t^e V}(x) = rv(x)/e$.

The following proposition illustrates how strong a condition on R is the existence of a 1-fibered ideal.

PROPOSITION. *Let $(R, \underline{m})$ be a d-dimensional, normal, analytically unramified local domain. If R has a 1-fibered ideal, then R is analytically irreducible.*

PROOF: If v is the prime divisor corresponding to the 1-fibered ideal in R, let J be a normal 1-fibered ideal in R whose powers are valuation ideals for v. If $v(J) = e$, then $J^n = (t^e)^n V \cap R$, where t generates the maximal ideal of V, the valuation ring of v. Thus $t^j V \cap R \subseteq J^{n(j)} \subseteq \underline{m}^{n(j)}$, where $n(j) \to \infty$ as $j \to \infty$. Thus R is a subspace of V so $R^\wedge$ is contained in the domain $V^\wedge$.

REMARK: If R is assumed to be quasi-unmixed, then the proposition follows from the interesting result of Katz [**K**].

I conjecture that the converse to the proposition holds, namely that every analytically irreducible local domain has a 1-fibered ideal.[1] To put this conjecture in perspective, a little more history of the subject follows.

Muhly and Sakuma [**MS**] studied a property they called condition (N). A normal local domain $(R, \underline{m})$ is said to satisfy condition (N) if every prime divisor v of R has a 1-fibered ideal I_v. (Muhly and Sakuma called such ideals "asymptotically irreducible", a name motivated by a certain property these ideals possess in the more specialized setting which they were

[1] Recently, Dale Cutkosky constructed an example which disproves this conjecture.

studying.) Zariski [**ZS**] proved that every 2-dimensional regular local ring satisfies (N). Muhly [**M**] showed that there are 2-dimensional analytically irreducible local domains not satisfying (N). Lipman [**L**] proved that 2-dimensional local rational singularities satisfy (N). Göhner [**G**] showed that if R is 2-dimensional analytically normal and if $R^\wedge$ is semifactorial, i.e., R has torsion class group, then R satisfies (N). Göhner asks if condition (N) implies that $R^\wedge$ is semifactorial for a 2-dimensional analytically normal domain.

Cutkosky has recently answered Göhner's question in the affirmative if the residue field is algebraically closed. The non-algebraically closed case is still open.[2] In fact, there are many other open questions, such as whether every d-dimensional regular local ring satisfies (N) for $d > 2$.

To return to the question about whether every analytically irreducible local domain has a 1-fibered ideal, we note that we may assume that R is complete. For if $(R, \underline{m})$ is a d-dimensional analytically irreducible local domain then R has a 1-fibered ideal if and only if $\overline{R^\wedge}$, the integral closure of its m-adic completion does. This follows from the fact that there is a 1–1 correspondence between prime divisors of R and prime divisors of $R^\wedge$, cf. [**A2**] and [**G**]. Thus we may assume that R is a complete normal local domain and, as such, R is a finitely generated module over a complete regular local ring $(S, \underline{n})$. So one approach to the question comes via classical questions about splitting of valuations for the "going down" theorem gives the following lemma.

LEMMA. *Assume that R is integral over a regular local subring S. If there exists a prime divisor v of S which does not split in K and if v has a 1-fibered ideal in S, for example, if* $\dim S = 2$, *then R has a 1-fibered ideal.*

Thus the rather classical question arises: given a regular local ring $(S, \underline{n})$ with quotient field L and a finite extension K of L, does there exist a prime divisor of S which does not split in K? Neither the answer nor the question appear to be in literature, although there are interesting closely related results of Abhyankar [**A1**] and Abhyankar and Zariski [**AZ**]. For example, in the latter paper is the theorem that if S is equicharacteristic and K is separable over L, then infinitely many prime divisors do split.

Using classical splitting criteria involving the coefficients of minimal polynomials cf. [**O,E,A1**] it is not difficult to find examples where every prime

---

[2] Not any longer, for Cutkosky has recently settled this case also.

divisor of S splits in K, cf. [**Sy**]. Heinzer suggests a better question: given a complete normal domain $(R, \underline{m})$ with quotient field K does there exist some regular local ring S lying below R such that R is integral over S and such that S has a prime divisor which does not split in K?

In closing we mention two other approaches to this question which were discussed at the microprogram. One is Izumi's theorem, [**I**] and [**R2**], which gives various characterizations of analytically irreducible local domains by means of certain linear inequalities on prime divisors. A related approach is followed by Tomari and K. Watanabe, cf. [**TW**]. They know that if a local domain R has a filtration with associated graded ring having certain properties, then R has a 1-fibered ideal. They prove the existence of such a filtration in several specific instances.

Huneke's paper [**H**] is an excellent reference for the work of Zariski and Lipman on complete ideals in 2-dimensional regular local rings, as are Göhner's paper [**G**] for property (N) and the recent paper [**HL**] by Heinzer and Lantz for exceptional prime divisors.

REFERENCES

[A1] S. S. Abhyankar, *Nonsplitting of valuations in extensions of two dimensional regular local domains*, Math. Annalen **170** (1967), 87–144.

[A2] S. S. Abhyankar, *Local uniformation on algebraic surfaces over ground fields of characteristic* $p \neq 0$, Ann. of Math. **63** (1956), 491–526.

[AZ] S. S. Abhyankar and O. Zariski, *Splitting of valuations in extensions of local domains*, Proc. Nat. Acad. Sci. U.S.A. **41** (1955), 84–90.

[E] O. Endler, O., "Valuation Theory," Springer-Verlag, Berlin, 1972.

[G] H. Göhner, *Semifactoriality and Muhly's condition (N) in two dimensional local rings*, J. Algebra **34** (1975), 403–429.

[HL] W. Heinzer and D. Lantz, *Exceptional prime divisors of two dimensional local domains*, in "Proceedings of the Microprogram on Commutative Algebra," Springer-Verlag, New York, 1989.

[H] C. Huneke, *Complete ideals in two-dimensional regular local rings*, in "Proceedings of the Microprogram on Commutative Algebra," Springer-Verlag, New York, 1989.

[I] S. Izumi, *A measure of integrity for local analytic algebras*, Publ. RIMS, Kyoto Univ. **21** (1985), 719–735.

[K] D. Katz, *On the number of minimal prime ideals in the completion of a local domain*, Rocky Mtn. J. Math. **16** (1986), 575–578.

[L] J. Lipman, *Rational Singularities with applications to algebraic surfaces and unique factorization*, Publ. Math. Inst. Hautes Études Sci. **36** (1969), 195–279.

[M] H. T. Muhly, *On the existence of asymptotically irreducible ideals*, J. London Math. Soc. **40** (1965), 99–107.

[MS] H. T. Muhly and M. Sakuma, *Asymptotic factorization of ideals*, J. London Math. Soc. **38** (1963), 341–350.

[O] A. Ostrowski, *Untersuchungen zur arithmetischen Theorie der Korper*, Math. Zeit. **39** (1934), 269–404.

[R1] D. Rees, *Valuations associated with ideals II*, J. London Math. Soc. **31** (1956), 221–228.

[R2] D. Rees, *Izumi's theorem*, in "Proceedings of the Microprogram on Commutative Algebra," Springer-Verlag, New York, 1989.

[S] P. Samuel, *Some asymptotic properties of powers of ideals*, Ann. of Math. **56** (1952), 11–21.

[Sy] J. D. Sally, *Non-splitting of prime divisors*, Proc. Camb. Phil. Soc. **103** (1988), 251–256.

[TW] M. Tomari and E. Watanabe, *Filtered rings, filtered blowing-ups and normal 2-dimensional singularities with "star-shaped" resolution*, preprint.

[ZS] O. Zariski and P. Samuel, "Commutative Algebra," Vol. II, Van Nostrand, Princeton, 1960.

Department of Mathematics, Northwestern University, Evanston IL 60208

The author was partially supported by a grant from the NSF.

A Method for the Study of Artinian Modules, With an Application to Asymptotic Behavior

R. Y. SHARP

0. Introduction.

If N is a Noetherian module over the commutative ring R (throughout the paper, R will denote a commutative ring with identity), then the study of N in many contexts can be reduced to the study of a finitely generated module over a commutative Noetherian ring, because N has a natural structure as a module over $R/(0 : N)$ and the latter ring is Noetherian. For a long time it has been a source of irritation to me that I did not know of any method which would reduce the study of an Artinian module A over the commutative ring R to the study of an Artinian module over a commutative Noetherian ring. However, during the MSRI Microprogram on Commutative Algebra, my attention was drawn to a result of W. Heinzer and D. Lantz [**2**, Proposition 4.3]; this proposition proves that if A is a faithful Artinian module over a quasi-local ring (R, M) which is (Hausdorff) complete in the M-adic topology, then R is Noetherian. It turns out that a generalization of this result provides a missing link to complete a chain of reductions by which one can, for some purposes, reduce the study of an Artinian module over an arbitrary commutative ring R to the study of an Artinian module over a complete (Noetherian) local ring; in the latter situation we have Matlis's duality available, and this means that the investigation can often be converted into a dual one about a finitely generated module over a complete (Noetherian) local ring.

This paper is partly expository and partly a research paper: most of the first two sections is exposition, while the majority of the last two sections constitutes new research. The first section describes the initial part of the above-mentioned chain of reductions, and shows how one can, for some purposes, reduce the study of an Artinian module over an arbitrary commutative ring to the study of an Artinian module A over a quasi-local ring R with maximal ideal M, and points out that then A has a natural structure as a module over the M-adic (Hausdorff) completion $\hat{R}$ of R. Unfortunately, however, it is not clear (to me at least) that the natural completion topology on $\hat{R}$ is the $\hat{M}$-adic topology (where $\hat{M}$ is the maximal ideal of the quasi-local ring $\hat{R}$): one does not know that M is a finitely generated ideal of R. It is for this reason that the above-mentioned Heinzer-Lantz Theorem does not seem (to me at least) enough in itself to bridge the gap to commutative Noetherian ring theory, and it is why a generalization of that theorem is proved in Section 3.

Section 2 gives a brief description of those aspects of Matlis's duality theory [**7**] which are used in Sections 3 and 4.

The generalization of the Heinzer-Lantz Theorem which is proved in Section 3 is as follows. Suppose that R is a quasi-local ring having maximal ideal M which is (Hausdorff) complete with respect to a multiplicative filtration $\mathcal{F} = (Q_n)_{n\in\mathbb{N}}$ (we always use $\mathbb{N}$ (respectively $\mathbb{N}_0$) to denote the set of positive (respectively non-negative) integers) such that $M = Q_1$ and, for each $n \in \mathbb{N}$, the closure of M^n in the $\mathcal{F}$-topology is Q_n, so that, in particular, $M^n \subseteq Q_n$. Suppose that A is a faithful Artinian R-module such that

$$A = \bigcup_{n\in\mathbb{N}} (0 :_A Q_n).$$

It is proved in Section 3 that in these circumstances R is Noetherian and complete with respect to its M-adic topology. This result completes the reduction procedure begun in Section 1.

The ideas of the proof of the above theorem owe much to those of Heinzer and Lantz in [**2**], and, of course, their theorem is a particular case. Thus Section 3 can be regarded as providing an exposition of the Heinzer-Lantz Theorem (which, in my opinion, deserves to be more widely known); however, it does in addition produce a result which I believe to be new, and so constitutes new research.

The last Section 4 of the paper also constitutes new research. It uses the chain of reductions described in Sections 1 and 3 to obtain a new result, one which I had been unsuccessfully attempting to achieve for some time by more direct means. This result is described as follows.

Let I be an ideal of R (it is not assumed here that R is Noetherian). Let N be a Noetherian R-module. It is a result of M. Brodmann [**1**] that both the sequences of sets

$$(\mathrm{Ass}_R(N/I^nN))_{n\in\mathbb{N}} \text{ and } \left(\mathrm{Ass}_R(I^{n-1}N/I^nN)\right)_{n\in\mathbb{N}}$$

are ultimately constant; we let $\mathrm{As}^*(I, N)$ and $\mathrm{Bs}^*(I, N)$ (respectively) denote their ultimate constant values. (Of course, $\mathrm{Ass}_R(X)$, for a Noetherian

R-module X, denotes the set of prime ideals belonging to the zero submodule of X for primary decomposition.) Now $\mathrm{Bs}^*(I,N) \subseteq \mathrm{As}^*(I,N)$; S. McAdam and P. Eakin [**10**, Corollary 13] proved that, in the special case in which $N = R$ (this demands that the ring R itself be Noetherian),

$$\mathrm{As}^*(I,R) \setminus \mathrm{Bs}^*(I,R) \subseteq \mathrm{Ass}(R).$$

It is straightforward to adapt their argument to show that, more generally,

$$\mathrm{As}^*(I,N) \setminus \mathrm{Bs}^*(I,N) \subseteq \mathrm{Ass}(N),$$

and this adaptation is sketched briefly in Section 4 below.

Now let A be an Artinian R-module. There is a theory of "secondary representation" for Artinian R-modules which is in several respects dual to the theory of primary decomposition for Noetherian R-modules. Accounts of this dual theory are available in [**5**,**3**,**12**]; we follow I. G. Macdonald's terminology from [**5**]. This theory associates with the Artinian R-module A a finite set $\mathrm{Att}_R(A)$ (or $\mathrm{Att}(A)$) of prime ideals of R, referred to as the attached prime ideals of A. In [**13**], use was made of Kirby's Theorem 1 of [**4**] to show (without any Noetherian hypothesis on R) that the sequences of sets

$$(\mathrm{Att}_R(0:_A I^n))_{n\in\mathbb{N}} \text{ and } (\mathrm{Att}_R((0:_A I^n)/(0:_A I^{n-1})))_{n\in\mathbb{N}}$$

are ultimately constant. Denote their ultimate constant values by $\mathrm{At}^*(I,A)$ and $\mathrm{Bt}^*(I,A)$ respectively. It is easy to see that $\mathrm{Bt}^*(I,A) \subseteq \mathrm{At}^*(I,A)$. The work of McAdam and Eakin mentioned above raises the following question: is it the case that

$$\mathrm{At}^*(I,A) \setminus \mathrm{Bt}^*(I,A) \subseteq \mathrm{Att}(A)?$$

For some time, I have been trying to prove, by direct means, that this is so, but without success. The statement is proved in Section 4 below, by use of the reduction procedure described in Sections 1 and 3, Matlis's duality, and the work of McAdam and Eakin. Of course, this is a rather roundabout route, but it is the only one I have been able to find so far, and so it provides a good illustration of the use of the reduction procedure described in the first part of the paper.

Before we begin the mathematics, a few words about notation are appropriate. The symbol L will always denote a module over R, our commutative ring (with identity); I will always denote an ideal of R; a Noetherian R-module will often be denoted by N; and an Artinian R-module will often be denoted by A. Additional properties of R will only be assumed when these are explicitly stated, perhaps by means of some of the following terminology. The phrase "(R, M) is quasi-local" will mean that R has M as its unique maximal ideal; by "R is local" we shall mean that R is both quasi-local and Noetherian.

1. Reduction to the quasi-local case.

The ideas in this section are known to numerous people, but I am not aware of a place where they have been written down in a coherent fashion. The roots of the ideas go back at least to Matlis's paper [**8**], but the underlying rings in that paper are assumed to be Noetherian.

(1.1) REMARK: Let A be an Artinian R-module and let $0 \neq x \in A$. Then $R/(0 : x) \cong Rx$ as R-modules, and so $R/(0 : x)$ is an Artinian ring. Hence there are only finitely many maximal ideals, $M_1, \ldots, M_r$ say, of R that contain $(0 : x)$, and there exists $t \in \mathbb{N}$ such that $(M_1 \ldots M_r)^t x = 0$.

(1.2) DEFINITION. *Let M be a maximal ideal of R. The M-torsion submodule $\Gamma_M(L)$ of L is defined by*

$$\Gamma_M(L) = \bigcup_{n \in \mathbb{N}} (0 :_L M^n).$$

(1.3) LEMMA. *Let* $\operatorname{Max}(R)$ *denote the set of maximal ideals of R. The sum $\sum_{M \in \operatorname{Max}(R)} \Gamma_M(L)$ of submodules of L is direct.*

PROOF: Suppose that $M_1, \ldots, M_n$ are n distinct maximal ideals of R (where $n \geq 2$) and $x_i \in \Gamma_{M_i}(L)$ for $i = 1, \ldots, n$ are such that $\sum_{i=1}^{n} x_i = 0$. Choose $j \in \mathbb{N}$ with $1 \leq j \leq n$. Now there exists $h \in \mathbb{N}$ such that ${M_i}^h x_i = 0$ for all $i = 1, \ldots, n$. Since

$$x_j = -\sum_{\substack{i=1 \\ i \neq j}}^{n} x_i,$$

we have

$${M_j}^h + \bigcap_{\substack{i=1 \\ i \neq j}}^{n} {M_i}^h \subseteq (0 : x_j),$$

so that $x_j = 0$.

(1.4) PROPOSITION. *Let A be a non-zero Artinian R-module. Then there are only finitely many maximal ideals M of R for which $\Gamma_M(A) \neq 0$; if the distinct such maximal ideals are $M_1, \ldots, M_n$, then*

$$A = \Gamma_{M_1}(A) \oplus \cdots \oplus \Gamma_{M_n}(A).$$

PROOF: The first claim is a consequence of (1.3) because the submodule $\sum_{M \in \mathrm{Max}(R)} \Gamma_M(A) = \bigoplus_{M \in \mathrm{Max}(R)} \Gamma_M(A)$ of A must satisfy the descending chain condition on submodules. It therefore remains only to show that

$$A \subseteq \Gamma_{M_1}(A) + \cdots + \Gamma_{M_n}(A).$$

Let $0 \neq x \in A$. It follows from (1.1) that there is a subset $\{i_1, \ldots, i_r\}$ (with r elements) of $\{1, \ldots, n\}$ such that $(M_{i_1} \ldots M_{i_r})^t x = 0$ for some $t \in \mathbb{N}$. (Observe that if M is a maximal ideal of R different from $M_1, \ldots, M_n$ and $(MM_{i_1} \ldots M_{i_r})^t x = 0$, then $(M_{i_1} \ldots M_{i_r})^t x = 0$ because $\Gamma_M(A) = 0$.) We shall show that $x \in \sum_{i=1}^n \Gamma_{M_i}(A)$ by induction on r. This claim is clear when $r = 1$, and so we suppose that $r > 1$ and that the claim has been proved for smaller values of r.

Since ${M_{i_1}}^t + (M_{i_2} \ldots M_{i_r})^t = R$, there exist $a \in {M_{i_1}}^t$ and $b \in (M_{i_2} \ldots M_{i_r})^t$ such that $a + b = 1$. Then $x = ax + bx$, and $bx \in \Gamma_{M_{i_1}}(A)$, while $(M_{i_2} \ldots M_{i_r})^t ax = 0$; hence, in view of the inductive assumption, $ax \in \sum_{i=1}^n \Gamma_{M_i}(A)$. This completes the inductive step.

(1.5) REMARK: In the situation of (1.4), denote (temporarily) $\Gamma_{M_i}(A)$ by A_i (for $1 \leq i \leq n$). For $j \in \mathbb{N}$, we have

$$(0 :_A I^j) = (0 :_{A_1} I^j) \oplus \cdots \oplus (0 :_{A_n} I^j),$$

and a natural isomorphism

$$(0 :_A I^j)/(0 :_A I^{j-1}) \cong \bigoplus_{i=1}^n (0 :_{A_i} I^j)/(0 :_{A_i} I^{j-1}).$$

It now follows from [**5**, (4.2)] that $\mathrm{Att}_R(A) = \bigcup_{i=1}^n \mathrm{Att}_R(A_i)$,

$$\mathrm{Att}_R(0 :_A I^j) = \bigcup_{i=1}^n \mathrm{Att}_R(0 :_{A_i} I^j),$$

and

$$\mathrm{Att}_R((0:_A I^j)/(0:_A I^{j-1})) = \bigcup_{i=1}^{n} \mathrm{Att}_R((0:_{A_i} I^j)/(0:_{A_i} I^{j-1})).$$

It follows that, for some purposes, such as that which will be our concern in Section 4 below, we can assume that the Artinian module A satisfies $A = \Gamma_M(A)$ for some maximal ideal M of R.

We therefore analyze the situation in which A is an Artinian R-module and there is a maximal ideal M of R with the property that each element of A is annihilated by some power of M.

(1.6) LEMMA. *Let A be an Artinian R-module. Suppose that, for some maximal ideal M of R, we have $A = \Gamma_M(A)$. Let $s \in R \setminus M$. Then multiplication by s provides an automorphism of A; in fact, for each $x \in A$, there exists $b_{x,s} \in R$ such that $x = b_{x,s}\, sx$.*

PROOF: There exists $n \in \mathbb{N}$ such that $M^n x = 0$. Now $M^n + Rs = R$, and so there exists $c \in M^n$ and $b_{x,s} \in R$ such that $c + b_{x,s}\, s = 1$. Hence $x = b_{x,s}\, sx$.

(1.7) REMARK: Let A be an Artinian R-module. Suppose that, for some maximal ideal M of R, we have $A = \Gamma_M(A)$. Then it is immediate form (1.6) that A has a natural structure as an R_M-module; in fact, for $x \in A$ and $a \in R$, $s \in R \setminus M$, we have

$$(a/s)x = b_{x,s}\, ax,$$

where $b_{x,s} \in R$ is such that $b_{x,s}\, sx = x$. It follows from this that a subset of A is an R-submodule if and only if it is an R_M-submodule, so that A is an Artinian R_M-module.

(1.8) LEMMA. *Let the situation be as in (1.6) and (1.7). Let $P' \in \mathrm{Att}_R(A)$. Then $P' \subseteq M$. Also*

$$\mathrm{Att}_{R_M}(A) = \{PR_M : P \in \mathrm{Att}_R(A)\}.$$

PROOF: There exists a P'-secondary submodule S of A. Let $0 \neq x \in S$, and suppose that $P' \not\subseteq M$, so that there exists $a \in P' \setminus M$. Then there exists $n \in \mathbb{N}$ such that $a^n x = 0$, and there exists $h \in \mathbb{N}$ such that $M^h x = 0$. Hence

$$R = Ra^n + M^h \subseteq (0 : x),$$

a contradiction.

In the light of (1.7), it now follows that, when A is considered as an R_M-module, S is (an R_M-submodule and) $P'R_M$-secondary. Hence a minimal secondary representation for A as an R-module is automatically a minimal secondary representation for A as R_M-module, and $\operatorname{Att}_{R_M}(A) = \{PR_M : P \in \operatorname{Att}_R(A)\}$.

(1.9) REMARK: Let the situation be as in (1.6), (1.7) and (1.8). Observe that the condition "A is an Artinian R-module and $A = \Gamma_M(A)$" is inherited by all submodules and all homomorphic images of A. Furthermore, for each $j \in \mathbb{N}$, $(0 :_A I^j) = (0 :_A (IR_M)^j)$. It is therefore immediate from (1.8) that, for each $j \in \mathbb{N}$, we have

$$\operatorname{Att}_{R_M}(0 :_A (IR_M)^j) = \{PR_M : P \in \operatorname{Att}_R(0 :_A I^j)\}$$

and

$$\begin{aligned}\operatorname{Att}_{R_M}&\left((0 :_A (IR_M)^j)/(0 :_A (IR_M)^{j-1})\right)\\ &= \left\{PR_M : P \in \operatorname{Att}_R\left((0 :_A I^j)/(0 :_A I^{j-1})\right)\right\}.\end{aligned}$$

It follows that, for some purposes, such as that with which we shall be concerned in Section 4, we can assume that the Artinian module A is over a quasi-local ring.

We shall need to consider the effect of factoring out the annihilator of an Artinian module.

(1.10) REMARK: Let A be a non-zero Artinian R-module. It is clear that $(0 :_A I) = (0 :_A I + (0 : A))$ and, indeed, that

$$(0 :_A I^j) = (0 :_A (I + (0 : A))^j) \text{ for all } j \in \mathbb{N}.$$

Thus, when we are interested in properties of the R-modules $(0 :_A I^j)$ $(j \in \mathbb{N})$, we can in some situations assume that $(0 : A) \subseteq I$. Suppose that this is so, and set $\bar{R} = R/(0 : A)$, $\bar{I} = I/(0 : A)$. Of course, A can be regarded as a (faithful, Artinian) $\bar{R}$-module in a natural way; each attached prime of each submodule and each homomorphic image of A must contain $(0 : A)$; and it is easy to see that

$$\operatorname{Att}_{\bar{R}}(A) = \{P/(0 : A) : P \in \operatorname{Att}_R(A)\},$$

and, for all $j \in \mathbb{N}$,

$$\begin{aligned} \operatorname{Att}_{\bar{R}}(0 :_A \bar{I}^j) &= \{P/(0:A) : P \in \operatorname{Att}_R(0 :_A I^j)\}, \\ \operatorname{Att}_{\bar{R}}\left((0 :_A \bar{I}^j)/(0 :_A \bar{I}^{j-1})\right) & \\ &= \{P/(0:A) : P \in \operatorname{Att}_R\left((0 :_A I^j)/(0 :_A I^{j-1})\right)\}. \end{aligned}$$

Note also that if, in addition, (R, M) is quasi-local, then $A = \Gamma_M(A)$ by (1.4), and so $\bigcap_{i \in \mathbb{N}} M^i \subseteq (0 : A)$. Thus, if we denote $M/(0 : A)$ by $\bar{M}$, then the quasi-local ring $(\bar{R}, \bar{M})$ satisfies $\bigcap_{i \in \mathbb{N}} \bar{M}^i = 0$.

The next stage in our reduction procedure is the passage to the M-adic (Hausdorff) completion $\hat{R}$ of a quasi-local ring (R, M) which possesses an Artinian module.

(1.11) LEMMA. *Let A be an Artinian module over the quasi-local ring (R, M). Then A has a natural structure as a module over the M-adic (Hausdorff) completion $\hat{R}$ of R, and this is such that, if $\psi : R \to \hat{R}$ denotes the canonical ring homomorphism, for a sequence $(c_n)_{n \in \mathbb{N}}$ of elements of R such that $(\psi(c_n))_{n \in \mathbb{N}}$ converges to $\hat{c}$ in $\hat{R}$, and for $x \in A$, the sequence $(c_n x)_{n \in \mathbb{N}}$ is ultimately constant, and its ultimate value is $\hat{c}x$.*

PROOF: We have $A = \Gamma_M(A)$, by (1.4). For $\hat{c}$, $(c_n)_{n \in \mathbb{N}}$ and x as in the statement, there exists $h \in \mathbb{N}$ such that $M^h x = 0$, so that, since there exists $k \in \mathbb{N}$ such that $c_n - c_\ell \in M^h$ for all $n, \ell \geq k$, we have $c_n x = c_k x$ for all $n \geq k$. It is routine to check that the ultimate value of the sequence $(c_n x)_{n \in \mathbb{N}}$ is independent of the choice of sequence $(c_n)_{n \in \mathbb{N}}$ in R for which $\lim_{n \to \infty} \psi(c_n) = \hat{c}$, so that we can define this ultimate value to be $\hat{c}x$. It is also routine to check that this definition produces an $\hat{R}$-module structure on A.

(1.12) COROLLARY. *Let the situation be as in (1.11). Then a subset of A is an R-submodule if and only if it is an $\hat{R}$-submodule, so that A is an Artinian $\hat{R}$-module. Furthermore,*

$$\operatorname{Att}_R(A) = \{\psi^{-1}(Q) : Q \in \operatorname{Att}_{\hat{R}}(A)\}.$$

PROOF: The first claim is immediate from the definition of the $\hat{R}$-module structure on A given in (1.11). Let $A = S_1 + \cdots + S_r$ be a minimal secondary representation of A as $\hat{R}$-module, with $\sqrt{(0 :_{\hat{R}} S_i)} = Q_i$ for $i = 1, \ldots, r$. It is clear that S_i, as R-module, is $\psi^{-1}(Q_i)$-secondary, and so the final claim follows from [**5**, (2.1)].

(1.13) REMARKS: Let the situation be as in (1.11) and (1.12), and let $j \in \mathbb{N}$. Of course, $I\hat{R}$ denotes the ideal of $\hat{R}$ generated by $\psi(I)$.

(i) We have $(0 :_A I^j) = (0 :_A (I\hat{R})^j)$, and so, by (1.12),

$$\mathrm{Att}_R(0 :_A I^j) = \left\{\psi^{-1}(Q) : Q \in \mathrm{Att}_{\hat{R}}(0 :_A (I\hat{R})^j)\right\},$$

$$\mathrm{Att}_R\left((0 :_A I^j)/(0 :_A I^{j-1})\right)$$
$$= \left\{\psi^{-1}(Q) : Q \in \mathrm{Att}_{\hat{R}}\left((0 :_A (I\hat{R})^j)/(0 :_A (I\hat{R})^{j-1})\right)\right\}.$$

These observations will be relevant in Section 4.

(ii) Let $\hat{M}$ denote the maximal ideal of the quasi-local ring $\hat{R}$. Of course, $\hat{R}$ has a natural topology, defined in terms of a filtration $\mathcal{F} = (Q_n)_{n\in\mathbb{N}}$ of ideals of $\hat{R}$, simply by virtue of the fact that it is the (Hausdorff) completion of R with respect to the M-adic topology. Unfortunately, because we do not know that M is a finitely generated ideal of R, it is not clear (to me at least) that the $\mathcal{F}$-topology of $\hat{R}$ is the $\hat{M}$-adic topology. However, we can make the following points.

(a) The filtration $\mathcal{F} = (Q_n)_{n\in\mathbb{N}}$ of $\hat{R}$ is multiplicative;

(b) $Q_1 = \hat{M}$;

(c) for each $n \in \mathbb{N}$, we have $(\hat{M})^n \subseteq Q_n$, and, in fact, Q_n is the closure of $(\hat{M})^n$ in $\hat{R}$ with respect to the $\mathcal{F}$-topology;

(d) for each $x \in A$, there exists $n \in \mathbb{N}$ such that $Q_n x = 0$.

Suppose that $A \neq 0$. We can regard A as a (faithful, Artinian) module over the quasi-local ring $\hat{R}/(0 :_{\hat{R}} A)$, and the generalization of the Heinzer-Lantz Theorem which is presented in Section 3 below will show that $\hat{R}/(0 :_{\hat{R}} A)$ is actually Noetherian and complete with respect to its maximal ideal topology.

2. A review of some aspects of Matlis's duality.

In this section we give a brief review of Matlis's duality theorem, and point out some consequences which will be used in Sections 3 and 4.

(2.1) THEOREM (E. Matlis [7]). *Let (R, M) be a complete local ring, and let $E := E(R/M)$ denote the injective envelope of the R-module R/M. Let D denote the (additive, R-linear and exact) functor* $\mathrm{Hom}_R(\cdot, E)$ *(from the category $\mathcal{C}(R)$ of all R-modules and R-homomorphisms to itself). For each R-module L, let*

$$\mu_L : L \longrightarrow DD(L) = \mathrm{Hom}_R(\mathrm{Hom}_R(L, E), E)$$

be the natural R-homomorphism for which $(\mu_L(x))(f) = f(x)$ for all $x \in L$ and $f \in \mathrm{Hom}_R(L, E)$ (so that, as L varies through $\mathcal{C}(R)$, the μ_L constitute a morphism from the identity functor to the functor DD).

(i) *The R-module E is Artinian; for each $f \in \mathrm{Hom}_R(E, E)$, there is a unique $a_f \in R$ such that $f(x) = a_f x$ for all $x \in E$.*

(ii) *If N is a Noetherian R-module, then $D(N)$ is Artinian.*

(iii) *If A is an Artinian R-module, then $D(A)$ is Noetherian.*

(iv) *If the R-module L is either Noetherian or Artinian, then*

$$\mu_L : L \longrightarrow DD(L)$$

is an isomorphism.

(v) *If L is an R-module of finite length (and we use ℓ_R to denote lengths of R-modules), then $\ell_R(D(L)) = \ell_R(L)$.*

The classical reference for this work is Matlis [**7**]; another reference is Sharpe and Vámos [**14**, Chapter 5].

(2.2) TERMINOLOGY: Let the situation be as in (2.1). If L is either an Artinian R-module or a Noetherian R-module, then $D(L)$ is referred to as the *Matlis dual* of L.

We now give an application of Matlis's duality which we shall use in Section 3.

(2.3) COROLLARY. *Suppose that (R, M) is an Artinian local ring (and so complete), and let $E = E(R/M)$. Suppose that L is a faithful R-submodule of E. Then $L = E$.*

PROOF: The canonical exact sequence $0 \longrightarrow L \xrightarrow{\mu} E \longrightarrow E/L \longrightarrow 0$ induces (with the notation of (2.1)) an exact sequence

$$0 \longrightarrow D(E/L) \longrightarrow D(E) \xrightarrow{D(\mu)} D(L) \longrightarrow 0.$$

Now, by (2.1)(i), each element of $D(E) = \mathrm{Hom}_R(E, E)$ is given by multiplication by an element of R, and since L is a faithful R-module, $D(\mu)$ is injective. Therefore $D(E/L) = 0$. But, by (2.1)(i), E is Artinian, and so, by (2.1)(iv),

$$E/L \cong D(D(E/L)) = 0.$$

The following lemma will be helpful in Section 4.

(2.4) LEMMA. *Let the situation be as in (2.1); let N be a Noetherian R-module, let A be an Artinian R-module, and let $j \in \mathbb{N}$. Then*

(i) $D(N/I^jN) \cong (0 :_{D(N)} I^j)$ *and*

$$D(I^{j-1}N/I^jN) \cong (0 :_{D(N)} I^j)/(0 :_{D(N)} I^{j-1});$$

(ii) $D(0 :_A I^j) \cong D(A)/I^jD(A)$ *and*

$$D\left((0 :_A I^j)/(0 :_A I^{j-1})\right) \cong I^{j-1}D(A)/I^jD(A).$$

PROOF: The standard natural equivalence of functors

$$\operatorname{Hom}_R(\cdot \otimes_R N, E) \longrightarrow \operatorname{Hom}_R(\cdot, \operatorname{Hom}_R(N, E))$$

(from $\mathcal{C}(R)$ to itself) (see [**11**, Chapter 8, Theorem 4], for example) leads to a commutative diagram

$$\begin{array}{ccc}
0 & & 0 \\
\downarrow & & \downarrow \\
D(N/I^{j-1}N) & \xrightarrow{\cong} & (0 :_{D(N)} I^{j-1}) \\
\downarrow & & \downarrow \\
D(N/I^jN) & \xrightarrow{\cong} & (0 :_{D(N)} I^j) \\
\downarrow & & \downarrow \\
D(I^{j-1}N/I^jN) & & (0 :_{D(N)} I^j)/(0 :_{D(N)} I^{j-1}) \\
\downarrow & & \downarrow \\
0 & & 0
\end{array}$$

in which the left column is the exact sequence induced by application of the functor D to the canonical exact sequence

$$0 \longrightarrow I^{j-1}N/I^jN \longrightarrow N/I^jN \longrightarrow N/I^{j-1}N \longrightarrow 0,$$

the right column is the canonical exact sequence, and the two horizontal maps are isomorphisms. Part (i) follows from this, and (ii) is then an easy consequence, because, by (2.1)(iv), $L \cong D(D(L))$ whenever L is an R-module which is either Artinian or Noetherian.

Our last result in this section will show that, in the situation of (2.1), for a Noetherian R-module N, we have $\mathrm{Att}_R(D(N)) = \mathrm{Ass}_R(N)$. To establish this, we shall use the fact that, for a $P \in \mathrm{Spec}(R)$ and an Artinian R-module A, it is the case that $P \in \mathrm{Att}_R(A)$ if and only if A has a homomorphic image with annihilator exactly P. Although this is proved in [**6**, 1.4] in the case when R is Noetherian, and that suffices for our purposes here, this characterization of $\mathrm{Att}_R(A)$ for Artinian A is actually valid without the Noetherian assumption on R. As this characterization has not (as far as I am aware) appeared before, it is presented here.

(2.5) LEMMA. *Let $J = \sqrt{(0 : A)}$, where A is an Artinian R-module. Then there exists $n \in \mathbb{N}$ such that $J^n A = 0$. (There is no Noetherian assumption on R here.)*

PROOF: By [**4**, Lemma 3], there is a finitely generated ideal J_0 of R such that $J_0 \subseteq J$ and $(0 :_A J^j) = (0 :_A {J_0}^j)$ for all $j \in \mathbb{N}$. Since J_0 is finitely generated, there exists $n \in \mathbb{N}$ such that ${J_0}^n A = 0$. Hence $A \subseteq (0 :_A {J_0}^n) = (0 :_A J^n)$.

(2.6) COROLLARY. *(There is no Noetherian assumption on R in this result.) Let A be an Artinian R-module. Then $\mathrm{Att}_R(A) = \{P \in \mathrm{Spec}(R) : A$ has a homomorphic image with annihilator precisely $P\}$.*

PROOF: Let $P \in \mathrm{Att}_R(A)$. Thus there is a minimal secondary representation $A = S_1 + \cdots + S_r$ for A, with $\sqrt{(0 : S_i)} = P_i$ for $i = 1, \ldots, r$ and $P = P_1$. Then $T = A/(S_2 + \cdots + S_r)$ is P-secondary (and Artinian). By (2.5), there exists $n \in \mathbb{N}$ such that $P^n T = 0$. It follows that T/PT is a homomorphic image of A with annihilator equal to P.

The reverse inclusion can be proved by use of the argument under "$\Leftarrow$" in the proof of [**6**, 1.4].

(2.7) COROLLARY. *Let the situation be as in (2.1).*

(i) *Let N be a Noetherian R-module. Then $\mathrm{Att}_R(D(N)) = \mathrm{Ass}_R(N)$.*

(ii) *Let A be an Artinian R-module. Then $\mathrm{Ass}_R(D(A)) = \mathrm{Att}_R(A)$.*

PROOF: Note that, by (2.1), if L is either Noetherian or Artinian, then L and its Matlis dual $D(L)$ have the same annihilator.

Let $P \in \mathrm{Ass}_R(N)$, so that N has a submodule T with annihilator equal to P. Application of the exact functor D to the inclusion map $T \longrightarrow N$ shows that $D(N)$ has a homomorphic image with annihilator equal to P. Since $D(N)$ is Artinian by (2.1), it follows from (2.6) that $P \in \mathrm{Att}_R(D(N))$.

Thus $\mathrm{Ass}_R(N) \subseteq \mathrm{Att}_R(D(N))$, and a very similar argument shows that $\mathrm{Att}_R(A) \subseteq \mathrm{Ass}_R(D(A))$.

The result now follows from Matlis's duality (2.1), since $D(N)$ is Artinian, $D(A)$ is Noetherian, and there are isomorphisms

$$N \xrightarrow{\cong} DD(N), \quad A \xrightarrow{\cong} DD(A).$$

3. A generalization of the Heinzer-Lantz Theorem.

In [**2**, 4.3], Heinzer and Lantz proved that if A is a faithful Artinian module over the (Hausdorff) complete quasi-local ring (R', M'), then R' is Noetherian. In this section, we shall adapt their ideas to prove a seemingly more general result over a quasi-local ring (R, M) which is (Hausdorff) complete with respect to the topology induced by a multiplicative filtration $\mathcal{F} = (Q_n)_{n\in\mathbb{N}}$ of ideals of R for which $M = Q_1$ and, for each $n \in \mathbb{N}$, the closure of M^n with respect to the $\mathcal{F}$-topology is Q_n: we shall show that if there exists a faithful Artinian R-module A for which $A = \bigcup_{n\in\mathbb{N}}(0 :_A Q_n)$, then R is Noetherian.

(3.1) Definition: Suppose that (R, M) is quasi-local. The *socle* of L is the submodule $(0 :_L M)$.

Of course, the socle of L has a natural structure as a vector space over R/M; the dimension of this is finite when L is Artinian.

(3.2) Remark: Suppose that (R, M) is quasi-local and that A is a faithful Artinian R-module. Then $A = \Gamma_M(A)$ by (1.4), and so

$$\bigcap_{n\in\mathbb{N}} M^n \subseteq (0 : A) = 0.$$

(3.3) Remark: Suppose that (R, M) is quasi-local and that A is a non-zero Artinian R-module. Let $0 \neq x \in A$. Since $A = \Gamma_M(A)$ by (1.4), the element x is annihilated by some power of M. Let h be the least $i \in \mathbb{N}$ such that $M^i x = 0$. Then

$$0 \neq M^{h-1}x \subseteq (0 :_A M).$$

Hence A is an essential extension of its socle.

(3.4) LEMMA (Heinzer and Lantz). (See [**2**, p. 212].). *Suppose that* (R, M) *is quasi-local, and set* $k = R/M$. *Let* A *be a non-zero Artinian* R*-module such that* $\dim_k(0 :_A M) = s$. *Then there exist* s *submodules* $A_1, \ldots, A_s$ *of* A *such that* $0 = \bigcap_{i=1}^{s} A_i$ *and, for each* $j = 1, \ldots, s$, *the Artinian module* A/A_j *has 1-dimensional socle.*

PROOF (after Heinzer and Lantz): Let $x_1, \ldots, x_s$ form a basis for $(0 :_A M)$. For each $j = 1, \ldots, s$, let

$$\mathcal{T}_j = \{T : T \text{ is a submodule of } A \text{ and } x_1, \ldots, x_{j-1}, x_{j+1}, \ldots, x_s \in T \text{ but } x_j \notin T\}.$$

Then $Rx_1 + \cdots + Rx_{j-1} + Rx_{j+1} + \cdots + Rx_s \in \mathcal{T}_j$. By Zorn's Lemma, $\mathcal{T}_j$ has a maximal member A_j with respect to inclusion.

If $B := \bigcap_{i=1}^{s} A_i$ were not zero, its socle would, in view of (3.3), be a non-zero subspace of $(0 :_A M) = Rx_1 + \cdots + Rx_s$, so that there would exist $a_1, \ldots, a_s \in R$, at least one of them, a_j say, being a unit, such that

$$a_1 x_1 + \cdots + a_j x_j + \cdots + a_s x_s \in B = \bigcap_{i=1}^{s} A_i;$$

since $x_1, \ldots, x_{j-1}, x_{j+1}, \ldots, x_s \in A_j$, it would follow that $x_j \in A_j$, a contradiction. Hence $\bigcap_{i=1}^{s} A_i = 0$.

Choose $j \in \mathbb{N}$ with $1 \leq j \leq s$. Of course $x_j + A_j$ is a non-zero element of the socle of A/A_j. Suppose that $y \in A$ is such that $x_j + A_j$, $y + A_j$ are two linearly independent elements of this socle. Then $A_j \subset Ry + A_j$ (the symbol $\subset$ denotes strict inclusion) and $x_j \notin Ry + A_j$; hence $Ry + A_j \in \mathcal{T}_j$, and this contradicts the maximality of A_j. Hence $\dim_k(0 :_{A/A_j} M) = 1$.

We now consider some consequences of the hypotheses that will apply to the main result of this section.

(3.5) REMARKS: Suppose that (R, M) is quasi-local, and that $\mathcal{F} = (Q_n)_{n \in \mathbb{N}}$ is a multiplicative filtration of ideals of R for which $Q_1 = M$ and, for each $n \in \mathbb{N}$, the closure of M^n in the $\mathcal{F}$-topology is Q_n. Assume that R is (Hausdorff) complete with respect to the $\mathcal{F}$-topology. Let A be a non-zero Artinian R-module such that

$$A = \bigcup_{n \in \mathbb{N}} (0 :_A Q_n).$$

(i) If L is a homomorphic image of A, then $L = \bigcup_{n \in \mathbb{N}} (0 :_L Q_n)$.

(ii) Set $J = (0 : A)$. Write $\bar{R} = R/J$, $\bar{M} = M/J$ and, for each $n \in \mathbb{N}$, $\bar{Q}_n = (Q_n + J)/J$. Of course A can be viewed as a faithful Artinian $\bar{R}$-module in a natural way. Since

$$A = \bigcup_{n \in \mathbb{N}} (0 :_A \bar{Q}_n),$$

it follows that $\bigcap_{n \in \mathbb{N}} \bar{Q}_n = (0 :_{\bar{R}} A) = 0$. Let $\bar{\mathcal{F}}$ denote the multiplicative filtration $(\bar{Q}_n)_{n \in \mathbb{N}}$ of the quasi-local ring $(\bar{R}, \bar{M})$: we have that $\bar{Q}_1 = \bar{M}$, that, for each $n \in \mathbb{N}$, the closure of $\bar{M}^n$ in the $\bar{\mathcal{F}}$-topology is $\bar{Q}_n$, and that $\bar{R}$ is (Hausdorff) complete with respect to the $\bar{\mathcal{F}}$-topology.

The above observations will be helpful during the proof of the following theorem, which is a generalization of the Heinzer-Lantz Theorem. The ideas in the proof were inspired by some of those in Heinzer's and Lantz's paper [**2**].

(3.6) THEOREM. *Suppose that (R, M) is quasi-local and (Hausdorff) complete with respect to the topology induced by a multiplicative filtration $\mathcal{F} = (Q_n)_{n \in \mathbb{N}}$ of ideals of R which is such that $Q_1 = M$ and, for each $n \in \mathbb{N}$, the closure of M^n in the $\mathcal{F}$-topology is Q_n. Suppose that there exists a faithful Artinian R-module A such that*

$$A = \bigcup_{n \in \mathbb{N}} (0 :_A Q_n).$$

Then R is Noetherian and complete with respect to its M-adic topology.

PROOF: Set $k := R/M$. Let $\dim_k(0 :_A M) = s, \geq 1$. We first reduce to the case where $s = 1$. By Lemma (3.4), there exist submodules $A_1, \ldots, A_s$ of A such that $\bigcap_{i=1}^{s} A_i = 0$ and, for each $j = 1, \ldots, s$, the Artinian R-module A/A_j has 1-dimensional socle. For each $j = 1, \ldots, s$, set $J'_j = (0 : A/A_j)$, so that A/A_j is a faithful Artinian module having 1-dimensional socle over the quasi-local ring R/J'_j. It follows from (3.5) that, if the result had been proved in the special case in which $s = 1$, then we would be able to deduce that each R/J'_j is Noetherian, and since there is an R-monomorphism $R \longrightarrow \bigoplus_{i=1}^{s} R/J'_i$ because

$$\bigcap_{i=1}^{s} J'_i = \bigcap_{i=1}^{s} (A_i : A) = (0 : A) = 0,$$

it would follow that R is Noetherian; the full result would then follow easily from the Artin-Rees Lemma.

Thus we can, and do, assume henceforth that $s = 1$, so that $(0 :_A M) \cong k$. Set $E = E_R(k)$, the injective envelope of the R-module k. Since A is an essential extension of $(0 :_A M)$ by (3.3), we see that A can be identified with a submodule of E.

We now introduce some notation. For each $n \in \mathbb{N}$, we set

$$A_n = (0 :_A M^n), \quad J_n = (0 :_R A_n) = (0 :_R (0 :_A M^n)), \quad E_n = (0 :_E J_n).$$

It should be noted that, since $M^n \subseteq Q_n$, each element of Q_n is a limit (in the $\mathcal{F}$-topology) of a sequence of elements of M^n, and each element of A is annihilated by Q_j for some $j \in \mathbb{N}$, it follows that

$$A_n = (0 :_A M^n) = (0 :_A Q_n).$$

Consequently, for $r \in \mathbb{N}$ with $r \leq n$, we must also have

$$A_n = (0 :_A M^n) = (0 :_A M^{n-r}Q_r) = (0 :_A Q_n)$$

because $M^n \subseteq M^{n-r}Q_r \subseteq Q_n$. A further point to note is that, for $n, k \in \mathbb{N}$, we have $Q_n J_k \subseteq J_{k+n}$ because

$$\begin{aligned} Q_n J_k A_{k+n} = Q_n J_k (0 :_A M^{k+n}) &= Q_n J_k (0 :_A M^k Q_n) \\ &\subseteq J_k (0 :_A M^k) = 0. \end{aligned}$$

Let $n \in \mathbb{N}$. It is now convenient to introduce the further notation $R_n := R/J_n$. Note that A_n is a non-zero faithful Artinian module over the quasi-local ring $(R_n, M/J_n)$; since $(M/J_n)^n A_n = 0$, it follows that A_n is Noetherian, too. Therefore, since A_n is faithful as R_n-module, we see that R_n is a Noetherian ring. But $M^n \subseteq J_n$, and so $(M/J_n)^n R_n = 0$; therefore $(R_n, M/J_n)$ is an Artinian local ring.

Now the faithful Artinian R_n-module A_n is a submodule of $E_n = (0 :_E J_n)$; but the latter is an injective envelope of the residue field of R_n (in view of the natural equivalence of functors

$$\operatorname{Hom}_R(\cdot \otimes_{R_n} R_n, E) \longrightarrow \operatorname{Hom}_{R_n}(\cdot, \operatorname{Hom}_R(R_n, E))$$

of [**11**, Chapter 8, Theorem 4]); hence $A_n = E_n$ by (2.3).

Fix $k \in \mathbb{N}$, and let $n \in \mathbb{N}$ be such that $n \geq k$. Our immediate aim is to show that

$$J_k = J_n + M^k = J_n + M^{k-r}Q_r \text{ for each } r \in \mathbb{N} \text{ with } r \leq k.$$

Our argument will involve lengths of modules, and we use ℓ to denote such lengths. It will also be convenient to write $Q_0 = R$, and to deal with an integer r such that $0 \leq r \leq k$.

Since $J_n \subseteq J_n + M^{k-r}Q_r$ and $R_n = R/J_n$ is an Artinian local ring, it follows that $\ell_R(R/(J_n + M^{k-r}Q_r)) < \infty$. In fact

$$\begin{aligned}\ell_R(R/(J_n + M^{k-r}Q_r)) &= \ell_{R_n}(R_n/(J_n + M^{k-r}Q_r)/J_n) \\ &= \ell_{R_n}(0 :_{E_n} (J_n + M^{k-r}Q_r)/J_n)\end{aligned}$$

by Matlis's duality (2.1)(v) and (2.4)(i). But $A_n = E_n$, and so

$$\begin{aligned}\ell_R(R/(J_n + M^{k-r}Q_r)) &= \ell_{R_n}(0 :_{A_n} (J_n + M^{k-r}Q_r)/J_n) \\ &= \ell_R(0 :_{A_n} J_n + M^{k-r}Q_r) \\ &= \ell_R(0 :_A M^{k-r}Q_r)\end{aligned}$$

since if $x \in (0 :_A M^{k-r}Q_r) = (0 :_A M^k) = A_k$, then $x \in A_n$ since $n \geq k$ and so x is automatically annihilated by J_n.

We have therefore shown that, for all $n, k \in \mathbb{N}$ with $n \geq k$ and all $r \in \mathbb{N}_0$ with $r \leq k$,

$$\ell_R(R/(J_n + M^{k-r}Q_r)) = \ell_R(A_k) = \ell_R(R/J_k),$$

so that, since $J_n + M^{k-r}Q_r \subseteq J_k$ and all these lengths are finite,

$$J_n + M^{k-r}Q_r = J_k,$$

as claimed.

The next stage in the proof is to show that, for each $k \in \mathbb{N}$, we have that $Q_k = M^k$ and Q_k is finitely generated. Since $R_{k+1} = R/J_{k+1}$ is Noetherian and, by what we have just shown,

$$(Q_k + J_{k+1})/J_{k+1} = (M^k + J_{k+1})/J_{k+1},$$

there exist $t \in \mathbb{N}$ and elements $z_1, \ldots, z_t \in M^k$ whose natural images in R/J_{k+1} generate the above ideal. Our aim is to show that $z_1, \ldots, z_t$ generate Q_k.

Let $a_k \in Q_k$. Then there exist $a_{10}, \ldots, a_{t0} \in R$ and $a_{k+1} \in J_{k+1}$ such that

$$a_k = \sum_{i=1}^{t} a_{i0}\, z_i + a_{k+1}.$$

Suppose, inductively, that $n \geq 1$ and we have found $a_{k+n} \in J_{k+n}$ and, for each $j = 0, \ldots, n-1$, elements $a_{1j}, \ldots, a_{tj} \in Q_j$ such that

$$a_k = \sum_{i=1}^{t} \left(\sum_{j=0}^{n-1} a_{ij} \right) z_i + a_{k+n}.$$

Now $J_{k+n} = J_{k+n+1} + M^k Q_n$, by the paragraph before last in this proof, and so $a_{k+n} = \sum_{r=1}^{p} h_r \ell_r + c_{k+n+1}$, where $p \in \mathbb{N}$, $h_1, \ldots, h_p \in Q_n$, $\ell_1, \ldots, \ell_p \in M^k$ and $c_{k+n+1} \in J_{k+n+1}$.

For each $r = 1, \ldots, p$, there exist $\ell_{1r}, \ldots, \ell_{tr} \in R$ and $j_{r,k+1} \in J_{k+1}$ such that

$$\ell_r = \sum_{i=1}^{t} \ell_{ir} z_i + j_{r,k+1}.$$

It follows that

$$a_k = \sum_{i=1}^{t} \left(\sum_{j=0}^{n} a_{ij} \right) z_i + a_{k+n+1}$$

where, for each $i = 1, \ldots, t$,

$$a_{in} = \sum_{r=1}^{p} \ell_{ir} h_r \in Q_n,$$

and $a_{k+n+1} = \sum_{r=1}^{p} h_r j_{r,k+1} + c_{k+n+1} \in Q_n J_{k+1} + J_{k+n+1} \subseteq J_{k+n+1}$. The inductive step is therefore complete.

We have therefore constructed, for each $i = 1, \ldots, t$, a sequence $(a_{in})_{n \in \mathbb{N}_0}$ with $a_{in} \in Q_n$ for all $n \in \mathbb{N}_0$. Since R is complete with respect to the topology induced by the filtration $(Q_n)_{n \in \mathbb{N}}$, it follows that $\sum_{n=0}^{\infty} a_{in}$ converges, to $\hat{a}_i \in R$, say. Now for all $n \in \mathbb{N}$ we have

$$\hat{a}_i - \sum_{j=0}^{n} a_{ij} \in Q_{n+1} \text{ for } 1 \leq i \leq t,$$

and so, for all $n \in \mathbb{N}$,

$$\begin{aligned} \sum_{i=1}^{t} \hat{a}_i z_i - a_k = \sum_{i=1}^{t} \left(\hat{a}_i - \sum_{j=0}^{n} a_{ij} \right) z_i - a_{k+n+1} &\in Q_{n+1} M^k + J_{k+n+1} \\ &\subseteq J_{k+n+1}. \end{aligned}$$

Hence

$$\sum_{i=1}^{t} \hat{a}_i z_i - a_k \in \bigcap_{j=k+2}^{\infty} J_j = \bigcap_{j=k+2}^{\infty} (0 :_R (0 :_A M^j)) \subseteq (0 : A) = 0.$$

It follows that $Q_k = M^k$, and this is finitely generated. In particular, we have proved that $\mathcal{F} = (Q_n)_{n \in \mathbb{N}}$ is the M-adic filtration of R, and that M is finitely generated. It now follows from [**15**, p. 260, Corollary 4] that R is Noetherian, and so is a complete local ring.

(3.7) COROLLARY. *Let A be a non-zero Artinian module over the quasi-local ring (R, M). Regard A as a module over the M-adic completion $\hat{R}$ of R in the manner indicated in (1.11). Then $\hat{R}/(0 :_{\hat{R}} A)$ is Noetherian, and a complete local ring.*

PROOF: This now follows immediately from (1.12), (1.13), (3.5) and (3.6).

4. An application to asymptotic behavior.

In this section, we shall use the reduction procedure developed in Sections 1 and 3, together with Matlis's duality, to prove a new result: the conjecture which this result establishes had previously resisted all the author's attempts to prove it by more direct means. We begin by setting the scene.

(4.1) NOTATION: Let N be a Noetherian R-module. Brodmann [**1**] (essentially) proved that both the sequences of sets

$$(\mathrm{Ass}_R(N/I^n N))_{n \in \mathbb{N}} \text{ and } \left(\mathrm{Ass}_R(I^{n-1}N/I^n N)\right)_{n \in \mathbb{N}}$$

are ultimately constant; let $\mathrm{As}^*(I, N)$ and $\mathrm{Bs}^*(I, N)$ denote their ultimate constant values (respectively).

Let A be an Artinian R-module. In [**13**], it was proved that both the sequences of sets

$$(\mathrm{Att}_R(0 :_A I^n))_{n \in \mathbb{N}} \text{ and } \left(\mathrm{Att}_R\left((0 :_A I^n)/(0 :_A I^{n-1})\right)\right)_{n \in \mathbb{N}}$$

are ultimately constant; let $\mathrm{At}^*(I, A)$ and $\mathrm{Bt}^*(I, A)$ denote their ultimate constant values (respectively).

In the notation of (4.1), we have $\mathrm{Bs}^*(I,N) \subseteq \mathrm{As}^*(I,N)$ and (by (2.6), for example) $\mathrm{Bt}^*(I,A) \subseteq \mathrm{At}^*(I,A)$. In the special case in which R itself is Noetherian, McAdam and Eakin [**10**, Corollary 13] showed that

$$\mathrm{As}^*(I,R) \setminus \mathrm{Bs}^*(I,R) \subseteq \mathrm{Ass}(R),$$

and we adapt their argument to show, in (4.2) below, that

$$\mathrm{As}^*(I,N) \setminus \mathrm{Bs}^*(I,N) \subseteq \mathrm{Ass}(N).$$

In the light of this, it is natural to conjecture that

$$\mathrm{At}^*(I,A) \setminus \mathrm{Bt}^*(I,A) \subseteq \mathrm{Att}(A),$$

and our main result of this section is to establish that this is so.

(4.2) THEOREM. *Let N be a Noetherian R-module. Then*

$$\mathrm{As}^*(I,N) \setminus \mathrm{Bs}^*(I,N) \subseteq \mathrm{Ass}(N).$$

PROOF: By replacing R by $R/(0:N)$ we can, and do, reduce to the case where R is Noetherian. The result can now be established by making straightforward modifications to the argument used by McAdam and Eakin (as presented in [**9**, 2.2]) in the special case in which $N = R$. We therefore give only an outline of the proof. Let $P \in \mathrm{As}^*(I,N) \setminus \mathrm{Bs}^*(I,N)$. We must prove that $P \in \mathrm{Ass}(N)$.

Localization and completion enable us to reduce the problem to the case where (R,M) is a complete local ring and $P = M$, and so we assume that this is the case. Also, we can, and do, assume that I is proper.

Let $G := G_I(R) = \bigoplus_{i \in \mathbb{N}_0} G_i$, where $G_i = I^i/I^{i+1}$ for all $i \in \mathbb{N}_0$, be the associated graded ring of R with respect to I. Note that $G_I(N) := \bigoplus_{i\in\mathbb{N}_0} I^iN/I^{i+1}N$ is a Noetherian graded module over G. Consideration of the homogeneous submodule $(0 :_{G_I(N)} G_+)$ of $G_I(N)$, where $G_+ = \bigoplus_{i\in\mathbb{N}} G_i$, enables us to find an $\ell \in \mathbb{N}$ such that

$$\begin{aligned}
&(I^jN :_N I) \cap I^\ell N = I^{j-1}N && \text{for all } j > \ell,\\
&\mathrm{As}^*(I,N) = \mathrm{Ass}_R(N/I^jN) && \text{for all } j > \ell, \text{ and}\\
&\mathrm{Bs}^*(I,N) = \mathrm{Ass}_R(I^{j-1}N/I^jN) && \text{for all } j > \ell.
\end{aligned}$$

A simple inductive argument, involving the exact sequences

$$0 \longrightarrow I^jN/I^{j+t+1}N \longrightarrow I^{j-1}N/I^{j+t+1}N \longrightarrow I^{j-1}N/I^jN \longrightarrow 0,$$

now shows that

$$\mathrm{Bs}^*(I,N) = \mathrm{Ass}_R(I^{j-1}N/I^{j+t}N) \text{ for all } j > \ell \text{ and all } t \in \mathbb{N}_0.$$

For each $j > \ell$, $V_j := ((I^jN :_N M) + I^\ell N)/I^\ell N$ is a non-zero subspace of the finite-dimensional vector space $(I^\ell N :_N M)/I^\ell N$ (over R/M). Hence $\bigcap_{j=\ell+1}^{\infty} V_j \neq 0$: let $y \in (I^\ell N :_N M)$ be such that $y + I^\ell N$ is a non-zero element of $\bigcap_{j=\ell+1}^{\infty} V_j$. For each $j > \ell$, we can write $y = u_j + w_j$, where $u_j \in (I^jN :_N M)$ and $w_j \in I^\ell N$. We can then show that both the sequences $(u_j)_{j \geq \ell+1}$ and $(w_j)_{j \geq \ell+1}$ are Cauchy, and use (the completeness of N and) Krull's Intersection Theorem to see that $u = \lim_{j\to\infty} u_j$ satisfies $(0 : u) = M$.

We are now completely ready to prove the result "dual" to (4.2).

(4.3) THEOREM. *Let A be an Artinian R-module. Then*

$$\mathrm{At}^*(I,A) \setminus \mathrm{Bt}^*(I,A) \subseteq \mathrm{Att}(A).$$

PROOF: First use (1.5) to reduce to the case where $A = \Gamma_M(A)$ for some maximal ideal M of R. Then use (1.9) to reduce to the case where (R,M) is a quasi-local ring. Clearly we can assume that $A \neq 0$ and that I is a proper ideal. We then use (1.10), (1.11), (1.12), (1.13) and (3.7) to reduce to the case where (R,M) is a complete local ring.

We can therefore apply Matlis's duality, and we use the notation of (2.1). By (2.1)(iii), the R-module $D(A)$ is Noetherian. Hence, by (4.2),

$$\mathrm{As}^*(I,D(A)) \setminus \mathrm{Bs}^*(I,D(A)) \subseteq \mathrm{Ass}(D(A)).$$

But, by (2.7), $\mathrm{Ass}(D(A)) = \mathrm{Att}(A)$, and, in view of (2.4) and (2.1)(iv), if n is chosen sufficiently large that $\mathrm{As}^*(I,D(A)) = \mathrm{Ass}(D(A)/I^nD(A))$ and $\mathrm{At}^*(I,A) = \mathrm{Att}(0 :_A I^n)$, then

$$\begin{aligned}\mathrm{As}^*(I,D(A)) &= \mathrm{Ass}(D(A)/I^nD(A)) = \mathrm{Att}(D(D(A)/I^nD(A))) \\ &= \mathrm{Att}(0 :_A I^n) = \mathrm{At}^*(I,A);\end{aligned}$$

similarly, $\mathrm{Bs}^*(I,D(A)) = \mathrm{Bt}^*(I,A)$. The result follows.

Of course, one would like a more direct proof of this result. However, as was mentioned earlier, all my attempts to produce a direct proof have so far been unsuccessful.

Acknowledgements.

The author gratefully acknowledges the support of the United Kingdom Science and Engineering Research Council Grant No. GR/E/42389, which enabled him to attend the Microprogram on Commutative Algebra held at the Mathematical Sciences Research Institute, Berkeley, in June and July 1987.

He would also like to thank William Heinzer and David Lantz for several very helpful conversations at MSRI about their work in [**2**], and Peter Vámos for many conversations about Artinian modules, both in the recent past and several years ago.

REFERENCES

1. M. Brodmann, *Asymptotic stability of* $\mathrm{Ass}(M/I^nM)$, Proc. AMS **74** (1979), 16–18.
2. W. Heinzer and D. Lantz, *Artinian modules and modules of which all proper submodules are finitely generated*, J. Algebra **95** (1985), 201–216.
3. D. Kirby, *Coprimary decomposition of Artinian modules*, J. London Math. Soc. (2) **6** (1973), 571–576.
4. D. Kirby, *Artinian modules and Hilbert polynomials*, Quart. J. Math. Oxford (2) **24** (1973), 47–57.
5. I. G. Macdonald, *Secondary representation of modules over a commutative ring*, Symposia Mathematica **11** (1973), 23–43.
6. I. G. Macdonald and R. Y. Sharp, *An elementary proof of the non-vanishing of certain local cohomology modules*, Quart. J. Math. Oxford (2) **23** (1972), 197–204.
7. E. Matlis, *Injective modules over Noetherian rings*, Pacific J. Math. **8** (1958), 511–528.
8. E. Matlis, *Modules with descending chain condition*, Trans. AMS **97** (1960), 495–508.
9. S. McAdam, "Asymptotic prime divisors," Lecture Notes in Mathematics, vol. 1023, Springer-Verlag, Berlin, 1983.
10. S. McAdam and P. Eakin, *The asymptotic* Ass, J. Algebra **61** (1979), 71–81.
11. D. G. Northcott, "An Introduction to Homological Algebra," Cambridge University Press, 1960.
12. D. G. Northcott, *Generalized Koszul complexes and Artinian modules*, Quart. J. Math. Oxford (2) **23** (1972), 289–297.
13. R. Y. Sharp, *Asymptotic behaviour of certain sets of attached prime ideals*, J. London Math. Soc. (2) **34** (1986), 212–218.
14. D. W. Sharpe and P. Vámos, "Injective modules," Cambridge Tracts in Mathematics and Mathematical Physics, vol. 62, Cambridge University Press, 1972.
15. O. Zariski and P. Samuel, "Commutative Algebra, Vol. II," Graduate Texts in Mathematics, vol. 29, Springer-Verlag, Berlin, 1975.

Department of Pure Mathematics, University of Sheffield, Hicks Building, Sheffield S3 7RH, ENGLAND

Symmetric Algebras and Factoriality

WOLMER V. VASCONCELOS

Introduction.

The symmetric algebras over a commutative, Noetherian, ring R are among the simplest extensions of R: If an R-module E is given by a presentation

$$R^m \xrightarrow{\varphi} R^n \longrightarrow E \longrightarrow 0,$$

its symmetric algebra over R, $S(E) = \mathrm{Sym}_R(E)$, is the quotient of the polynomial ring $R[T_1, \ldots, T_n]$ by the ideal J generated by the 1-forms

$$f_j = a_{1j}T_1 + \cdots + a_{nj}T_n, \; j = 1, \ldots, m.$$

Despite this simplicity of description, there is considerable difficulty in ascertaining when $S(E)$ is an integral domain or, more generally, inherits *nice* depth properties from R. One source of the problem arises from the intractability of translating syzygetic properties of E into ideal theoretic properties of $S(E)$. There is, for instance, no adequate manner of finding the details of a projective resolution of the components of $S(E)$ from the resolution of E, except for exceptional cases. Instead certain indirect methods involving linkage and *ad hoc* complexes, dubbed approximation complexes, have been used to provide a measure of the necessary connection.

The aim of this paper is to continue to develop a computer-assisted approach to the study of ideal transforms of certain affine algebras derived from symmetric algebras initiated in [20]. Their interest lies in the fact that the modifications in question are factorial domains and symbolic blow-up rings whose structure is, in general, unknown. By focusing on a restricted family of cases certain predictions about the structure and finiteness of the transforms can be made. Using symbolic computation packages, a method is suggested for obtaining the defining equations of the rings beyond the horizon of those predictions. The methods of [20] are streamlined, and applied to new settings.

Unless stated otherwise, to simplify the discussion, we shall assume that R is a regular local ring or a polynomial ring over a field.

Among the several problems of a long standing that we shall be looking at, it is noteworthy to single out the following:

(a) The first question is: When is the symmetric algebra of the R-module E, $S(E)$, factorial? One set of conditions that has been identified is expressed in the sizes of the Fitting's ideals of E. Specifically, if E has a presentation as above, the notation $I_t(\varphi)$ will denote the ideal generated by the t-sized minors of φ. Consider the condition

$$(1) \qquad \mathcal{F}_k : \text{height}(I_t(\varphi)) \geq \text{rank}(\varphi) - t + 1 + k, \; 1 \leq t \leq \text{rank}(\varphi).$$

For $S(E)$ to be a domain requires $\mathcal{F}_1$, while E satisfies $\mathcal{F}_2$ if $S(E)$ is factorial. While these conditions are not enough, in general, they are useful in several interesting cases. They are sufficient conditions if E is a module of projective dimension at most one–that is, when $S(E)$ is a complete intersection–and imply that if $S(E)$ is factorial, then the projective dimension of E cannot be two. Other cases, in higher dimension, have been clarified as well (cf. [13]).

They lend support to the *first factorial conjecture*: If $S(E)$ is factorial then it must be a complete intersection. It can be shown that if the enveloping algebra of $S(E)$ is an integral domain–a condition somewhat stronger than factoriality–then $S(E)$ must indeed be a complete intersection (cf. [21]).

(b) If symmetric algebras that are factorial seem rare, there is a straightforward process that produces the *factorial closure* of any symmetric algebra $S(E)$. The setting is a sequence of modifications of the algebra $S(E)$, each more drastic than the preceding. Define:

(i) $D(E) = S(E)/\text{mod}\ R\text{-torsion}$ ($D(E)$ is a domain); (ii) $C(E) =$ integral closure of $D(E)$; (iii) $B(E) =$ the graded bi-dual of $S(E)$, that is if $S(E) = \oplus\text{Sym}_t(E)$, then $B(E) = \oplus\text{Sym}_t(E)^{**}$, where (**) denotes the bi-dual of an R-module. $B(E)$ is a factorial ring (cf. [6]; see also [13], [18]). These algebras are connected by a sequence of homomorphisms:

$$S(E) \longrightarrow D(E) \longrightarrow C(E) \longrightarrow B(E).$$

The algebra $D(E)$ is easy to obtain from $S(E)$. The other two algebras, $C(E)$ and $B(E)$, are a different matter. The body of evidence suggests the validity of the *second factorial conjecture*: $B(E)$ is a Gorenstein domain. The harder part of this question is, of course, whether $B(E)$ is Noetherian. The approach taken here to this and the next question is dependent on the structure.

The equality of the two last algebras is well–understood (cf. [13]):

THEOREM. *Let R be a normal, Cohen-Macaulay, universally Japanese domain.*

(a) *If E satisfies $\mathcal{F}_2$ then $B(E) = C(E)$.*

(b) *Conversely, if $S(E)$ is a domain and $B(E) = C(E)$ then E satisfies $\mathcal{F}_2$.*

We shall see later that the condition $S(E) = C(E)$ considerably simplifies the search for $B(E)$.

(c) If P is a prime ideal, denote its n–th symbolic power by $P^{(n)}$. The subring of the polynomial ring

$$R_s(P) = \sum P^{(n)} t^n \subset R[t],$$

is its symbolic power algebra–or symbolic blow-up ring. These algebras are not always Noetherian–counterexamples are shown to exist already in the ring of polynomials in three indeterminates (cf. [17]).

There is a tenuous connection between the finiteness of an algebra such as $B(E)$ and the symbolic blow–up of a codimension two prime ideal. It turns out that both (b) and (c) can be interpreted as problems in computing ideal transforms. This tool, used by Nagata [16] in his solution of the Hilbert's Fourteenth Problem, is a critical element of this study.

To deal with questions (b) and (c) we used in [20] a computer-assisted method, to successively add to $S(E)$ elements of ever higher degree while testing for equality between the two algebras. Here we refine the method and broaden the scope of the algebras for which one can make some predictions about the structure of the first approximations.

We shall now describe the contents of this paper. In section two, both the factorial closure of a symmetric algebra and the symbolic blow-up of a prime ideal are interpreted as ideal transforms. There is a feature making them rather special: They are both ideal transforms of $\mathbb{N}$–graded rings with respect to ideals generated by two homogeneous elements of degree zero. Furthermore, the ideal transform can be expressed as a directed union of Noetherian subalgebras each of which can be tested for equality with the full transform. With the ground thus set, we outline a method to compute the successive subalgebras–*the approximations*–using the Gröbner basis method of Buchberger. It was used to obtain the presentation of several algebras and get insight into the form of some approximations. In particular, Theorem was intuited from the calculations. An attempt to prove it by the computer was also made; it only partially succeeded.

Although arbitrary modules (over polynomial rings) can be considered, we looked for classes of modules for which some predictions concerning the subalgebras could be made. Noteworthy are modules which are free outside of the origin—e.g. modules of global sections of vector bundles in projective space in the case of the factorial closure or prime ideals which are complete intersections in the punctured spectrum of a local ring in the case of symbolic blow-ups. In section 2 we consider modules whose symmetric algebra $S(E)$ is a normal complete intersection. The main results are: (i) Theorem that says that if r is the degree where $S(E)$ and $B(E)$ begin to differ and if $B(r)$ is the subalgebra of $B(E)$, generated by the homogeneous elements of $B(E)$ of degree at most r, then several assertions can be made about the structure of $B(r)$, all derived from the explicit equations for a presentation of $B(r)$. These can be read off the presentation of the module E itself. It sets the background, for the next approximation, $B(r+1)$. This algebra has also a definite form whose details however have to be computed (cf. Theorem).

Algebras which are normal almost complete intersections are considered next. The details of $B(r)$ can also be written down from the original data by Theorem . Ultimately, what permits these descriptions is the fact that the Koszul homology modules of codimension two perfect ideals and codimension three Gorenstein ideals are Cohen-Macaulay.

Section four consists of two results. First, a formula that describes the symbolic square of an almost complete intersection. Then, a sketch aimed at linking the finiteness of the factorial closure of modules and the symbolic blow-up of certain codimension two prime ideals.

1. Ideal transforms.

Let I be an ideal of the Noetherian integral domain R, of field of fractions K. The *ideal transform* of I, $T_R(I)$, is the subring of all elements of K that can be transported into R by a high enough power of I:

$$B = T_R(I) = \bigcup_i I^{-i}.$$

In other words, B is the ring of global sections of the structure sheaf of R on the open set defined by I. The fundamental reference for this notion is [16]. We point out the following observation. If $I = I_1 \bigcap I_2$, and I_2 has grade at least two, then the ideal transforms of I and I_1 are the same. Furthermore, if R has the condition S_2 of Serre, we may even replace I by a subideal generated by two elements. As a matter of fact, in all the cases treated here the ideal I can always be taken to be generated by two elements, even when the condition S_2 is not present. This will represent an important computational simplification.

1.1. Factorial closure of a symmetric algebra.

Let R be Cohen-Macaulay factorial domain and let E be a module with a presentation:

$$R^m \xrightarrow{\varphi} R^n \longrightarrow E \longrightarrow 0.$$

Denote by $D(E)$ the quotient of $S(E)$ modulo the ideal of torsion elements. To obtain $B(E)$ one might as well apply the bi-dualizing procedure on $D(E)$. In particular we may assume that E is a torsion-free module. (More about this later.) According to [18], $B(E)$ can be described in the following manner. First, embed $D(E)$ into a polynomial ring $K[U_1, \ldots, U_e]$, K the field of fractions of R and e the rank of E. Then:

$$B(E) = \bigcap_{p \subset R} D(E)_p, \quad \text{height}(p) = 1. \tag{2}$$

PROPOSITION 1.1.1. *Let J be the Fitting ideal $I_{n-e}(\varphi)$ and put $M = JD(E)$. Then $B(E)$ is the M-ideal transform of $D(E)$.*

PROOF: Because E is assumed torsion free, J is an ideal of height at least two. If T denotes the ideal transform of $D(E)$ with respect to J, it is clear from the equation above that $T \subset B(E)$. Conversely, if $b \in B(E) \setminus D(E)$, denote by L the conductor ideal $D(E) :_R b$. Let Q be a minimal prime of L; if J is not contained in Q, the localization E_Q is a free R_Q-module and therefore $S(E_Q) = B(E_Q)$, so that $L_Q = R_Q$–which would be a contradiction. This shows that the radical of L contains J. □

COROLLARY 1.1.2. *There exists an ideal I generated by a regular sequence $\{f, g\}$ such that $B(E) = T_{D(E)}(I)$.*

PROOF: Let $\{f, g\}$ be any regular sequence contained in the ideal J (recall: height$(J) \geq 2$). It is clear that $T_{D(E)}(J) \subseteq T_{D(E)}(I)$. On the other hand, by Equation 2 any such transform must be contained in $B(E)$. □

1.2. Symbolic blow-ups.

Let R be an integral domain and let P be a prime ideal. The set of associated prime ideals of the collection of modules $\{R/P^n, n \geq 1\}$ is known to be finite (cf. [1]; see also [7]). If it reduces to P, $R_s(P)$, the symbolic blow–up, coincides with the ordinary Rees algebra of P. If not, pick f an element belonging to each of those primes except for P.

PROPOSITION 1.2.1. *Let R be an integral domain with the S_2 property of Serre. Let P be a prime ideal and let f be an element as above. Let g be an element of P such that $\{f, g\}$ is a regular sequence. Then $R_s(P)$ is the ideal transform of $R(P)$ with respect to $I = (f, g)$.*

PROOF: This follows from the choices of f and g, and the fact that the ideal transform $T_{R(P)}(f,g)$ is a graded ring that lives in $R[t]$, since $\{f, g\}$ is a regular sequence. □

1.3. The computation of ideal transforms.

In both of our examples of ideal transforms we could write the algebra B as

$$R \oplus B_1 \oplus B_2 \oplus \cdots,$$

where each B_i is a finitely generated R-module. We denote by $B(r)$ the subalgebra of B generated by its homogeneous components up to degree r. Note that B is also the ideal transform, with respect to the same ideal, of any of the $B(r)$'s ($r \geq 1$). In particular we have:

PROPOSITION. *$B = B(r)$ if and only if grade $I \cdot B(r) \geq 2$.*

This section consists of a series of observations on how to use the theory of Gröbner bases, as implemented by Buchberger's algorithm [4], to compute ideal transforms. It is not very clear whether similar methods could be used to deal with arbitrary ideal transforms. Our case is rather special in that each of the rings is a graded algebra, the ideal is generated by elements of degree zero, and the ideal transform lives naturally in a polynomial ring. We make use of each of these features. [21] and [22] discuss computations of a more general nature in commutative algebra.

We represent each $B(r)$ as a quotient $P(r)/J(r)$, where $P(r)$ is a graded polynomial ring. Thus $J(E) = J(1)$ is the ideal of 1-forms given by the presentation of the module E. We seek to construct $J(r + 1)$ from $J(r)$ and to test whether the equality $B = B(r)$ already holds.

To get started, let us indicate how to obtain the algebra $D(E)$. We assume the module E is given as the cokernel of the matrix $\varphi = (a_{ij})$, an $n\times m$ matrix. Let f be a nonzero element of R chosen so that the localization of E at f is a free module. If E has rank e, f can be taken in the n-e–th Fitting ideal of E; if E is a prime ideal, f may be an element of E itself. Let

$$L(E) = \bigcup_{t\geq 1}(J(E) \ : f^t);$$

then $D(E) = R[T_1, \ldots, T_n]/L(E)$ (see Example). Note that if E is an ideal, $D(E)$ is the corresponding Rees algebra. In particular, it provides a presentation of the module E_0 obtained by moding out the R-torsion of E.

If we are looking for the factorial closure, the double dual of E must be determined first (In the case of the symbolic blow-up, $D(P) = B(1)$.). Instead of using the method to be discussed later, at this point the following alternative procedure may be applied. From the preceding one may assume that E is a torsion-free R-module. In fact, we suppose that E is given as a submodule of a free module $RT_1 \oplus \cdots \oplus RT_e$, generated by the forms

$$g_k = b_{k1}T_1 + \cdots + b_{ke}T_e, \ k = 1, \ldots, s.$$

The ring $D(E)$ is then the subring of the polynomial ring $R[T_1, \ldots, T_n]$ generated by the forms g_k's over R. To obtain the bi-dual of E we found useful to compute the ideal transform of $\{f,\ g\}$, chosen earlier, with respect to the subring generated by the g_k–forms and all the products T_iT_j. This is somewhat cumbersome but effective.

There are three parts to obtaining $J(r+1)$ from $J(r)$. Assume that R is a polynomial ring in the variables $x_1, \ldots, x_d$, and that E is a module with a presentation as before.

Step 1. Let $\{f,\ g\}$ be a regular sequence as in Corollary . To determine whether new generators must be added to $B(r)$ in order to obtain $B(r+1)$, we must compute the degree *(r+1)*–component of the ideal transform $T_{B(r)}(f,g)$, that is

$$\bigcup_{t\geq 1} B(r)_{r+1} : (f,g)^t.$$

This module is nothing but B_{r+1}. It is convenient to keep track of degrees and work in the full ring $A = B(r)$. Observe that for any ideal $L = (f_1, \ldots, f_p)$ of A,

$$A : L = \bigcap_{1\leq i\leq p} (Af_1 :_A f_i)f_1^{-1}.$$

When applied to $L = (f,g)^t$, we obtain:

$$A(t) = A : (f,g)^t = (\bigcap_{1\leq i\leq t} (Af^t :_A g^i f^{t-i}))f^{-t} = (\bigcap_{1\leq i\leq t} (Af^i :_A g^i))f^{-t}.$$

Phrased in terms of the ring $P(r)$ this would be:

$$(\bigcap_{1\leq i\leq t} (L_{r+1}, f^i) :_{P(r)} g^i)f^{-t},$$

where L_{r+1} denotes the subideal of $J(r)$ generated by the elements of degree at most *r+1*. (Note: One can also use the full $J(r)$ instead of L_{r+1}, a move that is useful when $B(r)$ turns out to be $B(r+1)$.) Because each $B(r)_j$, $j \leq r$, is a reflexive module, $A(t)$ and $A(t+1)$ can only begin to differ in degree *r+1*. Stability is described by the equality:

$$f(\bigcap_{1\leq i\leq t} Af^i :_A g^i)_{r+1} = (\bigcap_{1\leq i\leq t+1} Af^i :_A g^i)_{r+1}.$$

Suppose $u_i = e_i f^{-t}, i = 1, \ldots, s$, generate $A(t)_{r+1}/A_{r+1}$; we now get to the determination of the equations for the u_i's.

Step 2. The brute force application of the Gröbner basis algorithm permits mapping the polynomial ring $P(r+1)$ onto the new generators to get the ideal $J(r+1)$. It does not work well at all. It is preferable making use of the fact that B is a rational extension of $B(r)$, by seeking the $B(r)$-conductors of each of the u_i. It provides for a number of equations of the form

$$A_{ji}U_i - B_{ji},\ j = 1, \ldots, r_i.$$

This is useful when $B(r)$ is normal and $B(r+1)$ is singly generated over it; the set above would be all that is needed. In general one needs one extra move.

As a matter of fact, it suffices to consider the R-conductors of the elements u_i whenever the next step is going to be used.

Step 3. Denote $P(r+1) = P(r)[U_1, \ldots, U_s]$ and let $L(r+1) = J(r)$ together with all the linear equations above. If $L(r+1)$ is a prime ideal, then it obviously equals $J(r+1)$. Let h be an element such that the localization $S(E_h) = B_h$, e.g. pick $h = f$ above. It is easy to see that

$$J(r+1) = \bigcup_t (L(r+1) :_{P(r+1)} h^t).$$

REMARK 1.3.2: In view of the complexity of Buchberger's algorithm, considerable experimentation was exercised. For instance, (i) the choice of f, g may be changed from one approximation to the next, and (ii) the well-ordering of the monomials has to be played with to permit the computation to go through. It always worked better when f and g were actual variables.

2. Complete intersections.

Let R be a regular local ring of dimension three, and let E be a reflexive module of projective dimension one, free on the punctured spectrum of R:

$$0 \longrightarrow R^m \xrightarrow{\varphi} R^n \longrightarrow E \longrightarrow 0.$$

We shall assume that $S(E)$ is a domain–equivalent here to the condition $\mathcal{F}_1$. If $\mathcal{F}_2$ does hold, $S(E)$ is a factorial domain. If $B(E) \neq S(E)$, we must have $m = d-1, d = \dim R$.

There is yet another additional condition that simplifies the computation of $B(E)$, the assumption that $S(E)$ be a normal domain. Normality for an algebra such as $S(E)$ can be expressed as follows (cf. [21]). Since $S(E)$ is a Cohen-Macaulay domain, by Serre's criterion, $S(E)$ is normal when the localizations $S(E)_P$, for height 1 primes P, are discrete valuation domains. We may assume that $\mathbf{p} = P \bigcap R \neq O$. Localizing at $\mathbf{p}$, we get $P_{\mathbf{p}} = \mathbf{p}S(E)_{\mathbf{p}}$ Change the notation so that R is a local ring of maximal ideal $\mathbf{p}$ and the presentation of E is minimal—that is, the entries of φ lie in $\mathbf{p}$. It follows that if $S(E)$ is normal the image of $J(E)$ in

$$\mathbf{p}/\mathbf{p}^2 \otimes R[T_1, \ldots, T_n]_{\mathbf{p}R[T_1,\ldots,T_n]},$$

must have rank equal to the embedding dimension of R minus one. In particular

$$\dim((I_1(\varphi) + \mathbf{p}^2)/\mathbf{p}^2) \geq \dim R - 1. \tag{3}$$

The method we shall employ does not use the integrality of $S(E)$, much less the normality of $S(E)$. These conditions permit however some early predictions.

The normality condition implies that we can find a system of parameters of $\mathbf{p}$ so that the ideal $I_1(\varphi) = (x_1, \ldots, x_{d-1}, x_d^a)$. From the matrix $\varphi = (a_{ij})$, we write the ideal of the definition of $S(E)$ in the following format:

$$J = (f_j = \sum_{i<d} x_i b_{ji} + x_d^a b_{jd},\ j = 1, \ldots, d-1),$$

where the b_{ji} are linear forms in the T_k's.

2.1. The first approximation.

Several Gorenstein rings appear, quite unexpectably, among the algebras $B(r)$. We shall attempt to explain this occurrence. Most occur as specializations of the defining ideals of some generic Rees algebras. The setting is as follows. Let R be a regular local ring and let I be an ideal. I is said to be *strongly Cohen-Macaulay* if the homology modules of the Koszul complex on a set (any will do) of generators of I are Cohen-Macaulay modules (see [12], [13]). One of the main results of [12] is:

THEOREM 2.1.1. *Let R be a regular local ring and let I be a strongly Cohen-Macaulay ideal. If I satisfies the condition $\mathcal{F}_1$, then the associated graded ring*

$$\mathrm{gr}_I(R) = \oplus_{i\geq 0} I^i/I^{i+1} \simeq \mathrm{Sym}(I/I^2)$$

is a Gorenstein ring.

A consequence is that the extended Rees algebra of I

$$A = R[It, t^{-1}]$$

is a Gorenstein algebra as well. This formulation provides for a presentation for A once the ideal I has been given by its generators and relations. Specifically, suppose $I = (x_1, \ldots, x_n)$ has a presentation

$$R^m \xrightarrow{\varphi} R^n \longrightarrow I \longrightarrow 0, \ \varphi = (a_{ij}).$$

We obtain A as the quotient of the polynomial ring $R[T_1, \ldots, T_n, U]$ modulo the ideal J generated by the 1-forms in the T_i's

$$f_j = a_{1j}T_1 + \cdots + a_{nj}T_n, \ j = 1, \ldots, m,$$

together with the linear polynomials

$$UT_i - x_i, \ i = 1, \ldots, n.$$

Since the Krull dimension of A is $\dim R + 1$, J is an ideal of height n. Let us indicate the cases we shall make use of.

(a) If X is a generic $n-1 \times n$ matrix (x_{ij}) and I is the ideal generated by the $n-1$-sized minors of X, then

$$J = (UT_1 - \Delta_1, \ldots, UT_n - \Delta_n, \ \Sigma_i x_{ji} T_i, \ j = 1, \ldots, n).$$

Note that by specializing $U = 0$, one obtains the ideal whose explicit resolution is given in [10].

(b) Let X be the generic, skew–symmetric matrix of order $n+1$, n even, $X = (x_{ij})$. Denote by I the ideal generated by the Pfaffians of X of order n. I is strongly Cohen-Macaulay and satisfies the condition on the local number of generators. J, in turn, is obtained as indicated above.

To return to the normal complete intersections, denote by $J(\varphi)$ the $(d-1) \times d$ matrix (b_{ji}), and by D the ideal generated by the $(d-1)$–sized minors of $J(\varphi)$. $D = (D_1, \ldots, D_d)$ is a Cohen-Macaulay ideal of height two. An important role is that of the ideal D *evaluated at* $\mathbf{p}$, that is, taken mod $\mathbf{p}$; it will be denoted by Δ. Note that normality requires $\Delta \neq 0$, while height $\Delta \leq 2$ by standard considerations. Δ can be interpreted as a kind of Jacobian ideal and its height is independent of the choices made.

In the remainder of this section, R will denote a regular local ring of dimension d, and $\mathbf{p}$ is its maximal ideal.

THEOREM 2.1.2. ([20, Theorem 2.2]) *Let $B(d-1)$ be the subalgebra of B obtained by adding to $S(E)$ the component of degree $d-1$ of B. Then:*

(a) *$B(d-1)$ is a Gorenstein ring.*

(b) *$B = B(d-1)$ if and only if* height $\Delta = 2$.

(c) *If* height $\Delta = 1$ *and Δ has no multiple component, or $a = 1$, then $B(d-1)$ is normal.*

PROOF: We begin by showing that $S(E)$ and B first differ in degree $d-1$. Since E is free on the punctured spectrum and $S(E) \neq B$, we must have in the resolution of E that $m = d-1$. The projective resolution of a symmetric power $S_t(E)$ is then (cf. [25]):

$$0 \to \wedge^m R^m \otimes S_{t-m}(R^n) \to \wedge^{m-1} R^m \otimes S_{t-m+1}(R^n) \to \cdots \to S_t(R^n) \to 0.$$

Because this complex is exact and E is free outside of $\mathbf{p}$, we have that $S_t(E)$ is a reflexive module for $t \leq d-2$, but not reflexive outside this range. Furthermore, a direct calculation shows that

$$\mathrm{Ext}^{d-1}(S_{d-1}(E), R) \;=\; R/I_1(\varphi).$$

Note that this measures the difference between B_{d-1} and S_{d-1}. Indeed, from the exact sequence

$$0 \longrightarrow S_{d-1} \longrightarrow B_{d-1} \longrightarrow C_{d-1} \longrightarrow 0$$

we obtain that

$$\mathrm{Ext}^d(C_{d-1}, R) = \mathrm{Ext}^{d-1}(S_{d-1}(E), R),$$

because B_{d-1} is a reflexive R-module. On the other hand, since the socle of $R/I_1(\varphi)$ is principal, we get, by duality, that $C_{d-1} = R/I_1(\varphi)$.

B_{d-1} is therefore obtainable from $S_{d-1}(E)$ by the addition of a single generator. We proceed to find this element. Denote by $D_1, \ldots, D_d$, the maximal minors of the matrix $J(\varphi)$. From the equations

$$f_j = 0,\ j = 1, \ldots, d-1$$

in $S(E)$, we obtain

$$x_i D_j \equiv x_j D_i$$

except that for $j = d$ we replace x_d by x_d^a. Let u be the element D_1/x_1 of the field of fractions of $S(E)$. By equation (2), $u \in B$.

We claim that $S(E)[u] = B(d-1)$, and that all of the asserted properties hold. First define the ideal $J(d-1)$ of the polynomial ring $P(d-1) = R[T_1, \ldots, T_n, U]$, generated by $J(E)$ and the polynomials

$$x_i U - D_i,\ i < d, \text{ and } x_d^a U - D_d.$$

By hypothesis $J(E)$ is a prime ideal of height $d-1$ and by normality one of the forms D_i has unit content. Thus the ideal $J(d-1)$ has height at least d. It is then a proper specialization of the ideal (a) of Theorem and is therefore a perfect Gorenstein ideal. Furthermore, since $\mathbf{p}P(d-1)$ is not an associated prime of $J(d-1)$, and E is free on the punctured spectrum of R, it follows easily that $J(d-1)$ is a prime ideal. This means that

$$P(d-1)/J(d-1) \simeq S(E)[u] \subseteq B(d-1).$$

We are now ready to prove the assertions. We begin by showing that u generates $B(d-1)$ over $S(E)$. Let v be the homogeneous generator of $B(d-1)$; we must have $u = bv + f$, where $b \in R$ and f is a form of $S(E)$ of degree $d-1$. We claim that b is a unit. Since the conductor of v, in degree 0, is $I_1(\varphi)$, the equations for v must be similar to those for u above, that is

$$x_i U - f_i,\ i < d, \text{ and } x_d^a U - f_d.$$

By normality, one of the D_i does not have all of its coefficients in $\mathbf{p}$; assume D_1 is such a form. $x_1 bU - D_1 + x_1 f$ must be a linear combination of the linear equations for v:

$$x_1bU - D_1 + x_1f = \sum_{i<d} b_i(x_iU - f_i) \ + \ b_d(x_d^aU - f_d)$$

with

$$(b_1 - b)x_1 + \sum_{1<i<d} b_ix_i \ + \ b_dx_d^a = 0.$$

But this is clearly impossible, unless b is a unit.

To prove (b) we use Proposition . As $B(d-1)$ is Cohen-Macaulay, we must have height $\mathbf{p}B(d-1) \geq 2$, a condition that is expressed by height $(\Delta) = 2$.

To make the notation simpler, in the proof of (c), we assume that $d = 3$, and label the variables $\{x,\ y,\ z\}$; the general case is similar.

The first condition in (c) is $I_2(J(\varphi)) = gK$ mod $\mathbf{p}$, where g is some 1-form in the T-variables, and $K \not\subseteq (g)$ mod $\mathbf{p}$. We must show that the localization of $B(d-1)$ at the minimal primes of $\mathbf{p}B(d-1)$ are discrete valuation domains. There are two cases to consider. (i) If height $(K, \mathbf{p}) \geq d+2$, $P = (g, \mathbf{p})$ is the only minimal prime of $(J(d-1), \mathbf{p})$. (ii) If height $(K, \mathbf{p}) \leq d+1$, $(J(d-1),\ \mathbf{p})$ has another minimal prime, $(K, \mathbf{p})$.

To verify normality we show that the image of $J(d-1)_P$ in the vector space $(P/P^2)_P$ has codimension 1.

We may write $D_i = gE_i + F_i$, where F_i has 'coefficients' in $\mathbf{p}$. g can be taken as an independent variable along with the x, y, z. We have the following matrix corresponding to the coordinates of the generators of $J(d-1)$ relative to the basis $\{x, y, z, g\}$ of $(P/P^2)_P$:

$$\begin{bmatrix} b_{11} & b_{12} & z^{a-1}b_{13} & 0 \\ b_{21} & b_{22} & z^{a-1}b_{23} & 0 \\ U - \partial F_1/\partial x & -\partial F_1/\partial y & -\partial F_1/\partial z & -E_1 \\ -\partial F_2/\partial x & U - \partial F_2/\partial y & -\partial F_2/\partial z & -E_2 \\ -\partial F_3/\partial x & \partial F_3/\partial y & z^{a-1}U - \partial F_3/\partial z & -E_3 \end{bmatrix}.$$

If $a = 1$, there exists a 3×3 minor which is a monic polynomial in U. On the other hand, if $a > 1$, then $(D_1, D_2) \subset \mathbf{p}$ and hence E_3 is not zero (mod P) by hypothesis and using the block consisting of the last 3 rows and columns 1, 2, 4 we get a non-vanishing 3×3 minor. □

EXAMPLE 2.1.3: (i) Our first example is the module in [18] whose transpose is the matrix:

$$\begin{bmatrix} x & y & z & 0 & 0 \\ 0 & 0 & x & y & z \end{bmatrix};$$

it has the associated matrix of 1-forms:

$$\begin{bmatrix} T_1 & T_2 & T_3 \\ T_3 & T_4 & T_5 \end{bmatrix}.$$

Because height $(\Delta) = 2$, $B(E) = B(2)$.

Note that at this stage only the leading forms of the entries are to be considered.

(ii) Consider

$$\begin{bmatrix} x & y & z & x^2 \\ y & z^2 & x^2 & z \end{bmatrix}.$$

In this case it is easy to see that height $(\Delta) = 2$, so that $B(E) = B(2)$. If we replace 'z' in the bottom row by 'z^2' we get the associated matrix of forms:

$$\begin{bmatrix} T_1 + xT_4 & T_2 & T_3 \\ xT_3 & T_1 & z(T_2 + T_4) \end{bmatrix}.$$

Now height $(\Delta) = 1$, so that $B \neq B(2)$. We shall see that $B = B(3)$.

2.2. The second approximation.

For convenience we stay with the hypothesis $d = 3$. Suppose we have a module E as above, for which $B \neq B(2)$. We begin by comparing $B(2)_3$ to B_3; in particular we will show that $B_3 \neq B(2)_3$ in all cases. (To contrast, see Example .)

Consider the exact sequence

$$0 \longrightarrow B(2)_3 \longrightarrow B_3 \longrightarrow C_3 \longrightarrow 0.$$

If we represent $B(2) = P(2)/J(2)$, taking into account that the ideal $J(2)$ is graded in the T's and U variables, it follows that

$$\operatorname{Ext}^3(C_3, R) \;=\; \operatorname{Ext}^1(J(2)_3, R).$$

We seek to determine this module. Given the presentation

$$0 \longrightarrow R^2 \xrightarrow{\varphi} R^n \longrightarrow E \longrightarrow 0,$$

the ideal $J(2)$ is generated by the maximal Pfaffians of the matrix

$$\begin{bmatrix} 0 & U & b_{11} & b_{12} & b_{13} \\ -U & 0 & -b_{21} & -b_{22} & -b_{23} \\ -b_{11} & b_{21} & 0 & z^a & -y \\ -b_{12} & b_{22} & -z^a & 0 & x \\ -b_{13} & b_{23} & y & -x & 0 \end{bmatrix}.$$

By [5] the resolution of this ideal is ($A = P(2)$):

$$0 \longrightarrow A(-4) \longrightarrow A(-3)^2 \oplus A(-2)^3 \longrightarrow A(-2)^3 \oplus A(-1)^2 \longrightarrow J(2) \longrightarrow 0.$$

The R-projective resolution of the degree 3 component of $J(2)$ is:

$$0 \longrightarrow R^2 \oplus A_1^3 \xrightarrow{\gamma} A_1^3 \oplus A_2^2 \longrightarrow J(2)_3 \longrightarrow 0.$$

We must compute $\operatorname{coker}(\gamma^*) = \operatorname{Ext}^1(J(2)_3,\ R)$. For this view γ as mapping 5-tuples into 5-tuples: Thus a 5-tuple $[x_1, x_2, x_3, x_4, x_5]$ is mapped to another of components:

$$\begin{aligned} &-Ux_2 - b_{11}x_3 - b_{12}x_4 - b_{13}x_5 \\ &Ux_1 + b_{21}x_3 + b_{22}x_4 + b_{23}x_5 \\ &b_{11}x_1 - b_{21}x_2 - z^a x_4 + yx_5 \\ &b_{12}x_1 - b_{22}x_2 + z^a x_3 - xx_5 \\ &b_{13}x_1 - b_{23}x_2 - yx_3 + xx_4 \end{aligned}$$

Note that [1, 0, 0, 0, 0] and [0, 1, 0, 0, 0] map to [0, U, 0, 0, 0] and [-U, 0, 0, 0, 0], respectively, so that we can express $\operatorname{Ext}^1(J(2)_3,\ R)$ as $\operatorname{coker}(\theta^*)$:

$$\theta\ :\ P_1^3 \longrightarrow P_1^3 \oplus P_2^2.$$

Here P_t stands for the degree t component (graded by the T-variables) of the polynomial ring $R[T_1, \ldots, T_n]$, and θ is the matrix:

$$\begin{bmatrix} -b_{11} & -b_{12} & -b_{13} \\ b_{21} & b_{22} & b_{23} \\ 0 & -z^a & y \\ z^a & 0 & -x \\ -y & x & 0 \end{bmatrix}$$

PROPOSITION 2.2.1. *$C_3 = R/J$, where J is an ideal (x, y, z^b), $1 \leq b \leq a$.*

In other words, C_3 has a description similar to that of C_2. By duality, it suffices to show that $\mathrm{Ext}^1(J(2)_3,\ R) = R/J$, with J as indicated. It will follow that B_3 is gotten from $B(2)_3$ by adding one generator. Unlike the previous approximation, when the conductor of the element u was (x, y, z^a), the R-conductor of the new generator may be a larger ideal.

PROOF: Decompose θ^* into $\alpha + \beta$, where α and β are the induced maps:

$$\begin{aligned} \alpha &: (P_1^3)^* \longrightarrow (P_1^3)^* \\ \beta &: (P_2^2)^* \longrightarrow (P_1^3)^* \end{aligned}$$

We first claim that coker (β) is a cyclic module. It suffices to show that the module $H = \mathrm{Hom}(\mathrm{coker}(\beta), R/\mathbf{p})$ is cyclic. We have the exact sequence

$$0 \longrightarrow H \longrightarrow (P_1^3)^\star \longrightarrow (P_2^2)^\star$$

where the mapping is the reduction mod $\mathbf{p}$ of the matrix of 1-forms. If we write each of its maximal minors as $\Delta_i = gh_i$, we obtain that H is generated by the vector (h_1, h_2, h_3).

Returning to coker (β), we may then pick a summand Re of $(P_1^3)^*$ so as to get the commutative diagram

$$\begin{array}{ccccc} (P_2^2)^* & \xrightarrow{\beta} & (P_1^3)^* & \longrightarrow & \mathrm{coker}(\beta) \\ & & {\scriptstyle\pi}\searrow & & \nearrow \\ & & & Re & \end{array}$$

Using the canonical identification of P_1^3 with its bi-dual, we write

$$\pi^*(e^*) = (H_1, H_2, H_3).$$

This vector reduces, mod $\mathbf{p}$, to (h_1, h_2, h_3). Note that π evaluates on $f = (f_1, f_2, f_3) \in (P_1^3)^*$ as

$$\pi(f) = (f_1(H_1) + f_2(H_2) + f_3(H_3)) \cdot e.$$

To prove our assertion it suffices to show that $\pi(\alpha((P_1^3)^*)) = I \cdot e$. Indeed, we have

$$\pi(\alpha(f_1, f_2, f_3)) = ((z^a f_2 - y f_3)(H_1) + (-z^a f_1 - x f_3)(H_2) + (y f_1 + x f_2)(H_3)) \cdot e,$$

which may be written

$$-(x(f_3(H_2) - f_2(H_3)) + y(f_3(H_1) - f_1(H_3)) + z^a(f_1(H_2) - f_2(H_1))) \cdot e.$$

By the assumption two of the forms $\{H_1, H_2, H_3\}$ are part of a basis of P_1. Suppose H_1 and H_2 have this property (although there may be lack of symmetry on the account of the exponent a, it will be clear that the choice does not play any role). We may pick H_1^* and H_2^* a dual pair to H_1 and H_2. To get $x \cdot e$ in the image, we map the vector $(0, 0, H_2^*)$. To get $y \cdot e$, we use $(0, 0, H_1^*)$. Having obtained these two elements in the image we easily get $z^a \cdot e$ as well. □

Before moving on to the description of $B(3)$, note the following consequences of the argument above. Reduce θ^* mod J and then dualize with respect to R/J, to obtain the exact sequence

$$0 \longrightarrow R/J\ (H_1',\ H_2',\ H_3') \longrightarrow (P_1^3)' \longrightarrow (P_2^2)'.$$

As the ideal generated by the forms H_1', H_2' and H_3' in the Cohen–Macaulay ring $R/J[T_1, \ldots, T_n]$ has height at least two, we may assume that, say, $\{H_1', H_2'\}$ is a regular sequence. From the equations

$$H_i' \cdot D_j' \;=\; H_j' \cdot D_i', \text{ where } D' \text{ is } D \text{ evaluated at J},$$

we get a decomposition

$$D' \;=\; h' \cdot (H_1',\ H_2',\ H_3').$$

Denote by g a lifting of h' to a form of degree 1 in $R[T_1, \ldots, T_n]$. $M = (x, y, z, g)B(2)$ is the only minimal prime of $\mathbf{p}B(2)$ of height one. If we denote by L the $B(2)$-conductor of the single generator v of $B(3)$ over $B(2)$—recall that $B(2)$ is a normal domain—it follows easily that L is precisely the ideal generated by J and g (note that $(D_1, D_2, D_3)B(2) \subset (x, y, z^a)B(2) \subset L$). Introduce a new variable V to map onto v; adding to $J(2)$ the equations of v over $B(2)$ we have the ideal of $P(2)[V]$:

$$J(3) = (J(2), xV - A_1, yV - A_2, z^bV - A_3, gV - A_4).$$

THEOREM 2.2.2. ([20, Theorem 3.2]) *Put* $\Delta = gK$. *If* height $(K) \geq 2$ *then:*

(a) *$B(3)$ is a Gorenstein ring, which is singly generated over $B(2)$; $J(3)$ is its defining ideal.*

(b) $B = B(3)$ *if and only if* height $(x, y, z, g, A_1, A_2, A_3, A_4) \geq 6$.

PROOF: (a) We first claim that $J(3)$ is a Gorenstein ideal. Indeed, we have the exact sequence:

$$0 \longrightarrow L(V - v) \longrightarrow B(2)[V] \longrightarrow B(3) \longrightarrow 0$$

where L is the image of (x, y, z^b, g) in $B(2)$. Note that L is a Gorenstein ideal so that $B(3)$ is Cohen-Macaulay. Applying $\mathrm{Hom}(-, B(2)[V])$, we get

$$0 \longrightarrow B(2)[V] \longrightarrow L^{-1}(V - v)^{-1} \longrightarrow W_{B(3)} \longrightarrow 0$$

where $W_{B(3)}$ denotes the canonical module of $B(3)$. Since $L^{-1} = (1, v)$, a simple calculation shows that $W_{B(3)}$ is principal. That $J(3)$ is a prime ideal follows in the same manner as $J(2)$: If P is a prime ideal of height four associated to $J(3)$ and containing $\mathbf{p}$ it must necessarily be $(\mathbf{p}, g)$—which is a contradiction as $(\mathbf{p}, g, A_1, A_2, A_3, A_4)$ has, by the normality of $B(2)$, height at least 5.

(b) Because $B(3)$ is a Cohen-Macaulay ring, the equality $B = B(3)$ is equivalent to saying that $\mathbf{p}B(3)$ is an ideal of height greater than one—in other words, height $(G = (\mathbf{p}, J(3))) \geq 6$. Since this ideal is contained in $Q = (x, y, z, g, A_1, A_2, A_3, A_4)$, height $Q \geq 6$. Conversely, let P be a prime minimal over G. If $g \in P$, $(A_1, A_2, A_3, A_4) \subset P$, and the ideal P contains Q. In the other case, $K \subset P$ so that P contains at least 5 independent linear forms besides the forms in V, so that height $P \geq 6$. □

REMARK 2.2.3:

(i) If $b = 1$, an application of the jacobian criterion will show that $B(3)$ is normal. Perhaps this is true in all cases.

(ii) In several other examples examined it was observed that whenever $\Delta = g \cdot K$ but g did not lie in K, then $B = B(3)$. Could this be general?

(iii) It is much less direct to predict the behavior of the higher $B(r)$; see Example .

(iv) Often there are ways of determining the height of the ideal Q without access to the coefficients A_i's.

(v) This theorem suggested itself while the scheme of section two was set up.

The following examples of non-homogeneous modules were examined using a package written by Gail Zacharias running in *MACSYMA*.

EXAMPLE 2.2.4: We go back to the Example (ii). Here $g = T_1$. The equations for $J(3)$ are those of $J(2)$ plus:

$$(xV+T_3U-T_1T_2T_4-T_1T_2^2, yV-((T_3T_4^2+T_2T_3T_4)z+(-T_1T_4^2-T_1T_2T_4)x-T_1^2T_4-T_3^3-T_1^2T_2), zV-T_1U+T_1T_3T_4-T_2T_3^2, T_1V-(U^2-T_1T_2T_4^2+(T_2^2T_3-T_1T_2^2)T_4+T_2^3T_3).$$

We verified that $x, y, V, T_2, U-T_1+T_3+T_4-z$ is a system of parameters for $B(3)$. Reduction mod this ideal showed a resulting ideal I such that length $(S/I) = 10$, while length $(I/I^2) = 46$. By [22], the ideal $J(3)$ is not in the linkage class of a complete intersection.

EXAMPLE 2.2.5: Let the module be defined by the matrix:

$$\varphi = \begin{bmatrix} x & y & z^2 & z^2 \\ 0 & x & y^2 & z^3 \end{bmatrix}$$

$J(3)$ is generated by $J(2)$ plus the forms

$$(xV-T_2^2T_4-z(-T_3T_4^2+2T_3^2T_4-T_3^3),\ T_2V+zU^2+T_1^2T_3T_4, yV+T_1T_2T_4+(T_3+T_4)Uz, zV+T_2U+T_1T_3T_4+T_1T_3^2).$$

Here height $(x, y, z, g = T_2, A_1, A_2, A_3, A_4) = 5$, so that $B \neq B(3)$. The next approximation, $B(4)$, has for its defining ideal:

$$J(4) = (J(3), xW - T_3T_4^3 + z(T_4^2 + T_3T_4)U - (3T_3^2 + T_1T_2)T_4^2 - 3T_3^3T_4 - T_3^4, yW + (T_4^2 + 2T_3T_4 + T_3^2)U + T_1^2T_4^2, zW - (T_3 + T_4)V - T_1T_4U, T_1T_3W + U^3 + V^2, (T_2 - zT_1)W + T_1T_3V + (T_3 + T_4)U^2 + T_1^2T_4U).$$

A system of parameters is $\{x, z, W, U-y, T_4-T_1\}$, and therefore $B = B(4)$.

EXAMPLE 2.2.6: The last example is the module with a presentation

$$\begin{bmatrix} x & z^2 & y & yz^2 \\ 0 & x & z^3 & y^2 \end{bmatrix}.$$

One has: $B(2) \subset B(3) = B(4) \subset B(5) = B$. The ideal of definition, $J(5)$, is (U, V and W have degrees 2, 3 and 5 respectively):

$$(J(2), xV-(z^2T_3^2+zT_2^2)T_4-T_3U-T_3^3, yV+zT_2U+T_1T_3^2, zV-T_3U+(T_1T_2+T_2T_3)T_4, T_2V+zU^2-zT_2T_4U+T_1^2T_3T_4, zW-UV-(T_1^2T_3+T_1^3)T_4^2,\ yW-T_1T_3V+T_2U^2-(T_1T_3+T_1^2)T_4Uz, xW-(-T_1^2T_4^3y+(T_3^2-T_1T_3)T_4U+(-T_2T_3^2-T_1^2T_2)T_4^2)z-(T_2T_4^2U-T_2^2T_4^3)y-(T_2T_4^2U-T_4U^2)x-T_3^2V-T_2^2T_4U+T_2^3T_4^2, T_2W+T_1^2T_4V+U^3-T_2T_4U^2, T_3W-V^2+(T_1+T_3)T_4U^2-(T_1+T_3)T_2T_4^2U).$$

3. Almost complete intersections.

Let E be a reflexive module such that its symmetric algebra is an almost complete intersection. This means, according to [21], that E has a projective resolution

$$0 \longrightarrow R \xrightarrow{\psi} R^m \xrightarrow{\varphi} R^n \longrightarrow E \longrightarrow 0.$$

We recall, cf. [21], that whether $S(E)$ is an integral domain is highly dependent on the Koszul homology of the ideal $I = I_1(\psi)$. In particular if I is strongly Cohen-Macaulay of codimension three, satisfies $\mathcal{F}_1$ and E^* is a third syzygy module, then $S(E)$ is a Cohen-Macaulay integral domain.

3.1. The first approximation.

We consider modules that are reflexive and free on the punctured spectrum of a regular local ring R. It implies that (with $\mathcal{F}_1$) $m = \dim R$. According to [21] this has the following consequence: If m is even then $S_{m/2}(E)$ is not a torsion-free R-module.

We focus on the case m odd. An application of the complexes of [25]–or through the approximation complexes–shows that $S_t(E)$ is reflexive for $t \leq (m-3)/2$ and torsion-free but not reflexive if $t = r = (m-1)/2$. We seek the equations for $B(r)$.

LEMMA 3.1.1. *$B(E)_r/S_r(E)$ is a cyclic R-module.*

PROOF: As in the proof of Theorem we consider the exact sequence

$$0 \longrightarrow S_r(E) \longrightarrow B(E)_r \longrightarrow C_r \longrightarrow 0,$$

and compute $\text{Ext}^{m-1}(S_r(E),\ R) = \text{Ext}^m(C_r,\ R)$. A direct application of the complexes of [25] shows that this module is isomorphic to $R/I_1(\psi)$. Because $I_1(\psi)$ is generated by a system of parameters, by duality it follows that C_r is cyclic. □

To get the equations for $B(r)$, first observe that the rows of the matrix φ are syzygies of the system of parameters $I_1(\psi) = (x)$. In particular $I_1(\varphi)$ is contained in (x). Denote by P the polynomial ring $R[T_1, \ldots, T_n]$. If $S(E)$ is a normal domain, we can write its defining ideal

$$J(E) = (f) = (x) \cdot J(\varphi).$$

Consider the exact sequence:

$$P^m \xrightarrow{J(\varphi)} P^m \longrightarrow C \longrightarrow 0.$$

Assume that $J(\varphi)$ has rank $m-1$; thus ker $J(\varphi)$ is cyclic, generated, say, by the vector of forms $(g) = (g_1, \ldots, g_m)$. Furthermore the associated prime ideals of C as an R-module are trivial–as E is free in codimension at most two–and therefore C is actually an ideal of P. It follows easily that C is a Gorenstein ideal and thus isomorphic to G, the ideal generated by the entries of (g).

We have come to the main result of this section.

THEOREM 3.1.2. *Let E be an R-module that is reflexive and free on the punctured spectrum of R, whose symmetric algebra is a normal domain.*

(a) *$B(r)$ is a Gorenstein ring singly generated over $S(E)$.*

(b) *Denote by $G(0)$ the ideal G evaluated at the origin. Then $B(r) = B$ if and only if height $G(0) \geq 2$.*

PROOF: Consider the equations

$$\begin{aligned} (x) \cdot J(\varphi) &= (f) \\ (g) \cdot J(\varphi) &= 0 \end{aligned}$$

Reading them mod (f), that is, in $S(E)$, when $J(\varphi)$ still has rank $m-1$ by the normality hypothesis, we get an element h in the field of fractions of $S(E)$, such that $h \cdot (x) \equiv (g)$. It follows from Equation 2 that h actually lies in B.

(a) Let $J(r)$ be the ideal of the polynomial ring $P[U]$

$$((f), (x) \cdot U - (g)).$$

$J(r)$ is by Theorem a Gorenstein prime ideal of height m.

(b) Because B is the ideal transform of $B(r)$ with respect to the ideal (x), these rings are equal if and only if the grade of $(x) \cdot B(r)$ is at least two. Since $B(r)$ is Cohen-Macaulay this translates into the condition above. □

REMARK: Of course, if $S(E)$ is normal, the rank of $J(\varphi)$ is at least $m-1$. It is likely to be $m-1$ in all cases $S(E)$ is a domain. If $S(E)$ is not a domain, then the rank of $J(\varphi)$ may be higher, as in one of the examples below.

EXAMPLE 3.1.3: Given a regular local ring (R,m) of dimension $n \geq 4$, Vetter [23] constructed an indecomposable vector bundle on the punctured spectrum of R, of rank $n-2$. Its module of global sections E has a presentation:

$$0 \longrightarrow R \longrightarrow R^n \xrightarrow{\varphi} R^{2n-3} \longrightarrow E \longrightarrow 0.$$

We look at $S(E)$; note that E satisfies $\mathcal{F}_1$. As $S(E)$ is an almost complete intersection, we can apply to it the theory of [21]: For n even, $S(E)$ cannot be an integral domain, while for n odd, and larger than 3, the approximation complex $Z(E)$ cannot be exact. For $n = 5$ the matrix φ is:

$$\begin{bmatrix} -x_2 & x_1 & 0 & 0 & 0 \\ -x_3 & 0 & x_1 & 0 & 0 \\ -x_4 & x_3 & -x_2 & x_1 & 0 \\ -x_5 & x_4 & 0 & -x_2 & x_1 \\ 0 & -x_5 & -x_4 & x_3 & x_2 \\ 0 & 0 & -x_5 & 0 & x_3 \\ 0 & 0 & 0 & -x_5 & x_4 \end{bmatrix}$$

The ideal of definition of $S(E)$ is $J(E) = [T_1, \ldots, T_7] \cdot \varphi$. An application of the *MACAULAY* program yielded that $J(E)$ is a Cohen-Macaulay prime ideal.

Finally we use the normality criterion of [21] to show that $S(E)$ is integrally closed. The Jacobian matrix of $J(E)$ with respect to the x-variables is:

$$J(\varphi) = \begin{bmatrix} 0 & T_1 & T_2 & T_3 & T_4 \\ -T_1 & 0 & -T_3 & -T_4 & T_5 \\ -T_2 & T_3 & 0 & T_5 & T_6 \\ -T_3 & T_4 & -T_5 & 0 & T_7 \\ -T_4 & -T_5 & -T_6 & -T_7 & 0 \end{bmatrix}.$$

Because rank $J(\varphi) = 4$, $S(E)$ is integrally closed. Note that the ideal $G(O)$ has height three, so $B(E) = B(2)$.

For $n = 6$ the module is defined by the matrix

$$\begin{bmatrix} -x_2 & x_1 & 0 & 0 & 0 & 0 \\ -x_3 & 0 & x_1 & 0 & 0 & 0 \\ -x_4 & x_3 & -x_2 & x_1 & 0 & 0 \\ -x_5 & x_4 & 0 & -x_2 & x_1 & 0 \\ -x_6 & x_5 & -x_4 & x_3 & -x_2 & x_1 \\ 0 & -x_6 & x_5 & 0 & -x_3 & x_2 \\ 0 & 0 & -x_6 & x_5 & -x_4 & x_3 \\ 0 & 0 & 0 & -x_6 & 0 & x_4 \\ 0 & 0 & 0 & 0 & -x_6 & x_5 \end{bmatrix}.$$

As remarked earlier, $S(E)$ cannot be an integral domain. A computation, as indicated in section 2, shows that $D(E)$ is defined by the linear forms arising out of this matrix plus the polynomial:

$$T_5^3+2T_4T_5T_6+T_3T_6^2+T_4^2T_7-2T_3T_5T_7-T_2T_6T_7+T_1T_7^2-T_3T_4T_8+T_2T_5T_8-T_1T_6T_8+T_3^2T_9-T_2T_4T_9-T_1T_5T_9.$$

It was further verified that $B(E)=D(E)$, and that it is a Cohen-Macaulay ring.

MACAULAY was also used to show that for $n=7$, $S(E)$ is not an integral domain. Since $S_3(E)$ is *not* a reflexive module, $B(E)$ will not have a description as the previous case. ($B(E)$ was not computed.) □

4. Prime ideals.

We touch briefly on the use of the method outlined in section 2 for the computation of symbolic blow-ups. The main result here is a formula describing the second symbolic power of almost complete intersections. The relationship between the finiteness of the factorial closures of rank two modules and the symbolic blow-up of codimension two prime ideals is examined.

4.1. The symbolic square of an ideal.

Let R be a regular local ring of dimension d and let I be a prime ideal of height g. Denote by $I^{(2)}$ the I-primary component of I^2. One way to express the equality–or difference–between I^2 and $I^{(2)}$ is through the module I/I^2: $I^2=I^{(2)}$ if and only if I/I^2 is a torsion-free R/I-module. Let us seek to determine the torsion of this module.

Let $I = (x_1, \ldots, x_n)$; there exists an exact sequence

$$H_1 \xrightarrow{\psi} (R/I)^n \longrightarrow I/I^2 \longrightarrow 0$$

where H_1 is the first Koszul homology module of I. Let us assume that H_1 is a Cohen-Macaulay module–which will make the sequence exact on the left—and further that $\dim(R/P) = 1$. The long exact sequence of the functor $\mathrm{Ext}(-, R)$ yields:

$$\mathrm{Ext}^{d-1}((R/I)^n, R) \longrightarrow \mathrm{Ext}^{d-1}(H_1, R) \longrightarrow \mathrm{Ext}^d(I/I^2, R) \longrightarrow 0.$$

On the other hand, the sequence

$$0 \longrightarrow I^{(2)}/I^2 \longrightarrow I/I^2 \longrightarrow I/I^{(2)} \longrightarrow 0$$

gives rise to an isomorphism $\mathrm{Ext}^d(I^{(2)}/I^2, R) = \mathrm{Ext}^d(I/I^2, R)$, since $I/I^{(2)}$ is a torsion-free R/I-module and therefore Cohen-Macaulay by the dimension condition. If we now let W denote the canonical module of R/I, $W = \mathrm{Ext}^{d-1}(R/I, R)$, and use the duality in the Koszul homology ([12] surveys these aspects), we get the exact sequence

$$W^n \xrightarrow{\alpha} \mathrm{Hom}(H_1, W) \longrightarrow \mathrm{Ext}^d(I^{(2)}/I^2, R) \longrightarrow 0.$$

PROPOSITION 4.1.1. *Let I be an almost complete intersection of height $d-1$. Then $I^{(2)}/I^2 = \mathrm{Ext}^d(R/I_1(\varphi), R)$, where $I_1(\varphi)$ denotes the ideal generated by the entries of the first order syzygies of I.*

PROOF: In this case $W = H_1$, a Cohen-Macaulay module, so that the target of α is $\mathrm{Hom}(H_1, W) = R/I$. Note that α is the W-dual of ψ and that ψ is given explicitly by: For a first order syzygy of I, $z = (x_1, \ldots, x_n)$, $[z] \in H_1$, $\psi([z]) = (\overline{x_1}, \ldots, \overline{x_n}) \in (R/I)^n$. Thus the image of α is the content of H_1 as a submodule of $(R/I)^n$. The assertion then follows by duality. □

REMARK 4.1.2: It is not difficult to show that the formula for $I^{(2)}/I^2$ remains valid without the assumption on the dimension of R/I as long as we assume that I is Cohen-Macaulay. If $d = 3$ this formula is proved in [14] using the resolutions of [25]. Villarreal [24] found a class of ideals for which the module $I^{(2)}/I^2$, while not necessarily cyclic as above, is nevertheless self-dual.

COROLLARY 4.1.3. *Let I be a prime ideal of a regular local ring (R, m). If I is a normal, almost complete intersection of dimension one, then $I^{(2)}/I^2$ is cyclic. Furthermore if $I \subset \mathrm{m}^2$, then the Cohen-Macaulay type of I is at least* $\dim R - 1$.

PROOF: Because the symmetric and Rees algebras of I coincide [12] and it is Cohen-Macaulay, by the normality criterion of [21] we not only must have $I_1(\varphi) = (y_1, \ldots, y_{d-1}, y_d^a)$ for some regular system of parameters (y), which proves the first assertion, but also the rank of the Jacobian matrix at the origin must be $d-1$. Picking a generating set for I such that $\{x_1, \ldots, x_{d-1}\}$ is a regular set, as $x_i \in \mathrm{m}^2$, this implies that there must be at least $d-1$ syzygies that contribute to the non-vanishing of the Jacobian ideal. □

A case where the method of section 2 works reasonably well is that of prime ideals defining monomial curves in 3-space. These curves, given parametrically by

$$\begin{cases} x = t^a \\ y = t^b \\ z = t^c \end{cases}$$

where $\gcd(a, b, c) = 1$, have for equations, according to [11], the 2×2 minors of a matrix

$$\begin{bmatrix} x^{a_1} & y^{a_2} & z^{a_3} \\ z^{b_1} & x^{b_2} & y^{b_3} \end{bmatrix}.$$

It follows that $I^{(2)}/I^2$ is always cyclic–which had already been observed in [14]. There is one feature of these ideals that despite the lack of normality behave as in section 3. To wit, if the all exponents in one row of the matrix above are not strictly larger than the exponents of the other row (i.e. $a_1 > b_2$, etc.) then Theorems and apply and in particular $I^{(3)}/I \cdot I^{(2)}$ is cyclic.

EXAMPLE 4.1.4: Begging the reader's indulgence, here is the defining ideal of the symbolic power algebra of the monomial curve (7, 9, 10). (The notation u_{ij} denotes one generator of degree i.)

$(u_{71}^2 + T_1 u_{33} u_{51}^2 - T_3^2 u_{41}^3, -u_{52} u_{71} + T_1 u_{33}^2 u_{51} + T_3 u_{32} u_{41}^2, u_{33} u_{71} + u_{51} u_{52} + T_1 T_3 u_{41}^2, -u_{32} u_{71} + T_1^2 u_{33} u_{51} - T_1 T_2 u_{41}^2 + u_{31} u_{33} u_{41}, u_{31} u_{71} + u_{32}^2 u_{41} + T_1 u_{33}^3, u_{21} u_{71} + T_1 T_3^2 u_{33}^2 - T_1^3 u_{32} u_{33} + u_{32}^3, T_3 u_{71} - u_{31} u_{51} + 2T_1 u_{32} u_{41}, T_2 u_{71} - T_3 u_{31} u_{41} - 2T_1^2 u_{33}^2, T_1 u_{71} + u_{32} u_{51} - T_3 u_{33} u_{41}, u_{52}^2 + (T_1^3 u_{33} + u_{32}^2) u_{41} + T_1 u_{33}^3, -u_{33} u_{52} -$

$u_{31}u_{51} + T_1u_{32}u_{41}, u_{32}u_{52} + T_3u_{31}u_{41} + T_1^2u_{33}^2, -u_{31}u_{52} + T_2u_{32}u_{41} + T_1T_3u_{33}^2, -u_{21}u_{52} - T_1^3T_3u_{32} - T_3u_{31}^2 + T_1^4u_{31}, T_3u_{52} - T_1T_2u_{41} + u_{31}u_{33}, T_2u_{52} + T_1^2T_3u_{33} + u_{31}u_{32}, T_1u_{52} + T_3^2u_{41} - u_{32}u_{33}, -u_{31}u_{32}u_{51} + (T_3u_{31}u_{33} + T_1u_{32}^2)u_{41} + T_1^2u_{33}^3, -u_{21}u_{51} + T_3u_{31}u_{33} - T_1^4u_{33} + T_1u_{32}^2, u_{31}^2u_{51} + (T_3u_{32}^2 - 2T_1u_{31}u_{32})u_{41} + T_1T_3u_{33}^3, T_3u_{51} - T_1^2u_{41} - u_{33}^2, (T_1^3u_{33} + u_{32}^2)u_{51} - T_1^2T_2u_{41}^2 + T_1u_{31}u_{33}u_{41} - T_3u_{32}u_{33}u_{41}, T_2u_{51} + T_3^2u_{41} - 2u_{32}u_{33}, T_1^2u_{31}^2u_{41} - T_1T_3u_{31}u_{32}u_{41} + T_1T_3^2u_{33}^3 + (T_1^3u_{32} + u_{31}^2)u_{33}^2 + u_{32}^3u_{33}, T_1u_{31}u_{51} - T_3^2u_{33}u_{41} + T_1^2u_{32}u_{41} + u_{32}u_{33}^2, T_1T_2u_{32}u_{41} + T_3^2u_{31}u_{41} + T_1^2T_3u_{33}^2 - u_{31}u_{32}u_{33}, (T_1^2u_{31}^3u_{33} + T_1T_3u_{32}^4 - T_1^2u_{31}u_{32}^3)u_{41} + T_1T_3^2u_{31}u_{33}^4 + (T_1^2T_3u_{32}^2 + 2T_1^3u_{31}u_{32} + u_{31}^3)u_{33}^3 + u_{31}u_{32}^3u_{33}^2, -T_1T_2u_{31}u_{41} + T_1T_3^2u_{33}^2 + (T_1^3u_{32} + u_{31}^2)u_{33} + u_{32}^3, T_1T_3^3u_{33} + T_2u_{32}^2 + T_1^3T_3u_{32} + T_3u_{31}^2 - T_1^4u_{31}, (T_2u_{32}^2 + T_3u_{31}^2)u_{41} + (T_1T_3u_{32} + T_1^2u_{31})u_{33}^2, (T_3^3 + T_1^2T_2)u_{41} - T_3u_{32}u_{33} - T_1u_{31}u_{33}, xu_{41} - T_3u_{32} - T_1u_{31}, -u_{21}u_{33} + T_1T_2u_{32} + T_3^2u_{31}, -T_2u_{33} + T_3u_{32} - T_1u_{31}, zu_{33} + T_3u_{21} + T_1T_2^2, yu_{33} - T_1u_{21} + T_2T_3^2, -xu_{33} + T_3^3 + T_1^2T_2, u_{21}u_{32} - T_2T_3u_{31} + T_1^2T_3^3 + T_1^4T_2, T_2^2u_{32} + u_{21}u_{31} + T_1T_3^4 + T_1^3T_2T_3, zu_{32} - xT_1^2T_3 + T_2u_{21}, yu_{32} + xT_1^3 + T_2^2T_3, xu_{32} - T_1u_{21} - T_2T_3^2, (T_3^2u_{31}u_{33} + T_1T_3u_{32}^2 - T_1^2u_{31}u_{32})u_{41} + T_1^2T_3u_{33}^2 - u_{31}u_{32}u_{33}^2, u_{21}u_{41} + T_1^2T_3u_{33} - u_{31}u_{32}, yu_{51} + T_3^2u_{32} - 2T_1T_3u_{31} + 2T_1^5, zu_{51} + 2T_1T_2u_{32} + T_3^2u_{31} - 2T_1^4T_3, yu_{52} - T_2T_3u_{31} - T_1^2T_3^3 + T_1^4T_2, T_1^2u_{31}^4u_{41} + T_1T_3^2u_{31}^2u_{33}^3 + (T_1T_3^2u_{32}^3 + T_1^2T_3u_{31}u_{32}^2 + 2T_1^3u_{31}^2u_{32} + u_{31}^4)u_{33}^2 + (T_1^3u_{32}^4 + 2u_{31}^2u_{32}^3)u_{33} + u_{32}^6, zu_{52} + T_2^2u_{32} + T_1T_3^4 - T_1^3T_2T_3, T_3^2u_{31}^2u_{41} + T_1T_3^2u_{32}u_{33}^2 + T_1^2T_3u_{31}u_{33}^2 + T_1^3u_{32}^2u_{33} + u_{32}^4, (T_3^2u_{32} - T_1T_3u_{31})u_{41} - T_1^3u_{33}^2 - u_{32}^2u_{33}, -T_2T_3u_{41} + T_1^3u_{33} + u_{32}^2, T_2^2u_{41} + T_1T_3^2u_{33} + u_{31}^2, T_1T_3u_{31}^3u_{41} + (T_1T_3^2u_{32}^2 + T_1^2T_3u_{31}u_{32} + T_1^3u_{31}^2)u_{33}^2 + T_1^3u_{32}^3u_{33} + u_{31}^2u_{32}^2u_{33} + u_{32}^5, -zu_{41} - T_2u_{31} + 2T_1^2T_3^2, yu_{41} + T_2u_{32} + 2T_1^3T_3, (T_3u_{31}^2u_{33} + T_3u_{32}^3 - T_1u_{31}u_{32}^2)u_{41} + (T_1T_3u_{32} + T_1^2u_{31})u_{33}^3, -zu_{31} + xT_1T_3^2 + T_2^3, yu_{31} + xT_1^2T_3 + T_2u_{21}, xu_{31} - T_3u_{21} + T_1T_2^2, xT_1T_3^3 + xT_1^3T_2 + u_{21}^2 + T_2^3T_3, zu_{21} + yT_2^2 - x^2T_1T_3, zT_2T_3 - yu_{21} - x^2T_1^2, zT_3^2 - yT_1T_2 + xu_{21}, z^2T_3 + y^2T_2 + x^3T_1, zT_1 + yT_3 + zT_2, xu_{51} - T_3^2u_{33} - 2T_1^2u_{32}, xu_{52} - u_{21}u_{33} + 2T_3^2u_{31}, xu_{71} - u_{21}u_{51} - T_1^4u_{33} + 3T_1u_{32}^2, yu_{71} + (T_2u_{31} + 2T_1^2T_3^2)u_{33} - 2T_1^4u_{32} + T_1u_{31}^2, zu_{71} + 2T_1^3T_2u_{33} + 2T_1^3T_3u_{32} + T_3u_{31}^2.)$

This ideal is Cohen-Macaulay. On the other hand, according to [19], the symbolic blow-up of a codimension two prime ideal is always quasi-Gorenstein, so that it must be Gorenstein.

There is a straightforward way of computing the symbolic powers of a prime ideal I. Let f be an element of R such that the localization I_f has the property that its symbolic and ordinary powers coincide. Then

$$I^{(n)} = \bigcup_{s \geq 1} I^n : f^s.$$

This forbidding expression is however just $(I^n, 1 - Tf) \bigcap R$, where T is an indeterminate over R. In contrast, $I^n : f$ is given by

$$(I^nT,(1-T)f)\bigcap R/f.$$

We have found the two computations comparable in *MACSYMA*.

Computing high symbolic powers is highly wasteful; this is the reason for emphasizing the search for fresh generators.

4.2. Bourbaki sequences.

This section will be used for a number of observations regarding the relationship between the finiteness of the factorial closure of modules and the symbolic blow-up of height two prime ideals.

This connection arises out of "Bourbaki sequences". Let us recall this notion (cf. [3]). Given a normal domain R and a finitely generated torsion-free R-module E, there exists an exact sequence

$$0 \longrightarrow F \longrightarrow E \longrightarrow I \longrightarrow 0,$$

where F is a free submodule of E, and I is an ideal. Furthermore, if R is a localization of a polynomial ring $k[x_1,\ldots,x_n]$ at the origin, k infinite, and E is reflexive then I can be taken to be prime of height two; for a detailed discussion, see [3], [8] and [15].

If the issue of actually obtaining $I = P$ (prime) from E or the converse (much easier) is skirted, a rudimentary connection between the two algebras, $B(E)$ and $R_s(P)$, can be sketched as follows.

The exact sequence of modules induces the exact sequence of symmetric algebras:

$$0 \longrightarrow F\circ S(E) \longrightarrow S(E) \longrightarrow S(P) \longrightarrow 0.$$

Since the R-torsion part of $S(E)$ maps onto the R-torsion of $S(P)$, we get a surjective homomorphism of integral domains

$$0 \longrightarrow (F) \longrightarrow D(E) \longrightarrow D(P) \longrightarrow 0.$$

If $S(P) = D(P)$, one can show that $S(E) = D(E)$, and that the generators of F give rise to a regular sequence on $S(E)$. Furthermore, if $S(P)$ is normal, then $S(E)$ is normal as well.

In general, it is an easy matter to see that there exists a regular sequence $\{f,\ g\}$, generating the ideal J of before, so that $B(E)$ and $R_s(P)$ are both ideal transforms of $D(E)$ and $D(P)$, respectively, with respect to J. Let us set up the local cohomology calculation associated to this ideal, that is, apply to this sequence the functors $\mathrm{Hom}_R(J^n, -)$:

$$0 \longrightarrow \mathrm{Hom}_R(J^n, (F)) \longrightarrow \mathrm{Hom}_R(J^n, D(E)) \longrightarrow \mathrm{Hom}_R(J^n, D(P)) \longrightarrow$$

$$\mathrm{Ext}^1_R(J^n, (F)) \longrightarrow \mathrm{Ext}^1_R(J^n, D(E)) \longrightarrow \mathrm{Ext}^1_R(J^n, D(P)) \longrightarrow 0.$$

Taking limits, we get:

$$0 \to [F] \to B(E) \xrightarrow{\varphi} R_s(P) \to H^2_J(F) \to H^2_J(D(E)) \xrightarrow{\psi} H^2_J(D(P)) \to 0,$$

where $[F]$ has the obvious meaning. Let us consider the special case where rank $(E) = 2$. Since $B(E)$ is a factorial domain, and $[F]$ is a divisorial ideal, it must be $B(E)e$, where e is the generator of F. If further $S(P) = D(P)$, ψ may be identified to multiplication by the irreducible element e. Note that if $S_t(E)$ is reflexive for $t \leq r$, then φ is surjective up to degree $t \leq r+1$.

REFERENCES

1. M. Brodmann, *Asymptotic stability of* Ass($M/I^n M$), Proc. Amer. Math. Soc. **74** (1979), 16–18.
2. P. Brumatti, A. Simis and W. V. Vasconcelos, *Normal Rees algebras*, J. Algebra **112** (1988), 26–48.
3. W. Bruns, E. G. Evans and P. Griffith, *Syzygies, ideals of height two and vector bundles*, J. Algebra **67** (1980), 429–439.
4. B. Buchberger, *An algorithmic method in polynomial ideal theory*, in "Recent Trends in Mathematical Systems Theory," N. K. Bose, ed., D. Reidel, Dordrecht, 1985, pp. 184–232.
5. D. Buchsbaum and D. Eisenbud, *Algebra structures for finite free resolutions, and some structure theorems for ideals of codimension 3*, Amer. J. Math. **99** (1977), 447–485.
6. D. L. Costa and J. L. Johnson, *Inert extensions for Krull domains*, Proc. Amer. Math. Soc. **59** (1976), 189–194.
7. P. Eakin and S. McAdam, *The asymptotic* Ass, J. Algebra **61** (1979), 71–81.
8. E. G. Evans and P. Griffith, *Local cohomology modules for local domains*, J. London Math. Soc. **19** (1979), 277–284.
9. R. Hartshorne and A. Ogus, *On the factoriality of local rings of small embedding dimension*, Comm. Algebra **1** (1974), 415–437.
10. J. Herzog, *Certain complexes associated to a sequence and a matrix*, Manuscripta Math. **12** (1974), 217–247.
11. J. Herzog, *Generators and relations of abelian semigroups and semigroup rings*, Manuscripta Math. **3** (1970), 153–193.
12. J. Herzog, A. Simis and W. V. Vasconcelos, *Koszul homology and blowing-up rings*, in "Proc. Trento Commutative Algebra Conference," Lectures Notes in Pure and Applied Math., vol. 84, Dekker, New York, 1983, pp. 79–169.
13. J. Herzog, A. Simis and W. V. Vasconcelos, *On the arithmetic and homology of algebras of linear type*, Trans. Amer. Math. Soc. **283** (1984), 661–683.
14. C. Huneke, *The primary components of and integral closures of ideals in 3-dimensional regular local rings*, Math. Annalen **275** (1986), 617–635.
15. M. Miller, *Bourbaki's theorem and prime ideals*, J. Algebra **64** (1980), 29–36.
16. M. Nagata, "Lectures on Hilbert's Fourteenth Problem," Tata Institute, Bombay, 1964.
17. P. Roberts, *A prime ideal in a polynomial ring whose symbolic blow-up is not Noetherian*, Proc. Amer. Math. Soc. **94** (1985), 589–592.
18. P. Samuel, *Anneaux gradués factoriels et modules réflexifs*, Bull. Soc. Math. France **92** (1964), 237–249.
19. A. Simis and N. V. Trung, *Divisor class group of ordinary and symbolic blow-ups*, Math. Z. (to appear).
20. W. V. Vasconcelos, *On the structure of certain ideal transforms*, Math. Z. (to appear).
21. W. V. Vasconcelos, *On linear complete intersections*, J. Algebra **111** (1987), 306–315.
22. W. V. Vasconcelos, *Koszul homology and the structure of low codimension Cohen-Macaulay ideals*, Trans. Amer. Math. Soc. **301** (1987), 591–613.
23. U. Vetter, *Zu einem Satz von G. Trautmann über den Rang gewisser kohärenter analytischer Moduln*, Arch. Math. **24** (1973), 158–161.

24. R. Villarreal, *Koszul homology of Cohen-Macaulay ideals*, Ph. D. Thesis, Rutgers University (1986).
25. J. Weyman, *Resolutions of the exterior and symmetric powers of a module*, J. Algebra **58** (1979), 333–341.

Department of Mathematics, Rutgers University, New Brunswick NJ 08903

This research was partially supported by NSF Grant DMS-8741451.

Noetherian Rings of Bounded Representation Type

ROGER WIEGAND

Let R be a reduced Noetherian ring with minimal prime ideals $P_1, \ldots, P_m$. The *rank* of the torsion-free R-module M is the m-tuple $\rho(M) = (r_1, \ldots, r_m)$, where r_i is the dimension of M_{P_i} as a vector space over R_{P_i}. We say R has *bounded representation type* provided there is an integer N such that each indecomposable finitely generated torsion-free R-module has rank less than or equal to $(N, \ldots, N)$. The least such N will be denoted by $\beta(R)$, and if no such N exists we write $\beta(R) = \infty$.

It follows from [**B1**, 1.2, 1.4] that rings of bounded representation type have Krull dimension at most one. In this paper we further restrict our attention to what Haefner and Levy [**HL**] call ring-orders. These are the one-dimensional, reduced, Noetherian rings for which the normalization $\tilde{R}$ is finitely generated as an R-module. Our goal is to characterize the ring-orders of bounded representation type. It is not clear how much of this work would go through without the assumption of finite normalization, nor is it clear what might be salvaged for rings with nilpotents.

Classically, representation type has been studied for ring-orders whose additive group is finitely generated. We call such a ring-order a *classical order*. For these rings bounded representation type is, by the Jordan-Zassenhaus Theorem [**SE**, Theorem 3.9] equivalent to *finite representation type*: There are only finitely many isomorphism classes of finitely generated torsion-free R-modules. In general, the two concepts are not equivalent. For example, each Dedekind domain has $\beta(D) = 1$, but the class group can be infinite.

The classical orders of finite representation type are exactly those satisfying the conditions DR of Drozd and Roiter [**DR**]:

DR1) The normalization $\tilde{R}$ is generated by three elements as an R-module;
DR2) The Jacobson radical of the R-module $\tilde{R}/R$ is cyclic.

This result appears in the 1967 paper [**DR**] of Drozd and Roiter, and a similar characterization can be found in Jacobinski's paper [**J**]. A different treatment, by Green and Reiner [**GR**], appeared in 1978 and corrected an error in [**J**], but still dealt only with classical orders.

In the last few years several authors, e.g., [**A**,**AR**,**AV**,**BGS**,**DW**,**EH**,**K**], have studied the maximal Cohen-Macaulay modules (= torsion-free modules in the one-dimensional case) over the local rings of hypersurface singularities (over an algebraically closed field). Also, Greuel and Knörrer [**GK**] have determined exactly which curve singularities (in characteristic 0) have finite representation type. Again, the corresponding local rings are exactly those that satisfy DR. Residue field extension (under the embedding of R in its normalization $\tilde{R}$) seems to be the main stumbling block in obtaining a general theory.

Using the approach of Drozd and Roiter, I show in Section 2 that every ring-order with bounded representation type satisfies DR. For the converse, treated in Section 3, I was unable to adapt their proof. Instead, I reduce the problem to a situation where one can appeal to the matrix reductions of Green and Reiner. Along the way, I am forced to impose a very mild but annoying (and probably unnecessary) separability condition: that the inclusion $R \to \tilde{R}$ induce no purely inseparable residue field extensions of degree 2 or 3. I prove that DR is equivalent to bounded representation type for ring-orders satisfying this separability condition, and to finite representation type for the semilocal ones.

It would be nice to find a more conceptual proof that DR implies finite representation type, but so far the difficult matrix reduction of [**GR**] appear to be unavoidable. In this paper I make no attempt to redo or modify these reductions. Otherwise, the paper is essentially self-contained. Consequently, many results are proved in some detail, even though similar results appear in the literature. Section 2 in particular relies heavily on the construction of Drozd and Roiter, although the arguments given here are a little cleaner. It seems worthwhile to make these results and methods accessible in the general context of ring-orders. The alternate approach — indicating the required modifications to existing proofs — would have resulted in at most a modest saving of space.

Two ways in which my approach differs from the classical approach are the systematic representation of modules by pullbacks, and avoidance of the completions of rings and modules. My (1.6) and (1.8) replace L. S. Levy's "Package Deal Theorem" (unpublished notes) and A. Jones's argument [**Jo**] that finite representation type descends from the completion of a local ring-order.

The last two sections are all new. In Section 4 I prove a theorem reminiscent of the Brauer-Thrall conjectures: If R is a local ring-order with infinite residue field and if DR fails, then there are infinitely many non-isomorphic indecomposable torsion-free modules of constant rank 1. Section 5 deals with the global problem, and presents a scheme for describing all the indecomposables over ring-orders that are locally of finite representation type.

1. Preliminaries.

All modules are assumed to be finitely generated. Our basic tool will be the conductor square

$$\begin{array}{ccc} R & \longrightarrow & \tilde{R} \\ \downarrow & & \downarrow \\ R/\underline{c} & \longrightarrow & \tilde{R}/\underline{c} \end{array} \tag{1.1}$$

Here $\tilde{R}$ is the normalization of the ring-order R, and $\underline{c} = \{r \in R \mid r\tilde{R} \subseteq R\}$ is the conductor ideal. Often it will be useful to replace R by a particular semilocal ring-order whose conductor square has the same bottom line as that of R:

1.2 **Singular semilocalization.** Call a maximal ideal $\mathcal{M}$ of the ring-order R *singular* if $R_{\mathcal{M}}$ is not a regular local ring (= discrete valuation ring or field). Equivalently, $\mathcal{M} \supseteq \underline{c}$. Let $S = R - (\mathcal{M}_1 \cup \cdots \cup \mathcal{M}_s)$, where the $\mathcal{M}_i$ are the singular maximal ideals, and let R_{sing} denote the ring of fractions $S^{-1}R$. Then, assuming $R \neq \tilde{R}$, R_{sing} is a semilocal ring-order, all its maximal ideals are singular, and the bottom line of its conductor square is the same as the bottom line of (1.1). We also use the notations M_{sing}, f_{sing}, etc., for localized modules and homomorphisms.

1.3. PROPOSITION. *Let R be a ring-order with $R \neq \tilde{R}$. Then $\beta(R) = \beta(R_{\text{sing}})$.*

PROOF: Clearly $\beta(R) \geq \beta(R_{\text{sing}})$, since each torsion-free R_{sing}-module comes from a torsion-free R-module. To prove the reverse inequality, we may assume R is connected (has no idempotents $\neq 0, 1$) and that $\beta(R_{\text{sing}}) = N < \infty$. Each minimal prime ideal of R survives in R_{sing}, so it makes sense to compare ranks over the two rings. Let F be any torsion-free R-module, and choose a torsion-free R-module G of rank $\leq (N, \ldots, N)$ and an R-homomorphism $f : F \to G$ such that f_{sing} is a split surjection. (See [**Bo**,

Ch. II, Prop. 19].) Let G_0 be the image of f. The induced map $f_0 : F \to G_0$ splits locally (hence globally by [**B1**, 1.9]), since $(f_0)_{\mathcal{M}} = f_{\mathcal{M}}$ for $\mathcal{M}$ singular and $(G_0)_{\mathcal{M}}$ is free for $\mathcal{M}$ non-singular.

In what follows, most of the action takes place on the bottom line of the conductor square (1.1). Usually we'll forget about R and $\tilde{R}$ and work instead with the following set-up:

1.4 **Artinian pairs** Let A be an Artinian subring of a ring B, and assume B is finitely generated as an A-module. Such a pair (A, B) is called an *Artinian pair*. By an (A, B)-module we will mean a pair (V, W) consisting of a projective B-module W and an A-submodule V of W such that $BV = W$. A morphism from (V_1, W_1) to (V_2, W_2) is simply a B-module homomorphism $f : W_1 \to W_2$ such that $f(V_1) \subseteq V_2$. Thus we have the notions of isomorphism of (A, B)-modules, the endomorphism ring $\mathrm{End}(V, W)$ of an (A, B)-module (V, W), and direct-sum decompositions of (V, W), via idempotents in $\mathrm{End}(V, W)$. We say the pair (A, B) has finite representation type provided there are only finitely many indecomposable (A, B)-modules up to isomorphism.

The bottom line of the conductor square (1.1) is an Artinian pair. We will show that the *semilocal* ring-order R has finite representation type if and only if the pair $(R/\underline{c}, \tilde{R}/\underline{c})$ has finite representation type. In the classical situation this is usually done by first going to the completion and then killing the conductor. While this approach is still available in our more general context, it seems easier to bypass the completion, since the bottom line of (1.1) contains all the complete local information we need. (It is shown in the appendix of [**WW2**] that for a local ring-order $R \neq \tilde{R}$ containing a field, the completion $\hat{R}$ can be recovered from the bottom line of (1.1).) The next few results establish the tools we need to pass between R and its associated Artinian pair.

1.5. Proposition. *The Krull-Schmidt Theorem holds for direct-sum decompositions of modules over an Artinian pair.*

Proof: The category of (A, B)-modules is additive, and given an idempotent $\phi \in \mathrm{End}(V, W)$, it factors as $(V, W) \xrightarrow{p} (\phi V, \phi W) \xrightarrow{i} (V, W)$, and $pi =$ identity. Further, the endomorphism ring of an indecomposable module is Artinian and, lacking idempotents, must be local. The Krull-Schmidt Theorem of [**B2**, Ch. I, 3.6] now applies.

If R is a ring-order, every torsion-free R-module M gives rise to an $(R/\underline{c}, \tilde{R}/\underline{c})$-module $M_{\text{art}} = (M/\underline{c}M, \tilde{R}M/\underline{c}M)$. We need to know which $(R/\underline{c}, \tilde{R}/\underline{c})$-modules arise in this fashion. The criterion, as in the case of completions, is determined by the rank function.

We say the module (V, W) over the Artinian pair (A, B) has *constant rank* r provided $W \cong B^{(r)}$. In general, B has orthogonal idempotents $\varepsilon_i = \varepsilon_i(W)$ with sum 1 such that $W\varepsilon_i \cong (B\varepsilon_i)^{(i)}$. (These idempotents do not necessarily decompose the *pair* (V, W).)

1.6. PROPOSITION. *Let R be a ring-order. The following are equivalent, for an $(R/\underline{c}, \tilde{R}/\underline{c})$-module (V, W):*

(a) *If P and Q are maximal ideals of R containing $\underline{c}$ and lying in the same connected component of* spec R, *then* $\dim_{\tilde{R}/P} W/PW = \dim_{\tilde{R}/Q} W/QW$.
(b) *Each idempotent $\varepsilon_i(W)$ lifts to an idempotent of $\tilde{R}$.*
(c) *$W \cong F/\underline{c}F$ for some projective $\tilde{R}$-module F.*
(d) *$(V, W) \cong M_{\text{art}}$ for some torsion-free R-module M.*

PROOF: It is easy to see that (a), (b) and (c) are equivalent and are implied by (d). Assuming (c), choose an isomorphism $\sigma : W \to F/\underline{c}F$, and define M by the pullback diagram

$$\begin{array}{ccccc} M & & \longrightarrow & & F \\ \downarrow & & & & \downarrow \\ V & \longrightarrow & W & \xrightarrow{\sigma} & F/\underline{c}F \end{array}$$

By [**W1**, Lemma 2.1], $M_{\text{art}} \cong (V, W)$.

1.7. PROPOSITION. *Let R be ring-order, and assume $R = R_{\text{sing}}$. Let M and N be torsion-free R-modules. If $M_{\text{art}} \cong N_{\text{art}}$, then $M \cong N$.*

PROOF: We have a pullback diagram

$$\begin{array}{ccc} M & \longrightarrow & \tilde{R}M \\ \downarrow & & \downarrow \\ M/\underline{c}M & \longrightarrow & \tilde{R}M/\underline{c}M \end{array}$$

with an analogous diagram for N. By hypothesis there is an isomorphism $\alpha : \tilde{R}M/\underline{c}M \to \tilde{R}N/\underline{c}N$ carrying $M/\underline{c}M$ onto $N/\underline{c}N$. It will suffice to lift

α to an isomorphism $\beta : \tilde{R}M \to \tilde{R}N$. Now $\tilde{R}M$ and $\tilde{R}N$ are projective $\tilde{R}$-modules, so we can choose an $\tilde{R}$-homomorphism β lifting α. Since $\underline{c}$ is contained in the Jacobson radical of $\tilde{R}$, β is surjective. But $\tilde{R}M$ and $\tilde{R}N$ have the same rank function, and it follows that β is an isomorphism.

1.8. **Lifting finite representation type.** Now we apply these results to the notion of finite representation type, using an idea due to Alfredo Jones [**Jo**]. Assume that R is a ring-order, that $R = R_{\text{sing}}$, and that the Artinian pair $(R/\underline{c}, \tilde{R}/\underline{c})$ has finite representation type. Let $X_1, \ldots, X_t$ be a list of the non-isomorphic indecomposable $(\tilde{R}/\underline{c}, \tilde{R}/\underline{c})$-modules. For each torsion-free R-module M, write $M_{\text{art}} = \oplus_j X_j^{(n_j)}$, and let $\nu(M)$ be the t-tuple $(n_1, \ldots, n_t)$ of non-negative integers. (This is well-defined, by (1.5).) Let $\mathcal{S}(R)$ be the set of all t-tuples of the form $\nu(M)$, M ranging over the non-zero R-modules.

1.8.1. LEMMA. *Let R be a ring-order with $R = R_{\text{sing}}$, and let M be a torsion-free R-module. Then M is indecomposable if and only if $\nu(M)$ is a minimal element of $\mathcal{S}(R)$.*

PROOF: Obviously M is indecomposable if $\nu(M)$ is minimal. Conversely, if $\nu(M) > \nu(N)$ we have $M_{\text{art}} \cong N_{\text{art}} \oplus X$ for some nonzero $(R/\underline{c}, \tilde{R}/\underline{c})$-module X. The criterion (a) of (1.6) shows that $X \cong U_{\text{art}}$ for some torsion-free R-module U. By (1.7), $M \cong N \oplus U$, and M decomposes.

1.9. THEOREM. *Let R be a ring-order, with $R = R_{\text{sing}}$. Then R has finite representation type if and only if $(R/\underline{c}, \tilde{R}/\underline{c})$ has finite representation type.*

PROOF: Suppose the pair $(R/\underline{c}, \tilde{R}/\underline{c})$ has finite representation type. By [**Jo**, Lemma 3], every subset of $\{0, 1, 2, \ldots\}^{(t)}$, in particular, $\mathcal{S}(R)$, has only finitely many minimal elements. By (1.7) and (1.8.1), R has finite representation type. Conversely, if R has finite representation type, let $Y_1, \ldots, Y_n$ be the indecomposable torsion-free R-modules. Write $(Y_k)_{\text{art}} = \oplus_j X_{jk}$, where the X_{jk} are indecomposable $(R/\underline{c}, \tilde{R}/\underline{c})$-modules. If, now, X is any indecomposable $(R/\underline{c}, \tilde{R}/\underline{c})$-module, we can find (by (1.6)) a complementary module X' such that $X \oplus X' \cong M_{\text{art}}$ for some torsion-free R-module M. Now decompose M, apply the art functor and use Krull-Schmidt (1.5), to deduce that X is isomorphic to some X_{jk}.

1.10. **Drozd-Roiter Conditions.** Let (A, B) be an Artinian pair, and let J be the radical of A. Let $\mu_A(M)$ denote the minimal number of generators for an A-module M. We say (A, B) satisfies DR provided

$$\text{DR1)} \quad \mu_A B \leq 3, \text{ and}$$

$$\text{DR2)} \quad \mu_A \left(\frac{JB + A}{A} \right) \leq 1.$$

1.10.1. REMARK: It is easily checked that the ring-order R satisfies the conditions DR of the introduction if and only if the pair $(R/\underline{c}, \tilde{R}/\underline{c})$ satisfies DR. (See [**GR**, p. 58].)

In order to build big indecomposables, we will need to define modules via their localizations. The result we need is proved in [**W3**] and is a minor modification of [**HL**, 1.6], but we will deduce it from a more general result.

1.11. LEMMA. *Let R be a commutative ring with only finitely many prime ideals, let $\mathcal{M}_1, \ldots, \mathcal{M}_n$ be the maximal ideals, and let F_i be a finitely presented $R_{\mathcal{M}_i}$-module for each i. If $(F_i)_P \cong (F_j)_P$ for each prime ideal $P \subseteq \mathcal{M}_i \cap \mathcal{M}_j$, there is a module F, unique up to isomorphism, such that $F_{\mathcal{M}_i} \cong F_i$ for each i.*

PROOF: Uniquess follows from faithfully flat descent, [**EGA**, 2.5.8]. To prove existence, we may assume inductively that there is an R-module G such that $G_{\mathcal{M}_i} \cong F_i$ for all $i > 1$. Choose $s \in \cap\{P \,\text{Prime} \mid P \nsubseteq \mathcal{M}_1]$, $t \in \cap\{Q \,\text{prime} \mid Q \nsubseteq \mathcal{M}_2 \cup \cdots \cup \mathcal{M}_n\}$, with $s + t = 1$. Then $R[s^{-1}] = R_{\mathcal{M}_1}$ and $R[t^{-1}] = (R - (\mathcal{M}_2 \cup \cdots \cup \mathcal{M}_n))^{-1} R$. If P is a prime ideal and $st \notin P$, then $P \subseteq \mathcal{M}_1 \cap \mathcal{M}_j$ for some $j \geq 2$, so $(F_1)_P \cong G_P$. It follows (by uniqueness) that $F_1[t^{-1}] \cong G[s^{-1}t^{-1}]$. The desired module F is defined by the pullback

$$\begin{array}{ccc} F & \longrightarrow & G[t^{-1}] \\ \downarrow & & \downarrow \\ F_1 & \longrightarrow & F_1[t^{-1}] \cong G[s^{-1}t^{-1}] \end{array}$$

1.12. PROPOSITION (Levy and Haefner). *Let R be a ring-order and $\{\mathcal{M}_i\}$ its set of maximal ideals. Suppose F_i is a torsion-free $R_{\mathcal{M}_i}$-module for each i, and that $(F_j)_P \cong (F_k)_P$ if P is a minimal prime ideal in $\mathcal{M}_j \cap \mathcal{M}_k$ (that is, they have the same dimension as vector spaces). Then there is a torsion-free R-module F such that $F_{\mathcal{M}_i} \cong F_i$ for each i.*

PROOF: We may assume R is connected and that $R \neq \tilde{R}$. By (1.11) and (1.2) there is a torsion-free R-module F such that $F_{\mathcal{M}_j} \cong F_j$ for each singular maximal ideal $\mathcal{M}_j$. If $\mathcal{M}_k$ is a non-singular maximal ideal, then $F_{\mathcal{M}_k}$ is a free $R_{\mathcal{M}_k}$-module, and the only question is whether it has the right rank. But since R is connected, there is a singular maximal ideal $\mathcal{M}_j$ and a minimal prime $P \subseteq \mathcal{M}_j \cap \mathcal{M}_k$. The desired conclusion follows from compatibility of the family $\{F_i\}$.

2. Indecomposables of large rank.

In this section we prove that ring-orders with bounded representation type satisfy DR. In fact, we prove more:

2.1. THEOREM. *Let R be a connected ring-order, and assume DR fails for R. Then for every $n \geq 1$ there is an indecomposable torsion-free R-module of constant rank n.*

PROOF: Suppose we can prove (2.1) when R is local. Then in the general case we note that by (1.10.1) there is a maximal ideal $\mathcal{M}$ such that DR fails for $(R_{\mathcal{M}})$. Let F_1 be an indecomposable $R_{\mathcal{M}}$-module of constant rank n. By (1.12) there is a torsion-free R-module F such that $F_{\mathcal{M}} \cong F_1$ and $F_{\mathcal{N}}$ is free of rank n for each maximal ideal $\mathcal{N} \neq \mathcal{M}$. Since R is connected and F_1 is indecomposable, it follows easily that F is indecomposable.

Assuming now that R is local, we use (1.6) and (1.10.1) to pass to the bottom line of the conductor square. We are now in the following situation: (A, B) is an Artinian pair, A is local, B is a principal ideal ring, and DR fails. We want to build, for each n, an indecomposable (A, B)-module of constant rank n. We will need the following lemma, for which I know no direct proof:

2.2. LEMMA. *Let (A, B) be an Artinian pair, with B a principal ideal ring. If B is free as an A-module then A is a principal ideal ring.*

PROOF: We know that B has only finitely many indecomposable modules, and that this property characterizes Artinian principal ideals rings. (See, for example [**Wa**, Theorem 2].) Therefore it suffices to show that the property descends from B to A. Let Y_i be the indecomposable B-modules, and write $Y_i = \oplus_j X_{ij}$, where the X_{ij} are indecomposable A-modules. Given any A-module X, write $X \oplus_A B = \oplus_i Y_i^{(r_i)} = \oplus_{i,j} X_{ij}^{(r_j)}$. Since $X \oplus_A B \cong$ direct sum of copies of X, Krull-Schmidt implies that X is a direct sum of some of the X_{ij}.

Now we show how failure of DR leads to a few very specific cases where it is rather easy to build big indecomposables.

2.3. PROPOSITION. *Let (A, B) be an Artinian pair, where A is local with maximal ideal M and residue field k. Assume B is a principal ideal ring and that DR fails for (A, B). Then there is a ring C between A and B such that, setting $D = C/\mathcal{M}C$, we have either*

(a) $\dim_k D \geq 4$, *or*

(b) $D \cong k[X, Y)/(X^2, XY, Y^2)$ *as k-algebras.*

PROOF: Let $d = \dim_k B/\mathcal{M}B = \mu_A B$. We may assume $d \leq 3$. Since $1 \in A$ is part of a minimal generating set for the A-module B, we can't have $d = 1$, or else $A = B$ and DR would hold. Suppose $d = 2$; let $I = (0 : B/A)$, and set $\bar{A} = A/I$ and $\bar{B} = B/I$. Then $\bar{B}/\bar{A}$ is a faithful cyclic $\bar{A}$-module, that is, $\bar{B}$ is a free $\bar{A}$-module of rank 2. Since $\bar{B}$ is a principal ideal ring, so is $\bar{A}$ by (2.2). Therefore each submodule of $\bar{B}/\bar{A}$, in particular $\mathcal{M}(B/A)$, is cyclic. Again, DR holds, contradiction. Therefore $d = 3$ and $(\mathcal{M}B + A)/A$ is not cyclic. Set $C = \mathcal{M}B + A$ and $D = C/\mathcal{M}C$. It is easy to see that either b) or a) holds, depending on whether or not $(\mathcal{M}B + A)/A$ can be generated by two elements.

The next lemma, whose proof is left to the reader, will allow us to replace the original Artinian pair (A, B) by the more manageable pair (k, D).

2.4. LEMMA. *Let (A, B) be an Artinian pair, let C be a ring between A and B, and let I be a nilpotent ideal of C. Let $F : (A, C)$-mod $\to (A/(I \cap A)$, $C/I)$-mod and $G : (A, C)$-mod $\to (A, B)$-mod be the additive functors defined by $(V, W) \to ((V + IW)/IW, W/IW)$ and $(V, W) \to (V, B \otimes_C W)$ respectively. Then*

(2.4.1) *F is surjective on isomorphism classes.*

(2.4.2) *G is full and faithful, and is injective on isomorphism classes.*

(2.4.3) *Both functors preserve and reflect indecomposable objects.*

Using (2.3) and (2.4) and changing notation, we need only consider the following situation: We have a field k and a finite-dimensional k-algebra A. Assuming either (a) $\dim_k A \geq 4$ or (b) $A = k[x, y]$ with $x^2 = xy = y^2 = 0$, we want to produce, for every n, an indecomposable (k, A)-module of rank n. Here is a basic pathological construction that does the job unless k is the 2-element field:

2.5. CONSTRUCTION: Let a and b be elements of A such that $\{1, a, b\}$ is linearly independent over k. Fix $n \geq 1$ and let ν be the nilpotent n by n matrix with 1's on the superdiagonal and 0's elsewhere. Set $W = A^{(n)}$, its elements written as column vectors. Finally, let V be the k-subspace of W consisting of all elements of the form $u + av + b\nu v$, u and v ranging over $k^{(n)}$.

2.6. PROPOSITION (Drozd and Roiter, [**DR**, Prop. 2.2]). *Assume either i)* $a^2 = ab = b^2 = 0$, *ii)* $a^2 = b^2 = 0$ *and* $\{1, a, b, ab\}$ *is linearly independent, or (iii)* $\{1, a, a^2, b\}$ *is linearly independent over* k. *Then* (V, W) *is indecomposable.*

PROOF: Let $\phi \in \mathrm{End}(V, W)$, and identify ϕ with its left-multiplication matrix. Since $k^{(n)} \subseteq V$ and $\phi(V) \subseteq V$ we can write, for every $u \in k^{(n)}$, $\phi(u) = u' + au'' + b\nu u''$, with $u', u'' \in k^{(n)}$. Moreover, u' and u'' are uniquely and linearly determined by u, so there are n by n matrices σ, τ with entries in k, such that

$$\phi = \sigma + a\tau + b\nu\tau \tag{2.6.1}$$

Also, $\phi(a1 + b\nu)$ carries $k^{(n)}$ into V (where "1" denotes the n by n identity matrix), so we can write $\phi(a1 + b\nu) = \mu + a\rho + b\nu\rho$, where μ and ρ are n by n matrices over k. From (2.6.1) we get

$$a\sigma + b\sigma\nu + a^2\tau + ab(\nu\tau + \tau\nu) + b^2\nu\tau\nu = \mu + a\rho + b\nu\rho \tag{2.6.2}$$

In Case (iii), I claim that $\nu^i\tau\nu^j = 0$ for all $i, j = 0, \ldots, n$. To see this, multiply each term in (2.6.2) by ν^i on the left and ν^j on the right, obtaining an equation I'll call (E). Since the claim is true if either $i = n$ or $j = n$, we may assume inductively that $\nu^{i+1}\tau\nu^j = 0$ and $\nu^i\tau\nu^{j+1} = 0$. Then the ab and b^2 terms in (E) vanish, and by linear independence the a^2 term, $a^2\nu^i\tau\nu^j$, has to vanish too. This proves the claim, and putting $i = j = 0$ we get $\tau = 0$. Then (2.6.2) yields $\sigma = \rho$ and $\sigma\nu = \nu\rho$, that is, σ commutes with ν. But then σ is an upper triangular striped matrix (constant on each diagonal), whence $\sigma \in k[\nu]$. Since $k[\nu] \cong k[X]/(X^n)$, a local ring, (V, W) is indecomposable in Case (iii).

In Cases (i) and (ii), we get, from (2.6.2), $\mu = 0$, $\rho = \sigma$, $\sigma\nu = \nu\sigma$. Assume $\varphi^2 = \varphi$, and use (2.6.1) to get $\sigma^2 = \sigma$ and $\sigma\tau + \tau\sigma = \tau$ (from the "1" and "a" terms respectively). As before, $\sigma = 0$ or 1, and either case forces $\tau = 0$. Again, (V, W) is indecomposable.

We have now dealt with possibility (b) of (2.3), as well as most of the examples arising in case (a). What remains is the following special situation: A is a k-algebra of dimension at least 4, but $\{1, a, a^2\}$ is linearly dependent for every $a \in A$. Thus each element of A satisfies a monic polynomial of degree 2. We proceed with the analysis of this case, much of which is carried out in [**DR**]. *Assume, for the rest of this section, that A is of this form.*

2.7. LEMMA. *If $a \in \operatorname{rad} A$ then $a^2 = 0$.*

PROOF: Assume $a \neq 0$, and write $a^2 = ra + s$, with $r, s \in k$. Then $s \in (\operatorname{rad} A) \cap k = 0$, so $a(a - r) = 0$. But this means $a - r$ is a non-unit, that is, $r = 0$.

Now we can dispose of the case where $\dim_k(\operatorname{rad} A) \geq 2$: Choose linearly independent elements $a, b \in \operatorname{rad} A$. If $ab = 0$, we get big indecomposables by Case (i) of (2.6). If $ab \neq 0$, then I claim that $\{1, a, b, ab\}$ is linearly dependent. For, if not, we have $ab = r + sa + tb$, with $r, s, t \in k$, and on multiplying this equation by ab, by a, and by b, we get $r = t = s = 0$, contradiction. Now we are in Case (ii) of (2.6), and again there are big indecomposables.

From now on we assume $\dim_k \operatorname{rad} A \leq 1$. Write $A/\operatorname{rad} A = \bigoplus_{i=1}^{s} K_i$, where each K_i is a field extension of k, and let $e_1, \ldots, e_s$ be the corresponding orthogonal idempotents of A.

2.8. LEMMA. *If $s \geq 2$ then $K_i = k$ for each i.*

PROOF: Suppose $K_1 \neq k$, choose $y \in K_1 - k$, and lift y to an element $x \in e_1 A$. Write $x^2 = rx + s$, with $r, s \in k$. Multiplying by e_2, we learn that $s = 0$, from which it follows that $y = r$, contradiction.

Since $\dim_k A \geq 4$ and $\dim_k(\operatorname{rad} A) \leq 1$, (2.8) implies that $s \neq 2$. Suppose $s = 3$. Then $K_i = k$ for all i and $\dim_k \operatorname{rad} A = 1$. Renumbering if necessary, we have $e_1 \operatorname{rad} A = \operatorname{rad} A$ and $e_i \operatorname{rad} A = 0$ for $i = 2, 3$. Choose a non-zero element $x \in \operatorname{rad} A$, set $a = x + e_2$, and check that $\{1, a, a^2\}$ is linearly independent, contradiction. (Don't forget (2.7)!) We have shown that $s = 1$ or $s \geq 4$.

If $s \geq 4$, we can apply Dade's Theorem (see (2.9) below), but it is amusing to note that we only need to worry about the case where k is the 2-element field: If $|k| \geq 3$, choose an element $a \in A$ such that $e_1 a$, $e_2 a$ and $e_3 a$ have distinct residues modulo $\operatorname{rad} A$. Then $\{1, a, a^2\}$ is linearly independent, contradiction.

Finally, suppose $s = 1$, that is, A is connected, and $K = A/\operatorname{rad} A$ is a field extension of degree at least 3. If $[K : k] = 3$, however, K has a primitive element, which, when lifted to A, gives an element a for which $\{1, a, a^2\}$ is linearly independent. Therefore $[K : k] \geq 4$, and by (2.4) we can forget about the radical and assume A is a field extension of degree at least 4 over k. Let E be the separable subfield, and write $E = k[u]$. Then $[E : k] \leq 2$, since otherwise $\{1, u, u^2\}$ would be linearly independent over k. Since K/E is purely inseparable there is an element $v \in K$ with minimal polynomial $X^p - v^p$ over E, where $p = \operatorname{char}(k)$. Then $\{1, v, \ldots, v^{p-1}\}$ is linearly independent over E, hence over k, and it follows that $p = 2$, and $v^2 \in E$.

Suppose $[E : k] = 2$. If $v^2 \notin k$, set $a = v$; otherwise set $a = uv$ (where $E = k[u]$). Then $a \in K - E$ and $a^2 \in E - k$, from which it follows that $\{1, a, a^2\}$ is linearly independent over k, contradiction. Therefore $[E : k] = 1$, that is, A/k is purely inseparable of degree ≥ 4. Furthermore, no element is of degree 4 over k. Therefore $x^2 \in k$ for every $x \in A$. Choose $a, b \in A$ such that $\{1, a, b\}$ is linearly independent, and look at the (k, A)-module (V, W) of (2.5):

Given an endomorphism φ of (V, W) we get equation (2.6.2) as before. It is easy to see that $\{1, a, b, ab\}$ is linearly independent over k. Therefore, from (2.6.2), we get $\rho = \sigma$, $\nu\rho = \sigma\nu$, $\nu\tau + \tau\nu = 0$, and $\mu = a^2\tau + b^2\nu\tau\nu$. Therefore $\sigma\nu = \nu\sigma$ and $\nu\tau = \tau\nu$ (characteristic 2). Thus σ and τ are both in $k[\nu]$, which commutes. If, now, φ is idempotent, expand "$\varphi^2 = \varphi$" using (2.6.1), and equate the "a" terms of the two sides of the equation, getting $\sigma\tau + \tau\sigma = \tau$. But $\sigma\tau + \tau\sigma = 2\sigma\tau = 0$, and hence $\varphi = \sigma \in k[\nu]$. Again, $\varphi = 0$ or 1. The proof of Theorem 2.1 is now complete.

We conclude this section with a proof of Dade's Theorem, which was the forerunner of all these constructions of indecomposable modules.

2.9. THEOREM (Dade [**D**]; see also [**GR**, (4.5)]). *Let (K, A) be an Artinian pair. Assume K is local and A has at least 4 maximal ideals. Then, for every $n \geq 1$, there is an indecomposable (K, A)-module of constant rank n.*

PROOF: By (2.4) we may assume K is a field and $A = K \times K \times K \times K$. Let W and ν be as in the construction (2.5), and let V be the K-subspace of W consisting of all elements $(x, y, x + y, x + \nu y)$, where x and y range over $K^{(n)}$. Then AV contains $(x, 0, x, x)$ for all $x \in K^{(n)}$ and $(0, y, y, \nu y)$ for all $y \in K^{(n)}$, so $AV = W$. Let $\varphi = (\alpha, \beta, \gamma, \delta)$ be an endomorphism of (V, W), where $\alpha, \beta, \gamma, \delta$ are n by n matrices over K. Since $\varphi(V) \subseteq V$, there are matrices σ, τ, ξ, η such that

$$\varphi(x, 0, x, x) = (\sigma x, \tau x, (\sigma + \tau)x, (\sigma + \nu\tau)x) \text{ and}$$
$$\varphi(0, y, y, \nu y) = (\xi y, \eta y, (\xi + \eta)y, (\xi + \nu\eta)\eta)$$

for all $x, y \in K^{(n)}$. Writing $\varphi = (\alpha, \beta, \gamma, \delta)$ and comparing coefficients, one obtains 8 equations, which lead very easily to these conclusions: $\varphi = (\alpha, \alpha, \alpha, \alpha)$ and $\alpha\nu = \nu\alpha$. It follows as before that (V, W) is indecomposable.

3. Finite representation type.

We already know, by (1.3), that the ring-order R has bounded representation type if and only if its singular semilocalization has bounded representation type. In §2 we showed that bounded representation type implies the Drozd-Roiter conditions DR. Here we establish the converse under a mild separability hypothesis.

3.1. THEOREM. *Let R be a ring-order satisfying the Drozd-Roiter conditions DR, and assume that the field extension $R/(P \cap R) \to \tilde{R}/P$ is separable for every maximal ideal P of $\tilde{R}$. Then R has bounded representation type. If, further, R is semilocal, then R has finite representation type.*

In the classical case of a module-finite $\mathbb{Z}$-algebra, the separability condition is automatic. Also, the Drozd-Roiter conditions force every residue field extension to be of degree at most 3, so our only concern is imperfect residue fields of characteristic 2 or 3.

To prove the theorem, we may assume R is semilocal and that every maximal ideal is singular, by (1.3). (It is very easy to prove that a semilocal ring has finite representation type if its singular semilocalization does. See §5 for a more general result.) Further, by (1.9) and (1.10.1) we may work instead with the Artinian pair $(R/\underline{c}, \tilde{R}/\underline{c})$. Since $R/\underline{c}$ is a direct product of local rings, it will suffice to prove this:

3.2. PROPOSITION. *Let (A, B) be an Artinian pair satisfying DR. Assume A is local, B is a principal ideal ring, and each residue field of B is a separable extension of the residue field of A. Then (A, B) has finite representation type.*

PROOF: This is proved by Green and Reiner [**GR**] under the extra assumption that the inclusion $A \to B$ induce no residue field extension, that is, $A/(\mathcal{N} \cap A) = B/\mathcal{N}$, for every maximal ideal $\mathcal{N}$ of A. We will replace (A, B) by a larger Artinian pair with no residue field extension, and then use:

3.3. LEMMA. *Let (A, B) be an Artinian pair, let A' be a ring containing A, and let $B' = B \otimes_A A'$. Assume that A' is a free A-module of finite rank, and that $JA' = J'$, where J and J' denote the radicals of A and A', respectively.*

(1) *If (A, B) satisfies DR, so does (A', B').*
(2) *If (A', B') has finite representation type, so has (A, B).*

PROOF: For (2), just copy the argument in the proof of (2.2). To prove (1), we have $\mu_{A'} B' \leq \mu_A B$, so DR1 holds for (A', B'). Also, since $A'B = B'$, we have $\frac{J'B'+A'}{A'} = \frac{JA'B+A'}{A'}$, a homomorphic image of the cyclic A'-module $\frac{JB+A}{A} \otimes_A A'$.

Returning to the local situation of (3.2), we let J be the maximal ideal of A and $k = A/J$. Write $B/\operatorname{rad} B = \prod_{i=1}^{s} K_i$, where the K_i are separable field extensions of k. Since $\dim_k B/JB \leq 3$ and $JB \subseteq \operatorname{rad} B$, extension we know $s \leq 3$ and $K_i = k$ for all but at most one index i. So we assume K_1 is a proper extension of k, necessarily of degree 2 or 3. Let $\in A[X]$ be a monic polynomial such that $k[X]/(\bar{f}) \cong K_1$, where $\bar{f}$ is the reduction of f modulo J. Let $A' = A[X]/(f)$. Let us see what we have accomplished:

3.4. LEMMA. *With notation as above and as in (3.3), the following are true:*

(1) *A' is a free A-module of rank $\deg f$ ($= 2$ or 3).*
(2) *$JA' = J'$.*
(3) *$A'/J' \cong K_1$*
(4) *$\operatorname{rad} B' = B'(\operatorname{rad} B)$*
(5) *If K_1/k is Galois, then every residue field of B' is isomorphic to K_1.*
(6) *If K_1/k is not Galois (hence of degree 3), then B' has 2 maximal ideals, whose residue fields are K_1 and a separable quadratic extension of K_1.*
(7) *B' is a principal ideal ring.*

PROOF: (1) is clear, since f is monic. Clearly $JA' \subseteq J'$, and $A'/JA' \cong k[X]/(\bar{f}) \cong K_1$. Both (2) and (3) follow. To prove (4), (5) and (6), we note that $B' \cong B[X]/(f)$, so $B'/B'(\operatorname{rad} B) \cong (B/\operatorname{rad} B)[X]/(\bar{f}) \cong \prod_{i=1}^{S} K_i[X]/(\bar{f})$. For $i > 1$, $K_i[X]/(\bar{f}) = k[X]/(\bar{f}) \cong K_1$; and

$$K_1[X]/(\bar{f}) \cong \begin{cases} K_1 \times K_1 \text{ or } K_1 \times K_1 \times K_1, & \text{if } \bar{f} \text{ splits in } K_1 \\ K_1 \times L, \text{ where } [L : K_1] = 2, & \text{otherwise.} \end{cases}$$

The latter case occurs only when $s = 1$ and $\deg f = 3$, and of course L/K_1 is separable. Since B is a principal ideal ring, (4) implies that $\operatorname{rad} B'$ is a principal ideal of B'. It follows easily that B' is a principal ideal ring. (See, e.g., [**AM**, Prop. 8.8].)

3.5. PROPOSITION. *Let (A, B) as in (3.2). There is a local extension ring A' of A satisfying the following:*

1. *A' is a free A-module, say of rank d.*
2. *$B' =: A' \otimes_A B$ is a principal ideal ring.*
3. *(A', B') satisfies DR, and $A' \to B'$ induces no residue field extension.*
4. *$d = 1$ if $A \to B$ induces no residue field extension.*
5. *If K_1 is the (necessarily unique) residue field properly extending $k = A/J$, then the residue field of A' is the Galois closure of K_1/k (hence $d = 2$, 3 or 6).*

PROOF: Apply (3.4) once if K_1/k is Galois, and twice otherwise. By (3.3), (A', B') still satisfies (DR).

The matrix reductions of [**GR**] show that the pair (A', B') has finite representation type, and by (3.3) this descends to (A, B). This completes the proof of (3.2) and (3.1).

3.6. REMARKS: The only reason for all the fuss about separability is to ensure that B' be a principal ideal ring. Only the case $\mu_A B = 3$ causes any difficulty: If R is a ring-order whose associated Artinian pair has $\mu_A B = 2$ (see (1.10)), then R is a Bass ring (see [**LW**]) and every torsion-free module is a direct sum of ideals; if, further, R is semilocal, there are only finitely many non-isomorphic ideals. Without separability, the assumption that $\mu_A B = 3$ is not enough to guarantee that the construction outlined here will produce a principal ideal ring B'. In the presence of DR2, however, the separability assumption may well be unnecessary.

4. Lots of ideals.

In [**GK**] it is shown that over an algebraically closed field a curve singularity of infinite representation type has, for every n, infinitely many non-isomorphic indecomposable torsion-free modules of rank n. It is reasonable to conjecture that this holds for every local ring-order with infinite representation type and infinite residue field, but a different approach will be needed. The proof in [**GK**] depends on the fact that there are infinitely many rings between R and $\tilde{R}$. A simple example shows that this is not always true in general.

4.1 EXAMPLE: Let L/K be a simple field extension of degree ≥ 4, and define R by the pullback diagram:

$$\begin{array}{ccc} R & \longrightarrow & L[[X]] \\ \downarrow & & \downarrow \\ K & \longrightarrow & L \end{array} \tag{4.1.1}$$

Then $L[[X]] = \tilde{R}$ and (4.1.1) is the conductor square of R. (See [**WW1**, 3.1].) There are only finitely many rings between K and L, so there are only finitely many rings between R and $\tilde{R}$.

We do, however, get infinitely many non-isomorphic (fractional) ideals:

4.2. PROPOSITION. *Let R be a local ring-order with infinite residue field. If DR fails, then R has infinitely many non-isomorphic faithful ideals (= torsion-free modules of constant rank 1).*

PROOF: By (1.6), (1.7), (2.3) and (2.4) it is enough to show that if k is an infinite field and D is a finite-dimensional k-algebra satisfying (a) or (b) of (2.3), then there are infinitely many non-isomorphic (k, D)-modules of constant rank 1. If U and V are distinct rings between k and D, the (k, D)-modules (U, D) and (V, D) are easily seen to be non-isomorphic. Therefore we may assume there are only finitely many intermediate rings. The usual proof of the primitive element theorem shows that then D is generated by a single element, say x, as a k-algebra. This rules out (b) of (2.3), so we may assume $d =: \dim_k D \geq 4$.

For each $t \in k$ let I_t be the k-subspace of A spanned by 1 and $x + tx^2$. We will show that for fixed $t \in k$ there are at most two elements $u \in k$ for which the (k, A)-modules (I_t, D) and (I_u, D) are isomorphic. If the pairs above are isomorphic, there is a unit $\alpha \in D$ such that $\alpha I_t = I_u$. Then $\alpha \in I_u$ since $1 \in I_t$, say, $\alpha = a + b(x + ux^2)$, with $a, b \in k$. Computing $\alpha(x + tx^2)$, we get

$$ax + (at + b)x^2 + b(t + u)x^3 + butx^4 \in I_u \tag{4.2.1}$$

If $b = 0$ we have $at = au$, whence $t = u$ since $\alpha \neq 0$. Assume from now on that $b \neq 0$.

If $d \geq 5$, the displayed powers of x are linearly independent. Since the coefficient of x has to be 0, we have $u = -t$. If $d = 4$, write $x^4 = cx^3 + \dots$, and reduce (4.2.1) accordingly. The new coefficient of x^3 is $b(t + u + ctu)$, and for fixed t there is at most one value of u satisfying $t + u + ctu = 0$.

5. The global case.

Let M be a torsion-free module over a ring order R, and let genus (M) denote the set of isomorphism classes of R-modules locally isomorphic to M.

5.1 THEOREM. *Let $R \neq \tilde{R}$ be a connected ring-order, and assume R_{sing} has finite representation type. Choose torsion-free R-modules $M_1, \dots, M_n$ such that $\{(M_j)_{\text{sing}} : 1 \leq j \leq n\}$ is a complete set of representatives for the set of isomorphism classes of indecomposable torsion-free R_{sing}-modules. Then $\bigcup_j$ genus (M_j) is the complete set of isomorphism classes of indecomposable torsion-free R-modules.*

PROOF: First of all, each M_j is indecomposable. For, if $M_j = X \oplus Y$ then, say, $X_{\text{sing}} = 0$. It will suffice to show $X_{\mathcal{M}} = 0$ for every non-singular maximal ideal $\mathcal{M}$. Since R is connected and $R \neq \tilde{R}$ there exist a singular maximal ideal $\mathcal{N}$ and a minimal prime P such that $P \subseteq \mathcal{M} \cap \mathcal{N}$. Then $X_P = 0$, and since $X_{\mathcal{M}}$ is free we have $X_{\mathcal{M}} = 0$.

Now let M be any indecomposable R-module. As in the proof of (1.3) we obtain a map $f : M \to M_j$ for some j, such that f_{sing} is a split surjection. Then, with $N = Imf$, the induced map $M \to N$ splits, so $M \cong N$. It remains to be shown that N is locally isomorphic to M_j. Since $N_{\text{sing}} = (M_j)_{\text{sing}}$, we only need to worry about the non-singular maximal ideals, and then it's just a matter of checking that the ranks agree. Another application of the connectedness argument in the preceding paragraph does the job.

5.2. COROLLARY. *The ring-order R has finite representation type if and only if*

(a) *R_{sing} has finite representation type, and*

(b) *genus* (M) *is finite for every indecomposable torsion-free* R*-module* M.

PROOF: All conditions can be checked on each connected component of $\operatorname{Spec} R$, so we may assume R is connected. If $R \neq \tilde{R}$, we apply (5.1), whereas if $R = \tilde{R}$ (i.e., R is a Dedekind domain), both sides of the implication simply assert that R has finite class group.

The theorem and corollary would be rather useless without some handle on the size of the genera. Fortunately, there is a good theory, developed in [**W1,LW,WW1,WW2**] which allows one to compute the genus in many cases. For example, the results of [**WW1**] allow one to make genus (M) into a group (with operation $*$) in such a way that $X \oplus Y \cong (X * Y) \oplus M$. (When $M = R$ this is the familiar law of Steinitz: $I \oplus J \cong (IJ) \oplus R$.) For faithful M, L. S. Levy has worked out a "Mayer-Vietoris" sequence, which appears in J. Haefner's University of Wisconsin thesis [**H**]:

$$(5.3) \qquad 1 \to \Delta_M L \to (\tilde{R}/\underline{c})^* \to \text{genus}(M) \to \operatorname{Pic} \tilde{R} \to 1$$

Here Δ_M is the subgroup of the group of units $(\tilde{R}/\underline{c})^*$ consisting of determinants of automorphisms of $\tilde{R}M/\underline{c}M$ that carry $M/\underline{c}M$ into itself and L is the group of units that lift to units of R. The connecting map $(\tilde{R}/\underline{c}) \to \text{genus}(M)$ is given by the group action defined in [**W1**], and the map $\text{genus}(M) \operatorname{Pic} R$ takes X to $(\det(\tilde{R}X))(\det(\tilde{R}M))^{-1}$.

If the ring-order R is a domain and is finitely generated as an algebra over an infinite perfect field k, then condition (b) is equivalent to finiteness of $\operatorname{Pic}(R)$. Equivalently, $\operatorname{Pic}(R)$ is finite, R has at most one singular maximal ideal $\mathcal{M}$, and the completion of $R_{\mathcal{M}}$ is isomorphic to a restricted power series ring $F + K[[X]]$, where $K \supseteq F \supseteq k$ are finite algebraic extension fields. These assertions are proved in [**W2**].

ACKNOWLEDGMENTS: I am grateful to the National Science Foundation for partial support for this research and to the University of Wisconsin for its hospitality and support during the 1985–1986 academic year, when much of this research was accomplished. I thank L. S. Levy for introducing me to the problems discussed in this paper, for his interest and encouragement, and for countless fruitful discussions on representation theory.

References

[**A**] M. Auslander, *Rational singularities and almost split sequences*, Trans. Amer. Math. Soc. **293** (1986), 511–532.

[**AM**] M. F. Atiyah and I. G. Macdonald, "Introduction to Commutative Algebra," Addison-Wesley, London, 1969.

[**AR**] M. Auslander and I. Reiten, *Almost split sequences for rational double points*, Trans. Amer. Math. Soc. **302** (1987), 87–98.

[**AV**] M. Artin and J.-L. Verdier, *Reflexive modules over rational double points*, Math. Ann. **270** (1985), 79–82.

[**B1**] H. Bass, *Torsionfree and projective modules*, Trans. Amer. Math. Soc. **102** (1962), 319–327.

[**B2**] H. Bass, "Algebraic K-Theory," Benjamin, New York, 1968.

[**Bo**] N. Bourbaki, "Éléments de Mathématiques, Algèbre Commutative," Hermann, Paris, 1961.

[**BGS**] R.-O. Buchweitz, G.-M. Greuel and F.-O. Schreyer, *Cohen-Macaulay modules on hypersurface singularities II*, Invent. Math. **88** (1987), 165–182.

[**D**] E. C. Dade, *Some indecomposable group representations*, Ann. of Math. **77** (1963), 406–412.

[**DR**] Ju. A. Drozd and A. V. Roiter, *Commutative rings with a finite number of indecomposable integral representations*, (in Russian), Izv. Akad. Nauk. SSSR, Ser. Mat. Tom 31 (1967), 783–798. English Translation Math. USSR-Izvestija **1** (1967), 757-772.

[**DW**] E. Dieterich and A. Wiedemann, *The Auslander-Reiten quiver of a simple curve singularity*, Trans. Amer. Math. Soc. **294** (1986), 455–475.

[**EGA**] A. Grothendieck and J. Dieudonne, "Éléments de Géometrie Algébrique," Chap. IV, Partie 2, Publ. Math. IHES, vol. 24, Paris, 1965.

[**EH**] D. Eisenbud and J. Herzog, *The classification of homogeneous Cohen-Macaulay rings of finite representation type*, preprint.

[**GK**] G.-M. Greuel and H. Knörrer, *Einfache Kurvensingularitäten und torsionfreie Moduln*, Math. Ann. **270** (1985), 417–425.

[**DR**] E. Green and I. Reiner, *Integral representations and diagrams*, Michigan Math. J. **25** (1978), 53–84.

[**H**] J. Haefner, *Direct sum behavior of lattices over sigma-I rings*, J. Pure and Appl. Algebra (to appear).

[**HL**] J. Haefner and L.S. Levy, *Commutative orders whose lattices are direct sums of ideals*, J. Pure and Appl. Algebra (to appear).

[**J**] H. Jacobinski, *Sur les ordres commutatifs avec un nombre fini de réseaux indécomposables*, Acta Math. **118** (1967), 1–31.

[**Jo**] A. Jones, *Groups with a finite number of indecomposable integral representations*, Michigan Math. J. **10** (1963), 257–261.

[**K**] H. Knörrer, *Cohen-Macaulay modules on hypersurface singularities I*, Invent. Math. **88** (1987), 153–164.

[**LW**] L. S. Levy and R. Wiegand, *Dedekind-like behavior of rings with 2-generated ideals*, J. Pure and Appl. Algebra **37** (1985), 41–58.

[**SE**] R. G. Swan and E. G. Evans, Jr., "K-Theory of Finite Groups and Orders," Springer Lecture Notes in Math., vol. 76, 1970.

[**W1**] R. Wiegand, *Cancellation over commutative rings of dimension one and two*, J. Algebra **88** (1984), 438–459.

[**W2**] R. Wiegand, *Picard groups of singular affine curves over a perfect field*, Math. Z. (to appear).

[W3] S. Wiegand, *Ranks of indecomposable modules over one-dimensional rings*, J. Pure and Appl. Algebra (to appear).

[Wa] R. B. Warfield, Jr., *Decomposability of finitely presented modules*, Proc. Amer. Math. Soc. **25** (1970), 167–172.

[WW1] R. Wiegand and S. Wiegand, *Stable isomorphism of modules over one-dimensional rings*, J. Algebra **107** (1987), 425–435.

[WW2] R. Wiegand and S. Wiegand, *Decompositions of torsion-free modules over affine curves*, in "Algebraic Geometry," Bowdoin 1985, Proc. Symp. in Pure Math., vol. 46, Part 2, pp. 503–513.

Department of Mathematics, University of Nebraska, Lincoln NE 68588